Bioremediation

Bioremediation

Applied Microbial Solutions for Real-World Environmental Cleanup

EDITED BY

Ronald M. Atlas
and Jim Philp

WASHINGTON, D.C.

American Society for Microbiology
1752 N Street, N.W.
Washington, DC 20036–2904

Library of Congress Cataloging-in-Publication Data

Bioremediation: applied microbial solutions for real-world environmental cleanup / edited by Ronald M. Atlas, Jim Philp.
p. cm.
Includes bibliographical references and index.
ISBN 1–55581–239–2 (alk. paper)
1. Bioremediation. I. Atlas, Ronald M., 1946– II. Philp, Jim.

TD192.5.B55693 2005
628.5–dc22
2004028933

Printed in the United States of America

10 9 8 7 6 5 4 3 2 1

Address editorial correspondence to ASM Press, 1752 N St., N.W., Washington, DC 20036–2904, USA

Send orders to ASM Press, P.O. Box 605, Herndon, VA 20172, USA
Phone: 800-546-2416; 703-661-1593
Fax: 703-661-1501
Email: books@asmusa.org
Online: www.asmpress.org

CONTENTS

CONTRIBUTORS

Robert T. Anderson
Department of Microbiology, University of Massachusetts, Amherst, MA 01003

Ronald M. Atlas
Graduate School, University of Louisville, Louisville, KY 40292

Mark J. Bailey
NERC Centre for Ecology and Hydrology, Mansfield Road, Oxford OX1 3SR, United Kingdom

Selina M. Bamforth
School of Civil Engineering and Geosciences, University of Newcastle-upon-Tyne, Newcastle-upon-Tyne NE1 7RU, United Kingdom

Lewis R. Barlow
Carl Bro Group, Spectrum House, Edinburgh EH7 4GB, Scotland, United Kingdom

D. Andrew Barry
Contaminated Land Assessment and Remediation Research Centre, Institute for Infrastructure and Environment, School of Civil Engineering and Electronics, The University of Edinburgh, Edinburgh EH9 3JL, Scotland, United Kingdom

Lena Ciric
NERC Centre for Ecology and Hydrology, Mansfield Road, Oxford OX1 3SR, United Kingdom

Colin J. Cunningham
Contaminated Land Assessment and Remediation Research Centre, The University of Edinburgh, Edinburgh, Scotland, United Kingdom

Mike Griffiths
Mike Griffiths Associates, St. Non, Pleasant Valley, Stepaside, Narberth, Pembrokeshire SA67 8NY, United Kingdom

Barry M. Hartman
Kirkpatrick & Lockhart LLP, 1800 Massachusetts Avenue, N.W., Washington, DC 20036–1800

Jonathan R. Lloyd
The Williamson Research Centre for Molecular Environmental Studies, The Department of Earth Sciences, The University of Manchester, Manchester M13 9PL, United Kingdom

Lynne E. Macaskie
School of Biosciences, The University of Birmingham, Birmingham B15 2TT, United Kingdom

Mark Mustian
Kirkpatrick & Lockhart LLP, 1800 Massachusetts Avenue, N.W., Washington, DC 20036–1800

Jim C. Philp
Department of Biological Sciences, Napier University, Merchiston Campus, 10 Colinton Road, Edinburgh EH10 5DT, Scotland, United Kingdom

Roger C. Prince
ExxonMobil Research and Engineering Co., 1545 Route 22 East, Annandale, NJ 08801

Henning Prommer
Department of Earth Sciences, Faculty of Geosciences, University of Utrecht, P.O. Box 80021, 3508 TA Utrecht, The Netherlands and CSIRO Land and Water, Private Bag No. 5, Wembley WA 6913, Australia

Ian Singleton
School of Biology, University of Newcastle-upon-Tyne, Newcastle-upon-Tyne NE1 7RU, United Kingdom

Andrew S. Whiteley
NERC Centre for Ecology and Hydrology, Mansfield Road, Oxford OX1 3SR, United Kingdom

PREFACE

Bioremediation: Applied Microbial Solutions for Real-World Environmental Cleanup, as the title implies, describes the environmental applications of microorganisms to remediate contaminated soils and waters. Written for both academics and practitioners, the book provides detailed knowledge of bioremediation research and real-world applicability of that knowledge. The book makes much use of "how to" information, covering how to bioremediate from site assessment to project closure. It provides a truly international perspective, balancing American (United States) and European (largely United Kingdom) coverage and showing the challenges facing bioremediation under differing regulatory frameworks and against differing histories of environmental awareness and public demands for remediation. It provides contemporary examples of the application of bioremediation and establishes the needs for future research and development efforts.

Bioremediation, as defined by the U.S. Office of Management and Budget, involves techniques using biological processes to treat contaminated soil or groundwater. It is a field that combines basic microbiology, advanced biotechnology, and environmental engineering and does so within the context of public demands for clean waters and soils, evolving risk-based regulatory frameworks that govern performance criteria, and public concerns about microorganisms—especially the deliberate release of genetically modified microorganisms into the environment. Thus, bioremediation still is a developing field, one that will be driven by scientific and technological developments, as well as public policy developments.

Although still considered an innovative technique by the U.S. Environmental Protection Agency, bioremediation is increasingly being used to treat contaminated soils and waters. As it is an innovative technology, there is great interest in research and development as well as actual applications. This volume describes both the bioremediation technologies being applied and those being developed. Thus, it is relevant to industrial engineers and managers who must apply technologies today to remove pollutants from contaminated soils and waters and to academic researchers whose efforts will develop future bioremediation technologies that can be applied for cost-effective cleanup efforts.

Bioremediation begins by defining the role of bioremediation in solving environmental contamination problems—providing an overview of the scientific underpinnings of bioremediation, as well as the practical considerations for applying bioremediation to the real-world problem of environmental contamination. Here, the biodegradative capacities of microorganisms upon which the success of bioremediation rests are highlighted.

The next chapters cover the more practical challenges of bioremediation, from risk-based design criteria, to the legal/regulatory frameworks that are the drivers of environmental remediation, to the engineering approaches for modeling bioremediation projects. The breadth of scientific and engineering principles presented reflects the complexities of applying biotechnological solutions for environmental problems. Risk assessment and risk reduction are critical considerations in establishing the need for remediation and the applicability of bioremediation. Awareness of the interactions between the legal, scientific, and engineering communities is essential for the successful use of bioremediation. We draw attention to the differences in regulatory frameworks between the United States and the United Kingdom since the regulatory requirements serve as drivers for the technological needs of remediation projects.

A knowledge of the basic principles of the legal requirements is essential for understanding how and when bioremediation can be applied for the restoration of contaminated soils and waters. When contaminated sites undergo remediation, there is a clear requirement to protect the environment while ensuring that risks to human health and the environment are minimized. Modeling provides a way of assessing the critical parameters that impact bioremediation and predicting the likelihood of successfully meeting established performance criteria. Effective design of a bioremediation project necessitates the integration of interdisciplinary knowledge developed by microbiologists, geochemists, hydrogeologists, mathematicians, and engineers.

Once the real-world needs and scientific, legal, and engineering challenges for bioremediation have been established, the actual applications of bioremediation are explored. Through extensive use of practical examples where bioremediation has been applied, a balanced international perspective on the applicability of bioremediation is provided. The varied approaches to bioremediation are described: in situ or ex situ methods and ones which may involve biostimulation, i.e., stimulating microbial activities by optimizing environmental conditions, e.g., by adding nutrients or oxygen to increase the rates of biodegradation; bioaugmentation, i.e., adding microorganisms to increase the diversity of microorganisms capable of biodegrading the contaminants; or natural monitored attenuation, i.e., monitoring the natural biodegradative activities to see that removal of the contaminants occurs at rates needed to meet targets set to reduce risk to human health and the environment. Perspectives for each approach are included, highlighting achievements and discussing limitations that establish the needs for future research and development efforts. The book covers the range of environmental contamination problems for which bioremediation can be and has been applied—from organic contamination of soil and groundwater with a myriad of compounds ranging from chlorinated solvents to plastics, to marine oil spills, to soils contaminated with metals and radionuclides.

A wide range of examples of successful employment of bioremediation are presented. The importance of monitoring is established, and a number of methods for monitoring the chemical disappearance of contaminants and the activities of microorganisms in bioremediation are described. The monitoring tools needed to manage and to design more effective bioremediation technologies are explained. Finally, the applications of biotechnology for clean environmental products and processes, effectively defining the field of preemptive bioremediation, are depicted. In this context, biotechnology holds great promise for future environmental applications. As highlighted throughout this book, to be successful, bioremediation must be economically and technically competitive with other physical and chemical remediation technologies. Establishing competitive costs for environmental managers, efficacy for regulators, and predictability for engineers will be key in the ability of bioremediation to become more significant in the highly competitive remediation industry. Bioremediation is still very much an evolving technology; there is a need for more research and development to identify and overcome the limitations and a real need to establish a critical dialogue among scientists, engineers, and environmental managers to ensure that discoveries in the laboratory can be successfully applied in the field. As the real-world applications of bioremediation continue to expand, so will the examples of specific successful approaches and the value of our work.

RONALD M. ATLAS, Louisville, Kentucky
JIM C. PHILP, Edinburgh, Scotland

ENVIRONMENTAL POLLUTION AND RESTORATION: A ROLE FOR BIOREMEDIATION

Jim C. Philp, Selina M. Bamforth, Ian Singleton, and Ronald M. Atlas

1

BIOREMEDIATION AND ITS PLACE IN THE WORLD

The past century witnessed a vast increase in global pollution. Industrial development, population growth, urbanization, and a disregard for the environmental consequences of releasing chemicals into the environment all contributed to the modern pollution situation. A huge range of industries, including most notably the oil and gas industry, contributed to the problem (Table 1.1). There was a large increase in the diversity of organic compounds that are industrially produced and which were carelessly released into the environment. Consequently, in the natural environment today there are numerous chemical contaminants, which are toxic to biological systems, that have originated from both natural (biogenic and geochemical) and anthropogenic sources.

Since the 1970s (post-Rachel Carson's *Silent Spring* era), there have been a public demand for environmental practices that reduce pollution and a growing demand for restoration of contaminated sites. Protecting human health and the environment from industrial pollution is a societal mandate today. Given the toxicity and environmental concerns associated with chemical pollutants, much effort has been directed towards ways in which diverse contaminants can be removed from ecosystems. These techniques vary from simple physical removal (e.g., landfill) to more expensive treatments such as incineration. Often these treatments do not remove the contaminant or may leave behind a toxic residue which has to be disposed of in other ways. The ability of microbes to degrade organic contaminants into harmless constituents has been explored as a means to biologically treat contaminated environments; today, it is the subject of many research investigations and real-world applications—it is the basis for the emergent field of bioremediation.

The process of bioremediation, defined as the use of microorganisms to detoxify or remove pollutants, which relies upon microbial enzymatic activities to transform or degrade the offending contaminants, is an evolving method for the removal and destruction of many environmental pollutants. Bioremediation, especially when it can be carried out in situ, is a cost-effective means of removing many chemical

Jim C. Philp, Department of Biological Sciences, Napier University, Merchiston Campus, 10 Colinton Rd., Edinburgh EH10 50T, Scotland, United Kingdom. *Selina Bamforth*, School of Civil Engineering and Geosciences, University of Newcastle-upon-Tyne, and *Ian Singleton*, School of Biology, University of Newcastle-upon-Tyne, Newcastle-upon-Tyne NE1 7RU, United Kingdom. *Ronald M. Atlas*, Graduate School, University of Louisville, Louisville, KY 40292.

Bioremediation: Applied Microbial Solutions for Real-World Environmental Cleanup
Edited by Ronald M. Atlas and Jim C. Philp © 2005 ASM Press, Washington, D.C.

TABLE 1.1 Categories of major industrial land uses and capacity for soil contamination[a]

Land use and category of contamination	Type(s) of contamination[b]
Category 1: high	
Hazardous waste treatment	O, I
Fine chemicals	O
Coal gasification	O, I
Integrated iron and steel	O, I
Oil refining and petrochemical	O, I
Pesticides	O
Scrap yards	O, I
Pharmaceutical	O
Category 2: moderate	
Fertilizer production	I
Wood preservatives	O
Electrical equipment	O
Minerals	I
Power stations	I
Shipbuilding	O, I
Tires	O, I
Pulp and paper	O
Tanneries	O, I
Category 3: low	
Animal processing	O
Glass	I
Printing	O
Airports	O
Railways	O
Dry cleaners	O

[a] Adapted from the work of Young et al. (146).
[b] O, organic; I, inorganic.

pollutants that adversely impact human health or environmental quality. It is an acceleration of the natural fate of biodegradable pollutants and hence a natural or "green" solution to the problem of environmental pollutants that causes minimal, if any, additional ecological impacts. The end products of effective bioremediation, such as water and carbon dioxide, are nontoxic and can be accommodated without harm to the environment and living organisms.

Although bioremediation has been utilized for decades, this technology is still considered an innovative technique by the U.S. Environmental Protection Agency (EPA). Whereas biological treatments of human and animal wastes (sewage treatment and composting) are widely employed, bioremediation currently comprises only 10 to 15% of all remediation methods used for the treatment of contaminated soils and groundwaters in the United States; reliance on the microbial capacity to biodegrade pollutants lags behind use of the physical treatments, namely, incineration, thermal desorption, and solidification. One of the major reasons for the reluctance to rely on bioremediation has been the difficulty in establishing engineering parameters that ensure reliability. However, with greater understanding of microbial diversity and the development of bioengineering, bioremediation is taking its place as a cost-effective technique in integrated environmental restoration efforts.

Recognizing the economic and environmental benefits of bioremediation, the Organization for Economic Co-operation and Development (OECD) estimated that the growth of bioremediation would be from $40 billion per year in 1990 to over $75 billion at the current time (97). Others have also estimated significant growth in the bioremediation market in North America and Europe (Table 1.2). According to a report from the McIlvaine Company (news release, 1998 [http://www.mcilvainecompany.com/news]), the popularity of soil incineration and groundwater pump-and-treat techniques is waning while that of bioremediation is increasing. The authors estimated that by 2002 world annual expenditures

TABLE 1.2 World bioremediation markets 1994–2000[a]

Market	U.S. $ (millions)		
	1994	1997	2000
United States	160–450	225–325	375–600
Europe	105–175	180–300	375–600
Germany	70–100	100–150	250–350
The Netherlands	10–20	15–35	30–60
Scandinavia	10–20	15–35	30–60
United Kingdom	5–10	7.5–20	15–30
Other	10–25	42.5–60	50–100
Canada	15–35	30–50	50–100

[a] *Source:* Tata Energy Research Institute (http://www.cleantechindia.com/eicnew/successstories/oil.htm).

TABLE 1.3 Economics of bioremediation[a]

Method	Range of cost of remediation ($US/ton of soil)
Incineration	400–1,200
Washing	200–300
Bioremediation	20–200

[a] *Source:* Tata Energy Research Institute (http://www.cleantechindia.com/eicnew/successstories/oil.htm).

for soil and groundwater cleanup using bioremediation techniques would increase to $1.1 billion, up from $870 million in 1997. Asian expenditures were expected to grow from $200 million in 1997 to over $300 million in 2002. Yet despite the growth in biological treatments, the market share remains relatively low—well under 10% of the $25 billion site remediation market. The largest application of bioremediation is for cleanup of landfills and hazardous waste dumpsites. In the United States, military applications rank second, whereas in the Europe-Africa segment and in Asia petroleum-contaminated sites are the second largest application.

Bioremediation tends to be an attractive technology because its use to remove pollutants involves relatively little capital outlay and because it can be inexpensive compared to physical methods such as incineration for decontaminating the environment (Tables 1.3 and 1.4); as witnessed in the *Exxon Valdez* oil spill, in which simple washing of oiled rocks with water was costing over $1 million per day, physical cleanup can be extraordinarily expensive, whereas bioremediation of hundreds of miles of shoreline was accomplished for less than $1 million. The vast and diverse metabolic capacities of naturally occurring microorganisms (both bacteria and fungi)—which can be harnessed for degradation of pollutants, the potential creation of genetically engineered bacteria by modern biotechnology, and the rapid development of bioengineering as a reliable field for application to environmental problems—mean that there is great promise for future development of bioremediation technologies.

TABLE 1.4 Typical costs of land remediation techniques[a]

Remediation technique	Cost ($US/m^3)
Thermal treatment (on-site incineration)	178–715
Excavation and disposal	53–134
Soil washing	26–71
Engineering capping	26–62
Encapsulation with geomembranes	71–107
Solidification/stabilization	17–178
In situ chemical oxidation	71–152
Bioremediation	2–268

[a] *Source:* MSI Marketing Research for Industry (92).

THE EVOLUTION OF BIOREMEDIATION

Microorganisms, especially bacteria, have benefited from time, evolving into many types with exceedingly diverse metabolic capabilities which can be utilized in bioremediation. Bacteria have existed on the planet for perhaps 3 billion years or so. Given their small size, large surface-to-volume ratio, very high rate of growth and division, and genome plasticity, bacteria evolve quickly. Yet it is only within approximately the last century that a vast array of synthetic chemicals which appear to have no natural counterparts have become common environmental pollutants, e.g., pesticides, herbicides, biocides, detergents, and halogenated solvents. That the biosphere has not been catastrophically polluted by these chemicals is testimony to the ability of the bacteria to evolve. These ideas were formalized by Alexander (4) as the principle of microbial infallibility, that is, the principle that no natural organic compound is totally resistant to biodegradation provided that environmental conditions are favorable. Given that most synthetic compounds are very similar to naturally occurring counterparts, it is not surprising that they can be biodegraded by microbial metabolism. Some xenobiotic (man-made) compounds have molecular structures that are not readily recognized by existing degradative enzymes; such compounds resist biodegradation or are metabolized incompletely, with the result that some xenobiotic compounds accumulate in the environment. However, the finding that some naturally halogenated compounds do exist and indeed are microbially produced (96) means that there is further potential scope for harnessing the ability of microbes to transform such xenobiotics.

THE NEED FOR BIOREMEDIATION

The major reasons for the control of water and soil pollution and the consideration of bioremediation are first and foremost, public health concerns; second, environmental conservation; and finally, the cost of decontamination. Clean water is essential to agriculture and industry, and for most countries water is a finite resource which has to be kept free of pollutants. Maintenance of safe potable water supplies is a major health concern around the world. The arguments for uncontaminated soil are similar; even countries with very large areas of land tend to have their development concentrated in certain places. Having land so contaminated that it cannot be developed in those areas is a real financial and developmental burden. For many of the industrial and industrializing nations, space is at a premium. Also, there are documented cases in which contamination has caused public health problems (e.g., see the discussion of Minamata disease below). Contaminated land can also become a contaminated water problem if the contamination migrates to groundwater. For many countries, groundwater is a major source of drinking water. When the pollution traverses national boundaries, the issues become political. All of this speaks to the critical need for cost-effective technologies, such as bioremediation, to help remove environmental contaminants.

The petroleum industry is a major contributor of organic contamination to the natural environment, releasing hydrocarbon contaminants into the environment in a number of ways. Severe subsurface pollution of soils and water can occur via the leakage of underground storage tanks and pipelines, spills at production wells and distribution terminals, and seepage from gasworks sites during coke production. Seepages of gasoline from underground storage tanks have caused widespread soil and aquifer contamination, threatening the safety of various potable water supplies. Polycyclic aromatic hydrocarbons (PAHs), dioxins, and dibenzofurans formed during the incomplete combustion of organic materials such as coal, diesel, wood, and vegetation are major airborne contaminants; these complex and sometimes carcinogenic compounds ultimately are deposited from the air and contaminate terrestrial and aquatic systems. Oil spills, such as the 1967 *Torrey Canyon* spill in the English Channel, the 1978 *Amoco Cadiz* oil spill off the coast of Brittany, and the 1989 *Exxon Valdez* oil spill in Prince William Sound, Alaska, have the potential to cause major marine ecosystem damage.

The complex and diverse range of petroleum-derived organic compounds released from such spillages is of major environmental concern. These consist of aliphatic (e.g., alkanes and alkenes) and aromatic (e.g., benzene, toluene, ethylbenzene, and xylene [BTEX]) hydrocarbons and PAHs. Chemicals synthesized from petroleum products and other manufacturing processes such as polychlorinated biphenyls (PCBs), polychlorinated dioxins, and *s*-triazines are equally as recalcitrant in the natural environment as their precursors, if not more so, since they are foreign to biological systems. These man-made chemicals (xenobiotics) are of particular concern because of their environmental persistence and potential toxicity and carcinogenicity.

The town gas industry, which utilized the coal gasification process for production of municipal gas supplies, is a good example of a polluting industry. Before the widespread use of natural gas, coal gasification was used from about the 1850s to the 1970s throughout Europe and the United States to produce gas for towns and cities. During the entire history of its use, there was virtually no concern over contaminated land. Disposal of the contaminants was often careless. As a result, it produced widespread unchecked environmental contamination. The contaminants from the town gas industry are readily recognizable at many sites today; the locations within cities of the plants are well documented, and in many cases the buildings where coal gasification was carried out still exist. There are more than 1,500 such sites in the United States (60), and the United Kingdom has around 1,000 former gasworks and associated sites, representing some 7,000 acres of land and a history of almost two centuries of production (93); others are present throughout Europe. The residues from coal gasification consist of coal tars, which are composed primarily of PAHs (42). They also contain phenolic materials, the volatile organics of the aromatic hydrocarbons BTEX, and a range of toxic heavy metals and cyanide. Traditionally, these residues were buried on-site or left in the bottom of gas holders. Today they remain as environmental contaminants in need of site remediation.

Agricultural industries are also major sources of organic contamination to terrestrial and aquatic ecosystems. Agricultural practices which involve regular applications of pesticides, such as the phenylamides (e.g., *s*-triazines) and organochlorines, to increase crop yields result in the introduction into the environment of persistent contaminants that can cause health and ecological problems. Fertilizer applications also result in major contamination of surface water and groundwater from agricultural runoff. Phosphate runoff can cause significant eutrophication of lakes, and nitrate and nitrite seepage into groundwater is a human health threat; nitrate and nitrite contamination in the agricultural heartland of the United States, for example, is toxic to newborns, sometimes resulting in casualties. The nonpoint pollution from agricultural runoff can cause widespread environmental problems.

As part of the 1977 Amendments to the Clean Water Act, the U.S. EPA developed a list of the common industrial pollutants that pose an imminent threat to public health and the environment. The list contains 114 organic compounds and 13 metals out of a total of 129 pollutants which have become known as the priority pollutants. These priority pollutants, along with petroleum, have become the main focus of emerging remediation technologies. The major criteria used in selecting priority pollutants are as follows (50).

1. Human toxicity:
   (a) systemic toxicity (acute and chronic effects),
   (b) carcinogenicity and genotoxicity,
   (c) reproductive and developmental effects,
   (d) neurobehavioral toxicity, and
   (e) local effects (e.g., irritation).

2. Likely presence in significant concentrations on land affected by post- or current industrial use.

3. Toxicity in plants and animals.

4. Potential for bioaccumulation and biomagnification.

5. Mobility in the environment (e.g., stability, volatilization).

6. Environmental persistence.

7. Potential to explode or ignite.

8. Potential for damage to buildings.

Priority pollutants in the environment pose a number of hazards (Table 1.5), not only to

TABLE 1.5 Some effects of pollution and those affected

Effect	The affected[a]
Cancer	Humans
Lung disease	Humans
Heart disease	Humans
Hypertension	Humans
Teratogenic effects	Humans
Reproductive effects	Humans
Allergy	Humans
Biomagnification	Humans, environment
Acute toxicity (immediate)	Humans, environment
Chronic toxicity (delayed)	Humans, environment
Neurotoxicity	Humans, environment
Biodiversity loss	Environment
Phytotoxicity	Environment
Zootoxicity	Environment
Eutrophication	Environment
Deoxygenation	Environment
Explosion	Humans, infrastructure
Fire	Humans, environment, infrastructure
Corrosion	Infrastructure
Radiation	Humans, environment

[a] Environment, effects on animals, plants, and the ecosphere; infrastructure, effects on human-made objects.

people (direct human toxicity), but also to flora and fauna (ecotoxicity). With regard to human hazards, the risk normally is to a restricted group that may be exposed directly to the contaminants in sufficient concentrations to do harm. Accidents such as at Chernobyl, for example, pose an extremely high risk to the localized population that may be exposed to high levels of contaminants but a substantially lower risk to other, distant populations; this is characteristic of most pollutants where the significant impacts and health risks are highly localized. This focuses concern on specific contaminated sites and on sources of environmental pollutants that may spread contaminants more widely, thereby potentially impacting additional populations. Accordingly, bioremediation is most often used for site remediation and is especially useful when it can be applied in situ to reduce further environmental contamination.

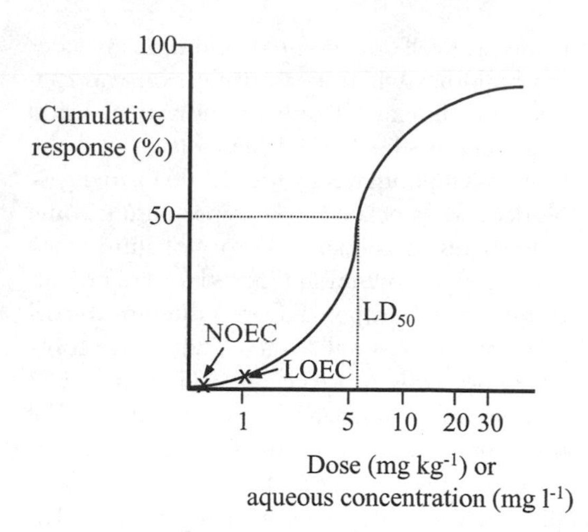

FIGURE 1.1 Cumulative dose-response curve in a lethality test. The typical curve is sigmoidal, and several important parameters can be derived from it. Probably the most widely used is the EC_{50}.

REDUCTION OF TOXICITY

A major aim of bioremediation, or any other remediation technology, must be the reduction of toxicity associated with the environmental contaminant, that is, the abatement of environmental impact. When the application of bioremediation is considered, one aim must be to minimize exposure to hazardous materials, thereby reducing toxicity. To study and compare toxic effects, there must be quantifiable effects (responses). It is necessary to distinguish between exposure and dose. Exposure is the concentration of the chemical in the air, water, or soil to which the test organism is exposed. Dosage is a more precise term, as it incorporates dose and frequency and duration of dosing. Increasing the concentration of a chemical in a toxicity test, or increasing the dose of the toxicant, produces a sigmoidal curve (Fig. 1.1), from which numbers relating to toxicity can be calculated. Central to considering the impact of chemical contaminants and the need for remediation is the dose-response relationship, which takes into account the potential exposure to the contaminant and the severity of the effect of such exposure. Site and risk assessment must account for toxicity of contaminants and possible environmental spread and human exposure.

Some of the most commonly encountered parameters used to describe toxicity are as follows: LC_{50}, the concentration producing 50% mortality in test organisms; LD_{50}, the dose producing 50% mortality in test organisms; IC_{50}, the concentration at which 50% of growth or activity is inhibited; EC_{50}, concentration at which 50% of the predicted effect is observed; LOEC, the lowest-observable-effect concentration, i.e., the concentration at which the lowest effect is seen; and NOEC, the no-observable-effect concentration, i.e., the maximum concentration at which no effect is observed.

Human and Animal Toxicity

The most convenient measure of human and animal toxicity historically has been mortality. Mortality tests are usually performed over very short periods, typically 96 h, and are thus known as acute tests, the endpoints being death, immobility, or cessation of growth or cellular functions. Acute toxicity normally refers to lethality as a result of exposure to a high concentration of a toxicant for a short period (thankfully not the typical concern of environmental pollution). Measurement of mortality with increasing concentrations of chemicals produces robust data that can be used to compare the toxicity of a wide range of chemicals without a need for knowledge of the underlying mechanisms of toxicity. Such data can also be used for risk assessment purposes. Table 1.6 gives toxicity data taken from the *Merck Index* (28) for a few pollutants to illustrate the range of toxic responses possible.

Although environmental contaminants can pose acute health risks, the concerns normally associated with contaminated land and water are with long-term effects. There are relatively few instances, typically ones associated with catastrophic events, in which one or more deaths occur immediately as a result of environmental pollution. As the science of toxicology has evolved, there has been increasing interest in sublethal responses associated with long-term exposure. With reference to contaminated land, water, and air, these sublethal responses are of more relevance. It is highly unlikely that humans will be exposed to concentrations of pollutants that will cause immediate death, although this is a situation frequently encountered by aquatic organisms. The long-term exposure to low levels of pollutants may have a deleterious effect not observed in an acute test, e.g., cancers from exposure to benzene or reproductive impairment from exposure to estrogen mimetics. This has led to the emergence of nonlethal toxicity tests, known as chronic tests. The test methodologies for chronic tests are highly variable and complicated; they must, for example, take into account the whole life cycle of the test organism.

Chronic toxicity refers to exposure to a low dose over a long period. Chronic toxicity, although not necessarily incurring death, can result in long-term debilitation of some form. Attempts have been made to quantify these long-term effects. Based on the average life span of a human, the corresponding exposures are acute if the exposure is for 1 day, subchronic if exposure is for 2 weeks to 7 years, and chronic if exposure is for more than 7 years (137). Bioremediation solutions can be used to reduce the impacts of environmental persistence of contaminants and thus to alleviate problems associated with chronic toxicity.

TABLE 1.6 Some toxic responses to common pollutants[a]

Compound	Toxicity (LD_{50} in rats)[b]	Human toxicity and effects
Benzene	3.8 ml/kg	Acute (from ingestion or inhalation): irritation of mucous membranes, restlessness, convulsions, excitement, depression. Death may follow from respiratory failure. Chronic: bone marrow depression and aplasia, rarely leukemia. Harmful amounts may be absorbed through skin. Benzene has been listed as a known carcinogen.
PCP	146 mg/kg	Ingestion causes increase, then decrease, of respiration, blood pressure, urinary output; fever; increased bowel action; motor weakness; collapse with convulsions and death. Causes lung, liver, and kidney damage, contact dermatitis.
Phenol	530 mg/kg	Ingestion of even small amounts may cause nausea, vomiting, circulatory collapse, tachypnea, paralysis, convulsions, coma, greenish or smoky-colored urine, necrosis of the mouth and gastrointestinal tract, icterus, death from respiratory failure, sometimes cardiac arrest. Average fatal dose is 15 g. Fatal poisoning may also occur by absorption following application to large areas.
TCDD	0.045 mg/kg	Extremely potent, low-molecular-weight toxin. Toxic effects in animals include anorexia, severe weight loss, hepatotoxicity, chloracne, teratogenicity, and delayed death. May reasonably be anticipated to be a carcinogen.

[a] From the *Merck Index* (28).

[b] Benzene was tested on young rats; the other compounds were tested on adult rats.

Most chronic tests rely upon the measurement of growth, which is an important indicator of the fitness of the individual. Growth alone, though, may be an inadequate measure, and a number of other measurements normally are taken. Reproduction is a sensitive endpoint for a chronic toxicity test. The effects may be temporary or permanent, and they range from delayed sexual maturity to complete loss of reproductive capability. The effects on the early developmental stages in the life cycle of the test organisms have also been studied. Morphological, biochemical, cytological, and behavioral changes have all been used as indicators and are being developed.

Standardized toxicological tests typically are used to determine the acute and chronic toxicities of chemicals. Chemicals are extracted and used on test organisms. Since it is typically unethical to run toxicity tests on humans, various surrogates are employed, ranging from bacteria to mice and nonhuman primates. To measure the toxicity of environmental contaminants and to exclude the possibility that some unsuspected toxic or mutagenic residue of pollutant is present, it is sometimes desirable to complement residue analysis with bioassays performed using actual environmental samples rather than extracts. Often, toxicity tests for environmental contaminants are initially run as screening tests using the microcrustacean *Daphnia*, bivalves (such as oyster larvae), fish (such as rainbow trout), and various bacteria, among other organisms, as indicators of toxicity to higher organisms, including humans.

As a rapid and convenient measure of acute toxicity, the reduction of light emission by luminescent bacteria (Microtox assay) can be used. Microtox measures the decrease in respiration, and subsequent light output, of the luminescent bacterium *Vibrio fischeri* as the toxic response. Bacterial bioluminescence is tied directly to cell respiration, and any inhibition of cellular activity (toxicity) results in a decreased rate of respiration and a corresponding decrease in the level of luminescence. The more toxic the sample, the greater the percent light loss from the test suspension of luminescent bacteria. Bacterial bioluminescence has proved to be a convenient measure of cellular metabolism and consequently a reliable sensor for measuring the presence of toxic chemicals in aquatic samples. While not approved by the U.S. EPA for compliance toxicity testing, the Microtox Acute Test has achieved official "Standards Status" in several countries, including an American Society for Testing and Materials standard (D-5660) in the United States; the final International Standards Organization draft (11348–3) entitled "Water Quality-Determination of the Inhibitory Effect of Water Samples on the Light Emission of *Vibrio fischeri* (Luminescent Bacteria) Test" also has been approved. Microtox testing can be a rapid, cost-effective tool in assessing toxicity of effluents, sediments, leachates, soils, sludges, groundwater, and surface water. Test data can be available within as little as 30 min, and numerous samples can be easily processed during a standard workday. Microtox is most useful when the bacterial response correlates well with the response of standard test species, which may be required in compliance tests, or indigenous species within the receiving water or water body of concern. Microtox data can be analyzed by calculating EC_{50}s based on dose-response curves. A published database that lists Microtox EC_{50}s (milligrams per liter) for over 1,200 chemicals is available.

One criticism of the Microtox assay is that the organism utilized in this assay is a marine organism and as such may not represent toxicity towards soil microbes. Other microbe-based bioluminescence assays using bacteria isolated from soil itself have been developed. These organisms include *Pseudomonas* and *Rhizobium* species (99). While these soil organism-based assays may be more relevant, they have not been widely used and hence lack comparative data to be used in standardizing data for regulatory purposes.

Mutagenic and carcinogenic contaminants are often assayed by utilizing the Ames test, which uses microorganisms as the test organisms (83). The Ames test procedure typically uses strains of the bacterium *Salmonella enterica* serovar Typhimurium for determining chemical mutagenicity. The serovar Typhimurium strains employed in the Ames test procedure are auxotrophs that require the amino acid histidine for growth. Several different strains are used, each specific for a type of mutation, such as frameshift, deletion, and so forth. The reason for using several strains is that they differ in their responses to different types of chemicals. For example, one strain may have greater permeability to large molecules than another and may hence be a better organism to use when large molecules are being tested. Often five strains are used in the test protocol.

In the Ames test, the auxotrophic bacteria are exposed to a concentration gradient of the chemical being tested on a solid growth medium that contains only a trace of histidine. The amount of histidine in the medium is just sufficient to support the auxotrophs long enough for the potential mutagenic chemicals to act. Normally, the test strain bacteria cannot grow sufficiently to form visible colonies because of the lack of histidine. Therefore, in the absence of a chemical mutagen, no colonies develop. If the chemical is a mutagen, many mutations will occur in the areas of high chemical concentration. It is likely that no growth will take place in these areas because of the occurrence of lethal mutations. At lower chemical concentrations along the gradient, fewer mutations will occur. Some of the mutants will be revertants to the prototrophs that do not require histidine. Since histidine prototrophs synthesize their own histidine, they grow and produce visible bacterial colonies on the histidine-deficient medium. The appearance of these colonies demonstrates that histidine prototrophs have been produced, and a high rate of formation of such mutants suggests that the chemical has mutagenic properties.

The Ames test is also useful to determine if a chemical is a potential carcinogen because there is a strong correlation between mutagenicity and carcinogenicity. Even though the Ames test does not actually establish whether a chemical causes cancer, determining whether a chemical has mutagenic activity is useful in screening large numbers of chemicals for potential carcinogenicity. In the Ames procedure, the chemical is incubated with a preparation of rat liver enzymes to simulate what normally occurs in the liver, where many chemicals are inadvertently transformed into carcinogens in an apparent effort by the body to detoxify the chemical. Following this activation step, various concentrations of the transformed chemical are incubated with the *Salmonella* auxotroph to determine whether it causes mutations and is a potential carcinogen. Further testing for carcinogenicity is done on the chemicals that test positive for mutagenicity.

These tests are useful in establishing whether soils or waters contain contaminants of health and/or environmental concern. In many cases, governmental bodies set regulatory standards for specific chemicals in an effort to control long-term effects. The primary concerns in setting acceptable exposure levels are carcinogenicity, teratogenicity, and direct toxicity. For purposes of considering the applicability of bioremediation, it is not necessary to differentiate among carcinogenic, teratogenic, and direct toxic effects. All can be considered forms of contaminant toxicity that may require remediation of the contaminated environment. The success of bioremediation can be assessed by demonstrating a reduction in the toxicities associated with environmental samples to levels that are considered safe.

Ecotoxicology

In considering ecotoxicity, it is important to recognize that the actual toxic effects of a chemical depend not only on the properties of the chemical itself but also upon several factors that impact possible exposure. Critical among these are bioavailability, which determines the concentrations of the chemical to which organisms actually are exposed, and biomagnification, which is a process through which chemicals can become concentrated at higher levels in a food web.

BIOAVAILABILITY

If a pollutant is present in soil or water but is not available to the biota, then it presents minimal risk. Accordingly, the concept of bioavailability is enshrined in the risk assessment approach to contaminated land and water. The bioavailability of pollutants in the environment is also an important factor affecting bioremediation and so is an essential area both for understanding and for further research. There are several definitions of bioavailability (79). In its simplest form, bioavailability can be defined as "the amount of contaminant present that can be readily taken up by living organisms, e.g., microbial cells" (80). Another definition that takes account of this and alludes to the risk assessment practice (78) is "a measure of the potential of a chemical for entry into ecological or human receptors." Although the importance of bioavailability for risk assessment and determination of the necessity of environmental remediation is undisputed, a number of factors influence bioavailability and quantification of bioavailability is extremely complex.

Solubility in water is one of several factors that significantly impacts bioavailability. Although the concentration of pollutant chemicals is usually rather low in natural water systems, it is in this aqueous phase that organisms most often contact toxic chemicals, making solubility an especially important consideration. Solubility of any solute in any solvent is mediated by forces of attraction, the main ones being (i) Van der Waals forces, (ii) hydrogen bonding, and (iii) dipole-dipole interactions. Charged or highly polar organic and inorganic molecules, hydrophilic molecules, are readily soluble in water. Conversely, nonpolar, hydrophobic molecules find solution in water a less energetically favorable state, and they have low water solubility. Benzene has moderate water solubility, but the addition of the polar hydroxyl group makes phenol much more water soluble.

The size and shape of a molecule are critical in determining its overall water solubility. Large molecules tend to have lower water solubility than small molecules because they have a higher molar volume. The addition of halogens to an organic molecule might be expected to increase the water solubility of the new molecule since halogens are electronegative (electron-withdrawing) species and would thus tend to increase the polarity of the molecule. However, in the increasingly chlorinated series of phenols, this is not the case: in fact, increasing chlorination very markedly decreases the solubility of chlorophenols. This is because the halogens increase the molecular volume of organic compounds (95, 145).

Other chemical functional groups affect water solubility in rather specific ways. The presence of polar functional groups does indeed tend to increase water solubility of molecules compared to that of unsubstituted hydrocarbons: alcohols, amines, ethers, ketones, and organic acids are good examples. Even with these examples, though, a general rule of increasing molecule size and decreasing water solubility still exists. The short-chain fatty acid acetic acid is much more water soluble than a long-chain fatty acid: the additions to the molecule are not more polar functional groups but simply more $C\text{-}H_2$ groups, making the molecule larger but less polar.

The situation is further complicated by more specific chemical reactions. The case of chlorophenols illustrates this point. Chlorophenols behave as weak acids when dissolved in water due to the loss of a proton from the hydroxyl group to leave a phenoxide ion. The more chlorines that are added to the ring, the more electrons are withdrawn from the electron-rich phenoxide ion. This lowers the phenoxide negative charge and therefore lowers its ability to hold a proton. Thus, the more chlorines on the ring, the more easily a proton is released and the greater the acidity. The ionized form is usually several orders of magnitude more water soluble than the neutral un-ionized form. The relative proportion of each species is highly dependent on pH.

Measuring Bioavailability. Measurement or estimation of bioavailability is problematic, as it is affected by many individual and interacting conditions relating to soil and water chemistry, pollutant chemistry and partitioning, and biological transformation and concentration factors (63). The techniques currently being researched can be categorized broadly as direct or indirect and biological or chemical (74), although truly direct chemical methods are not possible since only organisms can determine whether a chemical is bioavailable.

Direct biological techniques measure the actual amount of a chemical taken up by a target organism, and this may ultimately be the most accurate measure of bioavailability, although it is by no means routine; e.g., the determination of chemical levels in earthworms is complicated, and determining the concentrations acquired by humans, while possible in body fluids, generally is impossible in tissues except upon autopsy. The great strength of these direct techniques when they can be used, however, is that they integrate all the biotic and abiotic modifying factors of chemical bioavailability (74). However, for a fledgling bioremediation industry, reliance on direct biological methods may not be practical.

Fortunately, it is also possible to observe a response to a chemical in an organism without actually measuring the concentration of the chemical. A range of responses is possible, such as lethality, enzyme induction or inhibition, and reproductive effects. These are regarded as indirect biological methods because the effect in the organism may be quantifiable but the chemical concentration remains unknown. A promising, novel approach is the use of genetically modified microorganisms that can detect and quantify specific pollutants. This type of biosensor is based on the highly specific genetic control mechanisms used by microorganisms to ensure that specific proteins are expressed only when they are needed, for example, for the detoxification of a particular toxic substance. This control is exerted by inducible promoters, consisting of a specific DNA sequence upstream of the genes to be controlled, and a DNA-binding protein that either activates or prevents transcription in response to the presence or absence of the target compound.

Biosensors of this type are easily generated by fusing a controllable promoter to a reporter gene that generates a detectable signal when the promoter is activated (102). The most popular reporters are the bacterial *lux* system and the green fluorescent protein, which produce light, because light can be easily quantified and, due to its rarity in biology, there is no interference from the background biochemistry of the host organism. Such biosensors can be both highly specific and responsive to very low concentrations, and the presence of the microbial cell membrane makes this approach truly a measure of bioavailability. There are limitations to developing such biosensors for detection of organic pollutants, since the required specificity is often lacking. With heavy metals, specificity may not be a problem. Nevertheless, careful selection of the host strain is required for a variety of reasons (106). The bigger challenge to make this technology applicable to contaminated land is the extraction procedure. Making a bioluminescent biosensor respond to a soil sample is more difficult. Therefore, use of these sensors generally relies on extraction procedures, but there is little agreement over which extraction procedure will accurately reflect bioavailability, although methods which use the soil solution seem to hold the most promise (128).

An interesting approach that may overcome the extraction-related problem is a direct soil contact bioassay using the soil bacterium *Nitrosomonas europaea* (26). This is a novel approach that allows solid-phase contact, thereby providing information about toxicity of the bioavailable pools of adsorptive soil pollutants. The solid-phase contact assay permits in situ toxicity testing of soil. The *Nitrosomonas* strain used in this assay is a recombinant strain that contains *lux* bioluminescent reporter genes. This recombinant strain is mixed with a soil slurry or soil extract, after which bioluminescence is measured. Tests with linear alkene benzosulfonates show that the solid-phase contact assay allows direct interaction of the test microorganisms with bioavailable pools of the toxicants in soil and thus provides a very sensitive method for evaluating the in situ toxicity and assessing the risks of soil contaminants.

Chemical methods can also be employed to assess bioavailability. Indirect chemical methods usually involve the extraction of a fraction of the chemical (metals or organics) from a soil, the extractability being equated to bioavailability and defined by the chemical itself, the nature of the extractant(s), and the experimental conditions applied. The origins of this approach are in sequential extraction procedures (125), which attempt to quantify the differentiation of toxicants into those weakly bound and those strongly bound to the soil matrix. These tests often assume that the weakly bound toxicants are those that are more bioavailable, although evidence for this is incomplete. A recent advance in this technique is sequential accelerated solvent extraction (123), which reduces sample preparation time and maintains relatively constant extraction conditions. Another technique under development is selective supercritical fluid extraction (21).

Because bioavailability to humans often is the critical issue, a number of methods center around the human gastrointestinal tract and give an indication of likely uptake of inorganic contaminants from the digestive system (the physiologically based extraction test [PBET]) (109). To mimic human conditions, these methods incorporate gastric juices and enzymes with mixing at 37°C and use soil residence times similar to those found for children after ingestion of food. Analysis of all solutions produced in the PBETs is completed by analytical chemistry techniques under matrix-matched conditions. These techniques have been validated in vivo for arsenic and lead by PBET and for lead by SBET and have been used by the U.S. EPA to assay toxicity due to bioavailable heavy metals (29).

A reproducible soil extraction technique for estimating soil-associated organic contaminant bioavailability has been created by using cyclic oligosaccharides (42, 105). The α-, β-, and γ-cyclodextrins are cyclic oligosaccharides formed by 6, 7, or 8 α-linked glucose units, respectively (121). They have a hydrophilic shell and are highly water soluble but also have a hydrophobic cavity within which hydrophobic organic compounds can form inclusion complexes (14). Hydroxypropyl-β-cyclodextrin (HPCD) has higher water solubility than other cyclodextrins, and HPCD has been shown to increase the solubility of naphthalene and phenanthrene by 20- and 90-fold, respectively, in the presence of 50 g of HPCD liter^{-1} (10). The technique may be limited to compounds with a shape and size that are compatible with the HPCD cavity (42). However sophisticated such extraction techniques become, any such chemically based extraction technique is ultimately limited by the lack of an organism, i.e., by the lack of the real detector of true bioavailability. There is no universal chemical test for bioavailability since organisms themselves are the real detectors; to complicate matters even further, bioavailability is species specific (104).

Pollutant Partitioning and Bioavailability. In addition to solubility, bioavailability depends upon the localized distribution of the chemical within the environmental matrix. As soon as a chemical is introduced to the environment it becomes a pollutant, but the fate of that pollutant depends to a large degree on its chemistry, which determines how it will partition to various "compartments," or phases, of the environment. Partitioning is a general phenomenon that describes the tendency of a pollutant to exist between phases at equilibrium. The compartments of the environment of relevance to partitioning are air, water, soil and sediment, and biota. The concept of pollutant partitioning is important to risk-based strategies for the assessment of contaminated land and water. Once a pollutant is released into soil, there are a number of possible fates: soil sorption, leaching (to surface water or groundwater), volatilization to the atmosphere, plant root uptake and translocation, uptake to plant foliage, and uptake and accumulation in, or excretion from, animals. Very often a contaminant spill on land becomes a water pollution issue. As many countries are heavily reliant on aquifers as a source of potable water, the fate of pollutants becomes a health and safety issue, with ramifications for the technique of choice for cleanup.

Partitioning of chemical pollutants is governed by fundamental properties of chemicals and environmental factors associated with contaminated soils and waters. The term fugacity literally means the tendency to flee and refers to a molecule's driving force to escape from the compartment it currently exists in. With knowledge of all of a chemical's partition coefficients, it is possible to predict its relative concentrations in air, water, and soil at equilibrium. The gradient of fugacity between two compartments (e.g., soil and soil pore water) determines the potential for the chemical to move from one compartment to the other. Fugacity therefore has units of pressure, but this can be related to the chemical concentration through modeling, and thus the distribution through various compartments can be predicted.

Vapor pressure is the partial pressure of a chemical in a gas phase that is in equilibrium with the pure liquid or solid phase of the chemical. Vapor pressure is greatly influenced by temperature, and figures are usually quoted at a constant temperature (usually 20 or 25°C). If the volume of the containing vessel is also known, the vapor pressure can be used to calculate the concentration of the chemical in the headspace of the vessel from the ideal gas law:

$$\frac{n}{V} = \frac{P}{RT}$$

where P is vapor pressure (in atmospheres), V is volume, n is the number of moles of the chemical, R is the gas constant, and T is the absolute temperature.

Clearly, vapor pressure is the driving force for volatilization and transfer of the polluting chemical from soil or water into the atmosphere. Vapor pressures of organic compounds are often compared to mercury vapor pressure (atmospheric pressure is 760 mm Hg). Thus, the vapor pressure of benzene is 87.2 mm Hg at 25°C. Volatilization potential, however, can be misleading in the absence of information on the water solubility of the compound. A compound with a low vapor pressure may still have a strong tendency to escape if the water solubility is low. For example, 2,4,6-trichlorophenol has a much lower vapor pressure than 3-chlorophenol, yet its tendency to escape to air is greater. This can be explained when the water solubilities are compared: the much lower value for 2,4,6-trichlorophenol results in its having a Henry's constant about 1 order of magnitude higher than that of 3-chlorophenol, even though its vapor pressure is much lower. Indeed, for the purpose of studying partition phenomena, Henry's law constant is more valuable than vapor pressure. Henry's law constant is the ratio of the equilibrium concentration of a compound in air and its equilibrium concentration in water. It can therefore be regarded as the partition coefficient between air and water and is the best indicator of the tendency of a chemical to volatilize from water. Henry's law constant can be written as

$$H = \frac{P}{S}$$

where H is Henry's law constant (in atmospheres per cubic meter per mole), P is vapor pressure (in atmospheres), and S is water solubility (in moles per cubic meter).

Another critical factor impacting partitioning and bioavailability is the hydrophobicity of the chemical. Polarity and water solubility do not give direct measures of the hydrophobicity of a compound. A hydrophobic compound should preferentially dissolve in a lipophilic phase, and this is the basis for the octanol-water partition coefficient, K_{ow}, which is the ratio of the concentration of a chemical, at equilibrium, when dissolved in octanol and in water, when both phases are saturated with each other. The K_{ow} is a useful measure that relates to the bioavailability and biodegradability of compounds (Table 1.7).

Measurement of K_{ow} was devised by the pharmaceutical industry to study partitioning of drugs as it roughly equates to the partitioning between water and body fat. It has proven to be very useful in predicting how chemicals will partition in the environment (57). Large molecules with low polarity have a high K_{ow} and are more likely to partition to solids in the environment. Small, polar molecules dissolve in water rather than octanol, have a low K_{ow}, and are much less likely to partition to solids in the

TABLE 1.7 Suggested values of log K_{ow}, log H_c, and $t_{1/2}$ influencing the fate and behavior of organic pollutants in soils[a]

Fate or behavior	Classification
Adsorption potential	
log K_{ow} ≤2.5	Low
log K_{ow} >2.5 and ≤4.0	Moderate
log K_{ow} >4.0	High
Volatilization	
log H_c ≤–3 and log (H_c/K_{ow}) ≤–8	Low
log H_c >–3 and log (H_c/K_{ow}) ≤–8	Moderate
log H_c ≤–3 and log (H_c/K_{ow}) >–8	Moderate
log H_c >–3 and log (H_c/K_{ow}) >–8	High
Degradation	
$t_{1/2}$ ≤10 days	Nonpersistent
$t_{1/2}$ >10 days and ≤50 days	Moderately persistent
$t_{1/2}$ >50 days	Persistent
Leaching	
log K_{ow} >4.0 or $t_{1/2}$ ≤50 days and log K_{ow} >2.5 and ≤4.0.	Low
$t_{1/2}$ >50 days and log K_{ow} >2.5 and ≤4.0	Possible
$t_{1/2}$ ≤50 days and log K_{ow} ≤2.5	Possible
$t_{1/2}$ >50 days and log K_{ow} <2.5.	High

[a] $t_{1/2}$, half-life.

environment, so they are more likely to remain in an aqueous phase.

$$K_{ow} = \frac{\text{concentration of pollutant in water-saturated octanol (milligrams per liter)}}{\text{concentration of pollutant in octanol-saturated water (milligrams per liter)}}$$

The range of values for K_{ow} is very high, and tabulated values are more often expressed as log K_{ow}. As a molecule becomes larger through increasing chlorination, the water solubility decreases and the log K_{ow} increases.

Although typically the organic matter content of soil is in the range of 0.1 to 5%, this is where organic pollutants sorb, and not to the soil as a whole. The soil adsorption coefficient, K_{oc}, takes account of this. For a liquid organic pollutant

$$K_{oc} = \frac{\text{mass of pollutant sorbed to the soil organic matter (milligrams per gram)}}{\text{mass of pollutant in the aqueous phase (milligrams per milliliter)}}$$

Alternatively, K_{oc} (in cubic meters per gram) = s/C_{Ae} (48), where s is the mass of solute sorbed per unit of dry mass of soil (in milligrams per milligram) and C_{Ae} is the equilibrium concentration in the aqueous phase (in grams per cubic meter). The range of values for K_{oc} is very large, and therefore the use of log K_{oc} is preferred. The soil partition coefficient is greater the more hydrophobic the pollutant. However, it has been measured for far fewer chemicals than K_{ow}; a list of measured K_{oc} values is given by Watts (137). It might be expected to be a more realistic measure than K_{ow}, but as data are relatively rare for K_{oc} a series of correlation equations that relate K_{oc} to K_{ow} has been developed. If the mass fraction of organic carbon in the soil is known, K_{oc} can be estimated by

$$K_{oc} = 6.3 \times 10^{-7} f_{oc} \times K_{ow}$$

where f_{oc} is the soil organic fraction. As a first approximate, the K_{ow}, or the log K_{ow}, can be taken as a measure of the potential of a pollutant to sorb.

K_{ow} and log K_{ow} have become increasingly used as predictors of the behavior of pollutants in the environment and have been correlated with toxicity. The most common mode of action regarding toxicity of industrial pollutants is probably narcosis. Nonpolar narcosis results from the perturbation of cellular membranes (41) as a result of the entry of hydrophobic (lipid-soluble) molecules into the phospholipid bilayer of the membrane. Narcosis has been shown to be a nonspecific physiological effect which is independent of chemical structure, and log K_{ow} is highly correlated with nonpolar narcosis toxicity (39). Phenols have greater toxicity than is predicted by nonpolar narcosis (40). They act by polar narcosis, which probably results from the possession of a strong hydrogen-bonding group that makes the compound more polar. Nevertheless, there is still a strong correlation between toxicity of these compounds and log K_{ow}. Pentachlorophenol (PCP) has yet greater toxicity as a result of a further mechanism: its acidity makes it a weak acid respiratory uncoupler (WARU). WARUs possess an acid-dissociable moiety, for example, a hydroxyl group; a bulky hydrophobic moiety, such as an aromatic ring; and multiple electronegative groups, such as halogens (124). During weak acid uncoupling, ATP synthesis is inhibited, with no effect on the respiratory chain (114); i.e., the tight coupling between electron transport and oxidative phosphorylation is abolished, allowing respiration to proceed without control by phosphorylation.

In soils, the organic fraction of soil (humus) is largely responsible for the sorption of hydrophobic organic compounds. This has significant consequences for risk assessment, as partitioning into the soil organic phase decreases exposure to some receptors. Many of the PAHs are sorbed to organic matter. The degree of retention is directly correlated with the log K_{ow} of the compound and thus to its hydrophobicity. As the number of rings increases, the log K_{ow} increases: the persistence of PAHs in soil increases with their molecular weight. All except naphthalene have a log K_{ow} of >4 and are highly sorbed. One possible mechanism is partitioning into the organic matter (32), rather than simple adsorption. Once bound into the physical matrix in this way, these compounds are difficult to remove (Fig. 1.2), and the sorp-

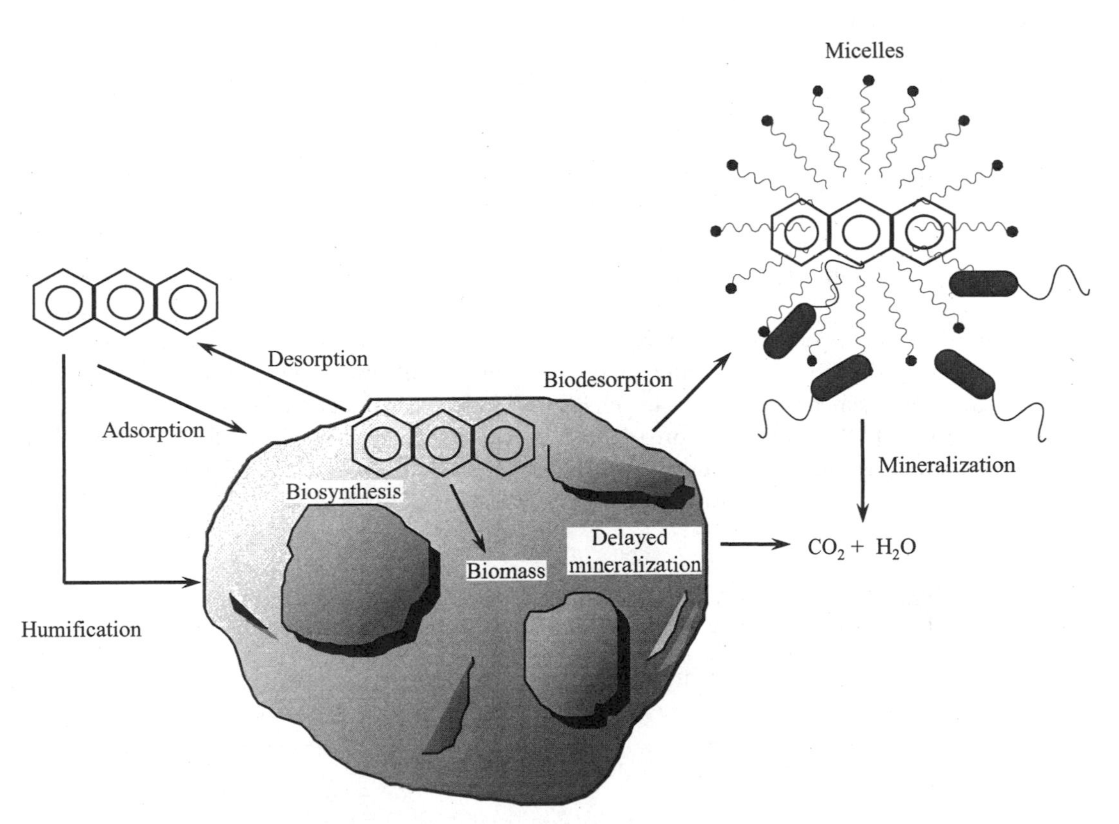

FIGURE 1.2 Fate of hydrophobic pollutants in soil, and the use of surfactants to try to improve desorption and bioavailability. With time, the pollutant becomes increasingly bound to the organic fraction of soil and is consequently more difficult to desorb; thus, it becomes less bioavailable and more difficult to biodegrade. Surfactants, including biosurfactants, may improve desorption and solubilization. There is strong evidence to suggest that biodegradation of such pollutants occurs in the aqueous phase.

tion process seems to become more severe with aging (65), a process with ill-understood mechanisms (33). Binding to humus may be irreversible (72). Such humification can be beneficial to prevent partitioning into water and uptake by living organisms, thereby detoxifying hydrophobic pollutants and dead-end metabolites, such as halocatechols.

BIOMAGNIFICATION

Biomagnification begins with bioaccumulation, which is the uptake of a chemical from the environment by a living organism that is able to concentrate that chemical. Bioaccumulation involves the uptake of the pollutant from the surrounding contaminated environment in low concentrations, either by active uptake (common for hydrophilic compounds) or by passive diffusion (common for hydrophobic compounds), and accumulation within the organism to much higher levels. Bioaccumulation occurs with many toxic pollutants, including metals, and the accumulation factors can be very high indeed, from very low levels of contamination. Bioaccumulation factors in hundreds of thousands have been recorded for animals and plants with organic and metallic pollutants.

Biomagnification involves progressive amplification through a food chain; whereas bioaccumulation is independent of trophic level, in biomagnification the concentration of the chemical increases as one moves up the food chain (84). Biomagnification is usually associated with hydrophobic organic compounds, such as organochlorine pesticides. Toxicants taken up by a living organism can either be metabolized to a more soluble form and excreted or be stored in tissues not directly involved in metabolism, e.g., hair and fat deposits. Biomagnification takes place when a predator consumes an organism containing a bioaccumulated toxicant. As the toxicant passes up the food chain, it has the potential to amplify at each trophic level. As a result of biomagnification, the ecotoxicological impacts of chemical pollutants often occur at the top levels of a food web. Occasionally, humans are at the top of the food chain: this is what was seen at Minamata, where mercury as methylmercury had passed up the food chain to the local people. Organochlorine pesticides are biomagnified within food chains, and the concentration factors in the top predators can be as much as 10^7 from water (122).

While polar, water-soluble hydrophilic chemicals are more readily available to organisms than nonpolar, hydrophobic chemicals, the latter are also lipophilic and hence more likely to be biomagnified. The potential for biomagnification can be estimated in a readily performed laboratory test that measures the partitioning between octanol and water. Generally, small, more polar molecules with low K_{ow}s have low tendencies to partition into fat deposits. Note that many of the hydrocarbon pollutants discussed can be described as nonpolar. As the PAHs increase in number of rings,

TABLE 1.8 Increasing ring number and molecular weight of the PAHs decreases water solubility and increases hydrophobicity and half-life in soil, thereby increasing persistence

PAH	No. of rings	Water solubility (mg/liter)	log K_{ow}	$t_{1/2}$ (days)[a]
Naphthalene	2	32	3.37	48
Anthracene	3	0.059	4.54	110
Chrysene	4	0.0033	5.86	230
Pyrene	4	0.13	5.18	200
Benzo[*a*]pyrene	5	0.0038	6.04	230
Benzo[*ghi*]perylene	6	0.00026	7.10	280

[a] $t_{1/2}$, half-life.

TABLE 1.9 Distribution between phases of some representative chemicals[a]

Phase	% Compound in phase[b]				
	PCP	Phenanthrene	Tetrachloroethane	DDT	HCB[c]
Air	5.8	76.2	99.9	0.4	7.5
Water	0.6	0.4	4.0×10^{-2}	5.0×10^{-2}	2.0×10^{-3}
Soil	81.0	20.2	2.0×10^{-2}	86.7	80.0
Sediment	11.9	3.0	3.0×10^{-3}	12.7	11.7
Aquatic-biota	2.0×10^{-3}	4.0×10^{-4}	4.0×10^{-7}	2.0×10^{-3}	2.0×10^{-3}
Vegetation	0.6	0.1	1.0×10^{-4}	0.6	0.6

[a] Adapted from the work of Connell et al. (36).
[b] Percentage distribution of the chemicals in all phases.
[c] HCB, hexachlorobiphenyl.

they become less water soluble and more hydrophobic and their log K_{ow} increases (Table 1.8). The higher the log K_{ow}, the greater the potential for biomagnification.

Table 1.9 shows how some important pollutants partition between the various environmental media that they contaminate. Of the group, phenanthrene and tetrachloroethane are relatively volatile (high Henry's law constant) and partition significantly into the air. Note that PCP, dichlorodiphenyltrichloroethane (DDT), and hexachlorobiphenyl all show about 80% partitioning to soil, which has a large influence on their bioavailability.

DEOXYGENATION

Whereas toxicity is a pollution effect with major implications for human health, deoxygenation is much more an ecotoxicological effect. Deoxygenation of water has serious consequences for the biota of surface water and contaminated soils. Groundwater and soil pore water deoxygenation slows natural attenuation of contamination to negligible levels. The problem of water deoxygenation stems fundamentally from the limited aqueous solubility of oxygen. Moreover, the solubility of oxygen in water is greatly influenced by temperature (Table 1.10). Even at saturation, figures of around 10 mg liter^{-1} are very low compared to the amount of oxygen available in air.

The organic chemicals present in domestic and industrial wastes are acted upon by bacteria attempting to biodegrade them oxidatively. When eutrophication occurs, the decomposition of the excessive concentrations of nutrients results in the metabolic consumption of the available oxygen. Therefore, a primary effect of discharge of pollution to a water body is depletion of oxygen. Similarly, in soils with limited porosity and hydraulic conductivity, oxygen depletion may be the principal limiting factor for natural attenuation. The situation can be illustrated by the Streeter-Phelps model developed in the 1920s (Fig. 1.3). Many variables have an influence on the dissolved oxygen (DO) concentration of water along a stretch of a river, e.g., varying temperatures, depths, photosynthetic activity, diurnal effects, and salinity effects. To properly model all of these effects and their interactions is an exceedingly difficult task and is also site specific. The simple model presented ignores such variables and is only at very best a first approximation to reality, but it serves to illustrate the two competing forces of oxygen depletion, caused by the waste, and reaeration (restoration of oxygen)

TABLE 1.10 Solubility of oxygen in water[a]

Temp(°C)	Chloride concn in water (mg/liter)			
	0	5,000	15,000	10,000
0	14.62	13.73	12.89	12.10
10	11.29	10.66	10.06	9.49
20	9.09	8.62	8.17	7.75
30	7.56	7.19	6.85	6.51

[a] Adapted from the work of Thomann and Mueller (126).

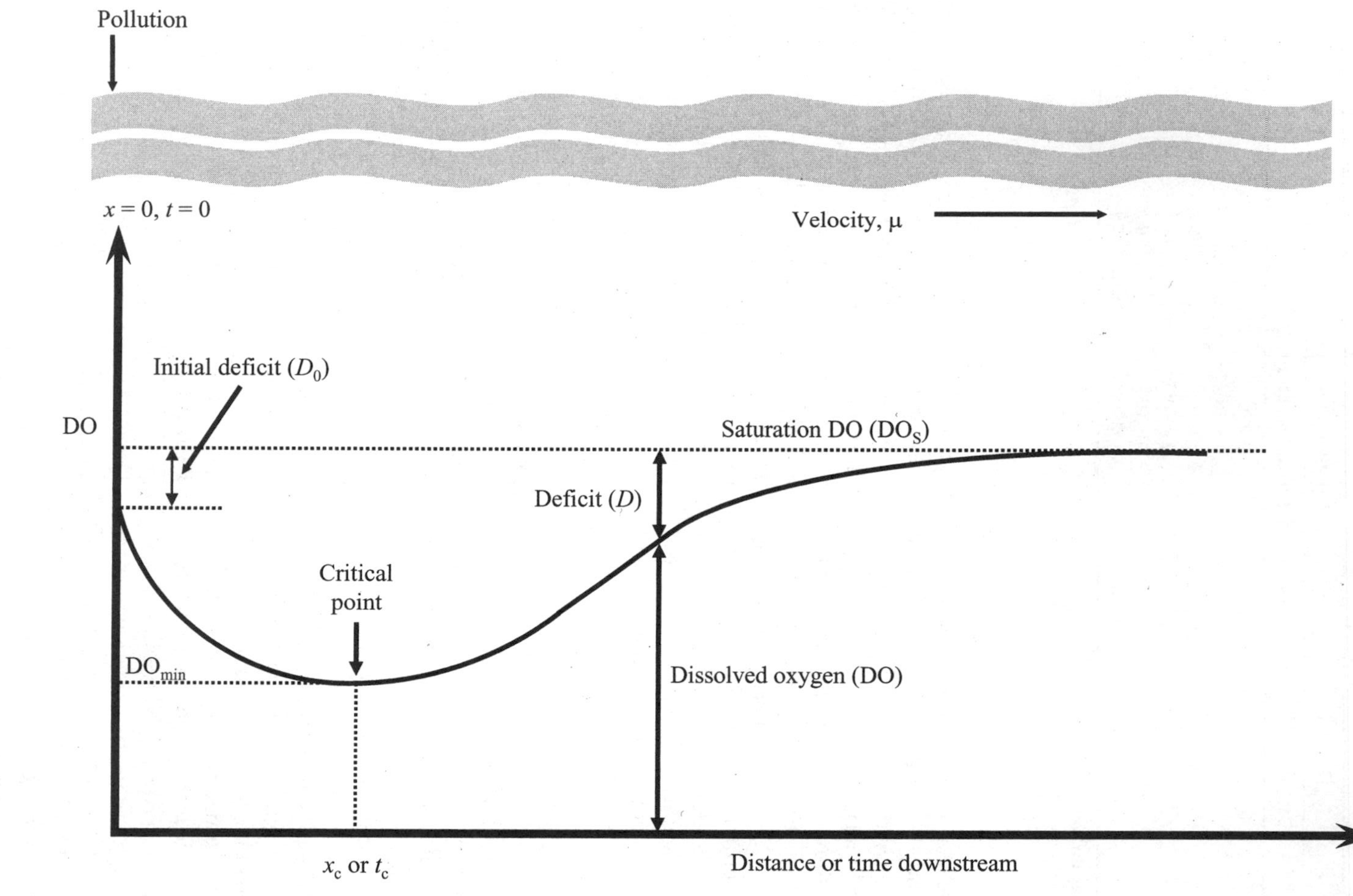

FIGURE 1.3 The oxygen sag curve showing decrease in oxygen concentration due to biodegradation of a pollutant in a waterway. As soon as a pollutant enters a river, it starts to deoxygenate water as a result of the biological oxygen demand it possesses. The two competing phenomena at play are the deoxygenation and the reaeration across the air-water interface. At some critical point downstream, the DO level reaches a minimum, after which the rate of reaeration exceeds the rate of deoxygenation, the DO starts to rise again, and the river recovers.

across the air-water interface. The model assumes a continuous discharge of waste at a given point on the river. It is also assumed that the water and wastes are uniformly mixed at any given cross-section. Finally, it is assumed that there is no dispersion of wastes in the direction of flow, i.e., that plug flow conditions prevail.

When an organic effluent enters the river, the competing forces of oxygen consumption (deoxygenation) and atmospheric reaeration quickly become apparent. In these early stages, the oxygen demand will often exceed the reaeration rate and the DO concentration will fall downstream of the outfall, creating an oxygen deficit. The rate of oxygen diffusion across the air-water interface is directly proportional to the oxygen deficit, so that if the rate of consumption lowers the oxygen concentration, the oxygen mass transfer rate will increase in the direction of air to water. Water with a measured DO concentration of 7 mg liter^{-1} and a saturation concentration of 9 mg liter^{-1} has an oxygen deficit of 2 mg liter^{-1}. If this water receives organic effluent and the DO concentration is lowered to 5 mg liter^{-1} (a dangerous level for many aquatic life forms), the oxygen deficit will be 4 mg liter^{-1}. Now oxygen will diffuse into the water twice as quickly.

At some point downstream, the rates of oxygen consumption and reaeration become equal and the oxygen concentration stops declining. This is the critical point. At the critical point, the oxygen deficit (*D*) is maximal and the DO concentration is minimal. Beyond this point, reaeration predominates and the DO concentration rises to approach saturation. Therefore, the river can be said to be recovering beyond the critical point.

The long tail characteristic of the recovery phase results from the oxygen mass transfer rate being directly proportional to the oxygen deficit. As the river recovers, the DO concentration increases; therefore, the oxygen deficit decreases and the rate of reaeration lowers. A more mathematical treatment of this situation is given by Masters (85). Note that if the organic effluent is strong, it is possible that the DO concentration can be reduced to zero, and an anaerobic zone is formed. This is easily recognized on a river by the foul smell, a mixture of anaerobic gases with the predominant smell being that of hydrogen sulfide. If the discharge environment is marine, the anaerobic conditions established soon mean that the most important electron acceptor is sulfate, as seawater contains some 2.7 g of sulfate kg^{-1} (118). This can lead to gross pollution by hydrogen sulfide, which not only is an extremely noxious and corrosive gas, but also is highly toxic to humans, flora, and fauna.

ENVIRONMENTAL POLLUTANTS AND THEIR BIODEGRADABILITY

The broadest classification of environmental pollutants is into two categories: organic and inorganic. Quantitatively, the organic pollutants of most concern are the hydrocarbons in their various forms. The most common are petroleum hydrocarbons (mixtures of *n*-alkanes; other aliphatics; mono-, di-, and polyaromatic compounds; heterocyclic aromatics; and other minor constituents), chlorinated solvents (e.g., trichloroethylene), surfactants, biocides (e.g., chlorinated phenols), and a host of other compounds specific to particular industries, e.g., nitroaromatics from munitions. A comprehensive encyclopedia (64) provides information on toxicity, carcinogenicity, and fate of the diversity of environmental pollutants. Fortunately, many of these pollutants are biodegradable by microorganisms in soils and waters (5, 136).

The biodegradability of environmental pollutants, and hence the degree of persistence of contaminants in natural environments, is influenced by various factors, most important of which are the chemical structure of the contaminant, the presence of a viable microbial population able to degrade the contaminant(s), and environmental conditions suitable for microbial biodegradative activities. Persistent contaminants are especially problematic if they are lipophilic, since they can be biomagnified as they move through the food web. Additionally, some pollutants can be metabolized to more toxic or carcinogenic compounds. PAHs, for example, are oxidized by mammalian enzymes (the P450 monooxygenase enzymes) to epoxides, which can subsequently form covalent adducts with DNA, resulting in mutations which may cause tumors. For bioremediation in soil after application of sewage sludge, these factors also greatly influence the bioavailability of such compounds in soils, the most important predictive factors being adsorption, volatilization, degradation, and leaching, which greatly influence biodegradability (139).

Crude Oil and Fuel Distillates

Over 2 billion tons of petroleum are produced annually worldwide. The production process, refining, storage, and distribution are all point sources of pollution of soil and water. Accidental spills at sea that result from tanker accidents are dramatic and make high-profile news stories, but quantitatively such spillages represent less than 10% of the total volume of petroleum hydrocarbon discharges into the environment. The low-level routine releases represent about 90% of the volume of hydrocarbon discharges, e.g., surface runoff and industrial effluents. In the marine environment, it is estimated that about 2 million tons of oil enter the seas annually. Only about 18% of this arises from refineries, offshore operations, and tanker activities (91). Crude oil contains a bewildering array of hydrocarbons, of the types mentioned above; a few representatives are shown in Fig. 1.4. Environmental contamination with petroleum introduces this myriad of hydrocarbons, causing a variety of problems (Box 1.1).

Hydrocarbons are diverse molecules that are extremely abundant in nature. However, the ability of plants and animals to degrade hydrocarbons is limited. It is within the bacteria, the filamentous fungi, and yeasts that hydrocarbon biodegradation is most common. Several reasons for the paucity of hydrocarbon biodegra-

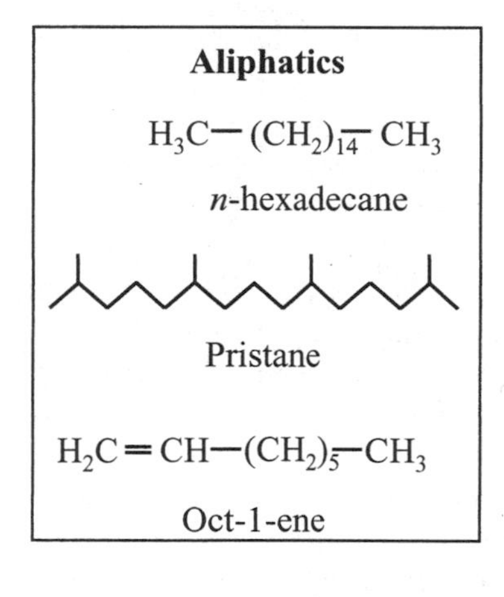

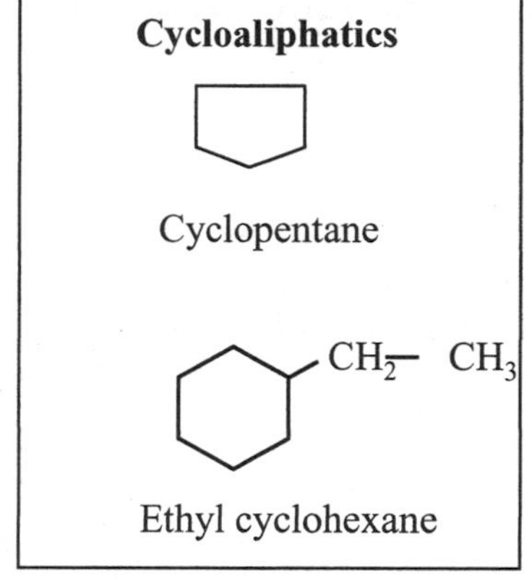

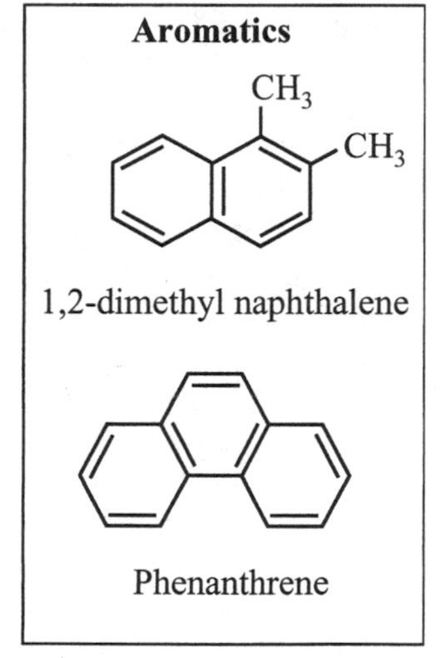

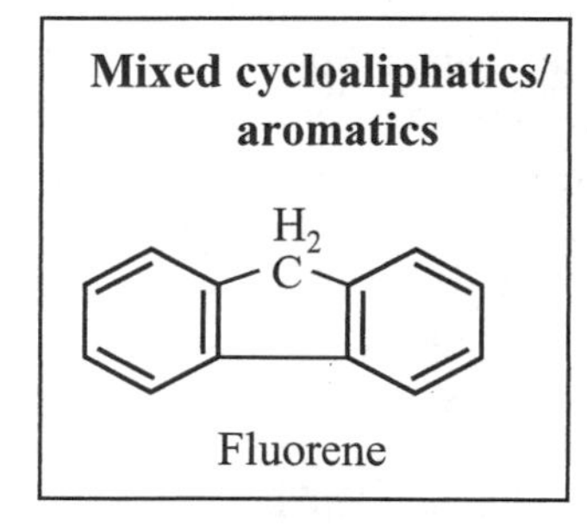

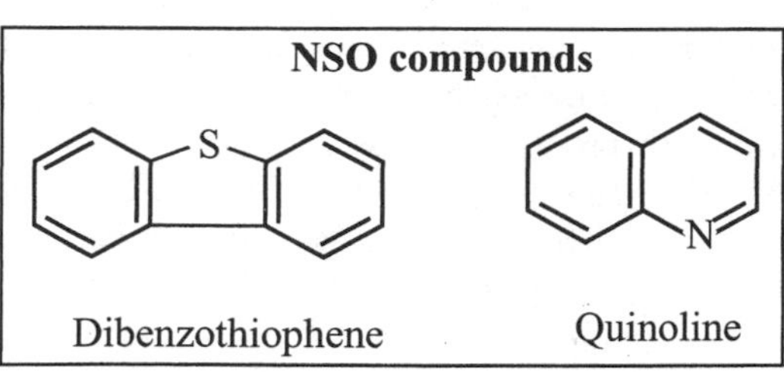

FIGURE 1.4 Chemical studies of various representative hydrocarbons found in crude oil.

BOX 1.1
Russian Oil Contamination

Currently, Russia is the world's second-largest crude oil producer, and the probability is that it still has large reserves undiscovered. It also has huge gas reserves. There are about 60,000 minor leaks from Russian pipelines every year. Russia has more than 1.5 million km of gas and oil pipelines, many of them dating from Soviet times and in poor condition. In eastern Siberia alone, the Security Council of the Russian Federation estimates that 3 to 10 million tons of crude oil leak each year. Arctic ecosystems have very little self-purification capacity, so oil pollution remains for long periods. Russia has 700 major accidents and spills (defined as those involving 25,000 barrels of oil or more) every year. There were a recorded 103 large-scale failures at oil and gas pipelines in the Russian Federation in 1991 to 1993, many of them in fragile Arctic and sub-Arctic areas.

Oil pipelines in areas like the Tyumen region and Khanty-Mansi autonomous district leak significant volumes of oil. There have been records of serious health problems from oil pollution in the more contaminated areas. Oil spills in Siberian rivers near the city of Nizhnevarovsk, for instance, have polluted drinking water and have been correlated with increased incidences of certain cancers.

The most severe problems, however, are in Chechnya, where an estimated 30 million barrels of oil have seeped into the ground from the region's black market oil industry. Since the collapse of the Soviet Union, thieves have tapped into the pipelines and have been stealing large quantities of oil from reserves at refineries. An estimated 15,000 "mini-refineries" have been built. These mini-refineries use only 50% of the oil (commercial refineries use 90%), with the rest dumped into the ground, contaminating water supplies, rivers, and fish. The Terek, Argun, and Sunzha river waters have at times been contaminated to levels 100 and 1,000 times above normal.

Much of the problem arises from the climate and geography of Russia, however. The severe cold of Siberian winters lasts for more than half of the year. Steel gets so cold that pipes can crack, and summer thaw means that pipelines can move on the ground. Without mechanisms to compensate, welds can break. Much of Siberia is swamp during summer, and this makes for the most challenging environment on earth in which to build an oil industry.

The good news is that it is estimated that about one-quarter of the world's registered engineers are Russian, and Russia has an enormously capable scientific infrastructure. Russia has an emerging bioremediation industry, and the development of oil and gas wealth has recently seen large investments, domestic and foreign, into the Russian oil fields.

The oil spill at Usinsk in the Komi Republic in 1994 was the largest pipeline spill on record. It is estimated that 130,000 tons of oil was spilled in this incident, which is more than three times the size of the *Exxon Valdez* spill. In addition, oil spills on land are much more difficult to deal with, as the relative ease of containment which can be achieved on water is not possible on land. The oil spread across 170 acres of streams and fragile bogs and marshland, creating a major environmental disaster in need of major remediation.

dation in the *Eukarya* can be identified. Many hydrocarbons are virtually insoluble in water, and thus their bioavailability is limited. Hydrocarbons are generally chemically inert but are subject to oxygen additions by various enzymes. Once oxidized, the alcohols and/or acids that are formed can undergo further metabolism and enter the central metabolic pathways.

In the United States, about 40% of the underground storage tanks have leaked and some 30,000 new releases are registered each year. About half of these facilities are used to store petroleum products. The U.S. EPA estimates that there are over 200,000 leaking underground storage tank sites in the United States. Much of the contamination is gasoline, which contains high concentrations of relatively soluble low-molecular-weight alkanes. Because these alkanes act as solvents to the lipid membranes of bacteria, gasoline is relatively toxic to microorganisms (115). Some of the contamination also comes from leaking diesel and jet fuel tanks. Diesel fuel is comprised largely of simple unbranched *n*-alkanes, with only around 4% polyaromatic compounds (58). The intermediate-chain-length alkanes present in diesel and jet fuel are highly biodegradable. The normal alkanes, or *n*-alkanes, have the general formula C_nH_{2n+2}, and they are the most biodegradable of the petroleum hydrocarbons.

Oxygenases, which catalyze the introduction of oxygen, are often the key enzymes involved in the initial attack on hydrocarbons. Dioxygenases catalyze the addition of both atoms of oxygen from molecular oxygen into the hydrocarbon substrate. Monooxygenases incorporate only one atom of molecular oxygen into the hydrocarbon substrate; the other atom of oxygen is reduced to water. Since these enzymes function as part oxygenase and part oxidase, they are also known as mixed-function oxidases.

The initial step in the oxidation of alkanes is via an alkane oxygenase-catalyzed reaction, which is carried out by either a P450 monooxygenase or a multiprotein monooxygenase (20). At least some alkane oxygenases are nonheme di-iron proteins. In the alkane monooxygenase reaction, the alkane is oxidized to an alcohol and then hydrolyzed to an aldehyde. This is subsequently converted via an oxidation reaction to fatty acids before entering the β-oxidation and tricarboxylic acid cycles. The monooxygenase system that catalyzes the reaction that many bacteria use to transform an alkane to an alcohol consists of three components: a membrane-bound monooxygenase, a soluble rubredoxin, and a rubredoxin reductase.

The alkane-oxidizing enzymes are encoded by both chromosomal and plasmid-borne genes, with the well-characterized OCT plasmid of *Pseudomonas putida* Gpo1, for example, encoding all the enzymes necessary for the degradation of *n*-alkanes to fatty acids (132). There are 10 genes involved in synthesizing the enzymes necessary for alkane oxidation; these can be found in two regions on the OCT plasmid, at the *alkBFGHJKL* operon and the *alkST* operon (131). The *alkB*, *alkF* and *alkG*, and *alkT* genes code for the production of the alkane hydroxylase, rubredoxins, and rubredoxin reductase, respectively. In turn, *alkJ* encodes alcohol dehydrogenase, *alkH* encodes aldehyde dehydrogenase, and *alkK* encodes acetyl coenzyme A production. It is thought that *alkL* may be involved primarily in alkane uptake, while *alkS* regulates the expression of the *alkBFGHJKL* operon. A model for alkane metabolism in *Pseudomonas oleovorans* proposed by van Beilen et al. (132) is illustrated in Fig. 1.5.

The aromatic hydrocarbons have a structure based on the benzene ring, C_6H_6. The benzene ring is ubiquitous in nature as a component of lignin and humus. It is found in the aromatic amino acids phenylalanine, tyrosine, and tryptophan. It is also present in fuel spills along with BTEX compounds (Fig. 1.6). BTEX compounds are a major concern as a result of fuel spills, and the term BTEX appears continuously in the bioremediation research and business literature.

All of the BTEX compounds are of concern as environmental pollutants. Benzene itself is an EPA category A carcinogen (human carcinogen, causing leukemia). The others all have severe acute effects of exposure. They are commonly found in gasoline and are highly volatile. Even so, they do adsorb to clay particles, allowing them to persist in soils. They also can leach into groundwater because their solubility is relatively high.

Due to leakages of underground storage tanks and pipelines, spills during transportation and from production sites, and seepage from surface contaminated sites such as gasworks, BTEX compounds have become major contaminants of the subsurface. Efforts have focused upon removing these contaminants before they enter aquifers and cause pollution of drinking water supplies. As there is very little oxygen present in the subsurface, it is important that knowledge of microbial metabolism of pollutants without oxygen is gained so that such abilities could be harnessed for bioremediation. Evidence now supports that anaerobic environments are host to a diverse array of microorganisms capable of metabolizing organic contaminants without the presence of molecular oxygen (25, 34, 35, 88, 107, 108, 147), including BTEX compounds (13, 19, 73).

The fused-ring PAHs have received much attention, both as pollutants and for their potential for biodegradation. Because of their widespread distribution, PAHs have been pol-

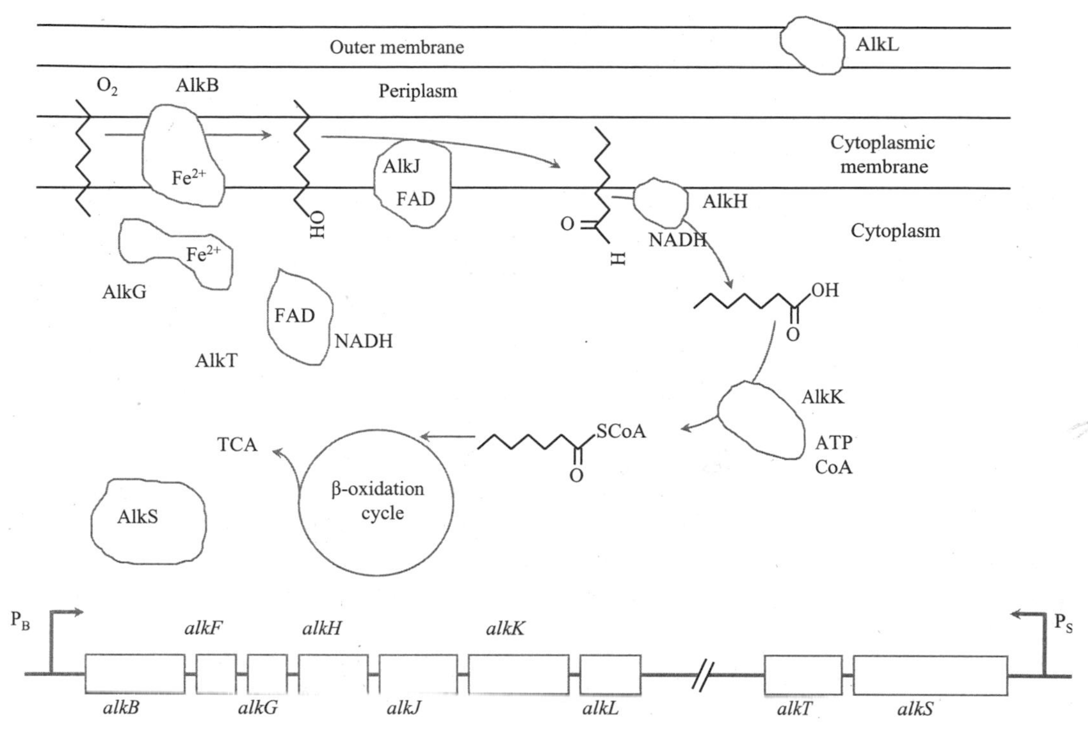

FIGURE 1.5 Model for alkane metabolism in *P. oleovorans*. (Top) Proposed location of various enzymes within the bacterial cell, along with the intermediates of *n*-alkane metabolism. (Bottom) Corresponding genes and operon arrangement. TCA, tricarboxylic acid; CoA, coenzyme A.

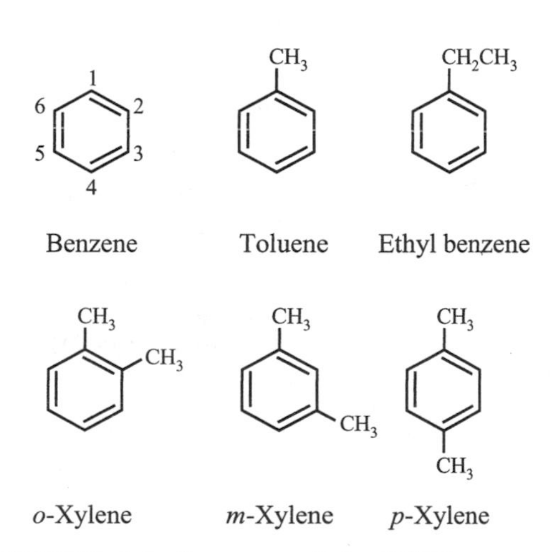

FIGURE 1.6 Structures of the BTEX group of compounds. The carbon atoms of the benzene molecule are numbered as an aid to the explanation of the nomenclature of aromatic compounds. For example, *o*-xylene is 1,2-dimethyl benzene.

lutants of great concern for many years. The simplest of the PAHs is naphthalene, and increasing the number of rings in the molecule decreases the aqueous solubility (Fig. 1.7), which has major implications for bioavailability and biodegradability. The group of compounds includes 16 priority pollutants, many of which are suspected carcinogens. Besides being present in petroleum, they are very common constituents of coal tars and are produced during high-temperature industrial processes such as oil refining and coke production in combined iron and steel plants.

Dioxygenases are involved at several stages in the pathways for catabolism of aromatic hydrocarbons. Dioxygenases involved in the initial attack on aromatic hydrocarbons bring about ring hydroxylation. These ring-hydroxylating dioxygenases, e.g., toluene dioxygenase, require reduced cofactors—NADH in catabolic pathways in addition to oxygen. They

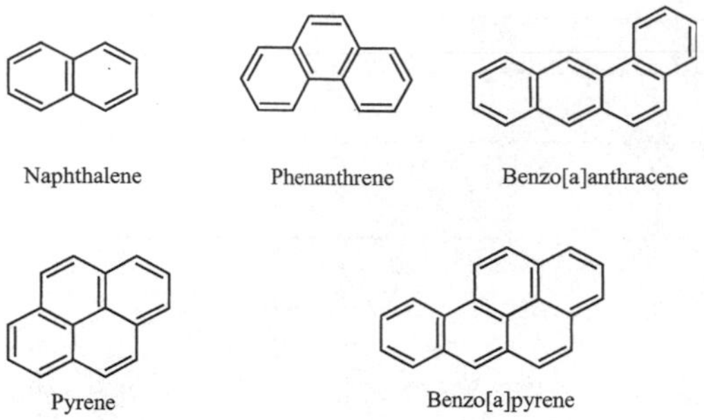

FIGURE 1.7 Structures of some representative PAHs.

dihydroxylate aromatic substrates to produce *cis*-diols. This is typical of the initial step in bacterial oxidation of aromatic hydrocarbons. Subsequently, other dioxygenases are involved in ring fission. These ring cleavage enzymes have no cofactor requirements and cleave the aromatic ring of hydroxylated aromatic substrates; for example, bacterial catechol 1,2-dioxygenase cleaves the dihydroxylated benzene ring to produce a *cis,cis*-muconic acid.

PAHs are of major concern as environmental contaminants, although less so in groundwater than in soil. This is because of their limited solubilities and relatively low concentrations in gasoline, which is a major source of groundwater contamination. The bacterial metabolism of naphthalene provides an example of the metabolism involved in the biodegradation of PAHs. The principal mechanism for the aerobic bacterial metabolism of naphthalene is via the oxidative action of the naphthalene dioxygenase enzyme, which forms the intermediate naphthalene dihydrodiols (30, 53, 89); these dihydrodiols are then dehydrogenated via the action of a suite of naphthalene dehydrogenase enzymes to form salicylic acid (Fig. 1.8). The conversion of naphthalene to salicylic acid

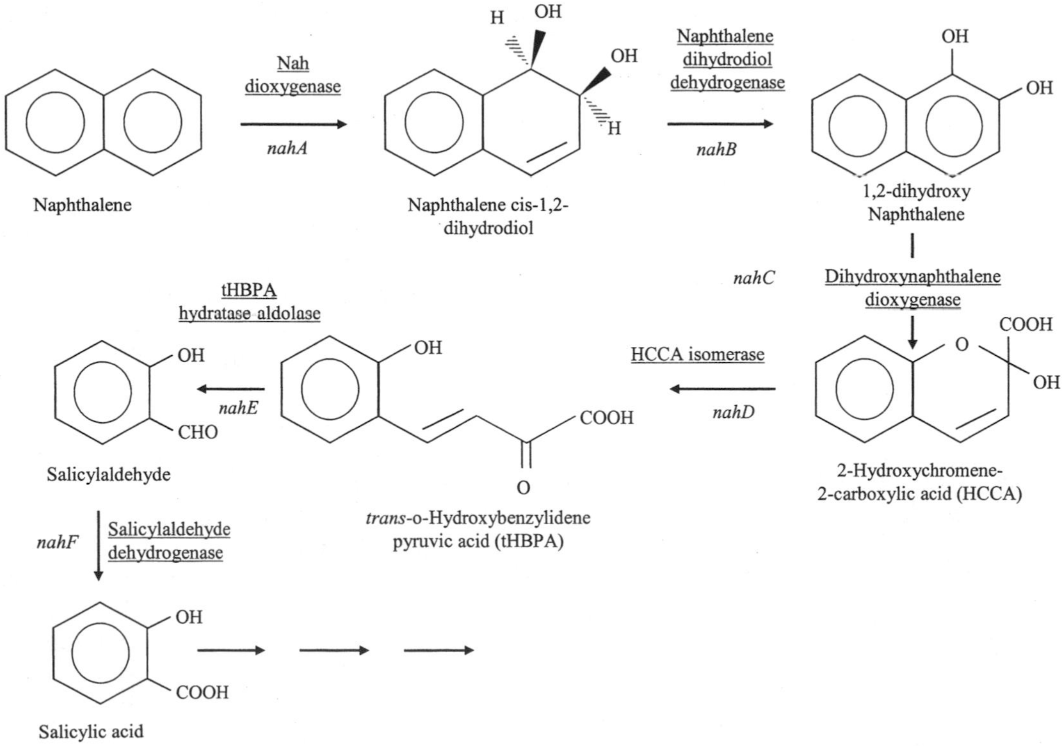

FIGURE 1.8 Initial steps in naphthalene metabolism in *Pseudomonas* spp. Enzymes (underlined) and genes (italic) are indicated.

constitutes the upper pathway of naphthalene metabolism. Salicylic acid can then be further metabolized via catechols, resulting in the end products carbon dioxide and water; this constitutes the lower pathway of naphthalene metabolism.

In most cases, the genes that encode naphthalene transformation are found on plasmids, the best studied of which is the NAH 7 plasmid of *P. putida* PpG7. This plasmid carries two operons: the *nah* operon, which encodes the initial stages of the metabolism of naphthalene (naphthalene to salicylic acid), and the *sal* operon, which encodes the transformation of salicyclic acid to catechols. The *nahR* gene produces a regulatory protein (a 36-kDa polypeptide) which is dependent on the presence of salicylate, and this protein induces the transcription of the *nah* and *sal* operons.

Fuel oxygenates are not petroleum hydrocarbons, but they are added to gasoline to promote more complete burning of the fuel by increasing the oxygen content. The most commonly used chemical in this sense is methyl *tertiary*-butyl ether (MTBE) (Fig. 1.9). Its use as a fuel oxygenate means that it is a chemical produced in a very high volume. MTBE is relatively soluble and volatile and appears to be relatively resistant to biodegradation (111). It therefore has the potential to be a widespread contaminant of groundwater, as it is highly mobile. It is, however, considered to be nontoxic at levels of exposure likely to be experienced (45).

Heterocyclic aromatics occur naturally in crude oil, and the aromatic organosulfur compounds (benzothiophene, dibenzothiophene) (Fig. 1.4) are of concern as contributors to the sulfur content of fuels. Biological desulfurization systems offer alternatives to hydrotreatment for the production of low-sulfur fuels (86).

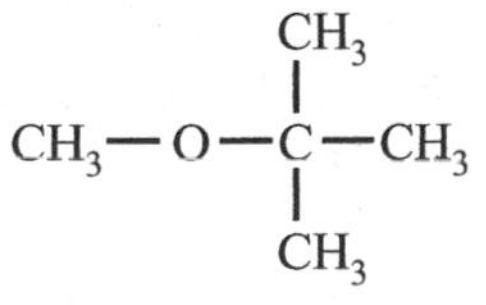

$$CH_3-O-C(CH_3)_2-CH_3$$

Methyl *tert*-butyl ether

FIGURE 1.9 Structure of the fuel oxygenate MTBE.

Haloalkenes and Haloalkanes

The most important haloalkenes are the chlorinated ethenes. Chlorinated ethenes and ethanes are common dry-cleaning agents. Halogenation of organic molecules generally makes them more resistant to aerobic biodegradation. Their oxidation state is raised and therefore electrophilic attack is energetically less favorable. Indeed, this is true for xenobiotic compounds carrying other substituents that confer electrophilicity on the molecule, e.g., nitro, azo, sulfo, and carbonyl groups. Thermodynamically, nucleophilic attack under reducing conditions is more likely (72). Increasing the degree of chlorination generally decreases the rate of biodegradation, i.e., the greater the degree of halogenation, the more resistant the molecule becomes. Increasing chlorination also leads to a marked drop in water solubility and corresponding increase in hydrophobicity, which have influence on biodegradability. The toxicity of the compound also adversely affects biodegradation; the interaction between toxicity and biodegradation is not well understood, but they are clearly linked and competing processes: at low concentrations a compound may be biodegradable, but biodegradation is arrested at higher concentrations by toxicity. If the concentration is very low for biodegradation, then this may not be enough to support microbial growth and reproduction, the concept of threshold first described by Jannasch (66) and subsequently elaborated upon in detail by Alexander (5).

Disinfection of drinking water by chlorination has had a long, successful pedigree. Several decades ago it was realized, however, that chlorination of water formed trace amounts of trihalomethanes, such as chloroform. Some of these compounds are suspected carcinogens, and strict limits are set for allowable concentrations in drinking water. Sulfate-reducing bacteria transform tetrachloroethene to trichloroethene (TCE) and *cis*-1,2-dichloroethene by anaerobic dehalogenation (11). The fully chlorinated but unsaturated tetrachloroethene (perchloroethene [PCE]) is subject to stepwise dechlorination to the toxic product vinyl chloride and eventually ethene (90). PCE degradation has been demonstrated in a methanogenic bacterial consortium growing on acetate (52, 134).

The haloalkene that causes the most concern is TCE (Fig. 1.10). It is widely distributed in soil and groundwater as a pollutant. Not only is it relatively resistant to biodegradation, but its aqueous solubility is only of the order of 1 mg liter^{-1}. Anaerobic TCE degradation occurs by reductive dechlorination (Fig. 1.11). In the commonly observed TCE transformation pathway, TCE is sequentially reduced to dichloroethene, vinyl chloride, and ethene. Extensive aerobic degradation of TCE by a methane-utilizing microbial consortium has been demonstrated (51, 76). The low specificity of methane monooxygenase allows the conversion of TCE to TCE epoxide, which subsequently spontaneously hydrolyzes to polar products (formic, glyoxylic, and dichloroacetic acids) utilizable by microorganisms.

Haloaromatics

For the same reasons as the haloaliphatics, the haloaromatic compounds are generally less biodegradable, certainly under aerobic conditions, than the nonhalogenated counterparts. Thus, they are more persistent in the environment. Moreover, there is great variety in the haloaromatics. Halogenated aromatic compounds are widely distributed in the environment as a result of their widespread use as herbicides, insecticides, fungicides, solvents, fire retardants, pharmaceuticals, and lubricants. Chlorinated compounds are used more frequently than fluorinated or brominated compounds. Several of these chemicals cause considerable environmental pollution and human health problems due to their persistence and toxicity. Anaerobic degradation of chlorophenols and chlorobenzoates is usually performed by complex methanogenic or sulfidogenic microbial consortia and in the presence of additional carbon sources.

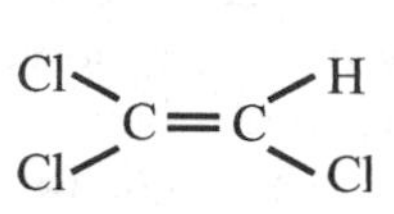

Cl, Cl C=C H, Cl

TCE

FIGURE 1.10 Structure of TCE.

Phenol was the first antiseptic, but chlorophenols possess far higher antimicrobial activity and acidity than phenol. Chlorinated phenols were subsequently developed as biocidal agents, especially di- and trichlorophenol and, above all, PCP. The fully substituted PCP (Fig. 1.12) is very recalcitrant, has limited aqueous solubility (except in alkali solution, where it forms a soluble sodium salt), and is persistent in the environment. It has long been used as a wood preservative and is a biocide of impressive effectiveness, being fungicidal, bactericidal, algicidal, insecticidal, molluscicidal, and herbicidal (100). Again, polychlorination is more readily dealt with by reductive dechlorination, although aerobic biodegradation is possible (112).

Other, less substituted chlorophenols have found uses in a variety of industries. They are used as intermediates in pesticide synthesis, as dyestuffs, and in the pharmaceutical industry. Many pesticide molecules are halogenated aromatics, with highly variable levels of biodegradability. Examples are DDT, 2,4-dichlorophenoxyacetic acid (2,4-D), and 2,4,5-trichlorophenoxyacetic acid (2,4,5-T) (Fig. 1.13), in which relatively small structural changes result in differing biodegradability. 2,4-D, for example, is biodegraded within days; 2,4,5-T differs only by one additional chlorine substitution in the *meta* position, yet this compound persists for many months. The additional substitution interferes with the hydroxylation and cleavage of the aromatic ring.

The haloaromatic groups are also found as part of other biocidal compounds, such as

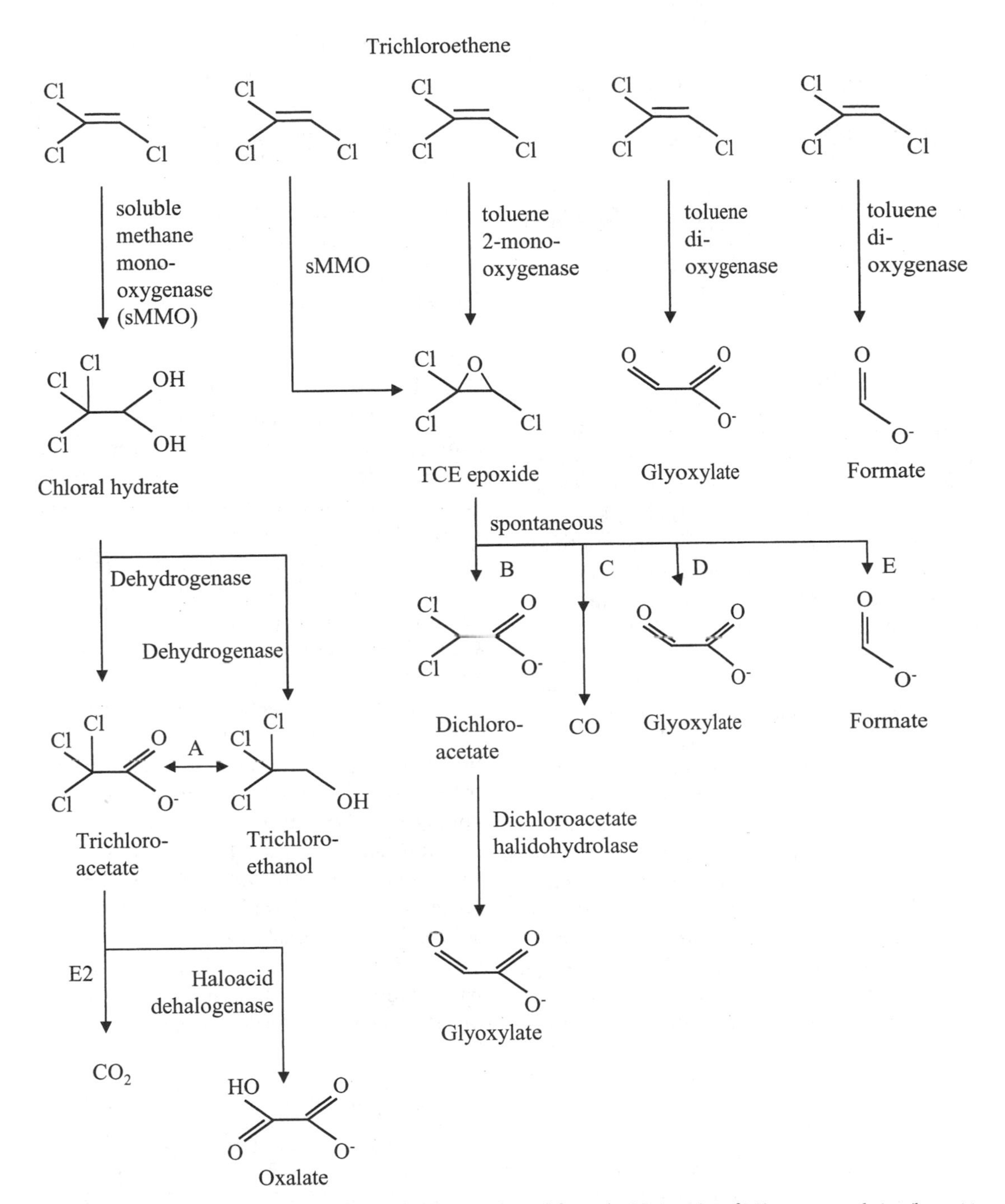

FIGURE 1.11 Biodegradation pathways for TCE. Adapted from the University of Minnesota website (http://umbbd.ahc.umn.edu/tce/tce_image_map.html).

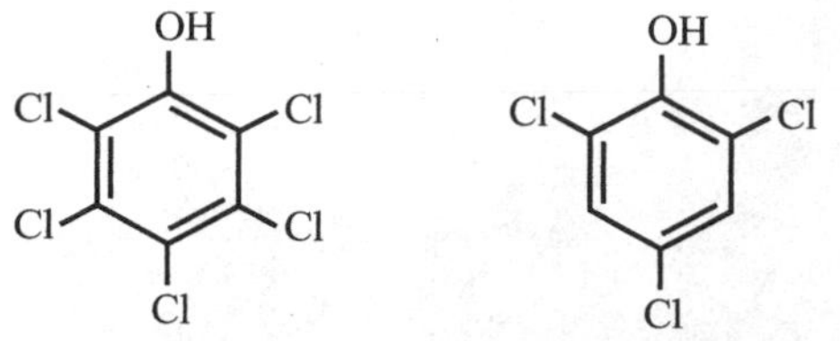

Pentachlorophenol 2,4,6-trichlorophenol

FIGURE 1.12 Structures of two examples of chlorophenols: PCP and trichorophenol.

the triazole fungicides (Fig. 1.14). The insolubility and hydrophobicity of some chloroaromatics make them useful industrial solvents. The chlorobenzenes exist as a group of 12 compounds. Historically, the chlorobenzenes were used in the synthesis of DDT, but now the most important use is in the synthesis of aniline dyes and a variety of pesticides. Biodegradation of numerous halogenated aromatic compounds under both aerobic and anaerobic conditions (55, 56, 94, 117) has been reported. Halobenzoates are used as substrates by various bacteria under aerobic conditions. Chlorophenols, particularly PCP, are used as preservatives for wood and canvas, as mentioned above. Other chlorophenols are used in the synthesis of pesticides, resins, dyes, and pharmaceuticals. Aerobically, PCP is converted by a bacterial monooxygenase to tetrachlorohydroquinone through the oxidative elimination of the chlorine *para* to the phenolic hydroxyl. Anaerobically, PCP is reductively dechlorinated stepwise to phenol. Phenol can be metabolized further anaerobically to methane and CO_2. Fungi transform PCP via pathways different from those of bacteria; e.g., in soil subject to bioremediation by a white rot fungus, methylation products including pentachloroanisole were produced (130).

Biodegradation of the herbicide 2,4-D occurs via a modified *ortho* cleavage pathway. This biodegradation pathway for 2,4-D is similar to that for other chlorinated aromatics, such as 3-chlorobenzoate (3CBA) (101). In this pathway, chlorinated aromatic compounds are degraded via a chloro-substituted catechol, which is *ortho* cleaved by a chlorocatechol 1,2-dioxygenase. The enzymes involved in the mineralization of chlorocatechols have wider substrate specificities than the ordinary *ortho* cleavage pathway enzymes, which is why it is called a modified *ortho* cleavage pathway (133). The genes for the modified *ortho* cleavage pathways are generally located on catabolic plasmids; their organization into operon structures differs substantially from that of the chromosomally encoded genes of the normal *ortho* cleavage pathway. Plasmid pJP4, which codes for the biodegradation of chlorinated haloaromatics, is an 80-kb IncP1 broad-host-range conjugative plasmid encoding most of the degradation pathway for 2,4-D, 2-methyl-4-chlorophenoxyacetic acid, and 3CBA and for resistance to mercuric chloride and phenylmercuric acetate (101). 3,5-Dichlorocatechol is oxidized via a series of steps to 2-chloromaleyl acetate by enzymes encoded by the *tfdCDEF* operon of pJP4. The chloromaleyl acetate is then ca-

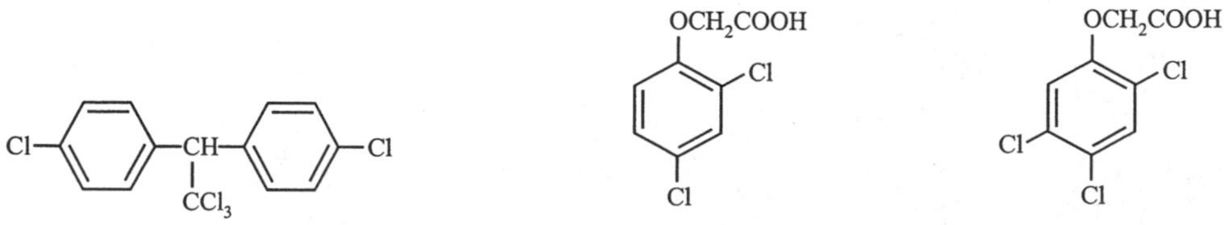

DDT
1,1,1-trichloro-2,2-bis(-p-chlorophenyl) ethane

2,4-D
(2,4-dichlorophenoxy) acetic acid

2,4,5-T
(2,4,5-trichlorophenoxy) acetic acid

FIGURE 1.13 Structures of some chloroaromatic biocides which vary greatly in their biodegradability.

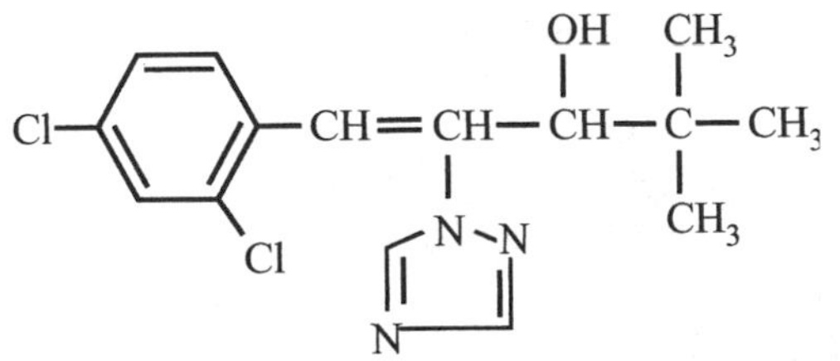

FIGURE 1.14 Structure of a triazole fungicide. Triazole fungicides exhibit their antifungal activity by inhibiting fungal ergosterol biosynthesis and are economically important agrochemicals since they have been widely used on crops such as wheat, barley, and orchard fruits.

tabolized by chromosomally encoded enzymes. Interestingly, microbial transformation of some organic contaminants such as the industrial solvent tetrachloroethylene (PCE) does not occur in oxygenated environments (69), with more rapid transformation of this contaminant under anaerobic conditions than aerobic conditions. Removal of chlorine from aromatics generally occurs much more readily in anaerobic environments (via reductive dechlorination), which is an important consideration when bioremediation operations for heavily chlorinated compounds are being designed.

PCBs

PCBs are mixtures of biphenyls with 1 to 10 chlorine atoms per molecule. This is a large family of compounds. They consist of two covalently linked, but not fused, benzene rings, with various degrees of chlorination of the carbon atoms. The family is large because of the many substitution possibilities, which number 210. The PCBs are oily fluids, and they have found widespread use as both insulators and coolants of transformers due to their low electrical conductivity and high boiling point. They have also been used as plasticizers.

PCBs, with their attendant chlorination (Fig. 1.15), are persistent, toxic, and carcinogenic (110). They are remarkably distributed in the environment, and they are able to biomagnify through food chains. Whereas most of the enumerated uses of PCBs would not seem to be conducive to widespread environmental contamination, PCB residues have in fact been detected in a large percentage of random environmental samples, and they accumulate in higher-trophic-level animals. Concentrations of PCBs over 1 ppm have been detected in one-third of the sampled U.S. population with no known occupational exposure. In some freshwater habitats, older individual predatory fish, such as trout and salmon, accumulate 20 to 30 ppm of PCBs, and predatory and fish-eating birds were found to contain, on occasion, several hundred parts per million of PCBs.

Although relatively resistant to biodegradation, a number of microorganisms that transform PCBs have been isolated. Degradation of PCBs typically is by cometabolism and is enhanced by the addition of less chlorinated analogs such as dichlorobiphenyl. Extensive degradation of some PCB congeners has been found in soils and aquatic waters and sediments. The specific congeners are differentially degraded, and various PCB products, according to their composition, exhibit different degrees of susceptibility to biodegradative

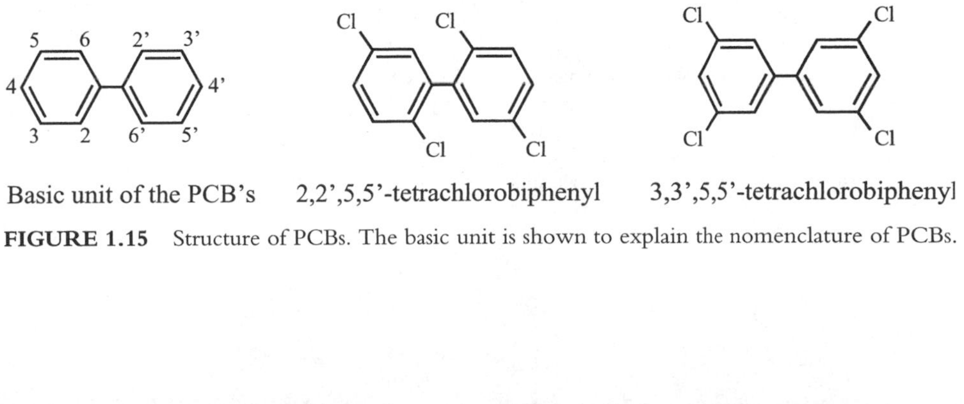

Basic unit of the PCB's   2,2',5,5'-tetrachlorobiphenyl   3,3',5,5'-tetrachlorobiphenyl

FIGURE 1.15 Structure of PCBs. The basic unit is shown to explain the nomenclature of PCBs.

transformations. Many PCBs (25) are transformed without the presence of molecular oxygen under one or more redox environments (nitrate, manganese, iron, and sulfate reducing conditions). Anaerobic PCB dechlorination is responsible for the conversion of highly chlorinated PCBs to lightly chlorinated *ortho*-enriched congeners. The products from this anaerobic process are readily degradable by a wide range of aerobic bacteria. The widespread anaerobic dechlorination of PCBs results in reduction of the potential risk due to the dioxin-like toxicity and carcinogenicity of PCB exposure.

Nitroaromatics

Nitroaromatic compounds are released into the biosphere almost exclusively from anthropogenic sources. Some compounds are produced by incomplete combustion of fossil fuels; others are used as synthetic intermediates, dyes, pesticides, and explosives. The most familiar of the nitroaromatics is trinitrotoluene (TNT) (Fig. 1.16), the universal high explosive. Nitro groups are also electron withdrawing, making them less susceptible to electrophilic attack. TNT may be found in gram-per-kilogram quantities in many contaminated soils (120) and is known to exert toxicological effects on a wide range of terrestrial and aquatic organisms, to be mutagenic, and to be a possible human carcinogen. Nitroaromatic biodegradation is at best slow and is best under reducing conditions (119).

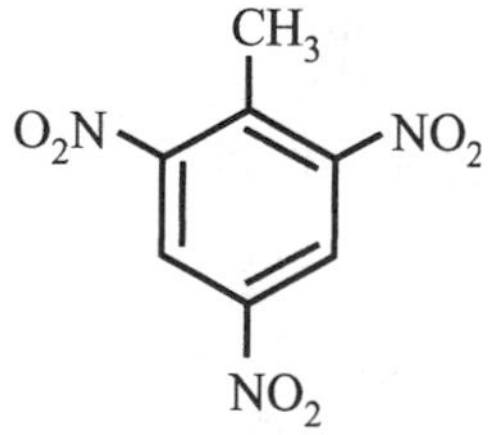

CH_3
O_2N NO_2
NO_2

FIGURE 1.16 Structure of TNT. Many military testing grounds are contaminated with TNT. Human exposure leads to a range of clinical conditions: anemia and abnormal liver, spleen enlargement, other harmful effects on the immune system, and skin irritation. There is evidence that TNT adversely affects male fertility, and TNT is listed as a possible human carcinogen.

Nitroaromatics are biodegraded under aerobic and anaerobic conditions (119). The anaerobic bacteria *Desulfovibrio* spp. can reduce nitroaromatic compounds, including 2,4,6-TNT to 2,4,6-triaminotoluene. Several strains of *Clostridium* spp. can catalyze a similar reduction, producing low-molecular-weight aliphatic acids. The fungus *Phanerochaete chrysosporium* mineralizes 2,4-dinitrotoluene and 2,4,6-TNT and shows promise as the basis for bioremediation strategies. A number of nitroaromatic compounds can serve as growth substrates for aerobic bacteria. Some bacteria can reduce the aromatic ring of dinitro and trinitro compounds with the elimination of nitrite. Monooxygenases can add a single oxygen atom and eliminate the nitro group from nitrophenols. Dioxygenases can insert two hydroxyl groups into the aromatic ring and precipitate the spontaneous elimination of the nitro group from a variety of nitroaromatic compounds. Reduction of the nitro group to the corresponding hydroxylamine is the initial reaction in the metabolism of nitrobenzene, 4-nitrotoluene, and 4-nitrobenzoate. These reactions have potential applications for the biodegradation of environmental contaminants.

Dioxins

Dioxins are among the most toxic chemicals known. Dioxin is a general term that describes a group of hundreds of chemicals that are highly persistent in the environment and also bioaccumulate rather efficiently. As knowledge stands, the most toxic compound is 2,3,7,8-tetrachlorodibenzo-*p*-dioxin, or TCDD (Fig. 1.17). The toxicities of other dioxins and chemicals like PCBs that act like dioxins are measured in relation to that of TCDD. TCDD is a category A carcinogen, meaning that it is a known human carcinogen. The major source of dioxin in the environment is incinerators that burn halogenated wastes. Dioxins are produced in paper mills that use chlorine bleaching in their process and are made during production of polyvinyl chloride. They are not used

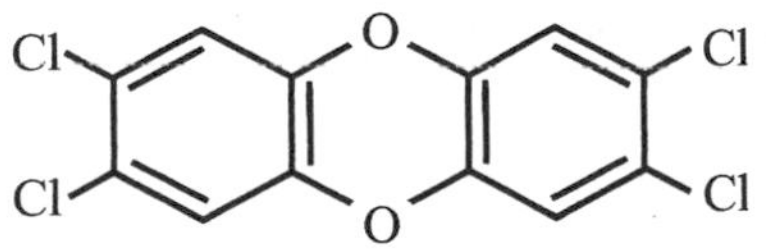

FIGURE 1.17 Structure of TCDD, one of the most toxic compounds known. In January 2001, its status as a suspected human carcinogen was changed to that of a known human carcinogen, based on sufficient evidence from a combination of epidemiological and mechanistic studies that indicated a causal relationship between exposure to TCDD and human cancer.

as industrial chemicals but are the inadvertent by-product of other industrial processes during the synthesis of chlorophenols and some chlorinated herbicides.

Since 1977, it has been recognized that dioxins are the major chlorinated toxic compounds in fly ash, which has been used in Europe in the construction of roads, dams, and bridges. The polychlorinated dibenzo-*p*-dioxins (PCDDs) belong to the most hazardous environmental pollutants, principally because of their molecular planarity and ability to bind to biological receptors (15). Aerobic cometabolism or transformation of PCDDs has focused mainly on congeners with four or fewer chlorine atoms (1, 3). The highly chlorinated forms are already highly oxidized. Therefore, oxidative, electrophilic attack is unlikely. These forms are more likely to be susceptible to nucleophilic attack, particularly reductive dechlorination. However, dioxins are neither toxic nor inhibitory to microorganisms and have no observable effect on soil respiration (23) or microbial activity and diversity (7).

Adriaens et al. (2) presented the first evidence for the reductive dechlorination of PCDDs and dibenzofurans by anaerobic microbial activity. They calculated half-lives of 1 to 4.1 years from laboratory microcosm experiments. However, real half-lives are much (perhaps orders of magnitude) longer as residual concentrations may never be biologically available. During reductive dechlorination of aged TCDD, the activity of a natural microbial consortium shifted from predominantly methanogenic to predominantly nonmethanogenic. The first phase of active TCDD dechlorination took place during active methanogenesis (15). Despite decreasing methanogenesis, dechlorination continued until the end of the experiment. Evidence for a dichotomous dechlorination pathway for highly chlorinated PCDD congeners by a microbial consortium was presented (Fig. 1.18):

1. a mixed *peri*-lateral dechlorination pathway for non-2,3,7,8-substituted congeners, and
2. a *peri*-dechlorination pathway for 2,3,7,8-substituted hepta- through penta- CDD isomers.

It was only in 1992 that the properties of a bacterium capable of mineralizing dibenzo-*p*-dioxin (DD) were reported (Fig. 1.19) (142). Chlorinated dioxins present considerably greater challenges for the reasons already outlined (mainly lower bioavailability and lack of sites for electrophilic attack). Until 1996, no aerobic enrichment cultures capable of utilizing chlorinated dibenzofurans and dibenzodioxins had been obtained. This changed when it was shown that *Sphingomonas* strain RW1 converted 1-chloro-DD (1-CDD) and 2-CDD to 3-chlorocatechol and 4-chlorocatechol, respectively, and catechol (Fig. 1.20) (140). Now both aerobic and anaerobic mechanisms of dioxin biodegradation are known (70).

The dioxygenation of monochlorinated DDs occurs on both the halogen-substituted aromatic nucleus and the nonsubstituted nucleus (54, 57). The accumulation of 3-chlorocatechol and 4,5-dichlorocatechol (from 2, 3-dichloro-DD) presents a significant problem for further biodegradation: 3-chlorocatechol is a potent inhibitor of *meta*-cleaving enzymes of catechol (16). 4,5-Dichlorocatechol is a strong inhibitor of chlorocatechol 1,2-dioxygenase. These compounds are therefore critical recalcitrant intermediates in the aerobic mineralization of dichloro- and possibly also trichloro- and higher congeners.

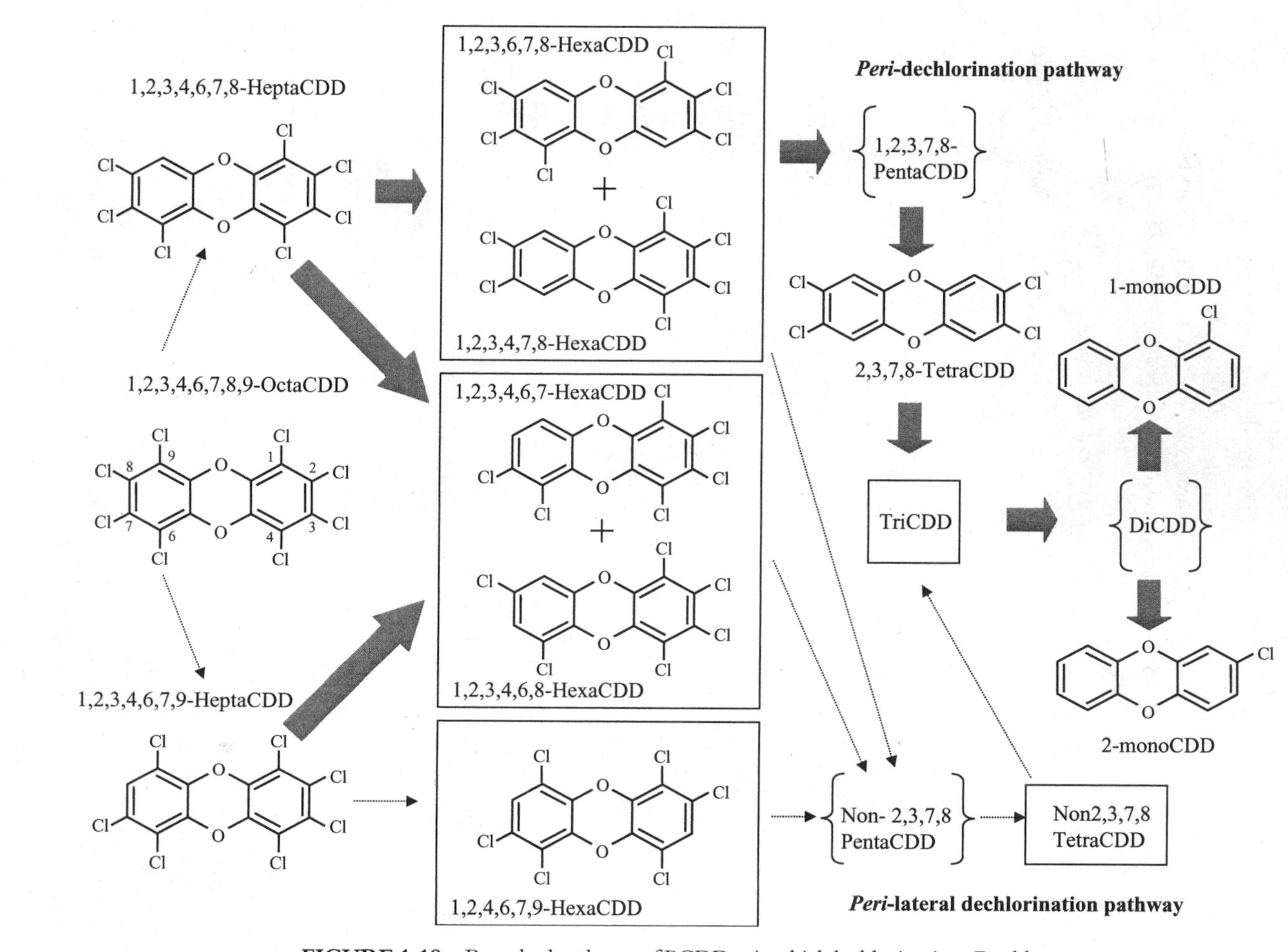

FIGURE 1.18 Branched pathway of PCDD microbial dechlorination. Dechlorination caused by activity of nonmethanogenic, non-spore-forming microbes (broad arrows) and intermediates found in trace concentrations (14) (braces) are indicated.

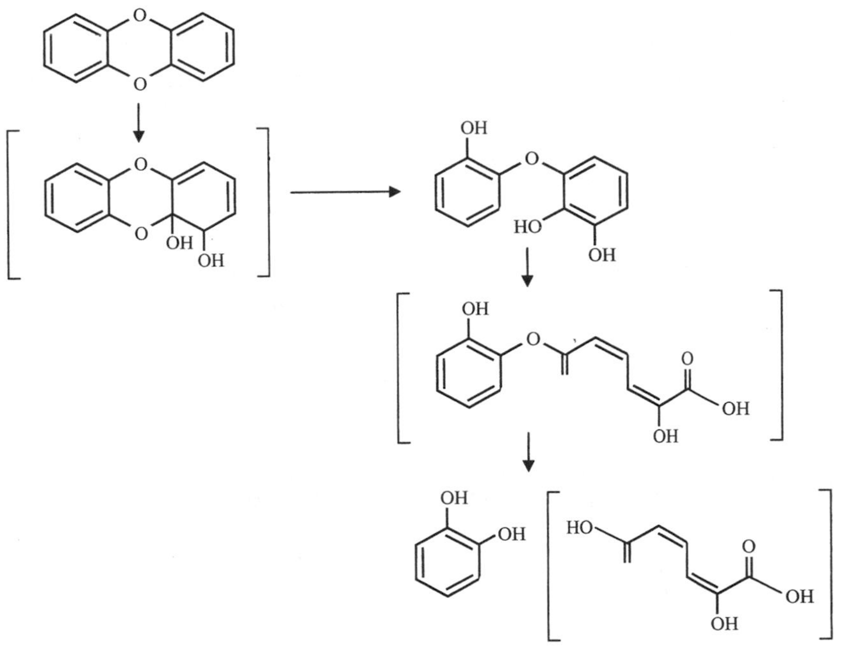

FIGURE 1.19 Metabolic pathway for the aerobic biodegradation of DD (142).

The genetics of dioxin biodegradation has been characterized (6, 71) (Fig. 1.21). Gene *fdx1* encodes a ferredoxin containing a [2Fe-2S] cluster of the putidaredoxin type, i.e., a ferredoxin which is related to ferredoxins typically acting as electron donors of monooxygenases. Gene *redA2* encodes a ferredoxin reductase with a high degree of homology to putidaredoxin reductase and to reductases acting with class IIB oxygenases. Genes *dxnA1* and *dxnA2* encode the terminal oxygenase. The α subunit shows homologies to those of the class IIB enzymes. Genes encoding enzymes catalyzing the second (extradiol cleavage of trihydroxybiphenyl or trihydroxybiphenyl ether) and the third (hydrolysis of the trihydroxybiphenyl ring cleavage product) catabolic reactions have also been localized and are usually on separated DNA frag-

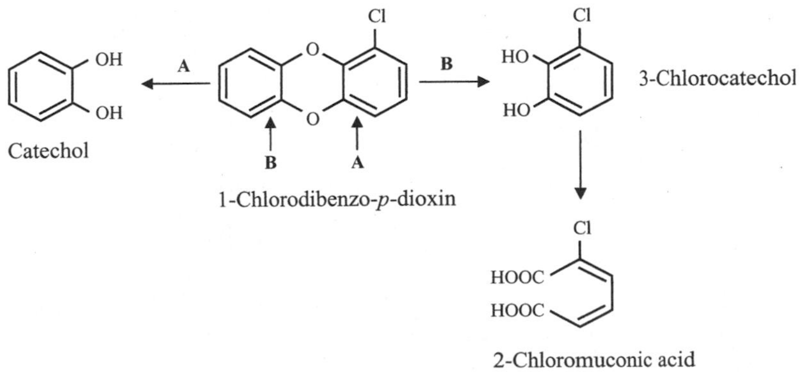

FIGURE 1.20 Metabolic pathway for the aerobic biodegradation of 1-CDD (140). A and B, sites of attack by the initial dioxygenase.

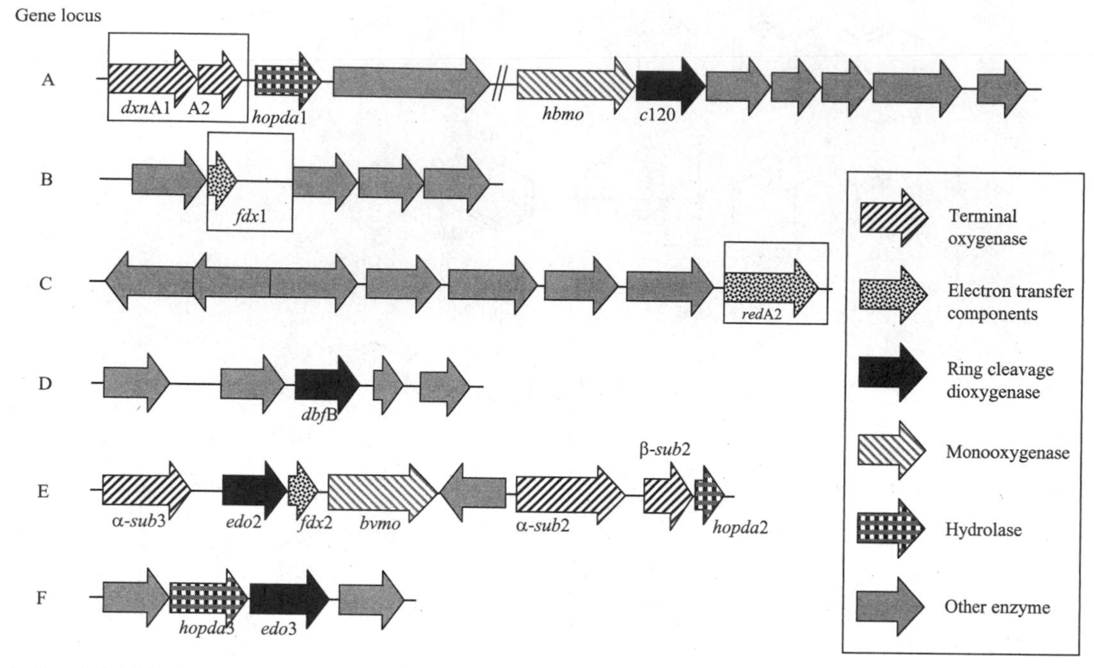

FIGURE 1.21 Gene organization for dioxin metabolism in *Sphingomonas* sp. strain RW1. Six fragments of the RW1 genome which are probably involved in dibenzofuran and dioxin degradation are shown. Genes coding for the initial dioxin dioxygenase (boxed) are indicated.

ments. Besides the gene for trihydroxybiphenyl dioxygenase, there are three genes encoding isofunctional extradiol dioxygenases. The gene sequences of *edo*2 and *edo*3 are most similar to those coding for ring cleavage enzymes with high activity against bicyclic substrates. Isoenzymes also seem to be present for the following hydrolase. Only one of the genes encoding such a hydrolase (*hopda1*) is clustered with other genes of the upper pathway (those encoding the terminal oxygenase). An analysis of the respective environments showed an association with sequences of additional pathways or parts of pathways, the function of which in DD and dibenzofuran degradation remains to be clarified.

ABSs

The alkylbenzyl sulfonate (ABS) compounds are included because they are widely used as anionic detergents and therefore have great potential for environmental contamination, purely through large-scale industrial and domestic usage (Fig. 1.22). Early "hard" detergents contained branched ABSs as the surfactants. In these early detergents, the alkyl portion of the ABSs had multiple methyl branching, which negatively influences bio-

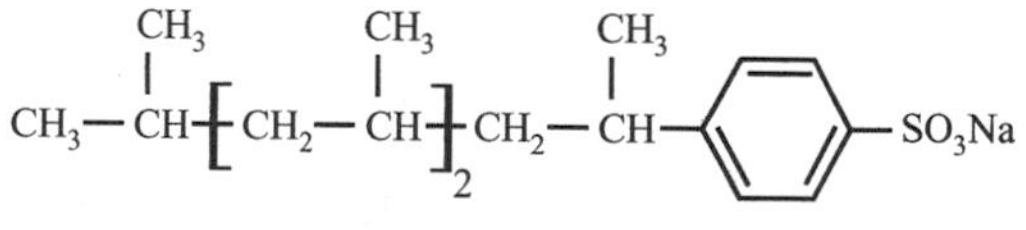

$$CH_3-CH(CH_3)-[CH_2-CH(CH_3)]_2-CH_2-CH(CH_3)-C_6H_4-SO_3Na$$

Nonlinear alkylbenzylsulfonate

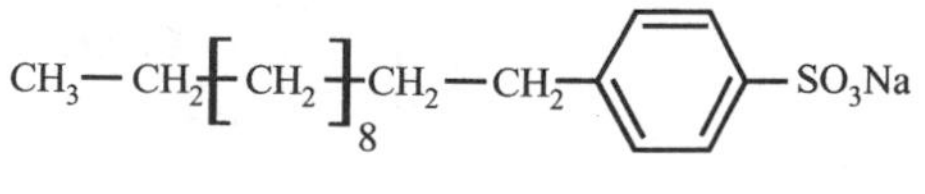

$$CH_3-CH_2-[CH_2]_8-CH_2-CH_2-C_6H_4-SO_3Na$$

Linear alkylbenzylsulfonate

FIGURE 1.22 Chemical structures of ABSs. The potential of these chemicals as surfactants can be seen from the possession of both charged hydrophilic groups and long-chain lipophilic groups.

degradation due to inhibition of β-oxidation. In contrast, linear ABSs are biodegradable in wastewater treatment plants. Nonlinear ABS is easier to manufacture and has slightly superior detergent properties, but the methyl branching of the alkyl chain interferes with biodegradation, since the tertiary carbon atoms block the normal β-oxidation sequence. ABS was specifically redesigned because its first form was resistant to biodegradation, and the industry switch to linear ABSs removed at a stroke the unsightly foaming and scum formation seen in receiving water bodies. Most importantly, the newly designed ABS is more easily biodegraded.

Plastics and Polymers

About 50 million metric tons of plastics are produced per year in the United States, much of which is for disposable goods and packaging material that wind up as environmental wastes and pollutants. Disposable goods and packaging material, about one-third of the total plastic production, have the largest environmental impact. More than 90% of the plastic material in municipal garbage consists of polyethylene, polyvinyl chloride, and polystyrene, in roughly equal proportions. Resistance to biodegradation of the polymers seems to be associated here with excessive molecular size. Biodegradation of long-chain C_{30} and higher *n*-alkanes declines with increasing molecular weight, and *n*-alkanes in excess of a molecular weight of 550 to 600 become refractory to biodegradation. If the molecular size of polyethylene is reduced to a molecular weight under 500, by pyrolysis, for example, the fragments are susceptible to biodegradation. Biodegradable polymers can be synthesized to replace or augment various plastics which have been accumulating in the environment because of resistance to microbial attack.

OPs

Organophosphates (OPs) were first used as nerve gas agents before being developed as pesticides. In their acute effects, they are anticholinesterases that block at neural connections between nerve and muscle and between synapses in the autonomic nervous system. They are in the first priority group of pesticides to be reviewed under the U.S. Food Quality Protection Act of 1996. Approximately 30 million kg of OPs is applied to approximately 60 million acres of U.S. agricultural crops annually. The pathways of exposure to humans are through direct contact by agricultural workers, airborne OPs in the atmosphere, residuals on crops, and runoff into groundwater. Human intake occurs through inhalation, absorption through the skin, and ingestion. The acute toxicity of compounds such as parathion (Fig. 1.23) is very high.

OPs have been used as pesticides and chemical warfare agents during the last 60 years. Sites contaminated with OPs are of concern when there is a high risk of contact with humans. OPs decompose in the environment over time to form relatively nontoxic compounds due to the action of light, water, and soilborne microorganisms. Bacteria that can biodegrade various organophosphorus pesticides, including dialkyl phosphates, dialkyl phosphorothioates, dialkyl phosphorodithioates, alkyl arylphosphonates, alkyl arylphosphonothioates, and alkyl alkyl-

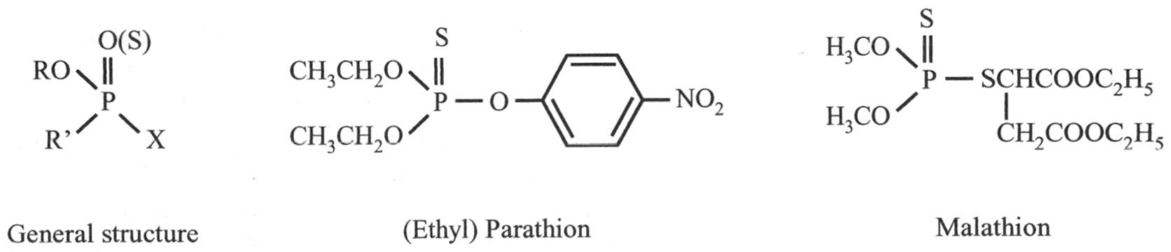

O(S)
RO, R', X, P

S
CH_3CH_2O, CH_3CH_2O, P—O—(C6H4)—NO_2

S
H_3CO, H_3CO, P—$SCHCOOC_2H_5$, $CH_2COOC_2H_5$

General structure

(Ethyl) Parathion

Malathion

FIGURE 1.23 Chemical structures of some common OPs. OPs are extremely toxic to humans, and some have had very widespread usage as pesticides.

phosphonates, have been isolated (37). *Agrobacterium radiobacter*, for example, can hydrolyze a wide range of OP insecticides (62). A gene encoding a protein involved in OP hydrolysis cloned from this bacterium had a high level of homology to other OP-hydrolyzing enzymes. Phosphotriesterases catalyze the hydrolytic detoxification of phosphotriester pesticides and chemical warfare nerve agents with various efficiencies (87). The proposed catabolic pathway for the breakdown of a typical phosphotriester, dimethyl paraoxon, to phosphate involves three hydrolytic steps. Phosphotriesterases catalyze the hydrolysis of paraoxon to dimethyl phosphate. Phosphodiesterases catalyze the hydrolysis of dimethyl phosphate to methyl phosphate. Phosphomonoesterases such as alkaline phosphatase catalyze the hydrolysis of methyl phosphate to phosphate.

Heavy Metals

Metals and their derivatives are a small group of elements. Only about 30 metals are used by industry, so their diversity is much less than that of the organics. Nevertheless, some metals are extracted in very large volumes, and several are highly significant environmental pollutants and toxicants. They vary in their toxicological effects. They also vary enormously in their bioavailability, depending on individual chemistry and pH, and in soils factors such as cation-exchange capacity greatly influence their mobility and availability. Various heavy metals other than mercury, including tin, cobalt, chromium, nickel, cadmium, and thallium, are used in metal alloys or as catalysts. Their mining, smelting, and ultimate disposal cause heavy-metal pollution problems. All these metals are substantially toxic to plants, animals, and many microorganisms. Radionuclides as environmental pollutants originate from atmospheric testing of thermonuclear weapons, uranium mining and processing, disposal of nuclear wastes, and the routine operation of nuclear power plants and accidents at such installations.

Nickel, zinc, copper, and boron are phytotoxic metals. Cadmium, lead, mercury, arsenic, beryllium, and copper are zootoxic. Arsenic is a metalloid, and its primary usage has been in agriculture, in formulation of herbicides, especially for the control of weeds in cotton fields. Sodium arsenite has been used as an insecticidal ingredient of sheep-dips. In industry, arsenic has found use in glass manufacture and a new role is in the semiconductor industry. Copper, which is released in significant amounts via smelting, is a category A carcinogen and has many acute effects on human health. In several countries, it is a major blight in drinking water. Cadmium is a superior plating metal. $Cd(OH)_2$ serves as the anode in nickel cadmium batteries, and other applications are found in the plastic and pigment industries. However, Cd is a highly toxic metal, which is bioaccumulated and is excreted with a half-life of 20 to 30 years. It causes a range of health effects from hypertension to cancer.

There is no better illustration of the hazards of environmental contamination than that provided by mercury. It is an important industrial element and has been used (is in use) in a wide variety of industries and products including miniature batteries, mercury vapor lamps, industrial catalysts, fungicides, and insecticides. Mercury is a serious threat to human health because of the way it acts on the central nervous system (CNS). The effects of mercury on the CNS include neurological damage, irritability, paralysis, blindness, insanity, chromosome damage, and birth defects (81). Methylmercury and other alkyl mercury species are lipid soluble and can penetrate the blood-brain barrier, attacking the CNS. Mercury species can spread throughout the entire body, bioaccumulating in lipids and organs such as the liver.

The risk of toxic chemicals catastrophically contaminating food was realized in the 1950s as a result of mercury poisoning. The city where it occurred is now a synonym for this type of risk: Minamata disease, which is discussed in detail by Ellis (46). At Minamata in Japan, discharge of inorganic mercury-bearing waste from a factory killed more than 100 people directly and ruined the lives of thousands of others. The factory discharged its waste to

Minamata Bay, which is a settling basin, not a flushing open coast. As a result, the many contaminants in the waste accumulated in the sediments rather than being diluted and dispersed. The inorganic mercury wastes, once incorporated into the anaerobic sediment of the bay, may have been methylated by microorganisms and may have added to the methylmercury contamination, though the contribution of biological methylation versus direct methylmercury discharge was not quantified, and the contribution from microbial methylation probably was minor in this incident. The environmental methylation of mercury can be attributed to the system responsible for the anaerobic generation of methane. Methylcobalamine (methyl-vitamin B_{12}) is able to transfer its methyl group to Hg^{2+} ions, yielding reduced vitamin B_{12}. The process may be enzymatically mediated, but it also proceeds spontaneously. Under environmental conditions, the predominant product is monomethylmercury ($Hg^{+}CH_3$). Thus, the release of mercury into the sediments of Minamata Bay started a novel chain of chemical change, biological uptake, and biomagnification that resulted in the concentration of methyl mercury in the flesh of shellfish and finfish. At the top of the food chain were the subsistence fishermen of the area (Fig. 1.24); much of their food was contaminated and, through biomagnification, a large number of people were exposed to sufficient toxin to poison them.

The heavy metals are the most perplexing inorganic pollutants for bioremediation. The heavy metals are those with a density of greater than 5 g cm^{-3}, of which there are about 40. Metals arise as pollutants from many industries, particularly ore extraction and manufacturing. For example, metals often have roles in catalysis. Metals can be conveniently subdivided into those with biological functions (e.g., copper, zinc, and cobalt) and those which are purely toxins (e.g., lead, mercury, and arsenic). Microbes cannot destroy metals. However, many can take them up and transform them to less

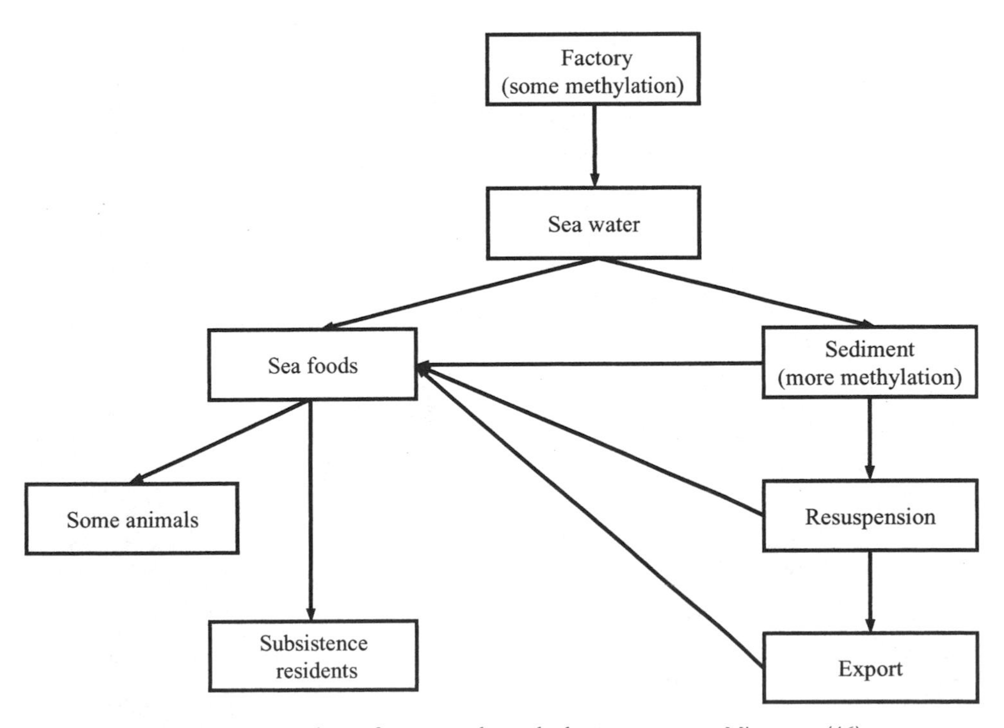

FIGURE 1.24 Flow of mercury through the ecosystem at Minamata (46).

hazardous oxidation states. Microorganisms are capable of transferring methyl groups onto various heavy metals and some metalloids, such as selenium and arsenic. Methylation processes result in altered toxicity and biomagnification of these elements. Arsenate can be methylated to mono-, di-, and trimethylarsines by some filamentous fungi, such as *Scopulariopsis brevicaulis*. The methyl group is donated here by either *S*-adenosylmethionine or methyl-cobalamine. The products are volatile and highly toxic compounds. A promising area is phytoremediation, the use of plants which can bioaccumulate heavy metals and thus move them out of contaminated environments. It is outside the scope of this chapter to discuss phytoremediation in more detail, but it is possible that a combined approach, i.e., the use of microbes to treat organics and the use of plants to remove metals, may be feasible. The potential of enhancing the activity of organic pollutant-transforming bacteria within plants (endophytic bacteria) is also being considered (12).

BIOCHEMICAL AND MOLECULAR ASPECTS OF BIOREMEDIATION

Given the limited reactivity of hydrocarbons and their chlorinated analogs, complex biochemical pathways are required for their biodegradation. This involves the synthesis of multiple enzymes and their ability to act in concert in the correct sequence to bring about conversion of the hydrocarbon to a central metabolite (Fig. 1.25). A common control strategy for the bacterium is to group the genes coding for the enzymes in operons. The operons can

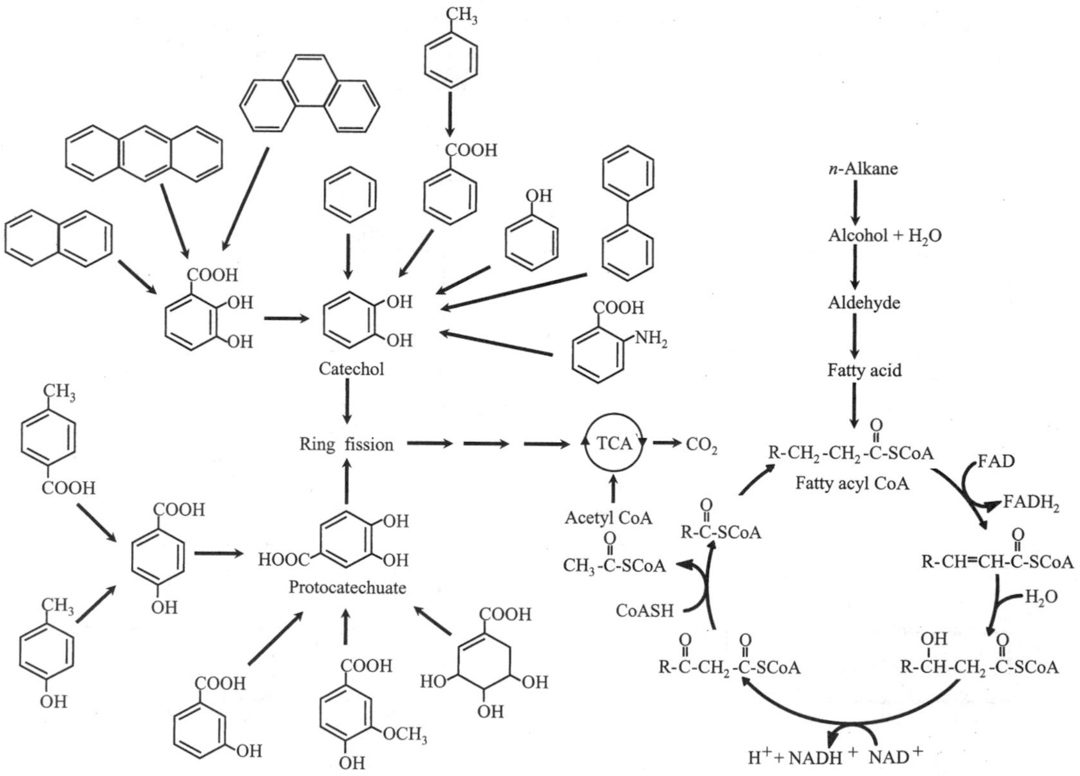

FIGURE 1.25 Bacterial metabolism of hydrocarbons to central metabolites. The remarkable economy of bacteria is illustrated: a vast number of hydrocarbons are converted to just two key intermediates, catechol and protocatechuate. From this point, ring fission occurs, and by a relatively few short steps the ring fission products are converted to central metabolites. TCA, tricarboxylic acid; CoA, coenzyme A.

be situated on the main chromosome, but commonly they are found on plasmids. Many such catabolic plasmids are self-transmissible via conjugation and can be transferred through soil bacterial populations, leading to spread of catabolic capabilities and evolution of new ones. Furthermore, many clusters of catabolic genes are located on transposons. Thus, novel catabolic capabilities can arise by the rearrangement of DNA sequences which is inherent in the activities of transposons.

Naturally occurring hydrocarbons exist in vast quantities, which has aided the evolution of biodegradation processes found in microorganisms. Hydrocarbons are also produced constantly by animals and plants and by microbial action. Benzene is commonly associated with petroleum. Lignin, the structural material of plants, and humus, the organic fraction of soils, also both contain the benzene ring, and together the quantity of these materials is vast. Pioneering work on the biochemistry of hydrocarbon biodegradation was being done in the 1960s. The first catabolic plasmids were discovered in the 1970s. It was soon realized that such catabolic plasmids were common in nature (141) and that the genus *Pseudomonas* was metabolically diverse in this respect. The scene was set for a large expansion in biodegradation research, and it became clear that not only was hydrocarbon biodegradation common in many bacteria, but also catabolic plasmids were widespread in many genera, both gram negative and gram positive. Knowledge of metabolic diversity was enhanced with the realization that complete reductive dechlorination by obligate anaerobes was possible. Anaerobic dissimilation of low-molecular-weight aromatics, such as benzene and toluene, follows a strategy similar to that of aerobic systems (61).

Natural environments contain mixed communities of microorganisms, and the importance of cometabolism of recalcitrant compounds is now known. In cometabolism, an organism growing on a particular substrate is able to gratuitously oxidize another substrate but is not able to assimilate it. However, the oxidation products of the second substrate are available for assimilation by other members of the community (9). As an important example, TCE, which is a common pollutant of groundwater, can be gratuitously transformed. Provision of either methane or toluene as a cosubstrate can be used for the bioremediation of TCE-contaminated groundwater.

One approach for providing the enzymatic capability to degrade diverse pollutants is to use genetic engineering to create microorganisms with the capacity to degrade a wide range of hydrocarbons. A hydrocarbon-degrading pseudomonad engineered by A. M. Chakrabarty was the first organism that the Supreme Court of the United States, in a landmark decision, ruled could be patented. However, considerable controversy surrounds the release of such genetically engineered microorganisms into the environment, and field testing of these organisms must be delayed until the issues of safety, containment, and potential for ecological damage are resolved. Given the current regulatory framework for the deliberate release of specifically genetically engineered microorganisms, it is unlikely that any genetically engineered microorganism would gain the necessary regulatory approval in time to be of much use in treating the oil spill for which it was created.

FACTORS AFFECTING BIOREMEDIATION

Bioavailability is extremely important to biodegradation of organic pollutants. It is frequently observed that the rate of removal of compounds from soils is very low even though the compounds are biodegradable. However, the substrates in these instances may not be in a form that is readily available to the microorganisms. Biodegradation of hydrophobic pollutants may take place only in the aqueous phase; for example, naphthalene (143) is utilized by pure cultures of bacteria only in the dissolved state. Bouchez et al. (24) similarly showed that phenanthrene biodegradation occurs only in the aqueous phase. The many observations of linear growth of bacteria and yeasts on slightly soluble substrates may be explained by the need for the substrates to be in the aqueous phase. In the case of the PAHs, it has been shown that linear growth is due to mass transfer limitation from the solid phase to the liquid phase (135).

The three main classes of hydrocarbons (aliphatic, alicyclic, and aromatic hydrocarbons) vary in their biodegradability according to size and solubility. It is believed that only molecules of hydrocarbons that are dissolved in the aqueous phase are available for intracellular metabolism (115). The rate at which a particular organic compound dissolves in water is critical to its biodegradability, as this governs the rate of transfer to the organism. The rate of transfer is determined by the equilibrium and actual concentration in the bulk phase and the aqueous phase. This is central to the concept of bioavailability as it relates to biodegradation. As a complicating issue in relation to bioavailability, there is also evidence that indicates the ability of microbes to access pollutants that are sorbed to surfaces and that microbial pollutant transformation rates observed in some cases cannot be explained by mass transfer processes or instantaneous desorption from surfaces alone (98). The application of surfactants to release hydrophobic pollutants, with the objective of increasing their bioavailability and biodegradability, has had mixed results (65, 77, 113, 127, 148, 149), and the effect with PAHs is not clear.

The transfer of knowledge from the research laboratory to field-scale bioremediation is not a simple matter. There are a host of factors in the field which are not difficult to control in the laboratory that have to be addressed. Bioremediation is a complex process, whose quantitative and qualitative aspects depend on the nature and amount of the pollutant present, the ambient and seasonal environmental conditions, and the composition of the indigenous microbial community. For bioremediation of waters, control of the environment may be relatively straightforward. The soil environment, however, is a different proposition. Decontamination of soils is especially complex due to the heterogeneous structure of soils, with mixed gaseous, solid, and liquid phases and associated interfaces.

It is an objective of the engineering professionals to ensure success by optimizing the physical and chemical factors that are important to the growth of the microorganisms. Many contaminated sites are characterized by poor nutrient concentrations (i.e., low levels of nitrogen and phosphorus), elevated or low temperatures, and a diverse range of cocontaminants, such as heavy metals, which can influence the process of bioremediation by inhibiting the growth of the pollutant-degrading microorganisms and therefore the efficacy of pollutant transformation and potential for bioremediation. Indeed, bioremediation treatments are often designed to overcome these limitations which may contribute to environmental persistence. Of particular importance are pH, redox potential, supply of oxygen, moisture, temperature, inorganic nutrients, cation-exchange capacity, pollutant bioavailability, and soil porosity.

Often the most important factor limiting rates of biodegradation in the environment is the availability of molecular oxygen. The initial enzymes of aerobic attack on hydrocarbons are oxygenases, which have an absolute requirement for molecular oxygen. Delivering air or oxygen to contaminated soils can be difficult for a number of reasons: the soil porosity may not be favorable, and therefore mass transfer from the gas phase to the aqueous phase will be limited. Also, the relatively low solubility of oxygen in water is a primary limiting factor. Most contaminated soils contain large populations of the appropriate microorganisms but can remain contaminated for decades or longer as a result of conditions that do not favor rapid biodegradation of complex pollutants.

Although anaerobic microorganisms have the potential to metabolize organic contaminants and do so in many field situations, oxygen is often an integral part in the oxidation of many organic pollutants, including hydrocarbons, as molecular oxygen is required to oxidize the carbon moiety (43). Fungal transformation of organic pollutants is also driven by oxygen since fungi are, in the main, aerobic organisms. It is important to understand whether the pollutant present will be transformed more rapidly under aerobic or anaerobic conditions and whether the cost of supplying oxygen to subsurface environments to remediate pollutants is met by an acceptable increase in transformation. DDT is a prime example of important decision making regarding oxygen availability. DDT is converted to DDE [1,1-dichloro-2,2-bis (*p*-chlorophenyl)ethylene] under anaerobic conditions. DDE is recalcitrant and has known toxic effects. However, under aerobic conditions, DDD [1,1-dichloro-2,2-bis(*p*-chlorophenyl)ethane], which is more readily broken down than DDE (68), is produced from DDT. Often a two-stage process is useful in completely transforming some pollutants; for example, better degradation of polychlorinated organic pollutants can be achieved by using a combination of anaerobic and aerobic treatments (17, 47, 49).

It is essential that contaminated sites be at the optimum temperature for bioremediation to progress successfully, since excessively high or low temperatures sometimes inhibit microbial metabolism. In addition, the solubility and bioavailability of a contaminant will increase as temperature increases, and oxygen solubility will be reduced, which will leave less oxygen available for microbial metabolism (82). Until recently, most research has focused upon the biodegradation of organic compounds at mesophilic temperatures; however, evidence suggests that the indigenous microbial communities of a contaminated site will adapt to the in situ temperature, such as those in Arctic and Antarctic soils. For example, a selection of *Rhodococcus* species that were isolated from an Antarctic soil were able to successfully degrade a number of *n*-alkanes at -2°C but were severely inhibited at a higher temperature (18). In addition, the PAHs naphthalene and phenanthrene were successfully degraded from crude oil in seawater at temperatures as low as 0°C (116).

The importance of isolating microorganisms able to metabolize organic contaminants at low temperatures was highlighted by Whyte and colleagues (138). These researchers discovered that the psychrotrophic *Rhodococcus* sp. strain Q15, which was isolated from a sediment in the Bay of Quinte, Ontario, Canada, was able to degrade a suite of alkanes from diesel oil in a temperature range of 0 to 30°C (138). As most contaminated sites will not be at the optimum temperature for bioremediation during every season of the year, these researchers suggest that this organism, along with other psychrophilic and psychrotrophic hydrocarbon-degrading microorganisms, may be of particular interest for the bioaugmentation of contaminated sites. In comparison to the psychrophiles and psychrotrophs, peroxidase enzymes produced by the ligninolytic fungi have an optimum activity between 50 and 75°C. This was demonstrated by Lau and colleagues (75), who found that $\geq$90% of a suite of PAHs was degraded during composting in spent-mushroom compost.

Previous work, for example, from the *Exxon Valdez* oil spill in 1989, which applied fertilizers to oil-contaminated beaches to accelerate the degradation of contaminating oil (103), has shown nutrient availability to be most commonly the limiting factor in bioremediation. In uncontaminated nonagricultural sites such as soils, mineral nutrients, including nitrogen, phosphate, and potassium (N, P, K), are very rarely limiting. However, when an environment is contaminated with organic compounds, the C:N ratio can increase dramatically and can become the rate-limiting factor in bioremediation (8, 27).

These nutrients can be supplied to the microorganisms by supplementing the contaminated material (soil or water) with a mineral fertilizer, for example, ammonium phosphate, or a mixture of other salts such as a combination of ammonium sulfate and calcium phosphate. In addition, organic material such as fish bones and manure can provide a source of nutrients. However, it is important to apply the correct balance and concentration of nutrients to the contaminated material, as the wrong ratio of carbon to nitrogen can result in little or no enhancement of microbial growth (8). This was shown in a creosote-contaminated soil, in which optimal biodegradation occurred at a C:N ratio of 25:1 relative to a C:N ratio of 5:1 (8).

Fungi may benefit from the addition of a readily degradable carbon source such as molasses, since many nonligninolytic fungi are unable to directly metabolize organic contaminants as their sole source of carbon but can cometabolize them when mineralizing other substrates (31). Another factor to consider when promoting fungal rather than bacterially mediated bioremediation is that ligninolytic enzyme synthesis can be inhibited in a high-level nutrient system (22).

It is therefore clear that before a contaminated site is supplemented with additional nutrients, a thorough site investigation should be conducted, not only to assess the nutrient levels on the site, but also to consider the nature of the pollutant and the intended method(s) of treatment. Excess nutrient loading of a contaminated site might initially result in an increase in microbial biomass and contaminant transformation rates, but long-term effects may be the inhibition of functionally important microorganisms, such as the ligninolytic fungi capable of transforming high-molecular-weight compounds.

Given the advances in molecular biology and the widespread availability of techniques for manipulating DNA, an evolution of bioremediation research is the construction of bacteria specifically designed for the treatment of specific wastes (129). A variety of strategies can be followed. Modification of the structure and regulation of catabolic operons and the copy number of plasmids can be used to increase expression with a view to making bioremediation occur more quickly. New genes are continuously being discovered, and they can be used to make new constructs, resulting in production of degradative pathways for new synthetic compounds. Biodegradation capacities not present at a particular site can be introduced. It is possible to make plasmids containing several operons, and they can be used to enhance the diversity of biodegradation.

The nature of some contamination means that despite having biodegradation capabilities, the natural populations may find the environment rather toxic. Heavy-metal resistance (Fig. 1.26) mechanisms often have a genetic basis (144), and the toxicity of solvents can be overcome by *cis-trans* isomerase (67) or efflux pump mechanisms (44). The transfer of genes for such mechanisms may be used to overcome toxicity of sites. The whole arena of genetic modification and release to the environment is, of

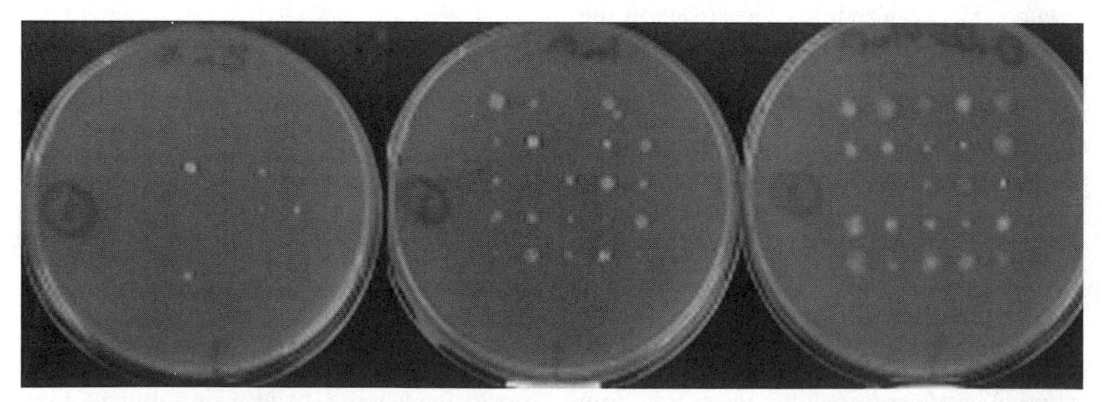

FIGURE 1.26 Replica plates of 25 strains of copper-resistant bacteria. From right to left, the plates contain increasing concentrations of copper nitrate (0.0001, 1.0, and 5.0 mM). As the concentration increases, some more sensitive strains are inhibited. Colonies resistant to these high levels of copper were green-blue in the presence of copper but cream colored in the absence of copper.

course, subject to strict control, as mentioned above.

CONCLUDING REMARKS

Overall, bioremediation is still very much an evolving technology. Whereas significant research in biodegradation processes has been in progress since the 1960s, its practical application to the bioremediation of contaminated soils on a large scale was only starting to be realized in the late 1980s. In 1990, as a technique for practical soil remediation it was of little significance. However, bioremediation in the field grew enormously in the early 1990s so that by 1995 it accounted for 15% of all site remediations (38). There are now many examples of success in the field, but a severe limitation is the need for costly site demonstrations as the technology remains worryingly inconsistent from one situation to another. There is clearly a need for more research and development to identify and overcome the limitations in specific circumstances.

REFERENCES

1. **Adriaens, P., and D. Grbic-Galic.** 1994. Reductive dechlorination of PCDD/F by anaerobic cultures and sediments. *Chemosphere* **29:** 2253–2259.
2. **Adriaens, P., Q. Fu, and D. Grbic-Galic.** 1995. Bioavailability and transformation of highly chlorinated dibenzo-*p*-dioxins and dibenzofurans in anaerobic soils and sediments. *Environ. Sci. Technol.* **29:**2252–2260.
3. **Adriaens, P., P. R.-L. Chang, and A. L. Barkovskii.** 1996. Dechlorination of chlorinated PCDD/F by organic and inorganic electron transfer molecules in reduced environments. *Chemosphere* **32:**433–441.
4. **Alexander, M.** 1965. Biodegradation: problems of molecular recalcitrance and microbial fallibility. *Adv. Appl. Microbiol.* **7:**35–80.
5. **Alexander, M.** 2001. *Biodegradation and Bioremediation*, 2nd ed. Academic Press, San Diego, Calif.
6. **Armengaud, J., B. Happe, and K. N. Timmis.** 1998. Genetic analysis of dioxin dioxygenase of *Sphingomonas* sp. strain RW1: catabolic genes dispersed on the genome. *J. Bacteriol.* **180:**3954–3966.
7. **Arthur, M. F., and J. I. Frea.** 1987. Microbial activity in soils containing 2,3,7,8-tetrachlorodibenzo-*p*-dioxin. *Environ. Toxicol. Chem.* **7:**5–13.
8. **Atagana, H. I., R. J. Haynes, and F. M. Wallis.** 2003. Optimization of soil physical and chemical conditions for the bioremediation of creosote-contaminated soil. *Biodegradation* **14:** 297–307.
9. **Atlas, R. M., and R. Bartha.** 1997. *Microbial Ecology: Fundamentals and Applications.* Benjamin/Cummings Science Publishing, Menlo Park, Calif.
10. **Badr, T., K. Hanna, and C. de Brauer.** 2004. Enhanced solubilization and removal of naphthalene and phenanthrene by cyclodextrins from two contaminated soils. *J. Hazard. Mater.* **112:** 215–223.
11. **Bagley, D. M., and J. M. Gossett.** 1990. Tetrachloroethene transformation to trichloroethene and *cis*-1,2-dichloroethene by sulfate-reducing enrichment cultures. *Appl. Environ. Microbiol.* **56:**2511–2516.
12. **Barac, T., S. Taghavi, B. Borremans, A. Provoost, L. Oeyen, J. V. Colpaert, J. Vangronsveld, and D. van der Lelie.** 2004. Engineered endophytic bacteria improve phytoremediation of water-soluble, volatile, organic pollutants. *Nat. Biotechnol.* **22:**583–588.
13. **Barbaro, J. R., J. F. Barker, L. A. Lemon, and C. I. Mayfield.** 1992. Biotransformation of BTEX under anaerobic, denitrifying conditions: field and laboratory observations. *J. Contam. Hydrol.* **11:**245–272.
14. **Bardi, L., A. Mattei, S. Steffan, and M. Marzano.** 2000. Hydrocarbon degradation by a soil microbial population with β-cyclodextrin as surfactant to enhance bioavailability. *Enz. Microb. Technol.* **27:**709–713.
15. **Barkovskii, A. L., and P. Adriaens.** 1996. Microbial dechlorination of historically present and freshly spiked chlorinated dioxins and diversity of dioxin-dechlorinating populations. *Appl. Environ. Microbiol.* **62:**4556–4562.
16. **Bartels, I., H.-J. Knackmuss, and W. Reinecke.** 1984. Suicide inactivation of catechol 2,3-dioxygenase from *Pseudomonas* MT-2 by 3-halocatechols. *Appl. Environ. Microbiol.* **47:**500–505.
17. **Beeman, R. E., and C. A. Bleckmann.** 2002. Sequential anaerobic-aerobic treatment of an aquifer contaminated by halogenated organics: field results. *J. Contam. Hydrol.* **57:**147–159.
18. **Bej, A. K., D. Saul, and J. Aislabie.** 2000. Cold-tolerant alkane-degrading *Rhodococcus* species from Antarctica. *Polar Biol.* **23:**100–105.
19. **Beller, H. R., A. M. Spormann, P. K. Sharma, J. R. Cole, and M. Reinhard.** 1996. Isolation and characterization of a novel toluene-degrading sulfate-reducing bacterium. *Appl. Environ. Microbiol.* **62:**1188–1196.
20. **Berthe-Corti, L., and S. Fetzner.** 2002. Bac-

terial metabolism of *n*-alkanes and ammonia under oxic, suboxic and anoxic conditions. *Acta Biotechnol.* **22:**299–336.
21. **Björklund, T. Nilsson, S. Bøward, K. Pilorz, L. Mathiasson, and S. B. Hawthorne.** 2000. Introducing selective supercritical fluid extraction as a new tool for determining sorption/desorption behavior and bioavailability of persistent organic pollutants in sediment. *J. Biochem. Biophys. Methods* **43:**295–311.
22. **Bogan, B. W., B. Schoenike, R. T. Lamar, and D. Cullen.** 1996. Expression of *lip* genes during growth in soil and oxidation of anthracene by *Phanerochaete chrysosporium. Appl. Environ. Microbiol.* **62:**3697–3703.
23. **Bollens, W. B., and L. A. Norris.** 1979. Influence of 2,3,7,8-tetrachlorodibenzo-*p*-dioxin on respiration in a forest floor and soil. *Bull. Environ. Contam. Toxicol.* **62:**648–652.
24. **Bouchez, M., D. Blanchet, and J.-P. Vandecasteele.** 1995. Substrate availability in phenanthrene biodegradation: transfer mechanism and influence on metabolism. *Appl. Microbiol. Biotechnol.* **43:**952–960.
25. **Boyle, A. W., C. J. Silvin, J. P. Hassett, J. P. Nakas, and S. W. Tanenbaum.** 1992. Bacterial PCB degradation. *Biodegradation* **3:** 285–298.
26. **Brandt, K. K., A. Pedersen, and J. Sorensen.** 2002. Solid-phase contact assay that uses a *lux*-marked *Nitrosomonas europaea* reporter strain to estimate toxicity of bioavailable linear alkylbenzene sulfonate in soil. *Appl. Environ. Microbiol.* **68:**3502–3508.
27. **Breedveld, G. D., and M. Sparrevik.** 2000. Nutrient limited biodegradation of PAHs in various soil strata at a creosote contaminated site. *Biodegradation* **11:**391–399.
28. **Budavari, S. (ed.).** 1989. *The Merck Index*, 11th ed. Merck & Co., Inc., Rahway, N. J.
29. **Cave, M., and J. Wragg.** 2000. Measurement of heavy metals: bioavailability and distribution in contaminated soils. *Earthwise* **15:**32–33.
30. **Cerniglia, C. E.** 1984. Microbial degradation of polycyclic aromatic hydrocarbons. *Adv. Appl. Microbiol.* **30:**31–71.
31. **Cerniglia, C. E., and J. B. Sutherland.** 2001. Bioremediation of polycyclic aromatic hydrocarbons, p. 136–187. *In* G. M. Gadd (ed.), *Fungi in Bioremediation.* Cambridge University Press, Cambridge, United Kingdom.
32. **Chiou, C. T., P. E. Porter, and D. W. Schmeddling.** 1983. Partition equilibria of nonionic organic compounds between soil and organic matter and water. *Environ. Sci. Technol.* **17:**227–231.
33. **Chung, N., and M. Alexander.** 2002. Effect of soil properties on bioavailability and extractability of phenanthrene and atrazine sequestered in soil. *Chemosphere* **48:**109–115.
34. **Coates, J. D., R. T. Anderson, and D. R. Lovley.** 1996. Oxidation of polycyclic aromatic hydrocarbons under sulfate-reducing conditions. *Appl. Environ. Microbiol.* **62:**1099–1101.
35. **Coates, J. D., J. Woodward, J. Allen, P. Philp, and D. R. Lovley.** 1997. Anaerobic degradation of polycyclic aromatic hydrocarbons and alkanes in petroleum-contaminated marine harbor sediments. *Appl. Environ. Microbiol.* **63:** 3589–3593.
36. **Connell, D. W., P. Lam, B. Richardson, and R. Wu.** 1999. *Introduction to Ecotoxicology.* Blackwell Science, Oxford, United Kingdom.
37. **Cook, A. M., C. G. Daughton, and M. Alexander.** 1978. Phosphorus-containing pesticide breakdown products: quantitative utilization as phosphorus sources by bacteria. *Appl. Environ. Microbiol.* **36:**668–672.
38. **Cookson, J. T., Jr.** 1995. *Bioremediation Engineering: Design and Application.* McGraw-Hill, New York, N.Y.
39. **Cronin, M. T. D., and J. C. Dearden.** 1995. Review: QSAR in toxicology. 1. Prediction of aquatic toxicity. *Quantit. Struct. Activ. Relat.* **14:** 1–7.
40. **Cronin, M. T. D., and T. W. Schultz.** 1997. Validation of *Vibrio fischeri* acute toxicity data: mechanism of action-based QSARs for non-polar narcotics and polar narcotic phenols. *Sci. Tot. Environ.* **204:**75–88.
41. **Cronin, M. T. D., and T. W. Schultz.** 1998. Structure-toxicity relationships for three mechanisms of action of toxicity to *Vibrio fischeri. Ecotoxicol. Environ. Safety* **39:**65–69.
42. **Cuypers, C., T. Pancras, T. Grotenhuis, and W. Rulkens.** 2002. The estimation of PAH bioavailability in contaminated sediments using hydroxypropyl-cyclodextrin and Triton X-100 extraction techniques. *Chemosphere* **46:** 1235–1245.
43. **Davies, J. I., and W. C. Evans.** 1964. Oxidative metabolism of naphthalene by soil pseudomonads: the ring fission mechanism. *J. Biochem.* **91:**251–261.
44. **de Bont, J. A. M.** 1998. Solvent-tolerant bacteria in biocatalysis. *TIBTECH* **16:**493–499.
45. **Duffy, J. S., J. A. Del Pup, and J. J. Kneiss.** 1992. Toxicological evaluation of methyl *tertiary* butyl ether (MTBE): testing performed under TSCA consent agreement. *J. Soil Contam.* **1:** 29–37.
46. **Ellis, D.** 1989. *Environments at Risk. Case Histories of Impact Assessment.* Springer-Verlag, Heidelberg, Germany.

47. **Evans, B. S., C. A. Dudley, and K. T. Klasson.** 1996. Sequential anaerobic-aerobic biodegradation of PCBs in soil slurry microcosms. *Appl. Biochem. Biotechnol.* **57–58:**885–894.
48. **Eweis, J. B., S. J. Ergas, D. P. Y. Chang, and E. D. Schroeder.** 1998. *Bioremediation Principles.* McGraw-Hill, Singapore, Singapore.
49. **Fathepure, B. Z., and T. M. Vogel.** 1991. Complete degradation of polychlorinated hydrocarbons by a two-stage biofilm reactor. *Appl. Environ. Microbiol.* **57:**3418–3422.
50. **Ferguson, C. C., D. Darmendrail, K. Freier, B. K. Jensen, J. Jensen, H. Kasamas, A. Urzelai, and J. Vegter (ed.).** 1998. *Risk Assessment for Contaminated Sites in Europe*, vol. 1. *Scientific Basis.* LQM Press, Nottingham, United Kingdom.
51. **Fogel, M. M., A. R. Taddeo, and S. Fogel.** 1986. Biodegradation of chlorinated ethanes by a methane-utilizing mixed culture. *Appl. Environ. Microbiol.* **51:**720–724.
52. **Galli, R., and P. L. McCarty.** 1989. Biotransformation of 1,1,1-trichloroethane, trichloromethane, and tetrachloromethane by a *Clostridium* sp. *Appl. Environ. Microbiol.* **55:**837–844.
53. **Habe, H., and T. Omori.** 2003. Genetics of polycyclic aromatic hydrocarbon metabolism in diverse aerobic bacteria. *Biosci. Biotechnol. Biochem.* **67:**225–243.
54. **Habe, H., J. S. Chung, J. H. Lee, K. Kasuga, T. Yoshida, H. Nojiri, and T. Omori.** 2001. Degradation of chlorinated dibenzofurans and dibenzo-*p*-dioxins by two types of bacteria having angular dioxygenases with different features. *Appl. Environ. Microbiol.* **67:**3610–3617.
55. **Häggblom, M. M.** 1992. Microbial breakdown of halogenated aromatic pesticides and related compounds. *FEMS Microbiol. Rev.* **103:**29–72.
56. **Häggblom, M. M., and P. W. Milligan.** 2000. Anaerobic biodegradation of halogenated pesticides: influence of alternate electron acceptors, p. 1–34. *In* J. M. Bollag and G. Stotzky (ed.), *Soil Biochemistry.* Marcel Dekker, New York, N.Y.
57. **Halden, R. U., and D. F. Dwyer.** 1997. Biodegradation of dioxins: a review. *Biorem. J.* **1:** 11–25.
58. **Heath, J. S., K. Kobis, and S. L. Sayer.** 1993. Review of chemical, physical and toxicological properties of components of total petroleum hydrocarbons. *J. Soil Contam.* **2:**221–234.
59. **Hemond, H. F., and E. J. Fechner-Levy.** 2000. *Chemical Fate and Transport in the Environment.* Academic Press, San Diego, Calif.
60. **Hoag, G. E., A. Dahmani, F. Nadim, C. S. Dulam, and E. Quinn.** 1998. Use of coal tar contaminated soil in road paving asphalt. *Land Contam. Reclam.* **6:**91–103.
61. **Holliger, C., and A. J. B. Zehnder.** 1996. Anaerobic biodegradation of hydrocarbons. *Curr. Opin. Biotechnol.* **7:**326–330.
62. **Horne, I., T. D. Sutherland, R. L. Harcourt, R. J. Russell, and J. G. Oakeshott.** 2002. Identification of an *opd* (organophosphate degradation) gene in an *Agrobacterium* isolate. *Appl. Environ. Microbiol.* **68:**3371–3376.
63. **Hund-Rinke, K., and W. Kördel.** 2003. Underlying issues in bioaccessibility and bioavailability: experimental methods. *Ecotoxicol. Environ. Safety* **56:**52–62.
64. **Irwin, R. S., M. van Mouwerkle, L. Stevens, M. D. Seese, and W. Basham.** 1997. *Environmental Contaminants Encyclopedia.* Water Resource Division, National Park Service, Fort Collins, Colo. http://www1.nature.nps.gov.
65. **Jain, D. K., H. Lee, and J. T. Trevors.** 1992. Effect of addition of *Pseudomonas aeruginosa* UG2 inocula or biosurfactants on biodegradation of selected hydrocarbons in soil. *J. Ind. Microbiol.* **10:**87–93.
66. **Jannasch, H. W.** 1967. Growth of marine bacteria at limiting concentrations of organic carbon in seawater. *Limnol. Oceanogr.* **12:**264–271.
67. **Junker, F., and J. I. Ramos.** 1999. Involvement of the *cis/trans* isomerase Cti in solvent resistance of *Pseudomonas putida* DOT-TIE. *J. Bacteriol.* **181:**5693–5700.
68. **Kantachote, D., I. Singleton, R. Naidu, N. C. McClure, and M. Mallaravapu.** 2004. Sodium application enhances DDT transformation in a long-term contaminated soil. *Water Air Soil Pollut.* **154:**115–125.
69. **Kao, C. M., Y. L. Chen, S. C. Chen, T. Y. Chen, W. S. Yeh, and W. S. Wu.** 2003. Enhanced PCE dechlorination by biobarrier systems under different redox conditions. *Water Res.* **37:**4885–4894.
70. **Kao, C. M., S. C. Chen, J. K. Liu, and M. J. Wu.** 2001. Evaluation of TCDD biodegradability under different redox conditions. *Chemosphere* **44:**1447–1454.
71. **Kasuga, K., H. Nojiri, H. Yamane, and T. Omori.** 1997. Genes of enzymes involved in the biodegradation of carbazole, dibenzofuran, fluorene, and dibenzo-*p*-dioxin by bacteria. *Water Sci. Technol.* **36:**9–16.
72. **Knackmuss, H.-J.** 1992. Potentials and limitations of microbes to degrade xenobiotics, p. 3–9. *In Proceedings of the International Symposium on Soil Decontamination Using Biological Processes*, 6 to 9 December, Karlsruhe, Germany.
73. **Langenhoff, A. A. M., D. L. Brouwers-Ceiler, J. H. L. Engelberting, J. J. Quist, J.**

G. P. N. Wolkenfelt, A. J. B. Zehnder, and G. Schraa. 1997. Microbial reduction of manganese coupled to toluene oxidation. *FEMS Microbiol. Ecol.* **22:**119–127.

74. **Lanno, R., J. Wells, J. Conder, K. Bradham, and N. Basta.** 2004. The bioavailability of chemicals in soil for earthworms. *Ecotoxicol. Environ. Safety* **57:**39–47.
75. **Lau, K. L., Y. Y. Tsang, and S. W. Chiu.** 2003. Use of spent mushroom compost to bioremediate PAH-contaminated samples. *Chemosphere* **52:**1539–1546.
76. **Little, C. D., A. V. Palumbo, and S. E. Herbes.** 1988. Trichloroethylene biodegradation by a methane-oxidizing bacterium. *Appl. Environ. Microbiol.* **54:**951–956.
77. **Liu, Z., A. M. Jacobson, and R. G. Luthy.** 1995. Biodegradation of naphthalene in aqueous nonionic surfactant systems. *Appl. Environ. Microbiol.* **61:**145–151.
78. **Loehr, R. C., and M. T. Webster.** 1997. Effects of treatment on contaminant availability, mobility, and toxicity, p. 137–386. *In* D. G. Linz and D. V. Nakles (ed.), *Environmentally Acceptable Endpoints in Soil.* American Academy of Environmental Engineers, Annapolis, Md.
79. **Madsen, E. L.** 2003. *Report on Bioavailability of Chemical Wastes with Respect to the Potential for Soil Bioremediation.* U.S. Environmental Protection Agency report EPA/600/R-03/076. U.S. Environmental Protection Agency, Washington, D.C.
80. **Maier, R.** 2000. Bioavailability and its importance to bioremediation, p. 59–78. *In* J. J. Valdes (ed.), *International Society for Environmental Biotechnology: Environmental Monitoring and Biodiagnostics.* Kluwer, Dordrecht, The Netherlands.
81. **Manahan, S. E.** 1994. *Environmental Chemistry,* 6th ed. Lewis Publishers, Boca Raton, Fla.
82. **Margesin, R., and F. Schinner.** 2001. Biodegradation and bioremediation of hydrocarbons in extreme environments. *Appl. Microbiol. Biotechnol.* **56:**650–663.
83. **Maron, D. M., and B. N. Ames.** 1983. Revised methods for the *Salmonella* mutagenicity test. *Mutat. Res.* **113:**173–215.
84. **Mason, C. F.** 1991. *Biology of Freshwater Pollution.* Longman Scientific and Technical, Harlow, Essex, U.K.
85. **Masters, G. M.** 1991. *Introduction to Environmental Engineering and Science.* Prentice-Hall Inc., Englewood Cliffs, N. J.
86. **McFarland, B. L., D. J. Boron, W. Deever, J. A. Meyer, A. R. Johnson, and R. M. Atlas.** 1998. Biocatalytic sulfur removal from fuels: applicability for producing low sulfur gasoline. *Crit. Rev. Microbiol.* **24:**99–147.
87. **McLoughlin, S. Y., C. Jackson, J.-W. Liu, and D. L. Ollis.** 2004. Growth of *Escherichia coli* coexpressing phosphotriesterase and glycerophosphodiester phosphodiesterase, using paraoxon as the sole phosphorus source. *Appl. Environ. Microbiol.* **70:**404–412.
88. **Meckenstock, R. U., E. Annweiler, W. Michaelis, H. H. Richnow, and B. Schink.** 2000. Anaerobic naphthalene degradation by a sulfate-reducing enrichment culture. *Appl. Environ. Microbiol.* **66:**2743–2747.
89. **Mishra, V., R. Lal, and R. Srinivasan.** 2001. Enzymes and operons mediating xenobiotic degradation in bacteria. *Crit. Rev. Microbiol.* **27:** 133–166.
90. **Mohn, W. M., and J. M. Tiedje.** 1992. Microbial reductive dehalogenation. *Microbiol. Rev.* **56:** 482–507.
91. **Morgan, P.** 1991. *Biotechnology and Oil Spills.* Shell Selected Papers Series PAC/233. Shell International Petroleum Corporation, London, United Kingdom.
92. **MSI.** 2002. *Contaminated Land Treatment UK.* MSI Data Report. MSI Marketing Research for Industry Ltd., Chester, United Kingdom. http://www.marketresearch.com.
93. **Munro, S., S. Wallace, P. Kirby, and P. Walker.** 1995. Meeting the environmental challenge: managing our former gasworks sites. *Land Contam. Reclam.* **3:**4–5.
94. **Neilson, A.** 1990. A review: the biodegradation of halogenated organic compounds. *J. Appl. Bacteriol.* **69:**445–470.
95. **Nirmalakhandan, N. N., and R. E. Spreece.** 1988. Prediction of aqueous solubility of organic compounds based on molecular structure. *Environ. Sci. Technol.* **22:**328–338.
96. **Oberg, G.** 2002. The natural chlorine cycle—fitting the scattered pieces. *Appl. Microbiol. Biotechnol.* **58:**565–581.
97. **Organization for Economic Cooperation and Development.** 1998. *Biotechnology for Clean Industrial Products and Processes: Towards Industrial Sustainability.* Report of *Ad hoc* Task Force chaired by A. T. Bull. Organization for Economic Cooperation and Development, Paris, France. http://www.oecd.org.
98. **Park, J. H., Y. C. Feng, P. S. Ji, T. C. Voice, and S. A. Boyd.** 2003. Assessment of bioavailability of soil-sorbed atrazine. *Appl. Environ. Microbiol.* **69:**3288–3298.
99. **Paton, G. I., G. Palmer, M. Burton, E. A. S. Rattray, S. P. McGrath, L. A. Glover, and K. Killham.** 1997. Development of an acute and chronic ecotoxicity assay using *lux*-marked *Rhizobium leguminosarum biovar trifolii. Lett. Appl. Microbiol.* **24:**296–300.

100. **Paulus, W.** 1993. *Microbicides for the Protection of Materials*. Chapman and Hall, London, United Kingdom.
101. **Perkins, E. J., M. P. Gordon, O. Caceres, and P. F. Lurquin.** 1990. Organization and sequence analysis of the 2,4-dichlorophenol hydroxylase and dichlorocatechol oxidative operons of plasmid pJP4. *J. Bacteriol.* **172:**2351–2359.
102. **Philp, J. C., C. French, S. Wiles, J. M. L. Bell, A. S. Whiteley, and M. J. Bailey.** 2004. Wastewater toxicity assessment by whole cell biosensor, p. 165–225. *In* D. Barceló (ed.), *Handbook of Environmental Chemistry*, vol. 5. *Water Pollution: Emerging Organic Pollutants in Wastewaters.* Springer Verlag, Berlin, Germany.
103. **Pritchard, P. H., J. G. Mueller, J. C. Rogers, F. V. Kremer, and J. A. Glaser.** 1992. Oil spill bioremediation: experiences, lessons and results from the *Exxon Valdez* oil spill in Alaska. *Biodegradation* **3:**315–335.
104. **Reid, B. J., K. C. Jones, and K. T. Semple.** 2000. Bioavailability of persistent organic pollutants in soils and sediments—a perspective on mechanisms, consequences and assessment. *Environ. Poll.* **108:**103–112.
105. **Reid, B. J., K. C. Jones, and K. T. Semple.** 2000. Non-exhaustive cyclodextrin-based extraction technique for the evaluation of PAH bioavailability. *Environ. Sci. Technol.* **34:** 3174–3179.
106. **Rensing, C., and R. M. Maier.** 2003. Issues underlying use of biosensors to measure metal bioavailability. *Ecotoxicol. Environ. Safety* **56:** 140–147.
107. **Rockne, K. J., J. C. Chee-Sandford, R. Sanford, B. P. Hedlund, J. T. Staley, and S. E. Strand.** 2000. Anaerobic naphthalene degradation by microbial pure cultures under nitrate reducing conditions. *Appl. Environ. Microbiol.* **66:** 1595–1601.
108. **Rockne, K. J., and S. E. Strand.** 2001. Anaerobic biodegradation of naphthalene, phenanthrene and biphenyl as a denitrifying enrichment culture. *Water Res.* **35:**291–299.
109. **Ruby, M. V., A. Davis, R. Schoof, S. Eberle, and C. M. Sellstone.** 1996. Estimation of lead and arsenic bioavailability using a physiologically based extraction test. *Environ. Sci. Technol.* **30:**422–430.
110. **Safe, S.** 1989. Polychlorinated biphenyls (PCBs): mutagenicity and carcinogenicity. *Mutat. Res.* **220:**31–47.
111. **Salanitro, J. P., L. A. Diaz, M. P. Williams, and H. L. Wisniewski.** 1994. Isolation of a bacterial culture that degrades methyl *t*-butyl ether. *Appl. Environ. Microbiol.* **60:**2593–2596.
112. **Salkinoja-Salonen, M. S., R. Hakulinen, R. Valo, and J. Apajalahti.** 1983. Biodegradation of recalcitrant organochlorine compounds in fixed film reactors. *Water Sci. Technol.* **15:** 309–319.
113. **Scheibenbogen, K., R. G. Zytner, H. Lee, and J. T. Trevors.** 1994. Enhanced removal of selected hydrocarbons from soil by *Pseudomonas aeruginosa* UG2 biosurfactants and some chemical surfactants. *J. Chem. Technol. Biotechnol.* **59:** 53–59.
114. **Schultz, T. W., and M. T. D. Cronin.** 1997. Quantitative structure-activity relationships for weak acid respiratory uncouplers to *Vibrio fischeri*. *Environ. Toxicol. Chem.* **16:**357–360.
115. **Sikkema, J., J. A. de Bont, and B. Poolman.** 1995. Mechanisms of membrane toxicity of hydrocarbons. *Microbiol. Rev.* **59:**201–222.
116. **Siron, R., E. Pelletier, and H. Brochu.** 1995. Environmental factors influencing the biodegradation of petroleum hydrocarbons in cold seawater. *Arch. Environ. Contam. Toxicol.* **28:**406–416.
117. **Slater, J. H., A. T. Bull, and D. J. Hardman.** 1995. Microbial dehalogenation. *Biodegradation* **6:**181–189.
118. **Snoeyink, V. L., and D. Jenkins.** 1980. *Water Chemistry*. John Wiley & Sons, New York, N.Y.
119. **Spain, J. C.** 1995. Degradation of nitroaromatic compounds. *Annu. Rev. Microbiol.* **49:**523–555.
120. **Spiker, J. K., D. L. Crawford, and R. L. Crawford.** 1992. Influence of 2,4,6-trinitrotoluene (TNT) concentration on the degradation of TNT in explosive-contaminated soils by the white rot fungus *Phanerochaete chrysosporium*. *Appl. Environ. Microbiol.* **58:**3199–3202.
121. **Szejtli, J.** 1982. Cyclodextrins and their inclusion complexes, p. 95–109. *In Proceedings of the First International Symposium on Cyclodextrins*, 30 September to 2 October, Akademiai Kiado, Budapest. D. Reidel Publishing, Dordrecht, The Netherlands.
122. **Tanabe, S., H. Tanake, and R. Tatsukawa.** 1984. Polychlorobiphenyls, total DDT and hexachlorohexane isomers in the western North Pacific ecosystem. *Arch. Environ. Contam. Tox.* **13:** 731–738.
123. **Tao, S., L. Q. Guo, X. J. Wang, W. X. Liu, T. Z. Ju, R. Dawson, J. Cao, F. L. Xu, and B. G. Li.** 2004. Use of sequential ASE extraction to evaluate the bioavailability of DDT and its metabolites to wheat roots in soils with various organic carbon contents. *Sci. Tot. Environ.* **320:** 1–9.
124. **Terada, H.** 1990. Uncouplers of oxidative phosphorylation. *Environ. Health Perspec.* **87:** 213–218.
125. **Tessier, A., P. G. C. Campbell, and M. Bisson.** 1979. Sequential extraction procedure for

the speciation of particulate trace metals. *Anal. Chem.* **51:**844–851.

126. **Thomann, R. V., and J. A. Mueller.** 1987. *Principles of Surface Water Quality Modelling and Control.* Harper & Row, New York, N.Y.
127. **Tiehm, A.** 1994. Degradation of polycyclic aromatic hydrocarbons in the presence of synthetic surfactants. *Appl. Environ. Microbiol.* **60:**258–263.
128. **Tiensing, T., S. Preston, N. Strachan, and G. I. Paton.** 2001. Soil solution extraction techniques for microbial ecotoxicity testing: a comparative evaluation. *J. Environ. Monit.* **3:**91–96.
129. **Timmis, K. N., R. J. Steffan, and R. Unterman.** 1994. Designing microorganisms for the treatment of toxic wastes. *Annu. Rev. Microbiol.* **48:**525–557.
130. **Tuomela, M., M. Lyytikainen, P. Oivanen, and A. Hatakka.** 1999. Mineralization and conversion of pentachlorophenol (PCP) in soil inoculated with the white-rot fungus Trametes versicolor. *Soil Biol. Biochem.* **31:**65–74.
131. **Van Beilen, J. B., S. Panke, S. Lucchini, A. G. Franchini, M. Rothlisberger, and B. Witholt.** 2001. Analysis of *Pseudomonas putida* alkane degradation gene clusters and flanking insertion sequences: evolution and regulation of the *alk* genes. *Microbiology* **147:**1621–1630.
132. **van Beilen, J. B., M. G. Wubbolts, and B. Witholt.** 1994. Genetics of alkane oxidation by *Pseudomonas oleovorans. Biodegradation* **5:** 161–174.
133. **van der Meer, J. R., W. M. de Vos, S. Harayama, and A. J. B. Zehnder.** 1992. Molecular mechanisms of genetic adaptation to xenobiotic compounds. *Microbiol. Rev.* **56:**677–694.
134. **Vogel, T. M., and P. L. McCarty.** 1985. Biotransformation of tetrachloroethylene to trichloroethylene, dichloroethylene, vinyl chloride, and carbon dioxide under methanogenic conditions. *Appl. Environ. Microbiol.* **49:**1080–1083.
135. **Volkering, F., A. M. Breure, A. Sterkenburg, and J. G. van Andel.** 1992. Microbial degradation of polycyclic aromatic hydrocarbons: effect of substrate availability on bacterial growth kinetics. *Appl. Microbiol. Biotechnol.* **36:** 548–552.
136. **Wackett, L. P., and C. D. Hershberger.** 2001. *Biocatalysis and Biodegradation: Microbial Transformation of Organic Compounds.* ASM Press, Washington, D.C.
137. **Watts, R. J.** 1997. *Hazardous Wastes: Sources, Pathways, Receptors.* John Wiley and Sons Inc., New York, N.Y.
138. **Whyte, L. G., J. Hawari, E. Zhou, L. Bourbonnière, W. E. Inniss, and C. W. Greer.** 1998. Biodegradation of variable-chain-length alkanes at low temperatures by a psychrotrophic *Rhodococcus* sp. *Appl. Environ. Microbiol.* **64:** 2578–2584.
139. **Wild, S. R., A. J. Beck, and K. C. Jones.** 1995. Predicting the fate of non-ionic organic chemicals entering soils following sewage sludge application. *Land Contam. Reclam.* **3:**181–190.
140. **Wilkes, H., R.-M. Wittich, K. N. Timmis, P. Fortnagel, and W. Franke.** 1996. Degradation of dibenzofurans and dibenzo-*p*-dioxins by *Sphingomonas* sp. strain RW1. *Appl. Environ. Microbiol.* **62:**361–371.
141. **Williams, P. A., and M. J. Worsey.** 1976. Ubiquity of plasmids in coding for toluene and xylene metabolism in soil bacteria: evidence for the existence of new TOL plasmids. *J. Bacteriol.* **125:**818–828.
142. **Wittich, R.-M., H. Wilkes, V. Sinnwell, W. Franke, and P. Fortnagel.** 1992. Metabolism of dibenzo-*p*-dioxin by *Sphingomonas* sp. strain RW1. *Appl. Environ. Microbiol.* **58:**1005–1010.
143. **Wodzinski, R. S., and D. Bertolini.** 1972. Physical state in which naphthalene and dibenzyl are utilised by bacteria. *Appl. Microbiol.* **23:** 1077–1081.
144. **Wuertz, S., and M. Mergeay.** 1997. The impact of heavy metals on soil microbial communities and their activities, p. 607–642. *In* J. D. van Elsas, J. T. Trevors, and E. M. H. Wellington (ed.), *Modern Soil Microbiology.* Marcel Dekker Inc., New York, N.Y.
145. **Yalkowsky, S. H., and S. C. Valvani.** 1979. Solubilities and partitioning. 2. Relationships between aqueous solubilities, partition coefficients and molecular surface areas of rigid aromatic hydrocarbons. *J. Chem. Eng. Data* **24:** 127–129.
146. **Young, P. J., S. Pollard, and P. Crowcroft.** 1997. Overview: context, calculating risk and using consultants, p. 1–24. *In* R. E. Hester and R. M. Harrison (ed.), *Contaminated Land and Its Reclamation.* Thomas Telford Publishing, London, United Kingdom.
147. **Zhang, X., E. R. Sullivan, and L. Y. Young.** 2000. Evidence for aromatic ring reduction in the biodegradation pathway of carboxylated naphthalene by a sulphate-reducing consortium. *Biodegradation* **11:**117–124.
148. **Zhang, Y., and R. M. Miller.** 1992. Enhanced octadecane dispersion and biodegradation by a *Pseudomonas* rhamnolipid surfactant (biosurfactant). *Appl. Environ. Microbiol.* **58:** 3276–3282.
149. **Zhang, Y., and R. M. Miller.** 1995. Effect of rhamnolipid (biosurfactant) structure on solubilization and biodegradation of *n*-alkanes. *Appl. Environ. Microbiol.* **61:**2247–2251.

SUSPICIONS TO SOLUTIONS: CHARACTERIZING CONTAMINATED LAND

Lewis R. Barlow and Jim C. Philp

2

DEFINING CONTAMINATED LAND

The public's perception of contaminated land has a tendency to be shaped by catastrophic accidents, such as the infamous Love Canal in the United States, where the dumping of toxic wastes into a waterway endangered the health of residents of an estate subsequently developed on the area, or the Lekkerkerk incident in The Netherlands, in which 800 residents were evacuated from homes that had been built on a former refuse tip near Rotterdam (45). High-profile incidents such as these have led to radical solutions—typically evacuation of residents and the excavation of contaminated material, its disposal to a landfill, and backfilling of the void thus created with a suitably clean replacement material. However, contaminated land is a problem of such highly variable significance that if this same conservative approach were to be applied to the huge number of contaminated sites now recognized, the cost of remediation would be unsupportable, not to mention technically infeasible. A successful approach to bioremediation involves site characterization that takes into account careful consideration of appropriate microorganisms, their survivability, and their response to various contaminants. Reliable information on the presence, type, and extent of contamination at a site is therefore vitally important.

Legislative Drivers

The investigation of a potentially contaminated site is usually triggered by one of two events. Either the site is subjected to scrutiny though a statutory nuisance (regulatory) regimen, or the site is selected for redevelopment and that redevelopment must conform to local area planning requirements. The former may be said to be a reactive approach and the latter proactive. In either case, the overall goal of the investigation is likely to be the same, i.e., the characterization of the contamination status of a site to an acceptable level of confidence in as efficient a manner as possible.

The multicultural, multigovernmental nature of the problem makes it difficult to assess the global scale of contaminated land; however, several countries have made attempts to quantify the number of sites in their territories. Even a consistent definition of contaminated land has evaded international consensus. Depending on definition, the number of sites for one nation

Lewis R. Barlow, Carl Bro Group, Spectrum House, Edinburgh EH7 4GB, Scotland, United Kingdom. *Jim C. Philp*, Department of Biological Sciences, Napier University, Merchiston Campus, 10 Colinton Road, Edinburgh EH10 5DT, Scotland, United Kingdom.

Bioremediation: Applied Microbial Solutions for Real-World Environmental Cleanup
Edited by Ronald M. Atlas and Jim C. Philp

can be variable, and international comparisons throw up some interesting anomalies.

Within the European Union (EU), there are three distinct groups of states with different attitudes towards contaminated land (8).

1. The "concerned" states, all with particular problems and all having contaminated land policy, all of which, except one, have traditionally had policy goals of multifunctionality.

2. The "less concerned" states, which have environmental concerns, but where the perceived high cost of protection and remediation is seen as an important consideration. All of the states in this group have pursued the policy of suitability for use.

3. The "unconcerned group," with relatively weak environmental protection and who await EU guidance. All have a policy goal of suitability for use for waste sites.

The rigorous stance taken toward environmental contamination by The Netherlands has produced some staggering statistics: a likely number of contaminated sites of 110,000 and a likely number of suspect sites of 600,000. The corresponding figures for Germany are 50,000 and 200,000 (52). This highlights the need for international consensus on definition, as it is unlikely that The Netherlands would have so many more sites than Germany.

Given that the United Kingdom was the cradle of the industrial revolution, for such a small nation it has a large number of contaminated sites. In the United Kingdom, a provision was made in the Environmental Protection Act 1990 for regulations requiring local authorities to compile registers of potentially contaminated land within their boundaries. However, surveyors and property developers were alerted to the possibility that this was likely to have a serious negative influence on land and property prices. The backlash against registers at the time was such that the government abandoned the idea in March 1993 (53). However, the concept of registers reappeared in the Environment Act (1995), which retrospectively inserted the United Kingdom legislation on contaminated land into the 1990 Environmental Protection Act. This legislation imposes upon local authorities the duty to identify contaminated land within their areas. Having done so, the regulators are expected to encourage the landowners and polluters to carry out remedial work on a voluntary basis. Failure to cooperate will result in the serving of a Remediation Notice to the Appropriate Person, with the property being registered and running the risk of blight. It has been estimated that there are about 150,000 disused or underutilized sites, but many of these are unlikely to be contaminated (53). An estimate of United Kingdom contaminated land suggests that between 100,000 and 220,000 ha of land is contaminated, representing between 0.4 and 0.8% of the total land area (43).

While Superfund sites in the United States are easily identifiable (52), since the U.S. Environmental Protection Agency (EPA) and others maintain several lists of sites, there is no comprehensive register as such. To some extent, the reason is the same as in the United Kingdom: fear of reduction in land value due to contamination blight. The likely number of suspect sites is 35,000, and the likely number of contaminated sites is 1,200. However, the total number of listed and unlisted sites in 31 large cities in the United States is over 500,000 (51).

Risk Assessment

Our current understanding of the magnitude of contaminated land problems makes it inappropriate to utilize the standards and guidelines for dealing with the problem that were drawn up 2 decades ago at national levels in many countries, such as the United Kingdom and The Netherlands. The magnitude of the problem, the cost of remediation, and scientific questions about the real threat posed by contaminated land to human health have driven the development of risk-based environmental assessment criteria for soils and groundwater. There has been a move from a mandatory requirement that the land be suitable for multifunctional uses towards a "fitness-for-use" standard, that is, towards the philosophy that the level of remediation required at a site should be based on the envisaged end use of the land. By this risk-based approach, land to be used for housing developments with domestic gardens would have to be remediated to a higher standard than land designated for use as a retail park. In this example, the potential exposure to residual contaminants in a garden would be greater than that at a retail park (where the ground surface is typically paved over), hence the requirement for the cleanup standards to be more stringent. Precisely how stringent these standards need to be may be quantified though a process of risk assessment.

PRINCIPLES OF RISK ASSESSMENT

Risk assessment with respect to contaminated land serves two main purposes. First, it can be used to measure the degree of significance of contamination at a site. Second, the level of cleanup required in order to make a site suitable for its intended use may also be determined.

Several terms used in the field of contaminated land risk assessment require working definitions. Toxicity is the potential of a material to produce injury in biological systems. A hazard is the nature of the adverse effect posed by the toxic material. Risk is a combination of the hazardous properties of a material with the likelihood of it coming into contact with sensitive receptors under specific circumstances. This type of assessment also considers potential pathways by which hazards may reach receptors. Risk, then, is a statistical entity.

When the basic principles of risk assessment are applied to a potentially contaminated site in the United Kingdom, a site may be statutorily "contaminated land" only if the local authority can establish the presence of a significant pollutant linkage, which must include a hazard, a pathway, and a receptor (48). A critical point about this United Kingdom legal definition of contaminated land is that if no pathway can be found to link a hazard to a receptor, then the site cannot constitute contaminated land (9). In the above example of domestic gardens, the receptors (sometimes referred to as targets) could be infants who may inadvertently ingest soil in gardens. Most risks to human health and the environment can be described in terms of single or multiple source-pathway-receptor scenarios.

This type of pollutant linkage analysis allows remediation strategies to be designed in a more realistic and cost-effective manner. The result for most sites is likely to be a less conservative remedial strategy than one based purely on multifunctionality, i.e., restoring a site to greenfield conditions.

There are four key steps in the process of assessing the risks associated with pollutant linkages. These are hazard (source) identification, exposure assessment, toxicity assessment, and risk characterization.

Hazard Identification. This is the stage at which the chemicals present on a site are anticipated, along with their characteristics, e.g., their concentrations, water solubility, and toxicity. Due to the likelihood of many tens or even hundreds of potential contaminants being present at a site, the hazard identification stage usually focuses on known contaminants of concern.

Exposure Assessment. Exposure assessment is the estimation of pollutant dosages for receptors, based upon the use of the site and the conditions therein. There are multiple facets to these calculations. Among the factors to be considered are exposure duration and frequency, mean body weight, and future population growth or decline.

Toxicity Assessment. This is the acquisition of toxicity data, such as dose-response data, and its evaluation for each contaminant for both carcinogens and noncarcinogens.

Risk Characterization. This is an assignment of the level of risk to each pollutant linkage. For many contaminated sites, the best that can be reasonably expected at an initial desk-based stage is a qualitative risk estimate, such as insignificant, low, medium, or high. The amount of data required for quantitative risk characterization may be beyond all but the most rigorously characterized sites.

The U.S. EPA has adopted detailed methodologies for achieving this process (59). The overall procedure is summarized in Fig. 2.1. Recently published guidelines by the United Kingdom Department of the Environment, Food and Rural Affairs (DEFRA) for environmental risk assessment and management take a slightly different approach (17) but with similar outcomes in mind. The DEFRA process has five stages.

1. Hazard identification.

2. Identification of consequences.

3. Estimation of the magnitude of consequences, including spatial and temporal scale, and time to onset of consequences.

4. Estimation of the probability of the consequences: probability of the hazard occurring, of the receptors being exposed to the hazard, and of harm resulting from exposure.

5. Evaluation of the significance of the risk.

Risk assessment then can be seen to be a decision-making tool. The objective of risk management is to break the hazard-pathway-receptor linkage. When taken alongside other considerations such as cost and technical feasibility, the outcome is remedial technology selection and implementation, which completes the chain of contaminated land management.

Risk assessment strategies across the EU are dealt with in detail by Ferguson and Kasamas

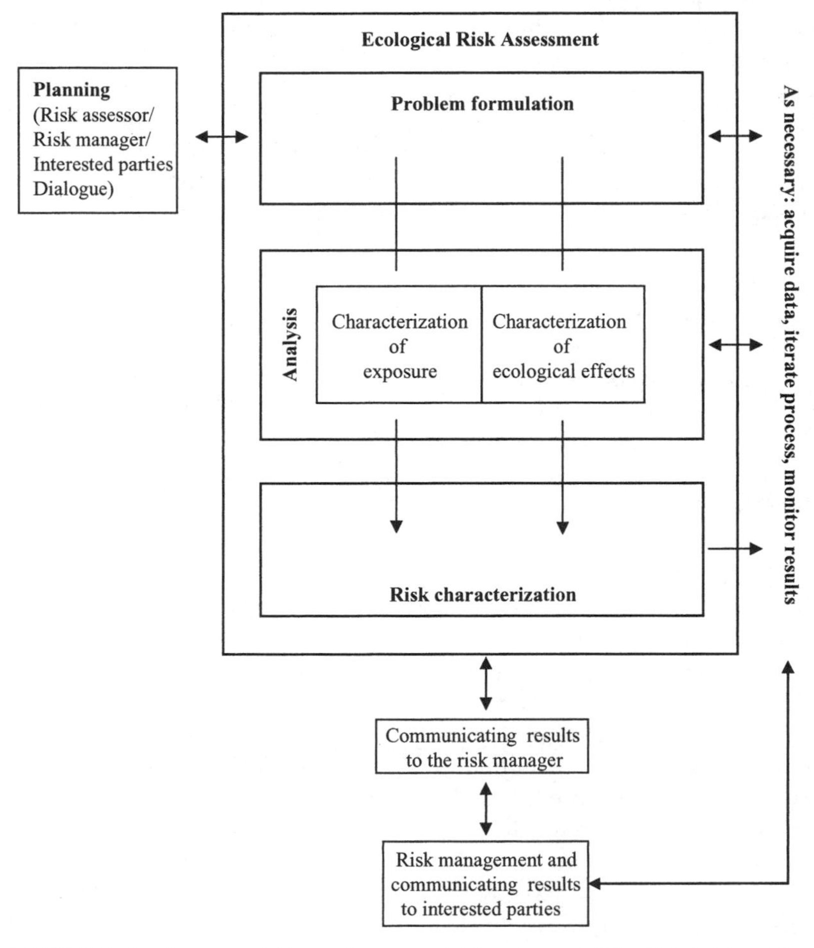

FIGURE 2.1 Framework for environmental risk assessment (after U.S. EPA guidelines [59]).

(23). A further British reference on risk assessment for contaminated land is the work of Cairney (5). Late in 1999, the United Kingdom Environment Agency published its own methodologies (36).

With the principles of risk assessment thus understood, the methods by which these principles can be applied to assessment of contaminated land may now be considered. There are three broad categories of assessment technique of increasing complexity:

- General statements
- Qualitative assessment
- Quantitative assessment

General Statements. General statements concerning risk usually are suitable for use only at an early stage of site assessment and are based on desk study information rather than actual physical data. They usually comprise simple statements such as the nature of contaminants expected to be present given the historical use of the site. The type and location of sensitive receptors and resources that may be sensitive to the site as well as the general significance of the identified hazards to these targets may be considered. General statements usually highlight the need for further investigation into potential hazards, pathways, and receptors.

Qualitative Risk Assessment. When applied to potentially contaminated sites, the purpose of a qualitative risk assessment is to express the significance or degree of real risk (as opposed to perceived risk) by use of a systematic and structured approach. Qualitative assessment is applicable to both desk study and intrusive investigations and is based on a systematic assessment of site-specific critical factors using professional judgment and expertise in addition to guidelines and standards. The causal chain of hazard-pathway-target is again the basis for qualitative risk assessment.

The degree of risk is dependent on both the nature of the impact on the receptor and the probability of this impact occurring. The underlying principle of site remediation is to eliminate or modify one or more of the above factors so that the risk is reduced to meet site-specific requirements. Formulation of the remedial objectives and strategy will essentially identify whether the source and/or pathway should be the focus of remedial objectives or whether protection of the receptor is a more viable option.

The risk assessment must be based on a dynamic assessment of the site for not just the existing, measured conditions but also foreseeable changes in any of the critical factors. For example, gas generation, groundwater level fluctuations, flooding, removal of surface pavement, or development could radically change conditions from those at the time of investigation. The long-term leaching potential for soil contaminants and gas generation from organic materials are two common causes for remedial work in advance of development even where existing water quality and soil gas concentrations do not necessarily pose a risk until development has taken place.

Quantitative Risk Assessment. As opposed to qualitative risk evaluation, the aim of a quantified risk assessment is to assign values for the existing and future deleterious effects associated with exposure. These values can be compared with acceptable health risks, which tend to vary from country to country.

Quantitative risk assessment requires high-quality data or a predetermined factor of safety and is often applied where a site is suspected to pose excessive human health risks. Risk assessments of this kind can be worked backwards by using an acceptable health risk as a starting point. The risk assessment procedure is then inverted until acceptable soil cleanup levels are obtained.

One of the reasons that quantified risk assessments are so data-intensive is that not only direct pathways need to be considered. Indirect contact can occur when contaminants are transported through soil, groundwater, surface water, uptake or adsorption by plants, dusts, or aerosols. Current understanding of the complex interactions between chemicals in the subsurface is poor. Also, most contaminated ground has previously been used for industrial or chemical works, and the presence of made ground and foundations usually causes a large degree of uncertainty in the various fate, attenuation, and transport processes that affect the movement of contaminants (58).

Quantitative risk assessment is a site-specific process, and the results and decisions drawn from a risk assessment are often dictated by the end use of a site. Many possible receptors exist, although in the majority of cases currently in the United Kingdom the primary driver for local authorities has tended to be the protection of human health.

The risk to human health posed by contaminants on a site is dependent on the concentration of the contaminant and the means of exposure, e.g., skin contact, inhalation, or ingestion. Essentially, the exposure from a certain contaminant can be quantified from the following equation or permutations of it (21):

$$\text{Exposure} = \frac{\text{(soil intake rate) (exposure time) (resorption rate) (contaminant concentration)}}{\text{body weight}}$$

where exposure or absorbed dose is the daily mass of contaminant absorbed per day, divided by the body weight of the receptor (in milligrams per kilogram of body weight per day); soil intake rate is the daily amount of soil a receptor is exposed to (in grams); exposure time is the number of days of exposure to the contaminant; resorption rate is the toxicokinetics-based empirical value quantifying the daily transfer of contaminants from the intake medium into the systemic circulation; contaminant concentration is the concentration of contaminant in the uptake medium (in milligrams per gram of soil); and body weight is the mass of receptor (in kilograms).

An understanding of the fate and transport of contaminants is crucial if a meaningful risk assessment is to be obtained. This analysis can be very complicated, since the number and types of processes affecting contaminants during transport are governed by both inherent contaminant characteristics and environmental conditions. Understanding of these complex dynamic processes requires the best approximation of the environmental chemistry of contaminants (e.g., biodegradability and hydrophobicity) and the environment at the site (e.g., geology and geochemistry).

SITE CHARACTERIZATION

A successful approach to bioremediation involves site characterization that takes into account careful consideration of appropriate microorganisms, their survivability, and their response to various contaminants. Reliable information on the presence, type, and extent of contamination at a site is therefore vitally important.

Since 2001, site investigation best practice in the United Kingdom has been set out by BS 10175: *Investigation of Potentially Contaminated Sites* (4). In the United States, similar guidance is given in the EPA document *Guidance for Scoping the Remedial Design* (56). This section should establish the reasons that good site investigation practice is necessary for environmental impact determination leading to selection of appropriate remediation strategies, especially where bioremediation is to be considered.

The following steps form the backbone of good contaminated land site investigation practice.

Identify Information Gaps. A desk study should be undertaken to decide if enough information exists to carry out a satisfactory risk assessment to a required degree of confidence. If not, the objectives of a further investigation need to be defined.

Site Safety. The potentially hazardous nature of the site will require consideration from the outset in order to identify any safety measures needed to protect personnel or the environment.

Nature of the Investigation. Once the objectives of a further investigation have been established, a decision on the nature of the investigation necessary to obtain suitable data must be made.

Sampling Locations and Depths. The locations and depths at which samples are to be collected and the number of locations required must be considered.

Chemical Analyses. The specification of what analyses should be carried out on the samples obtained during the investigation will be determined through research on the history of the site and consideration of the conditions encountered on-site.

Sampling Methods. The methods by which samples are to be obtained, preserved, and transported to the chemical testing laboratory require consideration.

The information obtained should allow the selection of remedial options and determination of whether bioremediation is feasible for the problems identified, what form this may take, and whether further pilot studies are necessary to facilitate this remedial method. The steps to reaching this crucial stage are outlined here.

Conceptual Site Modeling

The purpose of desk-based research (the desk study) is to establish baseline contamination conditions at a site and identify the need for any further work. It should be both thorough and conservative. A basic question is asked: "Do we have sufficient information about this site to adequately characterize its contamination status?"

Desk studies may vary in investigative depth, depending on site-specific requirements, although the items discussed here are considered to be the minimum requirements for a competent contamination desk study. Key information sought at the desk study stage includes:

- the history of the site, including previous owners, occupiers, and uses;
- a site visit, during which any visual evidence of potential contamination, site conditions, and nearby features are recorded;
- local geology and hydrogeology, including the presence and quality of groundwater and surface waters;
- the above- and below-ground layout of the site and its historical development;
- any history of mining, including shafts and worked seams;
- nearby waste disposal tips, abandoned pits, and quarries;
- information on previous investigations at the site;
- processes used on the site, including their locations, raw materials, products, waste, and methods of disposal; and
- nearby sensitive receptors, e.g., water courses, houses, parks, and areas of ecological sensitivity.

Once the relevant data sources have been searched, the information should be reviewed and evaluated to identify potential environmental issues and any areas of concern.

An interpretative desk study, i.e., one that not only provides factual data but also involves professional interpretation, will normally include a conceptual site model. This is a key component of the overall risk assessment process and is used as a holistic tool to consider potential sources of impact, potential migration pathways along which identified contaminants could migrate, and potential receptors which may become exposed. The presence of all three is known as a pollutant linkage.

In order for a site to constitute contaminated land in the United Kingdom, a pollutant linkage of significance between the contamination source and a sensitive receptor via an appropriate environmental pathway must be identified. The degree of significance of a pollutant linkage depends on a number of factors, including the hazardous nature of the source, the type of pathway, and the sensitivity of the target.

The conceptual site model is used as the framework on which to gauge the levels of risk associated with a site. Gaps in the model can be identified and may form the basis for further investigation by intrusive means. For example, a potential pollutant linkage in a conceptual site model may be the contamination of groundwater from a leaking underground storage tank. Desk study research may have identified the presence of the underground tank, although whether the tank is actually leaking may be unknown. The significance of this possibility must be considered in the conceptual model, to the point where the professional must decide whether there exists sufficient information to characterize the site conditions or whether intrusive site investigation work is required. This could take the form of groundwater monitoring to detect the potential presence of leaked fluid from the tank, or trial pitting to investigate the tank itself. Many investigative options are possible, the design of which should be based on the suspicions raised from the desk study process. Box 2.1 shows a general illustrative example of a conceptual model.

BOX 2.1
Leaking Underground Fuel Storage Tank Model

The following conceptual model (Box Figure 2.1.1) and source-pathway-receptor matrix (Box Table 2.1.1) are from the Institute of Petroleum (29). In the analysis of a leaking underground fuel storage tank (an exceedingly common occurrence), there are many pathways by which the pollutants can reach receptors, and naturally there are several possible receptors.

In the conceptual model, all possible combinations should be identified, but the matrix can be used to delineate which are the critical pathways and receptors and which ones pose insignificant risk.

The example is a great simplification. Each case will be site specific with respect to geology, hydrogeology, geography (human population density is critical), and other factors. As a result, the source-pathway-receptor matrix can become complex. Once complete, the matrix saves time and effort since a number of insignificant risks can be identified and ignored.

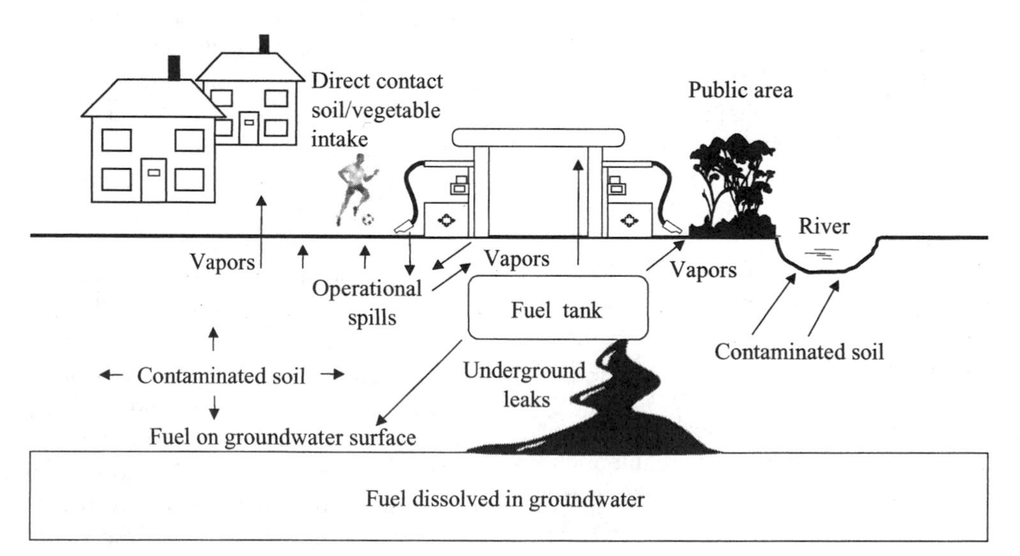

BOX FIGURE 2.1.1 Risk assessment for a gas station with a leaking underground storage tank (after Institute of Petroleum guidelines [29]).

BOX TABLE 2.1.1 Source-pathway-receptor for leaking underground storage tank[a]

Primary source	Secondary source	Hazard[b]	Transport mechanism	Pathway	Exposure medium	Receptor
Fuel tank	None	Dizziness, CNS depression, potential carcinogenicity	Vapor, through air	Vapor inhalation	Air	Humans
	None	Vegetative dieback, leaf function damage	Vapor, through unsaturated zone	Vapor absorption	Air	Vegetation adjacent trees
	None	Derogation of groundwater quality	Product loss and vertical migration to water table	Groundwater dissolution	Water	Groundwater aquifer
	None	Derogation of surface water quality	Product loss and dissolution in groundwater	Base flow and discharge to adjacent surface water body	Water	Adjacent river
	Contaminated soils	Dizziness, CNS depression	Vapor, through unsaturated zone	Vapor inhalation	Air	Human (recreation)
	Contaminated soils	Skin irritation, contact dermatitis when extreme	Contact with contaminated soil	Dermal contact at surface	Soil	Human (recreation)
	Contaminated soils	Flammability	Vapor, through unsaturated zone	Vapor buildup in basement	Air	Humans (residential)
	Contaminated soils	Flammability	Vapor, through unsaturated zone	Vapor buildup in basement	Air	Property
	Contaminated soils	CNS depression, asphyxiation	Vapor, through unsaturated zone	Vapor buildup in basement	Air	Humans (residential)
	Contaminated soils	Derogation of surface water quality	Bulk fluid, through unsaturated zone	Free product flow to adjacent river	Water	Adjacent river
	Free product on water table	Derogation of soil quality	Evaporation to overlying soils	Vapor phase	Soil vapor	Soil
Fuel pump	None	Derogation of soil quality	Spillage and percolation through cracked pavement	Leaching	Soil	Soil
	None	Various, potential carcinogenicity	Vapor, through air	Inhalation	Air	Humans (customers)

(Continued)

BOX TABLE 1 *Continued*

Primary source	Secondary source	Hazard[b]	Transport mechanism	Pathway	Exposure medium	Receptor
Operational spills	None	Vegetative dieback	Vapor, through unsaturated zone	Vapor absorption	Soil gases	Home-grown produce
	None	Various, potential carcinogenicity	Vapor, through unsaturated zone	Consumption of contaminated produce	Vegetable produce	Humans (residential)
	None	Various, potential, carcinogenicity	Vapor, through air	Inhalation	Air	Humans (customers)
	Contaminated soils	Dizziness, CNS depression	Vapor, through unsaturated zone	Vapor inhalation	Air	Humans (customers)

[a] Based on reference 29.
[b] CNS, central nervous system.

Investigative Techniques

If a desk study has identified a significant possibility of a pollutant linkage at a site but there is insufficient information to accurately characterize the risk associated with this linkage, further site investigation may be required.

The planning phase of a site investigation involves designing a program of works in a phased manner to acquire the key information needed to understand the site and refine the conceptual model. The focus of an investigation is usually the potential contamination source but will always depend on any knowledge gaps identified in the desk study. In either case, the three media that are usually the subject of a contaminated land investigation, either directly or indirectly, are soil, water, and gas.

If bioremediation is identified as a potential remedial option at the desk study stage, other pertinent characteristics may be of relevance at this point. However, the selection of bioremediation as a viable remedial option usually occurs once more quantitative data are available, i.e., after at least the first phase of intrusive investigation.

In determining which media are to be sampled and to which analyses they are to be subjected, it is important to consider the context within which these samples will be assessed. A plethora of environmental standards and guidelines exist for soils, waters, and gases, which can vary widely from country to country. Fundamentally, samples will be tested for determinants that are suspected to be of environmental significance, whether that may be to humans, crops, buildings, or other potential receptors.

There exist a number of techniques that aid the contamination status of a site to be determined. These techniques fall into two broad categories: intrusive or nonintrusive.

Nonintrusive techniques have the general advantage of offering large amounts of data within a short timescale without causing significant disturbance. These data often require significant interpretation and analysis but may prove useful in the early stages of a phased investigation and often enable areas of a site that may require further investigation at a later stage to be highlighted.

Conductivity surveys involve the use of a varying electromagnetic field to induce a current, which creates a secondary field, the strength of which is proportional to ground conductivity. Conductivity surveys allow rapid reconnaissance and can be used to interpret variations in groundwater quality and the presence of buried metallic objects. They can be affected by local "noise," for example, buried and overhead cables, pipes, or fences. Their reported use in contaminated land and water assessment has been extremely limited. A ground electromagnetic conductivity meter was used to investigate incidences of nitrate contamination of well water at dairies (19).

Electrical resistivity surveys measure apparent resistivities along a linear array of electrodes to produce an image-contoured two-dimensional cross-section. Electrical resistivity of soils depends on various factors, including soil type, water content, saturation, and pore fluid properties (67). Electrical resistivity surveys are relatively easy to carry out and provide a good resolution of resistive layers, and evidence has suggested that they are well suited to measurement and delineation of subsurface contamination (66). Residual petroleum on soil particle surfaces results in a permanent increase in soil resistivity (11). Naudet et al. (38) have suggested that both redox (indicative of biodegradation) and electrical resistivity tomography (indicative of mineralization) maps can be used to optimize the positioning of pumping wells for remediation. However, surveys are difficult or impossible to use on paved areas.

Ground-penetrating radar (GPR) can be used to detect buried tanks or near surface targets such as plastic pipes, metallic objects, voids, and mines through measurement of reflected microwave frequency radiation pulsed into the subsurface with an antenna. Equipment is drawn over the ground surface on a grid pattern. This technique requires expert processing and interpretation to properly characterize made ground and can suffer signal interference through reinforced concrete and from adjacent foundations.

GPR has been used predominantly for environments with low electrical conductivity, such as freshwater aquifers or dry sandy soils (47). At radar frequencies, it is possible to distinguish between a water-saturated medium and a non-aqueous-phase liquid (NAPL)-saturated medium (6). However, light non-aqueous-phase liquid (LNAPL)- and dense non-aqueous-phase liquid (DNAPL)-saturated media have very similar electromagnetic properties, and the type of contaminant can be better distinguished by acoustic properties. The amplitude of the reflection from water can discriminate the contaminated zone from the noncontaminated area in the case of LNAPL floating on the water table (41), and information on the thickness of the hydrocarbon layer can be gained.

GPR has been used during actual remediation projects. A four-dimensional monitoring GPR survey was used to outline the LNAPL plume in the subsurface of an impacted gas station in Brazil (10). Low reflectivity corresponded to hydrocarbon in the vapor phase in the vadose zone. Pumping of groundwater reduced the plume in the vadose zone. A steady restoration of reflectivity suggested a decrease of LNAPL saturation along the GPR survey. Osterreicher Cunha et al. (42), however, reported inconclusive results in the use of GPR during a bioventing project. If the difficulties can be resolved, GPR would be an excellent candidate technique for monitoring in situ bioremediation due to the difficulties in obtaining representative samples. The uses of GPR in particular (33), and electrical and electromagnetic methods more generally (46), in environmental applications have been reviewed.

Magnetic profiling involves the measurement of the earth's total magnetic field intensity using one or more sensors and is a rapid reconnaissance method for ferrous targets. This technique can also be affected by cultural "noise" and temporal variations in the magnetic field.

Seismic refraction is the measurement of compression and/or shear waves which have been critically refracted along an acoustic boundary and radiated back to the surface. The seismic signal is detected by use of an array of geophones. The techniques can be used for estimation of the thickness and depth of lithological units with different densities or for establishing the depth of groundwater table or vertical boundaries, such as edges of old backfilled quarries. Data production is slow and requires careful use in a culturally noisy environment, for example, with moving traffic or operating drilling rigs. The technique has been applied to determination of the depth and geometry of a landfill lower boundary (34); this is a difficult task, as field methods generally lack the necessary depth resolution. Although results demonstrated that tomographic refraction may be an efficient and cost-effective means of studying the very shallow subsurface (<20-m depth), complementary geological and other geophysical data were still required.

Infrared photography can highlight distressed vegetation resulting from contaminated ground or landfill gases and can be carried out by using remotely controlled model aircraft. Results from these surveys need to be interpreted with great care, as camera angle can be affected by pitch and roll of the aircraft and by the appearance of shadows.

Intrusive site investigation involves the disturbance of the ground to some extent in order to obtain samples for quantitative analysis or simply to enable below-ground strata to be viewed. The most commonly used intrusive techniques are trial pits, boreholes, and probing techniques. These and several other techniques are described below.

Trial pits and trenches are usually formed by a back-acting excavator, although they can also be hand dug. A suitably wide bucket which allows a good view of the excavation but minimizes the amount of material excavated should be chosen. Trial pitting facilitates the detailed examination of ground conditions and is a rapid and inexpensive investigative technique. However, the investigation depth is limited by the size of the machine, generally approximately 4.5 m, and the method is not suitable for sampling below water.

Cable percussion boreholes are most commonly used to sample soils and groundwater at depth and to monitor gas levels in the ground. They enable the installation of permanent monitoring wells and have less potential for adverse effects on health and safety and the above-ground environment than trial pits. Undisturbed samples can be collected, and integrated sampling for contamination, geotechnical sampling, and gas/water sampling may be undertaken. Boreholes are more costly and time-consuming than trial pits and hand augering, and there exists the potential for contaminating underlying aquifers and groundwater flow between strata within an aquifer unless the boreholes are properly cased.

Driven-tube or "window" samplers consist of a hollow metal tube, usually containing a plastic sleeve, which is driven into the ground with a hydraulic or pneumatic hammer. Undisturbed samples of the complete soil profile can be recovered from depths down to approximately 10 m in optimal ground conditions. There is poor sample recovery in noncohesive granular material, and obstructions such as bricks are difficult to penetrate.

Hand augering is a very basic technique with many designs available for different soil types, conditions, and sampling requirements. It allows the examination of soil profiles and collection of samples at preset depths, although only limited depths can be achieved if obstructions are present and its ease of use is very dependent on soil type.

Hollow-stem auger boreholes use a continuous-flight auger with a hollow central shaft. The technique forms a fully cased hole, avoiding potential problems of cross-contamination arising with cable percussion techniques, and can be used for installation of water and ground gas monitoring wells. It is less suited for deeper boreholes than cable percussion unless large rigs are used.

Cone penetration techniques can be used in conjunction with down-hole monitoring equipment to provide on-site screening, for example, a remote laser-induced fluorescence meter for organic compounds. There is a high mobilization cost for the most powerful equipment, and recovery is poor in noncohesive granular material.

Spike holes can be made with a small-diameter bar that is driven to form a hole and then removed to allow gas or vapor monitoring. It is an inexpensive and quick investigative technique that allows assessment of immediate hazards. The effectiveness of this technique is heavily dependent on ground conditions, and the method has a limited depth of penetration (approximately 1 m). Greater detail on intrusive investigation techniques is given by Patata and Mastrolilli de Angelis (44).

Prior to the undertaking of any fieldwork, the potential risks to the health and safety of all persons who may be affected by the investigation and the possible impacts of the investigation on the surrounding environment must be considered. Where risks are identified, appropriate precautions must be taken to minimize the risks.

Exposure to a hazard should be controlled, where possible, through the use of measures other than personal protective clothing or equipment. A health and safety plan describes the precautions that should be followed by site personnel in order to minimize their exposure to potential hazards. Where insufficient information is available to assess a risk, it should be assumed that control and monitoring are necessary.

Soil Characterization

Characterization of the soil media at a site requires sampling and usually chemical analysis (Fig. 2.2). The more sensitive the receptors or the greater the hazard, the greater the degree of confidence in the outcome of the risk assess-

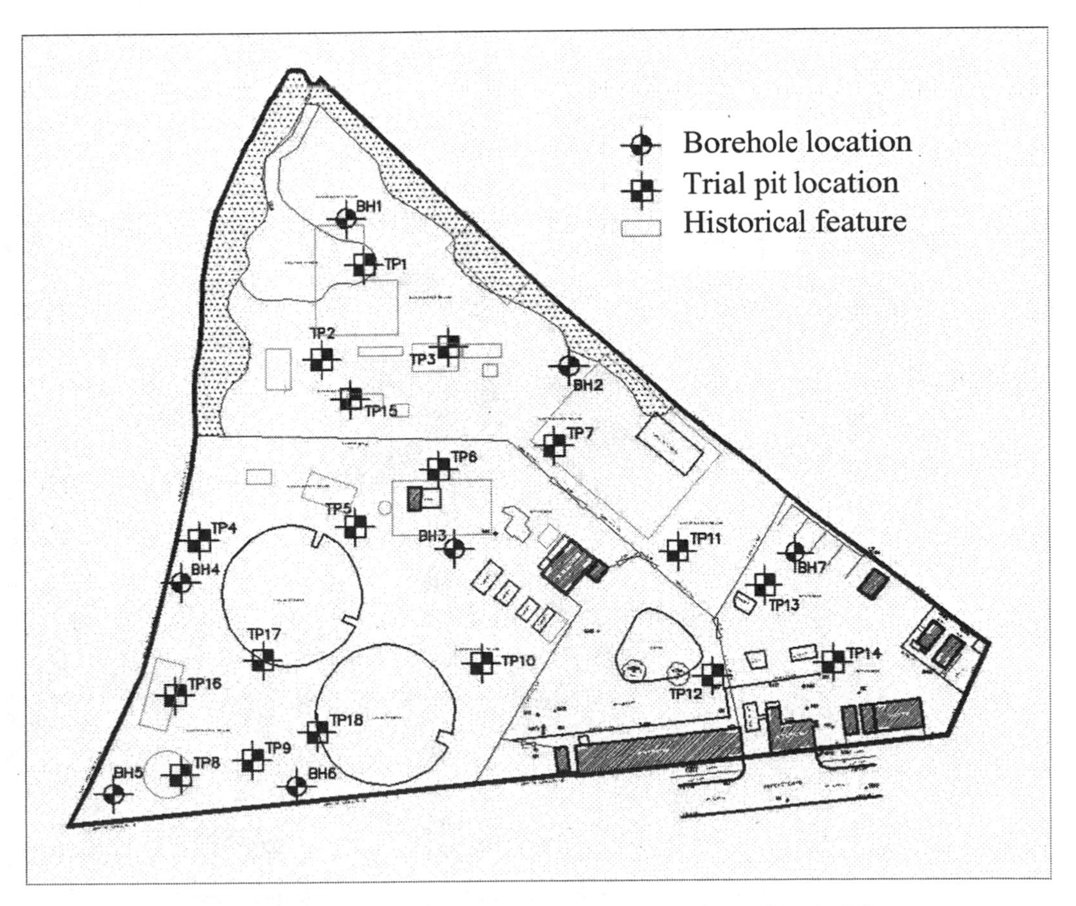

FIGURE 2.2 An example of investigation design, considering historical features.

ment and the subsequent risk management that is required. In such cases, a greater number of sampling locations and samples will be needed. Other factors, such as accurate delineation of an area of contamination, also necessitate more intensive sampling.

Targeted sampling involves sampling at locations selected on the basis of the conceptual model and that are known, or suspected to be, sources or areas of contamination. Potential point sources of contamination may include past or present storage tanks (above and below ground), below-ground fuel supply pipe work, drains, backfilled pits and waste disposal areas, or handling areas where spills of hazardous materials could have occurred.

Nontargeted sampling is usually carried out by using a regular pattern of sample locations. The reliability of interpolation between sample locations depends on variations in soil characteristics. For example, in well-stratified sediments, vertical variations in concentration will normally be much greater than horizontal variations, so that interpolation horizontally will be much more reliable than vertical interpolation.

If there are any regular topographical patterns on the site, for example, ditches at regular intervals or systematic undulations of the terrain, the sampling pattern should not coincide with the topography in a way that could introduce a bias or systematic error in the samples. This can be avoided by careful selection of the base or starting point of the sampling grid and, where necessary, by careful selection of the grid spacing.

The sampling depths should take into account the nature of the proposed development. For example, services and strip foundations are typically installed to a depth of approximately 1 m, but main sewers can be installed at much greater depths. The samples should be collected to represent a specific depth or narrow band of strata (Fig. 2.3).

Samples of natural strata, if uncontaminated, can indicate the local, background chemical

FIGURE 2.3 Scheduling of soil samples prior to laboratory analysis. The labeling procedure, often overlooked, should be viewed as an essential part of the overall procedure and should be given requisite attention to detail.

conditions and can be of assistance when the extent of contamination migration and/or the degree of remediation that is appropriate is being determined. Soils taken from beneath made ground can be subsoils and can differ in composition from the topsoils that would be naturally associated with them.

The spacing between sampling locations should be determined according to the conceptual model, the stage of the investigation, and the requirements of the risk assessment. The sampling density should depend upon the confidence and robustness required of decisions that will be based on the information obtained. Thus, the area and depth of interest will be related to the contaminants present, the pathways and the receptors, and the smallest area that might be of concern.

A higher-density sampling grid can also be necessary when a high level of confidence is required for the outcome of a risk assessment. Lower-density sample locations may be acceptable on large sites, subject to the chosen spacing providing adequate data and being consistent with the objectives of the investigation.

Soil samples have to be labeled so that identification is clear and unambiguous. The containers should show the sampling date, site and project identification, identification number, sampling depth, and processing following sampling, if necessary. Additional information, such as particulars about a sample location, conditions, or difficulty with sample acquisition, should be recorded in a field notebook.

The actual sampling tool, of course, must be clean, and between samplings must be cleaned scrupulously with appropriate solvents, which range from water to acids to polar organic solvents that cause no further pollution. The sampling tool must have a diameter of around twice the maximum particle diameter to be sampled. Thus, if the material to be sampled contains very coarse material such as bricks, then a laboratory spatula is a ridiculous sampling tool. Equally, drilling techniques that acquire only small samples will generate unrepresentative samples. Obtaining unrepresentative samples is one of the two biggest flaws encountered in sampling programs. If the material to be sampled contains stones of 1 kg, then the collection of 1 kg of a soil sample is clearly unrepresentative. It can be envisioned that, on occasion, samples of more than 50 kg would be required. The other major flaw in sampling programs is cross-contamination, which calls for proper cleaning to eliminate it. Detailed advice on soil sampling is given by the U.S. EPA (57).

The care taken to obtain representative samples in the field will be nullified if the sample is improperly subsampled in the laboratory (28). Gerlach et al. (26) compared five methods for splitting soil samples in the laboratory and showed that grab sampling is a poor method for subsampling particulate samples in terms of overall accuracy levels. The riffle splitter produced the smallest overall inaccuracy levels. Such is the significance of laboratory subsampling that the U.S. EPA has issued guidance on it (25). The problem stems from the inherent heterogeneity of soil.

Water Characterization

The emergence of contaminant hydrogeology as a subdiscipline within hydrogeology has been accompanied by the ability to measure concentrations at the level of parts per billion and below. Exacting low-level detection limits immediately posed demands and questions as to a corresponding ability to eliminate extraneous influences.

Diffuse source contamination of groundwater can be the result of a number of diffuse inputs from within a site but can also result from off-site sources. When there is no clearly defined source, groundwater monitoring wells should be installed on a nontargeted basis. The monitoring wells should be used to determine the direction of groundwater flow and the water quality upon entering and leaving the site. On a small site, this will probably require at least four monitoring locations. Wells should be located on the site so that they can be triangulated. Wells in a line will not provide adequate information to establish the direction of groundwater flow (36).

Monitoring well locations should be determined on the basis of the information available and the need for further information about the source and the migration of contaminants. The monitoring wells should therefore be located on a targeted basis. Subsequent monitoring wells can then be located on the basis of the information from the initial installations and therefore again should be installed on a targeted basis.

Wherever practicable, a groundwater monitoring well should be installed directly below the potential source. However, such installations can allow contaminants to migrate vertically. An alternative position is to install the monitoring well at the outer down-gradient edge of the potential source.

Where a contamination source is known or suspected, a groundwater monitoring well should be installed up-gradient of the source and a minimum of two should be installed down-gradient of the source. These monitoring wells can also be used to determine the direction of groundwater flow and the quality of the groundwater flowing onto the site. Further monitoring wells should be considered, depending upon the objectives and phase of the investigation, for example, at progressive distances down the hydraulic gradient from the source of contamination.

In the designing of a groundwater monitoring program, consideration should be given to the nature of the likely contaminants. If contaminants which were not anticipated are encountered, this may lead to a supplementary investigation to allow the installation of specific monitoring wells to address the contamination encountered. Groundwater monitoring will include addressing contaminants in solution such as metals and organic compounds (for example, phenols) and also the possibility of hydrophobic materials (for example, NAPLs) that may be present as free product.

Where volatile NAPLs are likely to be present, information on the potential location of contamination and migration plume can be obtained by carrying out soil gas monitoring. This information can be used to determine the location of monitoring wells. If NAPLs are present, consideration should be given to the effects of solubility, sorption, degradation, metabolites, and potential for migration.

Where liquids that are less dense than water (LNAPLs) are present, at least one borehole should be screened over a depth range that spans the level of the water table so that they can be more easily detected and the thickness of the LNAPL layer can be measured. DNAPLs will move to the base of the hydrological unit and can collect on, or be deflected by, lenses of low-permeability material.

Investigations for DNAPLs are difficult and require monitoring of wells that fully penetrate the aquifer and are screened at the base and at points where low-permeability material is present. Separate wells formed to different depths can be necessary at monitoring locations due to the difficulty of forming adequate seals in nested wells.

Where a monitoring well installation passes through low-permeability strata, routes allowing dispersal of contamination into underlying groundwater can be created. In such situations, a larger-diameter hole should be formed down to the low-permeability strata and an impermeable plug of bentonite-cement grout, with a minimum thickness of 1.0 m, should be inserted. Contamination can adversely affect the plug, and if necessary, suitable alternative material should be selected. This may necessitate some preliminary trials to confirm that the selected material is effective. The plug should be allowed to set before the borehole is continued by forming a smaller-diameter hole. In this way, a seal to prevent the downward migration of contamination can be created.

When all monitoring work has been completed and there is no further need for the monitoring wells, they should be sealed by grouting with suitable material, ensuring that the grouting is effective above and below the water table.

Despite the complexity of NAPL behavior in the subsurface and the importance of medium heterogeneity and the interactions between the various fluids, it is surprising to find that relatively few specialized NAPL site characterization techniques have been developed and thoroughly tested. The majority of routine site investigations continue to rely on standard techniques such as drilling, and monitoring and sampling of fluid levels in conventional piezometers. Guidelines on groundwater sampling are given by Yeskis and Zavala (65).

Other techniques that have been applied include soil gas surveys, coring, and geophysical techniques. Despite these efforts, significant additional work is required to develop and refine specialized techniques for determining the location and saturation of NAPLs in the subsurface. Modeling NAPL flow and remediation requires dependable field data for calibration and validation and as model inputs. Some of the more commonly used techniques applicable to this discussion are considered below.

Drilling and the subsequent installation of monitoring wells are the most common site investigation techniques at NAPL-contaminated sites (1, 62). These methods deliver direct information on the presence or absence of contaminants and allow broad delineation of the lateral, and sometimes the vertical, extent of contamination.

Gas Characterization

Soil gas samples are similar to water samples in that they can be representative of a large zone. Nevertheless, the sampling strategy differs from that used for waters because of the greater ability of soil gases to migrate in all directions within the ground (16).

Where there is the possibility of soil gas contamination (for example, on or adjacent to areas of landfill, alluvial ground, solvent or fuel storage, mining, buried dock sediment, and/or peat), it is necessary to determine the composition and migration potential of the soil gas. Degradation of organic matter can give rise to both methane and carbon dioxide and to a variety of trace gases, depending on the ground conditions and the nature of the material.

Volatile organic compounds (VOCs) can have associated vapors, the concentrations of which can vary in the soil gas above different parts of a plume but which can be used to indicate the location of the plume. When data from driven-tube sampler holes, cable percussion boreholes, and monitoring wells are interpreted, the strata penetrated should be taken into account, as smearing during the formation of the borehole for the installation can reduce the porosity of the ground and affect gas migration.

The detection and determination of gases can be done with instruments (either portable or laboratory based) or colorimetric gas detection tubes. Samples of soil gas can also be collected for analysis at a permanent laboratory. Portable instruments are used on-site for both landfill gases and VOCs. These may be nonspecific, e.g., flame ionization detectors or photoionization detectors (PIDs), or may be for the specific measurement of gases, such as methane, oxygen, and carbon dioxide.

Measurements of soil gas atmosphere in spike holes are subject to significant variation, depending upon the porosity of the ground and the weather conditions. Consequently, the results of the measurements from spiking should be interpreted with caution. A negative result does not necessarily mean the absence of a problem, as gas or volatiles could be present at a greater depth.

Concentrations can also build up when ground gases are confined, for example, in wet ground conditions when the soil pores become blocked at the ground surface. Installation of deeper monitoring points using boreholes is preferable.

Investigations for vapors associated with VOCs are usually part of a screening process, for example, to identify the location of a contaminant plume. The screening process is usually carried out using driven spikes or driven probes in conjunction with portable instruments. Screening may also be carried out in boreholes and driven boreholes during formation. Sample collection devices such as activated carbon tubes may be used to enable laboratory identification and analysis.

Screening for VOCs is usually carried out by using nonspecific instruments such as PIDs. PIDs can be fitted with lamps of different energies to vary the response to different groups of compounds. The greater the energy of the lamp, the greater the range of solvents causing a response. It can be necessary to obtain samples of the soil gas by adsorption onto a suitable medium or by using a gas syringe or sampling bag in order that laboratory analysis can be carried out to determine the composition and the contaminants present.

Monitoring of the soil gas profile during the formation of boreholes can provide useful information on the vertical distribution of VOC vapors and concentrations. Monitoring during installation can also give important safety information. Screening for vapors from VOCs tends to be limited by the depth to which the probeholes can penetrate, but the depth should be at least 1 m. When screening to establish the location of a migration plume is done, testing should be carried out at a consistent height above the water table to enable quantitative comparison of the results.

Sampling of VOCs for laboratory analysis has to be carefully designed to minimize losses through the two most common mechanisms, volatilization and biodegradation. The U.S. Army Corps of Engineers has prepared guidance for its staff to use sample collection and handling procedures that minimize VOC losses from solid samples (55).

QUANTIFYING THE PROBLEM

When a site investigation has been completed, a lag phase usually occurs while the samples obtained from the investigation are analyzed by the chemical testing laboratory. Average turnaround times of around 10 days are common in the United Kingdom, although premiums can be paid for more rapid reporting times. It is seldom heard of for a chemical testing laboratory to provide comment on the significance of the results of its tests. Rather, it is usually the responsibility of the site assessor (usually a specialist consultant) to gauge the real meaning of the contaminant concentrations reported by the laboratory. The simplest manner in which to carry out this type of assessment is to compare the results obtained with a set of robust, suitably recognized safety standards.

Given that the perception of contaminated land has changed over the last 20 years to a recognition of it as a widespread infrastructural problem of highly variable seriousness, it is not surprising that standards for cleanup have also evolved. It is this switch from rare-catastrophic to common-variable significance that has necessitated a move towards risk-based assessment.

Generic Screening Guidelines

In The Netherlands, the potential consequences of contaminated land have historically been taken rather more seriously, for a number of good reasons. It is a small, densely populated country with some large multinational companies. Agriculture is highly intensive, and of course the land is low-lying, and soil contamination can quickly become a serious threat to water resources. A much larger range of contaminant guideline concentrations is available, reflecting the demanding approach to cleanup that has been adopted there. The ALARA principle (as low as reasonably achievable) and the use of the best available techniques to control soil pollution have been employed. The underlying premise of the Soil Protection Act of 1987 is that soil pollution is not allowed (22). Given that it is impossible to prevent soil pollution completely, the Act tolerates emissions as long as they do not endanger soil multifunctionality.

Dissatisfaction with such approaches has led to the realization that cleanup standards have to be founded on risk-based assessment, and the fitness-for-use approach is central to risk assessment concepts (20).

The situation in the United States is more complex. The U.S. EPA Soil Screening Guidance is a tool to help standardize soil remediation at sites on the National Priority List. The outcome is soil screening levels (SSLs) for contaminants in soil that may be used for guidance purposes. SSLs are not national cleanup standards.

At sites where contaminant concentrations fall below SSLs, no further action is warranted under the Comprehensive Environmental Response, Compensation and Liability Act. Generally, where contaminant concentrations exceed the SSLs, further investigation, but not necessarily cleanup, is warranted (58). This resembles the "trigger-action" approach, and SSLs are risk-based concentrations derived from risk assessment procedures. However, the U.S. EPA lists generic SSLs for 110 chemicals, using default values that are conservative and likely to be protective for the majority of site conditions. Note that generic SSLs are not necessarily protective of all known human exposure pathways, reasonable land uses, or ecological threats.

The U.S. soil screening process is a seven-stage process.

1. Develop a conceptual site model based on historical records and available background.
2. Compare the soil component of the conceptual site model to the soil screening scenario.
3. Define data collection needs for soils to determine which site areas exceed SSLs.
4. Sample and analyze soil at the site.
5. Derive site-specific SSLs if needed.
6. Compare site soil contaminant concentrations to calculate SSLs.
7. Decide how to address areas identified for further study.

Essentially, SSLs are risk-based concentrations derived from equations combining exposure assumptions with EPA toxicity data.

In the United Kingdom, the Interdepartmental Committee on the Redevelopment of Contaminated Land (ICRCL) guidelines (30, 31) provided advice on only a limited range of contaminants. The guidelines, however, embody the concept of soil trigger, or threshold, and action concentrations for decision-making purposes. The trigger level indicates that some further investigation is required, and the action level means that remedial treatment is needed for the proposed end use of the land. These guidelines inadvertently became adopted as cleanup criteria and were formally withdrawn by DEFRA in 2003.

Due to the previous lack of coherent guidance and some dubiety over the scientific basis to the ICRCL soil guidance values, the United Kingdom Department of the Environment (now the Department of Environment, Transport and the Regions), with others, developed a framework for contaminated land risk assessment that included the derivation of new soil guideline values (SGVs) (17). These SGVs have been derived by using the Contaminated Land Exposure Assessment (CLEA) model, which calculates human exposure via 10 different exposure pathways (12–15). These guideline values are aimed to be used as a screening tool for establishing whether further action on a site is necessary. When guideline values are exceeded, the intention is that either a further phase of risk assessment or remedial action is triggered.

Site-Specific Risk Assessment—Focus on the CLEA Model

The source-pathway-receptor concept forms the backbone of the CLEA model, which is a generic model for determining potential exposures via pathways that may lead to direct risks to human health from specific site uses.

Although the CLEA model was initially developed for deriving SGVs for acceptable human health risks, further development has enabled specific risk assessments to be carried out. Importantly, it does not attempt to model the possible impact of contaminated soil on groundwater or surface water quality or on buried service and construction materials or temporary risks to site workers during redevelopment or other construction works.

The United Kingdom Environmental Protection Act 1990 defines contaminated land as follows (18):

> any land which appears to the local authority in whose area it is situated to be in such a condition, by reason of substances in, on, or under the land, that significant harm is being caused or there is a

significant possibility of such harm being caused; or pollution of controlled waters is being, or is likely to be, caused.

The guideline values derived from the CLEA model are being produced to help determine whether land is contaminated according to the above definition.

Table 2.1 lists the exposure pathways that can be considered in the CLEA model. The significance of each pathway depends on the contaminant of interest; for example, indoor inhalation of soil vapor is the predominant pathway for a contaminant such as benzene.

Particular to the CLEA model is the inclusion of time-dependent effects such as biodegradation and natural attenuation. This is more realistic than assuming that, for example, in 10 years' time the concentration of a benzene spillage at the soil surface will not have altered. A significant proportion of the benzene will have volatilized or migrated from the soil surface; therefore, an exposure scenario in 10 years will be different from the current situation.

Removal processes included in the model are photolysis, chemical reaction, volatilization, biodegradation, runoff, erosion, leaching, and crop off-take. Biodegradation is particularly important in determining risks from organic contaminants, but the rate of biodegradation is highly site specific and depends on the types and growth rates of microbial populations, which in turn are influenced by factors such as moisture content, temperature, pH, nutrient availability, and toxicity.

TABLE 2.1 Exposure pathways included in the CLEA model

Exposure pathway
Outdoor ingestion of soil
Indoor ingestion of soil
Consumption of home-grown vegetables
Ingestion of soil attached to vegetables
Skin contact with outdoor soil
Skin contact with indoor dust
Outdoor inhalation of fugitive dust
Indoor inhalation of fugitive dust
Outdoor inhalation of soil vapor
Indoor inhalation of soil vapor
Ingestion of drinking water from mains supply[a]
Skin contact with mains water during showering and bathing, etc.[a]
Inhalation of vapor during showering and bathing and from ambient vapors otherwise derived from mains water[a]

[a]Not computed when an estimate of ADI from background (nonsoil) sources is available.

Environmental models developed for organics that are particularly susceptible to biodegradation usually represent chemical and biological degradation as a single first-order process with a rate constant specified by the user. While most environmental models do not include this often highly significant process, the CLEA model does include this degradation scenario.

If the initial contaminant concentration at time zero is C_0 and first-order decay is assumed, then the concentration at time t is C_t, where

$$C_t = C_0 e^{(-kt)}$$

and the average concentration C_i over the time interval $(\Delta T)_i = t_i - t_{i-1}$ is given by:

$$\overline{C_i} = \frac{C_0}{(\Delta T)_i} \int_{t_{i-1}}^{t_i} e^{(-kt)} dt$$

This time-dependent function allows the risks to human health posed by a contaminated site to be calculated after a specified lag period between site investigation and first end-user access, which could be several years.

The CLEA model also enables the user to assess the health risks posed by background contamination in addition to those posed by the actual site. If this information is known by the user, it can be entered into the model, where it is subsequently extrapolated to create appropriate childhood intakes adjusted for lower dietary intakes and body weights.

The CLEA model can compute rough estimates of background exposure based on contaminant concentration in mains water supply if no other background concentration data are available. Intake via the following routes can then be calculated, bearing in mind that they will provide rough estimates of background exposure and are not necessarily as applicable for lipophilic or highly volatile substances:

- ingestion of drinking water from mains supply,
- skin contact with mains water during showering and bathing, and
- inhalation of vapor during showering and bathing and from ambient concentrations otherwise derived from mains water (e.g., cooking, washing machines, dishwashers).

Exposure Scenarios and Risks to Human Health

As yet, no comprehensive model that takes into account all receptors of potential concern on contaminated sites is available. Instead, there are partial models to quantify the risks posed to groundwater, surface water, ecosystems, and human health.

By taking the approach that a site should be suitable for its intended use, a number of scenarios have been defined to encompass the different activity patterns that may occur on a site. Studies of how people spend their time have been made by academic and commercial organizations around the world, although the results are not usually in the form required for developing exposure scenarios. Information in government publications and market research studies have been used for deriving exposure scenarios in the CLEA model, but many simplifying assumptions have had to be made.

Research into how people spend their time is usually carried out by completion of either questionnaires or diaries. Some surveys concentrate on time spent on specific activities, such as watching television; others give a broad picture of the way in which time is used. *Social Trends* (39) defines four broad categories of time use: work and travel, essential activities, sleep, and free time, with particular emphasis on the amount of time available for leisure activities. This information, which is published annually, gives a rough estimate of how and where people spend their time.

The risk characterization process summarizes and integrates the exposure assessment and toxicity assessment into a quantitative and qualitative expression of risk. Health risks are quantified in terms of carcinogenic and noncarcinogenic effects. Uncertainties and limitations in risk estimates are also identified in the risk characterization process. Once the risks posed by a site are quantified (forward calculation), the process may be reversed (back calculation) to obtain acceptable soil cleanup levels.

Carcinogenic effects are characterized by estimating enhanced probabilities that an individual will develop cancer over a lifetime of exposure, based on projected intakes from a given scenario and the dose-response information summarized in the toxicity assessment. Noncarcinogenic effects are characterized by comparing calculated intakes of substances, based on specific exposure scenarios, to reference doses and reference concentrations.

Noncarcinogenic risks are usually classified in terms of hazard indexes or hazard quotients (HQ). An HQ for a chemical is found after both exposure and toxicity information has been combined in the following way. Exposure information is used to calculate the average daily intake (ADI) of the contaminant in milligrams per kilogram per day. This is found by considering factors such as body weight, skin area, and exposure characteristics. The ADI is then compared with the tolerable daily intake (TDI), which is found through toxicological studies.

Clearly, if the ADI is higher than the TDI, then a risk is present, as the concentration of the contaminant is such that it will cause harm to human health. Division of the ADI by the TDI gives the HQ. If the HQ is very close to unity, it may be demonstrated that the contaminant still poses no risk, as the assumptions inherent in the human health risk assessment always tend to err on the conservative side.

The theoretical tolerable excess lifetime cancer risk typically used in the context of genotoxic carcinogens on contaminated sites ranges from 10^{-6} (e.g., Denmark) to 10^{-4} (The Netherlands) per substance, with the majority of countries preferring 10^{-5}. In the

United Kingdom, theoretical lifetime risks of around 10^{-5} to 10^{-4} are generally considered acceptable for the general public, bearing in mind that the real risk is unlikely to be higher and may well be very much lower. However, derivation of TDIs by dividing maximum tolerable risk by slope factor is not generally favored in the United Kingdom (22).

Public concern about exposure to potential carcinogens and, in particular, the belief that there is no safe level of exposure have been part of the motivation for improving risk characterization. The numerical excess risk criteria (e.g., 10^{-4} or 10^{-6}) for tolerable levels of carcinogenic contaminants mentioned above are frequently misunderstood. It is not valid to conclude from a target concentration of contamination based on a 10^{-6} risk criterion that 550 excess deaths from cancer will occur in the population. Equally, on the basis that one in four people in Britain die from cancer—a background rate of 250,000 in 1 million—it would be unwise to claim that those exposed to a "10^{-6}" concentration have a lifetime risk of cancer mortality of 250,001 in 1 million (21). Further work on the communication and understanding of risks associated with contaminated land was commissioned by the Scotland and Northern Ireland Forum for Environmental Research in 2000 (50).

Dealing with Uncertainty

In the past, there has been a tendency to calculate the plausible health risks posed by the worst-case exposure scenario associated with a contaminated site. This approach has led to some overly conservative soil cleanup values, and in recent years there has been a reaction against this type of risk assessment, typical of the 1980s.

Keenan et al. (32) drew attention to a U.S. EPA example of the theoretical cancer risk from dioxin exposure for a child living close to a hypothetical municipal waste incinerator. They comment that:

> At first review, the analysis seemed reasonable until one noted that the child ate about one teaspoonful of dirt each day, that his house was downwind of the stack, that he ate fish from a pond near the incinerator, his fish consumption was at the 95th percentile level, he drank contaminated water from the pond, he ate food grown primarily from the family garden, and he drank milk from a cow that had grazed on forage on the farm.

The above example, highlighted by DEFRA in 2002 (12), is clearly unhelpful in deriving realistic risks posed by contaminated ground, but the problem is compounded when conservative assumptions are present throughout a risk assessment procedure. For example, if human health risk depended only on the product of three independent factors, each of which was chosen at the 90th percentile level, the product would represent a 99.9th percentile risk. It should, however, be noted that TDIs are often derived from studies to determine safe levels of contaminant in food, not soil. Hence, chemicals tested for this scenario are likely to be more bioavailable; i.e., soil guidelines based on these TDIs are likely to be conservative (24).

U.S. EPA exposure assessment guidelines have endorsed the use of simulation distributions, such as those produced by Monte Carlo models (Fig. 2.4), since 1992 in order to produce more realistic estimates of exposure. CLEA follows this route and is therefore a stochastic model. As opposed to calculating single risk values or exposure durations, CLEA uses probability density functions to reflect the variability and/or uncertainties associated with exposure parameters. Through each pass of the Monte Carlo simulation, a single point value is randomly picked from a given probability density function to produce a single estimate of intake or risk.

In 1997, the U.S. EPA announced its policy to accept probablistic risk assessment (PRA). The primary advantage of PRA within the Superfund program is that it gives a quantitative description of variability in risk. These two components make a more complete risk characterization than is possible in the point estimate approach (7). PRA differs from the point estimate approach in that, instead of a single figure being chosen for each exposure factor,

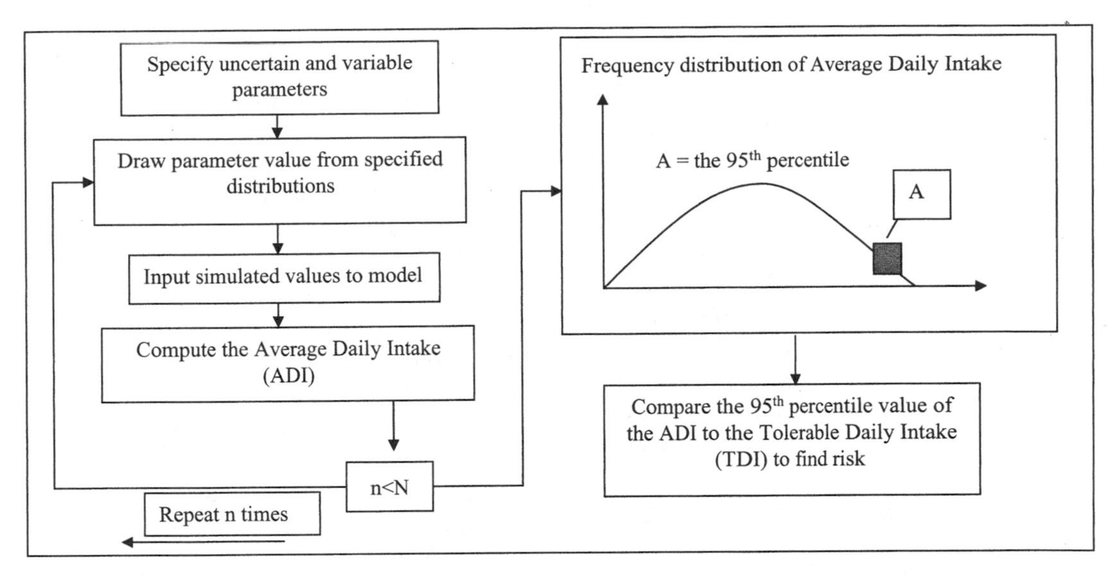

FIGURE 2.4 Procedure of Monte Carlo analysis (after the work of Ferguson and Denner [24]).

e.g., a mean weight for a person of 70 kg, a range of plausible values is chosen. The output of PRA is a range or distribution of risks experienced by the various members of the population of concern.

REMEDIAL PLANNING

The outcome of the investigative and risk assessment processes should be a report that is readily understandable by the client, the other members of the professional team, and the appropriate regulatory authorities. A high standard of presentation is therefore important, and the report should describe the various stages of the work undertaken, together with the findings obtained and any assumptions that may have been made.

Within the constraints of the site and the available information, the report should indicate the location of those parts of the site affected by contamination and identify the nature of these contaminants. Potential pathways and targets should be described, and the report should clearly define the options available to the client. The availability of contaminants, relative to actual or possible receptors, should be considered.

The decision to remediate is based on a knowledge of the nature and scale of the problems at a site; risks to the public, water bodies, and other sensitive receptors; demands of the local regulators; legal imperatives; aims of the site owner; and the economics of the situation. Once the goals and constraints of remediation have been identified and rationally evaluated, potential remedial options, including bioremediation, can be put forward. In general, the following evaluation process allows sound technical decisions on the feasibility of a particular bioremediation option to be made.

Economics of Remedial Technologies

Despite its undeniable promise, bioremediation still remains an innovative technology, not an established one, according to the U.S. EPA. Not technical success alone, but also economic competitiveness, will determine whether bioremediation can stay the course. There are two main considerations here. First of all, the market size for remedial technologies has to be established, and then the economics of bioremediation compared to that of other relevant technologies needs to be shown to be competitive.

It is notoriously difficult to quantify the size

of markets for biotechnology products and remedial technologies. For environmental biotechnology as a whole, as a new industry, the figures that have to be collected are fragmented and very difficult to gather. Despite these difficulties, rough estimates of the global market for biotechnology products have been calculated (54). The Organization for Economic Cooperation and Development (40) has estimated that the annual market for environmental biotechnology products is currently U.S. \$56 to \$120 billion. In context, the annual market for all industrial enzymes (e.g., detergent enzymes and baking and beverage enzymes) is of the order of \$1.5 billion. In diagnostics (e.g., DNA polymerase, alkaline phosphatase, and glucose oxidase), the total figure is \$0.15 to 0.2 billion. Contaminated soil remediation alone has been estimated to have an annual market of \$10 to \$25 billion. The total cost to the United States for all site cleanup has been estimated at between \$250 billion and \$1 trillion (49), the higher figure being for complete restoration of sites.

In 1998, Frost and Sullivan determined the European market for contaminated site remediation services to be worth U.S.\$1.02 billion. Landfilling still tends to dominate the market, which has had the effect of restraining new technology development. The European market is expected to rise to \$1.62 billion by 2005 as a result of market developments such as a concerted effort to reuse brownfield sites. Table 2.2 shows how the percentage of revenues by service has been distributed in Europe and how the future appears. The biological treatment sector looks due for steady growth in Europe, as some problems with acceptance are overcome.

Now that some figures for a market for bioremediation have been established, the next aspect is to determine how it can compete with other technologies on a cost basis. Bioremediation, as yet an innovative technology, does not have the economy of scale of established technologies which have been used in many full-scale projects. A detailed analysis of United Kingdom costs is given in Table 2.3.

Current figures show that prices for bioremediation services are falling. One Scottish bioremediation contractor, Environmental Reclamation Services, Ltd. Glasgow, quoted (2004) ex situ bioremediation costs for, e.g., 5,000 m^3 in windrows of 5,000-mg kg^{-1} diesel range organics, with a target of 1,000 mg kg^{-1}, of about £12 per m^3 (about \$20). At the time of writing, variation tended to be £10 to £30 per m^3. Compared to landfilling, in the United Kingdom ex situ bioremediation has therefore become cost-effective: legislative changes brought about by the

TABLE 2.2 Total contaminated site remediation service market within Europe

Yr	% of revenues obtained by:				
	Bioremediation	Physical/chemical treatment	Thermal treatment	Containment	Other
1995	17.3	53.0	16.2	8.2	5.3
1996	17.8	52.6	16.2	8.1	5.3
1997	18.3	53.1	15.2	8.1	5.3
1998	18.3	53.6	14.8	8.1	5.2
1999	19.7	53.2	14.0	8.0	5.1
2000	18.6	54.4	13.8	8.0	5.2
2001	18.1	55.1	13.6	8.0	5.2
2002	18.1	54.9	13.8	7.9	5.3
2003	18.2	54.6	13.8	7.9	5.5
2004	18.6	54.1	13.9	7.8	5.6
2005[a]	19.0	53.5	14.1	7.7	5.7

[a] Predicted values.

TABLE 2.3 1997 costs of remedial technologies in the United Kingdom[a]

Remediation method	Cost range/ton
Excavation/off-site disposal	$11–80 (£7–50)
Soil washing	$80–400 (£50–250)
Physicochemical washing	$80–275 (£50–170)
In situ stabilization/solidification	$100–190 (£60–010)
In situ electrokinetic techniques	$65–200 (£40–120)
Bioslurry	$80–130 (£50–80)
Biopiles	$25–75 (£15–45)
Land farming	$16–160 (£10–100)
Windrow turning	$8–100 (£5–60)
In situ biotreatments	$8–260 (£5–160)
Bioventing	$24–130 (£15–80)
Thermal treatment	$65–1,100 (£40–700)
Incineration	$80–1,900 (£50–1,200)
Cement/pozzolan solidification	$32–275 (£20–170)
Solvent extraction	$48–960 (£30–600)
Kiln-based vitrification	$48–800 (£30–500)

[a] Taken from reference 64.

EU Landfill Directive in mid-2004 have made the dumping of contaminated soil in landfill sites excessively expensive.

The U.S. EPA (61) has published a compendium of costs for six common remediation technologies: bioremediation, thermal desorption, soil vapor extraction (SVE), on-site incineration, groundwater pump-and-treat systems, and permeable reactive barriers. The information was gathered from approximately 150 projects, and an analysis was made to determine correlation between unit costs and quantity of soil or groundwater treated: this would indicate whether any of the technologies demonstrated economies of scale.

A correlation between unit cost and quantity treated was evident for four of the technologies—bioventing, thermal desorption, SVE, and pump-and-treat systems—and economics of scale were observed with all four, i.e., the unit cost decreased as larger quantities were treated. Of the four, bioventing showed the best correlation.

For other bioremedial technologies, however, no such quantitative correlation was observed. It was acknowledged that the unit costs for bioremediation are affected by factors including soil type and aquifer chemistry, site hydrogeology, type and quantity of amendments used, and type and extent of contamination.

Table 2.4 gives some comparative costs for remediation services in the United Kingdom and the United States. Since this table was compiled, the situation regarding landfilling of

TABLE 2.4 Comparative costs of remediation techniques in the United Kingdom and the United States[a]

Remediation approach	Cost/ton ($) in:	
	United Kingdom	United States
Bioremediation	9–286	26–65
Landfill	12–89	98–163
Incineration	89–2,150	163–523

[a] Based on the data of Wood (64) and Levin and Gealt (35).

contaminated soil has changed dramatically in the United Kingdom. In mid-July 2004, the implementation of the Landfill Directive classified soil contamination as nonhazardous (<10,000 mg of toxic material kg^{-1}, <1,000 mg of carcinogenic material kg^{-1}) or hazardous (>10,000 and 1,000 mg kg^{-1}, respectively). The cost of landfilling nonhazardous soil has gone to $21 per ton plus a landfill tax of $27 per ton (total, £27.00 per ton, or roughly $48.50). The cost for hazardous soil is £45 ($80) per ton plus a landfill tax of £15 ($27) per ton, making a total of £60, or roughly $107. However, another result of the implementation of the Landfill Directive is that the number of landfill sites in the United Kingdom that can accept contaminated soil has been greatly reduced, which has significantly increased the transportation costs. In some areas of the United Kingdom, it is not unlikely that contaminated soil would have to be transported 200 miles to a suitable landfill site, and the cost for transportation alone might amount to another $72 per ton. This has opened up much greater prospects for bioremediation since the transport cost is immediately removed from the economics.

Risk-Based Remedial Design

The outcomes of the risk assessment exercise feed into the next step in the contaminated land management process, which is selection of the remedial technology appropriate to nullifying the risks. It should be remembered that land remediation should be designed to break the pollutant linkage between source and receptor. In reality, several factors are likely to interplay in the eventual choice of technologies. In the United Kingdom, the four main factors that influence decision-making are as follows:

1. remedial objectives and treatment targets (for the purpose stated above);
2. environmental merit, e.g., incineration destroys contaminants but also destroys the soil;
3. cost-effectiveness; and
4. local factors, e.g., public opinion and local waste license regulations.

There now exist a large variety of treatment technologies, either established or innovative. It is beyond the scope of this book to look at all of these in detail. Therefore, a brief description of technologies classed as physical, chemical, thermal, or solidification is given. Landfill is not discussed, since it is not a remedial technology. Although in many countries it remains the cheapest option, it is merely a disposal option, as the anaerobic environment of the landfill is not conducive to biodegradation of many common contaminants.

Bioremediation has to compete successfully with these technologies, so for a screening exercise choices of which ones to compare it with have to be made. To narrow the field down, use has been made of U.S. EPA data on remedial technologies (60). The most frequently used established technologies in the United States are incineration (thermal), thermal desorption (thermal), solidification/stabilization (physical), and SVE (physical) and, for groundwater, pump-and-treat technologies. SVE and thermal desorption are interesting cases, as they were until recently classed as innovative technologies, but they have crossed the barrier to implementation and are now established. The EPA has defined innovative technologies as those whose use is limited by lack of data on cost and performance. They have only limited full-scale application, and in situ and ex situ bioremediation thus remains classed as innovative.

Bioremediation techniques together (for the fiscal year 1997) had been used in 11% of all Superfund remedial actions. For comparison, SVE had been used in 28%. For remedial treatment trains, in which two or more techniques are used in sequence, bioremediation was used in 6 of 17. Contaminated sites treated by bioremediation have mostly been BTEX (benzene, toluene, ethylbenzene, and xylene) containing. Tables 2.5 and 2.6 compare the applications of different techniques with respect to contaminant types, soil types, cost, and remediation times, and they contain data compiled from the work

TABLE 2.5 Comparative data on established remedial treatments[a]

Compound or parameter	Incineration	Thermal desorption	Solidification/stabilization		SVE
			Cement	Vitrification	
Organic					
Volatile			X		
Semivolatile			X		
Halogenated			X		
Nonhalogenated			X		
PAH					X
PCB		X			X
Dioxins/furans		X	X		X
Pesticide/herbicide		X	X		X
Inorganic					
Heavy metals	X				X
Nonmetals	X	X			X
Cyanides					X
Soil types	All	All	All but peats	All but peats	Porous
Corrosives		X	X		X
Explosives			X		X
Fuels					
Gasoline			X		
Diesel			X		
Jet fuel			X		
Kerosene			X		
Cost/ton ($)	80–1,920	64–1,120	32–272	48–812	16–144
Duration (mo)		1–13			5–10

[a] Data compiled from the work of Armishaw et al. (2), Martin and Bardos (37), the EPA (60), and Wood (64). X indicates that the method is not appropriate for use for the given substance.

of Armishaw et al. (2), Martin and Bardos (37), the U.S. EPA (60), and Wood (64).

FIELD TESTS

To establish the feasibility of a particular remedial technology, small-scale or laboratory experiments should be undertaken under carefully controlled conditions using actual media, e.g., soil and groundwater collected from the site and actual contaminants of concern. An example is a tray test, commonly used to evaluate the remedial effectiveness of thermal desorption. Contaminated soil is sampled, weighed, and placed in ovens at various temperatures for various periods, and the soil is retested after being heated to assess the likely effectiveness of heating at driving off organic compounds.

A further stage is to conduct pre-pilot-scale tests, which would be conducted in the field at very small scale, usually involving just one well, in the case of an in situ method. These tests provide simple yes-or-no answers, for example, to the question of whether this particular technology is applicable under these circumstances. If the answer is yes, the tests provide design parameters needed to create a pilot-scale test.

Before any bioremediation technology is undertaken, it is prudent to at least do a field pilot-scale test of the proposed method. This is highly recommended for any in situ or on-site technique, be it bioremediation, soil washing, or SVE.

A pilot test allows techniques, equipment, processes, and materials to be tested under field conditions but at a small scale. Invariably, the findings lead to some modification or adjust-

TABLE 2.6 Comparative data for bioremediation techniques[a]

Compound or parameter	Ex situ			In situ	
	Slurry bioreactor	Biopile/windrow	Landfarm	Pump treat	Bioventing
Organic					
Volatile					
Semivolatile					
Halogenated					
Nonhalogenated					
PAH					
PCB	X	X	X		X
Dioxins/furans		X	X		X
Pesticide/herbicide		X			X
Inorganic					
Heavy metals	X	X	X	X	X
Nonmetals	X	X	X	X	X
Cyanides	X	X	X		X
Soil types	All but peats	Porous	Porous	Porous	Porous
Corrosives	X			X	X
Explosives			X	X	X
Fuels					
Gasoline					
Diesel					
Jet fuel					
Kerosene					
Cost/ton ($)	80–130	8–110	16–160	8–260	24–290
Duration (mo)	Variable	3–30	1–24	3–48	12

[a] Data compiled from the work of Armishaw et al. (2), Martin and Bardos (37), the EPA (60), and Wood (64). X indicates that the method is not appropriate for use for the given substance.

ment in the final design of the remediation. Sometimes the pilot test results are such that the approach may be abandoned altogether. The rationale for a pilot test is simple: it provides insurance against failure at a larger scale. On smaller sites, the prepilot and pilot trials may be combined or the prepilot trial may be scaled up and the pilot trial dispensed with. On larger, more complex sites, both may be required.

Site investigation and characterization provide information that allows the professional to determine environmental impacts, proceed with risk analysis, determine liabilities, assess options, and plan remediation.

If bioremediation is deemed to be a suitable remedial technology at a site under investigation, a choice of which technology is most suitable has to be made. Even this step in the process can be complex; a simplistic summary is given in the form of a decision tree (Fig. 2.5).

Technology Briefs

The major competing, nonbiological technologies must be summarized in a form that can be compared to the bioremediation techniques. This has been done in the form of technology briefs below. Each of these gives the salient features of the technology under discussion and also some of the advantages and disadvantages of each. This is how the bioremediation technologies are described in chapter 5 but in greater detail.

Physical Treatment Systems. Physical treatment systems separate contaminants rather than destroy them. The principal techniques are

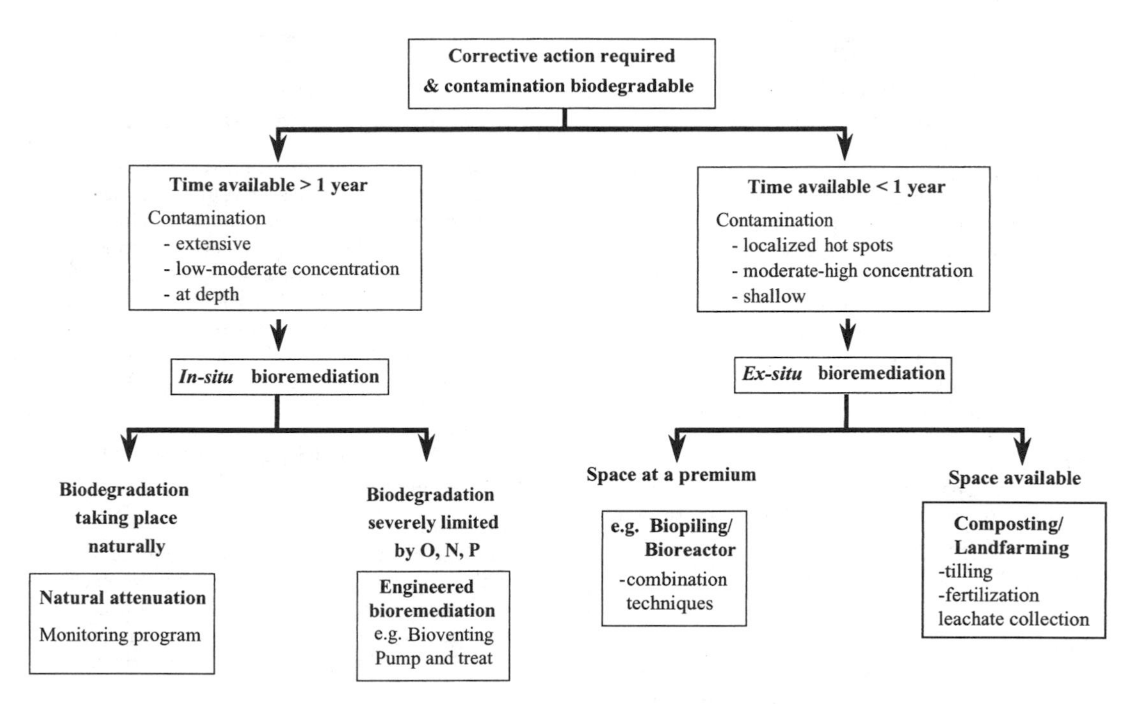

FIGURE 2.5 A simplified bioremediation decision tree. Whereas the early decisions based on time available tend to be clear-cut, later decisions on which individual biotechnology approach to use become more involved, and consequently the decision tree becomes more complex.

SVE, washing (water, surfactants), steam stripping, solid separation, and electroremediation. They can be considered to be preliminary treatments and thus are useful in remediation trains.

Chemical Treatment Systems. Chemical treatment systems either destroy contaminants or convert them to different, less toxic forms. They consist of oxidation (e.g., chlorine, hypochlorite, ozone), reduction, solvent extraction, neutralization, and precipitation. Chemical treatments have not achieved large popularity. Addition of solvents and oxidizing agents may add to the toxic burden of the system.

Thermal Treatment Systems. Thermal treatment systems heat the soils up to around 2,500°C. As such, they do destroy organic contaminants. Incineration can be performed in situ or ex situ. Although energy intensive, they are rapid, and most contaminants are destroyed completely, with some exceptions.

Solidification Treatment Systems. Solidification treatment systems use specific binders to convert contaminated soils, sludges, and waters into a solid to greatly reduce mobility and toxicity. Inorganic binder processes, e.g., cement encapsulation and vitrification, have greater potential for contaminated land treatment than organic binder systems: the organic polymers are too expensive.

SVE. SVE removes volatile and semivolatile contaminants from the unsaturated zone by applying a vacuum connected to a series of wells (Box 2.2). The contaminant groups most suited to SVE are VOCs and the more volatile fuels: it can be applied to volatile compounds with a Henry's law constant of >0.01 or a vapor pressure of >0.5 mm. SVE will not remove hydrophobic materials such as polychlo-

BOX 2.2
Technology Brief—SVE

Vacuum pumps or blowers induce a pressure gradient in the subsurface, resulting in an airflow field about an extraction well (27). These systems can be combined with groundwater pumping wells to remediate soil previously beneath the water table.

Gas-phase contaminants are removed via advective airflow entering the extraction wells. Volatilization is induced. High-vapor-pressure contaminants are removed first, and the soil progressively becomes enriched in less volatile compounds.

Advantages
Few moving parts; low operating costs; rapid installation; rapid results; no excavation; no disruption; ability to remediate under buildings; no toxic reagents; enhanced bioremediation if airflow is controlled.

Limitations
Contaminants with low vapor pressures have poor removal efficiencies; other techniques may be required (treatment train); there are constraints in soils with low air permeability; contaminants are not treated (toxicity unaltered).

rinated biphenyls (PCBs) or dioxins, which form firmly bound soil residues. An important limitation is the inability to treat soils of low porosity. It is therefore unsuited to the saturated zone.

Incineration. Incineration is the high-temperature thermal oxidation of contaminants to destruction. Incinerators come as a variety of technologies, including rotary kiln, fluidized bed, and infrared (37) (Box 2.3). The two most common incinerators used for contaminated soil are rotary kiln and fixed hearth, and the fluidized bed. Rotary kiln and fixed hearth are twin-chamber processes. The primary chamber volatilizes the organic components of the soil, and some of them oxidize to form carbon dioxide and water vapor. In the second chamber, high-temperature oxidation is used to attempt to completely convert the organics to carbon dioxide and water.

BOX 2.3
Technology Brief—Incineration

A typical incinerator system consists of several components: waste storage, preparation, and feeding; combustion chamber(s); air pollution control; residue and ash handling; and process monitoring.

Rotary kilns are the most common incinerators for waste materials (27). The rotary kiln is a cylindrical, refractory-lined reactor set at a slight angle (rake). As the kiln rotates, the waste moves through the reactor and is mixed by tumbling (63).

Advantages
Reported removal efficiencies of beyond 99%; detoxification; heat recycling.

Limitations
Very high initial capital investment; requirement of highly trained operatives; much greater complexity than other treatment technologies; emission of hazardous gases; de novo synthesis of dioxins and furans; adverse public perception; destruction of soil.

Fluidized Bed Incinerators. Fluidized bed incinerators, by contrast, are single-chamber systems containing fluidizing sand and a headspace above the bed. Fluidization with pressurized air creates high turbulence and enhances volatilization and combustion of the organics in contaminated soil, to form carbon dioxide and water.

Most of the reported limitations of soil incineration are operational problems. For example, there are specific feed size and material handling requirements that can impact on applicability or cost at specific sites. Volatile metals can exit the incinerator with the flue gases, entailing additional gas treatment facilities that add to the cost. Sodium and potassium form ashes that are aggressive to the brick lining. Above all, incineration is a costly, high-energy operation with poor public perception. It also destroys the soil.

Thermal Desorption. Thermal desorption is a high-temperature treatment that releases pollutants from soil to the gas phase (Box 2.4). The temperatures and residence times used in thermal desorption systems volatilize selected con

BOX 2.4
Technology Brief—Thermal Desorption

Thermal desorption is a lower-temperature thermal treatment (up to 600°C) which involves two processes: transfer of contaminants from the soil into the vapor phase (volatilization) and off-gas treatment, e.g., higher-temperature treatment to destroy (up to 1,400°C).

Advantages
Can be used for very small-scale projects; flexibility in operations (e.g., variable temperature, use of catalysts); organic material of soil may not be destroyed.

Limitations
Clay soils can greatly increase treatment cost; emissions; volatile metals cause operational problems.

taminants but do not oxidize or destroy them, as happens in incineration. Low-temperature thermal desorption can remove petroleum hydrocarbons from all soil types. An important feature of thermal desorption is that the treated soil is not destroyed; it will even support biological activity.

Perhaps the major operational problem encountered in thermal desorption treatment of contaminated soil involves particulates. All thermal desorption systems require treatment of the off-gas to remove particulates and organic contaminants. Dust and organic matter in the soil affect the efficiency of capture and treatment of off-gas.

Solidification. Unlike other remedial technologies, the objective of in situ solidification is to immobilize the contaminants within the contaminated matrix, rather than to remove them (Box 2.5). Immobilization is achieved with a binder such as Portland cement. In situ techniques use auger-caisson systems and injector heads to apply the solidification agents to soils.

As with all in situ techniques, sampling is problematic. In some situations, a large volume increase is inevitable, and this impacts greatly

BOX 2.5
Technology Brief—Solidification

Cement- and pozzolan-based solidification can be done both in situ and ex situ. In situ solidification can be performed by using an auger system that drills into the soil, combined with blades that mix. Solidification agents can be injected into the mixing zone. Ex situ treatments can be done in dedicated facilities using equipment borrowed from cement mixing. Soil and solidification agents can be added directly to a rotating drum.

In vitrification, soils are heated to melt them and produce a glass-like product. The high temperature results in the combustion of organic contaminants, and noncombustible contaminants are immobilized in the glassy matrix.

Advantages
Inorganics are permanently immobilized, especially with vitrification; vitrified material has long-term durability.

Limitations
Relatively high cost; vitrification is an energy-intensive process; soil structure is completely destroyed, especially during vitrification; organic contaminants are not well fixed by inorganic binders; cement solidification results in large volume increase, which increases the cost of subsequent landfilling.

on the applicability of solidification. Also, if the contaminants go to great depth, this limits applicability. Water contact and freeze-thaw add uncertainties to the integrity of long-term immobilization. The main target pollutants are metals, especially radionuclides, in the case of ex situ vitrification.

Case Studies

To help understand the applicability of risk assessment, site characterization, and remediation selection, we will discuss a few real-world case studies. These examples highlight the complexities of dealing with contaminated land and bringing science and practice together under critical public scrutiny.

The following case study example (Box 2.6) was chosen because it covers elements of most sections of this chapter: types and source of pollutants, cleanup standards, risk assessment,

BOX 2.6
The Millennium Dome Remediation Project

The former gas works at the Greenwich Peninsula, London, United Kingdom, was one of the largest gas works in Europe and was chosen as the site for the Millennium Dome. The site had also had a plant manufacturing sulfuric acid and ammonium sulfate fertilizer. There was a separate tar works that produced a range of organic chemicals, including benzene.

The extensive soil contamination required attention in two ways. There is a statutory obligation to protect groundwater, and the site had to be made fit for use, in keeping with United Kingdom policy.

Site Investigation

The objectives were to investigate the underlying geology and determine the chemical nature and concentrations of contaminants. Around 2,000 exploratory excavations were made, and 6,000 soil and water samples were taken for chemical analyses.

The contaminants were:

- coal tars, polyaromatic hydrocarbons (PAHs), phenols, benzene, and other VOCs
- cyanides, heavy metals
- sulfur compounds, ammonia
- mostly restricted to the top 3 to 4 m, above clay deposits.

Corresponding with the area of the future Millennium Dome was a site of more than 4 ha highly contaminated with VOCs, affecting both soil and perched water. Extensive clay deposits prevented groundwater contamination at the most polluted part of the site, namely, the tar production and distillation area. There were also widespread underground obstructions and tanks.

Risk Assessment

Extensive use was made of Geographic Information Systems in the risk assessment. There was no formal risk assessment process in the United Kingdom at the time, although ICRCL guidelines existed for gas works-contaminated soils. A semiquantitative risk assessment model was developed. This involved the preparation of a series of matrices for each hazard source, pathway, and potential receptor. A qualitative level of risk was assigned, leading to the identification of the most critical pollutant linkages. The identified risks were:

- existing and future effects on the adjacent river and the underlying Thames Gravels;
- the Millennium Dome visitors and staff, especially with regard to exposure to VOCs.

Risk Management: Remedial Strategy

Remedial strategy was developed as a two-phase process:

1. statutory remediation to nullify the most significant long-term sources of pollution, particularly to controlled waters; and
2. development remediation, depending on residual contamination after statutory remediation, with particular regard to fitness for use of the site.

The statutory remediation was complex, reflecting the nature of the site. Several techniques were involved.

1. SVE of 4.3 ha to remove VOCs such as benzene, using a dual-phase groundwater-SVE system.
2. Selective landfilling of untreatable waste: 250,000 m^3.
3. Screening of excavated material: 150,000 m^3.
4. Filling of excavated voids: 145,000 m^3.
5. Crushed concrete: 80,000 m^3.
6. Treatment of contaminated groundwater using pH adjustment, flocculation, rapid sand filtration, air stripping, biological treatment, and settlement: 50,000 m^3.
7. Soil washing to reduce volume for disposal: 30,000 m^3.
8. Removal of underground tanks: >150.

Project duration: 14 months.
Cost: $32 (£20) million.

SVE. Benzene was the principal contaminant of concern for SVE due to its mobility and carcinogenicity. It was observed that the vacuum system, as it created a large ingress of air through the shallow ground mass, stimulated biodegradation of VOCs. The SVE reduced the VOCs by 110 tons, including 12 tons of benzene. It was calculated that 90% of the removal resulted from incidental bioremediation due to the induced airflow into the unsaturated zone.

technology selection, and implementation. It was also a high-profile project, under great scrutiny as it was obvious that the end use of the remediated site was going to attract global attention. Another principle of soil remediation is evident: no one treatment is suitable for a large and complex site where the contaminants are patchy and diverse. Note that although bioremediation was not chosen as a primary remedial treatment, a significant proportion of the treatment ended up being biological. A more complete account of the project is given by Barry (3).

The second case study (Box 2.7) was selected because bioremediation was the primary remediation option selected rather than landfill—a comparative rarity in the United Kingdom. This illustrates the utility of careful risk assessment in remedial design. Had the design been based solely on treatment targets, then bioremediation may not have been considered, and the large financial savings would never have been realized. Risk assessment also showed that there was no need to treat all of the material, and this added considerably to the economics of the project.

BOX 2.7
Bioremediation of Coal Tar-Contaminated Soil

A former coal mine head in the United Kingdom included a coke works that had operated between 1900 and the early 1960s. Former coke works are often highly contaminated with coal tars and other such waste products. These chemicals are potentially harmful to people using the site in the future as well as groundwater resources and, in this case, a nearby wetland nature reserve.

Site Investigation

RJB Mining, the largest coal mining company in the United Kingdom, commissioned a site investigation to assess the nature and extent of the contamination. This showed that levels of coal tars and heavy metals at the site exceeded recommended government guideline values. An initial estimate suggested that 80,000 m^3 of soil and rubble needed to be treated.

The initial remedial options considered by RJB Mining were disposal to an off-site landfill at an estimated cost of $9.0 million (£5.6 million), and encapsulation in an on-site landfill at an estimated cost of $1.9 million (£1.2 million). Not only was disposal to a landfill prohibitively expensive, but transportation of the contaminated soil off-site would have had a major operational impact. Vehicle movements to and from the site were restricted (due to planning conditions), so transportation of the contaminated soil off-site would have resulted in a decrease in coal output. On-site encapsulation of the whole 80,000 m^3 was the preferable option but would have resulted in long-term liabilities to the company because of the waste management license conditions. The company would have been required to place $640,000 (£100,000) in trust to cover any long-term monitoring and further remedial works necessary.

Risk Assessment

At this time, RJB Mining engaged environmental consultants to assess the level of cleanup required using the Environment Agency's risk-based approach (36). The approach showed that the majority of the material could be buried on-site without requiring treatment or encapsulation. Risk-based cleanup target levels were derived for PAHs, heavy metals, phenols, and ammonia. Reduction of the high levels of PAHs in the soil was the primary objective of the bioremediation process.

By using this approach, after consultation with the Environment Agency and the Local Planning Authority, it was agreed that future users of the site would not be exposed to harmful contamination and that the only potential risk requiring mitigation was the underlying groundwater. RJB Mining then agreed on remediation targets with the Environment Agency that ensured that the site presented no risk to users and visitors and minimized the risk to groundwater.

Risk Management: Remedial Technology Selection

On this basis, 63,900 m^3 of soil and rubble was used as backfill, and only 16,100 m^3 of soil from

BOX 2.7 continues

BOX 2.7
Bioremediation of Coal Tar-Contaminated Soil *(continued)*

the site of the former coke works required treatment before being incorporated into backfill. As a result, RJB Mining engaged a specialist bioremediation company to undertake the treatment of the 16,100 m^3 of soil.

Bioremediation was used to treat 14,800 m^3 of this soil that was contaminated with biodegradable chemicals. The remaining 1,300 m^3 of soil was sent to an off-site landfill because of high levels of heavy-metal contamination. The treatment technology to be employed was windrows, an ex situ bioremediation technique.

Implementation

Treatment times varied, but each windrow of 500 to 600 m^3 of contaminated soil needed a maximum time of 12 weeks for PAH levels to reach the cleanup target (the average time was 6 to 7 weeks). The bioremediation project took a total of 30 weeks, starting in April 1999.

Summary

A summary of the initial options and the decisions with the relevant costs is shown in Box Table 2.7.1.

BOX TABLE 2.7.1 Summary of risk-based remediation at RJB Mining

Remediation method	Amt of soil (m^3)	Unit cost (\$ m^{-3})	Total cost (\$ [millions])
Initial options			
Disposal to landfill	80,000	11.2	9.0
On-site encapsulation	80,000	1.6	2.54[a]
Options after risk-based assessment			
Disposal to landfill	16,100 + 1,300	112 + 112[c]	1.8
Bioremediation	14,800[b]	32[d] + 16[e]	0.85

[a] Includes $0.64 million additional cost for insurance.
[b] Soil contaminated with biodegradable chemicals.
[c] Soil contaminated with heavy metals removed to landfill.
[d] Direct cost of treating the soil by bioremediation.
[e] Indirect costs including moving soil and preparing the treatment site.

With regard to economics, using bioremediation according to a risk-based approach to clean up the contaminated soil allowed RJB Mining to save around $8 million (£5 million) compared with disposal to an off-site landfill. The unit cost of bioremediation was less than half that for landfill, making it the preferred economic option for the soil requiring treatment.

CONCLUDING REMARKS

The design and implementation of an appropriate contamination investigation are essential in gaining a satisfactory understanding of both contamination issues and their possible solutions. The data produced should be reliable, reproducible, and representative of actual site conditions, both at the site and in the chemical testing laboratory.

The principal objectives of a contamination investigation are to determine the nature of contaminants present on the site and their likely behavior, spatial distribution, and volumetric extent. However, no investigation can guarantee to find all contaminants that may exist within a site or to precisely quantify their presence. The refined objective then becomes the assessment, within an agreed (and reasonable) degree of certainty, of the likelihood of any contamination being present and the significance of the risks that may be associated with these contaminants. It is risk assessment that is hence at the heart of any contaminated land investigation.

The late Colin Ferguson, a pioneer in contaminated land assessment in the United Kingdom in the late 20th century, had four fundamental rules of risk assessment. These were:

1. Question all of your results.
2. State all of your uncertainties.
3. You have to make a decision.
4. You must be able to sleep at night.

The ethical (not to mention legal) ramifications of incorrect contaminated land risk assessments can be highly significant, and specialist advice should be sought where any doubt exists in such areas.

REFERENCES

1. **American Petroleum Institute.** 1989. *A Guide to the Assessment and Remediation of Underground Petroleum Releases.* Publication no. 1628, 2nd ed. American Petroleum, Institute, Washington, D.C.
2. **Armishaw, R., R. P. Bardos, R. M. Dunn, J. M. Hill, M. Pearl, T. Rampling, and P. A. Wood.** 1992. *Review of Innovative Contaminated Soil Clean-up Processes.* Report LR 819 (MR). Warren Spring Laboratory, Warren, Ohio.
3. **Barry, D. L.** 1999. The Millennium Dome (Greenwich Millennium Experience Site) contamination remediation. *Land Contam. Reclam.* **7:** 177–190.
4. **British Standards Institute.** 2001. *Investigation of Potentially Contaminated Sites—Code of Practice.* BJ 10175. British Standards Institute, London, United Kingdom.
5. **Cairney, T.** 1995. *The Re-use of Contaminated Land. A Handbook of Risk Assessment.* John Wiley & Sons Ltd., Chichester, United Kingdom.
6. **Carcione, J. M., G. Seriani, and D. Gei.** 2003. Acoustic and electromagnetic properties of soils saturated with salt water and NAPL. *J. Appl. Geophys.* **52:**177–191.
7. **Chang, S. S.** 1999. Implementing probabilistic risk assessment in USEPA Superfund Program. *Hum. Ecol. Risk Manage.* **5:**737–754.
8. **Christie, S., and R. M. Teeuw.** 1998. Contaminated land policy within the European Union. *Eur. Environ.* **8:**7–14.
9. **Clifton, A., M. Boyd, and S. Rhodes.** 1999. Assessing the risks. *Land Contam. Reclam.* **7:**27–32.
10. **de Castro, D. L., and R. M. G. C. Branco.** 2003. 4-D ground penetrating radar monitoring of a hydrocarbon leakage site in Fortaleza (Brazil) during its remediation process: a case history. *J. Appl. Geophys.* **54:**127–144.
11. **Delaney, A. J., P. R. Peapples, and S. A. Arcone.** 2001. Electrical resistivity of frozen and petroleum-contaminated fine-grained soil. *Cold Regions Sci. Technol.* **32:**107–119.
12. **Department for Environment, Food and Rural Affairs.** 2002. *Contaminated Land Exposure Assessment Model (CLEA): Technical Basis and Algorithms.* Research and Development publication CLR10. The Environment Agency, Bristol, United Kingdom.
13. **Department for Environment, Food and Rural Affairs.** 2002. *Toxicological Reports for Individual Soil Contaminants.* CLR9, TOX1–10. The Environment Agency, Bristol, United Kingdom. http://www.environment-agency.gov.uk.
14. **Department for Environment, Food and Rural Affairs.** 2002. *Contaminants in Soils: Collation of Toxicological Data and Intake Values for Humans.* Research and Development publication CLR9. The Environment Agency, Bristol, United Kingdom.
15. **Department for Environment, Food and Rural Affairs.** 2002. *Assessment of Risks to Human Health from Land Contamination: An Overview of the Development of Soil Guideline Values and Related Research.* Research and Development publication CLR7. The Environment Agency, Bristol, United Kingdom.
16. **Department of the Environment.** 1996. *Waste Management Paper 27: Landfill Gas.* Her Majesty's Stationery Office, London, United Kingdom.
17. **Department of the Environment, Transport and the Regions.** 2000. *Guidelines for Environmental Risk Assessment and Management.* Revised departmental guidance. Her Majesty's Stationery Office, London, United Kingdom. http://www.defra.gov.uk.
18. **Department of the Environment, Transport and the Regions.** 2000. *Contaminated Land: Implementation of Part IIA of the Environmental Protection Act 1990.* DETR circular 02/2000. Her Majesty's Stationery Office, London, United Kingdom. http://www.defra.gov.uk.
19. **Drommerhausen, D. J., D. E. Radcliffe, D. E. Brune, and H. D. Gunter.** 1995. Electromagnetic conductivity surveys of dairies for groundwater nitrate. *J. Environ. Qual.* **24:** 1083–1091.
20. **Dutch Ministry of Housing, Spatial Planning and Environment.** 2000. Circular on target val-

ues and intervention values for soil remediation. DBO/1999226863. Dutch Ministry of Housing, Spatial Planning and Environment, The Hague, Netherlands.

21. **Ferguson, C. C.** 1996. Assessing human health risks from exposure to contaminated land: a review of recent research. *Land Contam. Reclam.* **4:** 159–170.
22. **Ferguson, C. C.** 1999. Assessing risks from contaminated sites: policy and practice in 16 European countries. *Land Contam. Reclam.* **7:**87–108.
23. **Ferguson, C. C., and H. Kasamas (ed.).** 1999. *Risk Assessment for Contaminated Sites in Europe,* vol. 2. *Policy Frameworks.* LQM Press, Nottingham, United Kingdom.
24. **Ferguson, C. C., and J. M. Denner.** 1998. Human health risk assessment using UK guideline values for contaminants in soils, p. 37–43. *In* D. N. Lerner and R. G. Walton (ed.), *Contaminated Land and Groundwater: Future Directions.* Engineering Geology Special Publication 14. Geological Society London, London, United Kingdom.
25. **Gerlach, R. W., and J. M. Nocerino.** 2003. *Guidance for Obtaining Representative Laboratory Analytical Subsamples from Particulate Laboratory Samples.* EPA/600/R-03/027. U.S. Environmental Protection Agency, Washington, D.C.
26. **Gerlach, R. W., D. E. Dobbs, G. A. Raab, and J. M. Nocerino.** 2002. Gy sampling theory in environmental studies. 1. Assessing soil splitting protocols. *J. Chemometrics* **16:**321–328.
27. **Grasso, D.** 1993. *Hazardous Waste Site Remediation. Source Control.* CRC Press, Boca Raton, Fla.
28. **Gy, P.** 1998. *Sampling for Analytical Purposes.* Wiley, New York, N.Y.
29. **Institute of Petroleum.** 1998. *Guidelines for the Investigation and Remediation of Retail Sites.* Portland Press, Colchester, United Kingdom.
30. **Interdepartmental Committee on the Redevelopment of Contaminated Land.** 1987. *Guidance on the Assessment and Redevelopment of Contaminated Land.* Guidance note 59/83. Department of the Environment, London, United Kingdom.
31. **Interdepartmental Committee on the Redevelopment of Contaminated Land.** 1990. *Notes on the Restoration and Aftercare of Metalliferous Mining Sites for Pastures and Grazing.* Guidance note 70/90. Department of the Environment, London, United Kingdom.
32. **Keenan, R. E., E. R. Algeo, E. S. Ebert, and D. J. Paustenbach.** 1993. Taking a risk assessment approach to RCA corrective action, p. 225–275. *In Developing Cleanup Standards for Contaminated Soil, Sediment, and Groundwater: How Clean is Clean?* Water Environment Federation, Alexandria, Va.
33. **Knight, R.** 2001. Ground penetrating radar for environmental applications. *Annu. Rev. Earth Planet. Sci.* **29:**229–255.
34. **Lanz, E., H. Maurer, and A. G. Green.** 1998. Refraction tomography over a buried waste disposal site. *Geophysics* **63:**1414–1433.
35. **Levin, M. A., and M. A. Gealt.** 1993. *Biotreatment of Industrial and Hazardous Waste.* McGraw-Hill, New York, N.Y.
36. **Marsland, P. A., and M. A. Carey.** 1999. *Methodology for the Derivation of Remedial Targets for Soil and Groundwater to Protect Water Resources.* Research and Development publication 20. The Environment Agency, Bristol, United Kingdom.
37. **Martin, I., and R. P. Bardos.** 1995. *A Review of Full Scale Treatment Technologies for the Remediation of Contaminated Soil.* EPP Publications, Richmond, Surrey, United Kingdom.
38. **Naudet, V., A. Revil, E. Rizzo, J. Y. Bottero, and P. Begassat.** 2004. Groundwater redox conditions and conductivity in a contaminant plume from geoelectrical investigations. *Hydrol. Earth Syst. Sci.* **8:**8–22.
39. **Office for National Statistics.** 2004. *Social Trends (34).* Office for National Statistics, London, United Kingdom.
40. **Organization for Economic Cooperation and Development.** 1998. *Biotechnology for Clean Industrial Products and Processes: Towards Industrial Sustainability.* Report of ad hoc task force chaired by A. T. Bull. Organization for Economic Cooperation and Development, Paris, France.
41. **Orlando, L.** 2002. Detection and analysis of LNAPL using the instantaneous amplitude and frequency of ground-penetrating radar data. *Geophys. Prospecting* **50:** 27–41.
42. **Osterreicher-Cunha, P., E. A. Vargas, J. R. D. Guimaraes, T. M. P. de Campos, C. M. F. Nunes, A. Costa, F. S. Antunes, M. I. P. da Silva, and D. M. Mano.** 2004. Evaluation of bioventing on a gasoline-ethanol contaminated undisturbed residual soil. *J. Hazard. Mater.* **110:** 63–76.
43. **Parliamentary Office of Science and Technology.** 1993. *Contaminated Land.* Parliamentary Office of Science and Technology, London, United Kingdom.
44. **Patata, L., and M. Mastrolilli de Angelis.** 1997. Field survey activities, p. 35–97. *In* P. Lecompte and C. Mariotti (ed.), *Handbook of Diagnostic Procedures for Petroleum-Contaminated Sites.* John Wiley and Sons Ltd., Chichester, United Kingdom.
45. **Pearce, F.** 1992. Sitting on a toxic time bomb. *New Scientist* **15(August):**12–13.
46. **Pellerin, L.** 2002. Applications of electrical and electromagnetic methods for environmental and

geotechnical investigations. *Surveys Geophys.* **23:** 101–132.

47. **Roth, K., U. Wollschlager, Z. H. Cheng, and J. B. Zhang.** 2004. Exploring soil layers and water tables with ground-penetrating radar. *Pedosphere* **14:**273–282.
48. **Rudland, D. J., R. M. Lancefield, and P. N. Mayell.** 2001. *Contaminated Land Risk Assessment: A Guide to Good Practice.* C552. Construction Industry Research and Information Association, London, United Kingdom.
49. **Russell, M., B. Colglazier, and M. English.** 1991. *Hazardous Waste Site Remediation: The Task Ahead.* Waste Management Research and Education Institute, University of Tennessee, Knoxville.
50. **Scotland and Northern Ireland Forum for Environmental Research.** 2000. *Communicating Understanding of Contaminated Land Risks.* SNIFFER project no. SR97(11)F. Scotland and Northern Ireland Forum for Environmental Research.
51. **Simons, R. A.** 1998. *Turning Brownfields into Greenbacks: Financing Environmentally Contaminated Real Estate.* Urban Land Institute, Washington, D.C.
52. **Soczo, E., and T. Meeder.** 1992. Clean-up of contaminated sites in Europe and the USA—a comparison. *In Eureco '92* (European Urban Regeneration Conference), Birmingham, United Kingdom.
53. **Syms, P. M., and R. A. Simons.** 1999. Contaminated land registers: an analysis of the UK and USA approaches to public management of contaminated sites. *Land Contam. Reclam.* **7:**121–132.
54. **ten Kate, K., and S. A. Laird.** 1999. *The Commercial Use of Biotechnology: Access to Genetic Resources and Benefit-Sharing.* Earthscan Publications Ltd., London, United Kingdom.
55. **U.S. Army Corps of Engineers.** October 1998. *USACE Sample Collection and Preparation Strategies for Volatile Organic Compounds in Solids.* U.S. Army Corps of Engineers, Washington, D.C.
56. **U.S. Environmental Protection Agency.** 1995. *Guidance for Scoping the Remedial Design.* EPA 540/R-95/025. U. S. Environmental Protection Agency, Washington, D.C.
57. **U.S. Environmental Protection Agency.** 1995. *Superfund Program Representative Sampling Guidance*, vol. 1. *Soil.* EPA 540/R-95/141. U.S. Environmental Protection Agency, Washington, D.C.
58. **U.S. Environmental Protection Agency.** 1996. *Soil Screening Guidance: User's Guide.* EPA/540/R-96/018. U. S. Environmental Protection Agency, Washington, D. C.
59. **U.S. Environmental Protection Agency.** 1998. *Guidelines for Ecological Risk Assessment.* EPA/630/R-95/002F. U. S. Environmental Protection Agency, Washington, D. C.
60. **U.S. Environmental Protection Agency.** 1999. *Treatment Technologies for Site Cleanup: Annual Status Report*, 9th ed. EPA-542-R99-001. U. S. Environmental Protection Agency, Washington, D. C.
61. **U.S. Environmental Protection Agency.** 2001. *EPA Remediation Technology Cost Compendium—Year 2000.* EPA-542-R-01-009. U. S. Environmental Protection Agency, Washington, D. C.
62. **Villaume, J. F.** 1985. Investigations at sites contaminated with dense non-aqueous phase liquids (NAPLs). *Ground Water Monit. Rev.* **5:**60–75.
63. **Wentz, C. A.** 1989. *Hazardous Waste Management.* McGraw-Hill, Singapore, Singapore.
64. **Wood, P. A.** 1997. Remediation methods for contaminated sites, p. 47–71. *In* R. E. Hester and R. M. Harrison (ed.), *Contaminated Land and Its Reclamation.* Thomas Telford Publishing, London, United Kingdom.
65. **Yeskis, D., and B. Zavala.** 2002. *Ground-Water Sampling Guidelines for Superfund and RCRA Project Managers.* EPA 542-S-02-001. U.S. Environmental Protection Agency, Washington, D.C.
66. **Yoo, G. L., and J. B. Park.** 2001. Sensitivity of leachate and fine contents on electrical resistivity variations of sandy soils. *J. Hazard. Mater.* **84:** 147–161.
67. **Yoo, G. L., M. H. Oh, and J. B. Park.** 2002. Laboratory study of landfill leachate effect on resistivity in unsaturated soil using cone penetrometer. *Environ. Geol.* **43:**18–28.

LEGAL AND REGULATORY FRAMEWORKS FOR BIOREMEDIATION

Barry Hartman, Mark Mustian, and Colin Cunningham

3

INTRODUCTION

Bioremediation is but one of many environmental cleanup technologies that is subject to governmental regulatory oversight. As with more conventional treatments, regulatory scrutiny of bioremediation generally will not be based upon the fact that bioremediation is being utilized, regardless of whether it involves biostimulation, bioaugmentation, or monitored natural attenuation. Instead, the scrutiny will be based upon the application of the technology to a specific problem, i.e., the nature and extent of environmental contamination and the necessary end point that must be reached so as to reduce risk and thus ensure protection of human health and environmental quality. In other words, the regulatory requirements are generally independent of the type of technology being proposed for a cleanup activity; instead, the regulatory requirements will be set to ensure that a specific end result is achieved.

It is not possible within the framework of this document to identify and evaluate all regulations and requirements that will impact any particular bioremediation activity. Instead, each specific application must be evaluated within the context of national and local laws and regulations that would apply to the specific situation. We have chosen to highlight the regulatory frameworks of the United States and the United Kingdom, as they represent the sorts of regulatory considerations that drive remediation efforts. The regulatory framework for bioremediation in the United States is advanced and governed by several key laws that establish the oversight mechanisms and parameters within which bioremediation must operate. The United Kingdom regulatory framework is developing in concert with that of the European Union, which is placing increased emphasis on environmental quality and the need for remediation of contaminated sites. Indeed, it is the regulatory requirement for cleanup of environmentally contaminated sites that is the ultimate driver for remediation efforts, and it is the standards established by the regulatory frameworks that establish the performance parameters that must be met by a bioremediation project.

Actually carrying out a bioremediation project requires a detailed evaluation of the

Barry Hartman and Mark Mustian, Kirkpatrick, Lockhart, Nicholson, Graham LLP, 1800 Massachusetts Avenue, N.W., Washington, DC 20036–1800. *Colin Cunningham*, Contaminated Land Assessment and Remediation Research Centre, The University of Edinburgh, Edinburgh, Scotland, United Kingdom.

Bioremediation: Applied Microbial Solutions for Real-World Environmental Cleanup
Edited by Ronald M. Atlas and Jim C. Philp

regulatory requirements at the beginning of a project, by persons with the necessary experience and training to ensure that the regulatory requirements are being met. Suffice it to say that the requirements for remediation of contaminated sites are highly dependent upon the specific remediation activity being undertaken. The coverage in this chapter should make the researcher, potential practitioner, site manager, and others aware of the various national and local programs and regulations that could impact a particular remediation project. Although it will not provide the reader all information necessary to successfully implement a bioremediation project, it will point toward the regulatory framework that must be considered. Beyond that, for pilot and real-world applications of bioremediation, it is strongly recommended that appropriate technical and legal resources be utilized to ensure that a successful, legally compliant result is achieved.

THE U.S. ENVIRONMENTAL REGULATORY PROCESS

Within the United States, almost any activity that has the potential to impact the natural environment or human health will likely require regulatory oversight and approval. This section will explain the basic principles of the American legal and regulatory system as it applies to bioremediation activities. The United States has developed overall regulatory programs which are aimed at different issues and problems. These programs address specific media (air, water, solid waste) and set up certain standard guidelines, regulations, and protocols. The purpose of this chapter is to look at bioremediation processes and technologies in the context of these regulatory programs. However, because of the size and diversity of the country, there is no standard protocol or procedure in place that will be applicable to any location and/or activity. Instead, it is necessary to understand the rules that are in place for the specific activity and specific place where you will be working.

Typically, both federal and state oversight of environmental programs is in place everywhere within the United States. In many places, this oversight extends to the local government level. There are, however, various programs in place which are applicable regardless of location and which are relatively consistent regardless of location. In general, these are the federal environmental laws that have been passed over the past 30 to 40 years. However, even the federal environmental laws are often implemented and regulated by state agencies, as the federal government has granted primacy to the state. In addition to federal laws implemented by the states, many states have their own laws and regulations, which cover the same regulatory area and which are oftentimes more stringent than federal law.

In most cases, any project that affects or impacts one or more environmental medium (air, water, or soil) will require regulatory approval. This approval can take various forms. In some cases, such as activities under the Clean Water Act (CWA), the approval would be in the form of a permit. In other cases, such as a cleanup under the Comprehensive Environmental Response, Compensation and Recovery Act (CERCLA) (42 U.S.C. §§9601 et seq. [2000]), approval would consist of a "record of decision" that identifies the treatment procedure to be carried out at the site. Regardless of the form of approval, it is necessary to obtain the proper approval prior to commencement of the project.

Many environmental laws include a requirement for an operating permit, which may be separate from the initial approval necessary for the selection of a technology, and construction of process systems. Operating permits impose conditions and emission or discharge limitations upon the operation of the system and require periodic monitoring and renewal of the permit. The limitations in an operating permit impose significant obligations on the permittee, and violations can result in both monetary penalties and potential civil and criminal liability. For example, negligent violations of an NPDES permit issued under 33 U.S.C. §1342 can result in penalties of up to $32,500 per day of violation and/or imprisonment for up to 1 year. See 33 U.S.C. §1319(c) and reference 9. Below, we identify some of the major areas of regulations, briefly explain the background on the laws and regulations in each area, and discuss how they tie into bioremediation activities. This chapter is intended to provide the reader a general overview and is not intended to provide substantive advice on the law or regulation being discussed.

CERCLA

CERCLA is the federal law that is most likely to interact with a bioremediation project. CERCLA is commonly known as Superfund, and it funds and regulates cleanup of hazardous waste sites throughout the United States. CERCLA, enacted in 1980, set up a trust funded by taxes on the chemical and petroleum industries and provided federal authority to respond to releases or threatened releases of hazardous substances that may endanger public health or the environment. The law authorized response action for both short-term removals, where releases or threatened releases required prompt response, and long-term remedial response actions that permanently and significantly reduce the dangers associated with releases of hazardous substances. CERCLA and the regulations enacted to implement CERCLA are extremely complex and involve many issues that are beyond the scope of this chapter. This chapter will provide a general walk-through of the steps that are taken during the course of a hazardous-site cleanup.

PA AND SI

The initial step in a cleanup is the preliminary assessment (PA) and site inspection (SI). The PA and SI are used by the Federal Environmental Protection Agency (EPA) to evaluate the potential for a release of hazardous substances from a site. The PA is a limited-scope investigation performed to distinguish between sites that pose little or no threat to human health and the environment and sites that may pose a threat and require further investigation. If the PA indicates that further investigation is required, an SI is performed.

The SI identifies sites that enter into the National Priorities List (NPL) site listing process and provides the data needed for hazard ranking system (HRS) scoring. During the SI, environmental and waste samples are collected to determine what hazardous substances are present. A determination of whether these substances are being released to the environment and whether these substances have reached nearby populations is also made.

HRS

The HRS is the mechanism used to place waste sites on the NPL. It uses the information from the PA and the SI to assess the relative potential of sites to pose a threat to human health or the environment. Information collected during the PA and SI is used to calculate an HRS score.

The HRS assigns numerical values to factors based upon the risk in three different categories:

- likelihood that a site has released or has the potential to release hazardous substances into the environment,
- characteristics of the waste (e.g., toxicity and waste quantity), and
- people or sensitive environments (targets) affected by the release.

Four pathways can be scored under the HRS:

- groundwater migration (drinking water),
- surface water migration (drinking water, human food chain, sensitive environments),
- soil exposure (resident population, nearby population, sensitive environments), and
- air migration (population, sensitive environments).

Scores are calculated for the pathways and combined to determine the overall site score. Sites with a score of 28.5 or greater are eligible for listing on the NPL.

NPL

Section 105(a)(8)(B) of CERCLA requires that the criteria provided by the HRS be used to prepare a list of national priorities among the known releases or threatened releases of hazardous substances, pollutants, or contaminants. The identification of a site for the NPL is intended to guide the EPA in:

- determining which sites warrant further investigation to assess the nature and extent of the human health and environmental risks associated with a site,
- identifying what CERCLA-financed remedial actions may be appropriate,
- notifying the public of sites the EPA believes warrant further investigation, and
- serving notice to potentially responsible parties that the EPA may initiate CERCLA-financed remedial action.

RI/FS

For sites on the NPL, a remedial investigation/feasibility study (RI/FS) is performed. The RI/FS is used to:

- characterize site conditions,
- determine the nature of the waste,
- assess risk to human health and the environment, and
- conduct treatability testing to evaluate the potential performance and cost of the treatment technologies being considered.

It is during the RI/FS phase of a CERCLA cleanup that bioremediation technologies would be evaluated, tested, and proposed. In order for bioremediation to be accepted in a CERCLA cleanup, the process must be able to reduce soil, groundwater, or surface water contaminant concentrations to below health-based levels. These levels are determined on the basis of the parameter of concern, the intended use of the water (drinking water, discharge to surface water, etc.), and other site-specific factors. The cleanup goals should meet chemical-specific "applicable or relevant and appropriate requirements" from other regulations, such as the Safe Drinking Water Act (SDWA) or the CWA.

ROD

The Record of Decision (ROD) is a public document that explains which cleanup alternatives will be used to clean up a Superfund site. The ROD is developed on the basis of the results from the RI/FS.

REMEDIAL DESIGN AND REMEDIAL ACTION

Remedial design is the phase in which the specifications and technologies are designed. The remedial action follows the design phase and involves the actual construction or implementation of the design.

CONSTRUCTION COMPLETIONS

The construction completion list is used to communicate the successful completion of cleanup activities. Sites qualify for the list when:

- any necessary physical construction is complete, whether or not final cleanup levels or other requirements have been achieved; or
- the EPA has determined that the response action should be limited to measures that do not involve construction; or
- the site qualifies for deletion from the NPL.

POSTCONSTRUCTION COMPLETION

Postconstruction completion activities provide for the long-term protection of human health and the environment and also involve optimizing remedies to reduce cost and/or increase effectiveness.

CWA

Historically, the United States has utilized a command-and-control strategy for addressing environmental risks. This strategy utilizes the top-down application of performance standards and permits to deal with specific environmental contaminants. A good example of this type of approach can be found in the passage of the CWA. Water protection efforts up until the early 1970s all centered on protection of stream uses and attainment of specific water quality standards within water bodies. The setting of goals and implementation procedures was to be determined on a state-by-state basis. This approach failed for a variety of reasons.

In 1971, the Senate Public Works Committee issued a report that spelled out the various problems with the nation's water pollution control program up to that point. The problems identified by the committee included:

- an almost total lack of enforcement under the 1948 Water Pollution Control Act abatement procedures;
- weaknesses in the permit program established under the Refuse Act in that it applied only to industrial polluters, authority was divided between two federal agencies, and procedures for issuing permits were slow and cumbersome;
- a lack of funding for construction projects;
- lack of research on treatment technologies; and
- the slow progress of states in establishing standards and setting up enforcement programs.

As a result of the weaknesses identified by this Senate Committee inquiry, the national approach to water pollution changed with the enactment of the 1972 Amendments to the Federal Water Pollution Control Act (FWPCA) (Federal Water Pollution Control Act Amendments of 1972, Public Law 92-500, 86 Stat. 816 codified as amended at 33 U.S.C. §§ 1251 to 1387 [2000]), commonly referred to as the CWA. The 1972 CWA contained numerous new provisions. The most important items were the new discharge permit requirements and the implementation of discharge limitations based upon the application of the best treatment technology available at the time, independent of the dischargers' impact upon the receiving stream. The purpose of the new technology-based regulatory approach was to quickly force the installation of the best water pollution control technology in the shortest time. In order to implement the regulatory program, it was necessary to create a permitting program. The CWA required the EPA to issue permits for all "point source" discharges into the nation's waters, upon the condition that the discharge met the requirements of the applicable sections of the CWA (33 U.S.C. § 1342[a][1] [(2000]). Each state was specifically granted the authority to administer its own permit program as long as that program complied with all applicable sections of the CWA (33 U.S.C. § 1342[b] [2000]).

SDWA

The SDWA is not directly related to bioremediation activities. It is unlikely that a bioremediation project would be driven directly by either the SDWA or the CWA. However, the standards and criteria derived from these statutes are often used to develop cleanup goals for bioremediation projects. In particular, cleanup standards for CERCLA projects will utilize either the drinking water maximum contaminant level criteria from the SDWA for determining cleanup standards for groundwater or the human health and aquatic life criteria developed in accordance with Section 303 of the CWA for determining cleanup standards for surface water discharges.

Brownfields Program

The Brownfields Program is an example of a new regulatory initiative that moves away from the command-and-control approach historically taken with environmental regulations. It is an attempt to create a public-private partnership to encourage reuse of properties that may potentially be contaminated with hazardous substances. The Brownfields Law (Small Business Liability Relief and Brownfields Revitalization Act, Public Law 107-118 [2002]), which is an amendment to CERCLA, was passed to provide mechanisms for funding and regulating cleanup of sites that may be less contaminated than those historically regulated under CERCLA. The intent is to encourage cleanup and reuse of old industrial and commercial sites which otherwise might never be redeveloped and to discourage the development of greenfield sites. The new law contains various provisions designed to reach that goal, including:

- funding for brownfield assessment and cleanups;
- definition of brownfield sites;
- identification of entities eligible for funding, including states, tribes, local governments, land clearance authorities, regional councils, redevelopment agencies, and other quasi-government entities;
- establishment of a program to provide training research and technical assistance to facilitate assessment and cleanup;
- provision of liability protection to adjacent property owners and bona fide purchasers of property;
- protection of innocent landowners;
- provision of funding for state response programs; and
- requirement of deferral of NPL listing if state or other party is cleaning up a site under a state program or if the state is pursuing a cleanup agreement.

Brownfield redevelopment regulation is relatively new at the federal level. However, it may develop into an area which could potentially make extensive use of bioremediation technology.

The law is intended to cover sites that, while contaminated by hazardous substances, are of relatively low risk. Furthermore, the program is intended to cover sites for which there is no viable responsible party and which will be assessed, investigated, or cleaned up by a person who is not liable for cleaning up the site. Because of the reduced up-front liability and greater flexibility for the schedules and cleanup standards, it is expected that this statute will encourage more innovative, cost-effective remediation strategies, which will likely include bioremediation strategies.

RCRA

The Resource Conservation and Recovery Act (RCRA) (42 U.S.C. §§6901 et seq. [2000]), enacted in 1976, implemented a comprehensive program to regulate the treatment, storage, and disposal of solid and hazardous waste within the United States. When enacted, the goals of RCRA were as follows:

- to protect human health and the environment from the hazards posed by waste disposal;
- to conserve energy and natural resources through waste recycling and recovery;
- to reduce or eliminate, as expeditiously as possible, the amount of waste generated, including hazardous waste; and
- to ensure that wastes are managed in a manner that is protective of human health and the environment.

RCRA has been significantly amended several times in order to handle evolving waste management needs. Currently, RCRA consists of 10 different subtitles (Subtitles A to J). The seven remaining subtitles outline general provisions; authorities of the administrator; duties of the Secretary of Commerce; federal responsibilities; miscellaneous provisions; research, development, demonstration, and information; and standards for medical-waste tracking. The three major subtitles contain the framework for the three major regulatory programs that comprise RCRA. The three major programs are hazardous waste management (Subtitle C), solid waste management (Subtitle D), and underground storage tank regulation (Subtitle I). These programs are discussed in more detail below.

RCRA SUBTITLE C—HAZARDOUS WASTE MANAGEMENT

Subtitle C of RCRA is intended to ensure that hazardous waste is managed safely. The federal government created what has been termed a "cradle-to-grave" system which tracks and regulates hazardous waste. Since the law was first passed, the EPA has developed a comprehensive regulatory framework for identifying, generating, transporting, storing, and disposing of hazardous wastes.

The Subtitle C regulations were developed to be prevention oriented, in other words, to prevent the release of hazardous wastes and constituents through a comprehensive and conservative set of management requirements. These management regulations, while arguably appropriate for tracking and regulating the ongoing generation of hazardous materials, are not intended to regulate the cleanup and remediation of previous releases of hazardous materials. These types of activities require a response-oriented program, such as that developed under CERCLA. However, many remediation activities do not fall under the CERCLA program. Historically, an RCRA permit was required before hazardous waste, including hazardous remediation wastes, could be treated, stored, or disposed of at a site. Obtaining a RCRA treatment, storage, or disposal permit is a lengthy, complex, and expensive procedure. Furthermore, many of the requirements associated with such a permit are not appropriate for a short-term remediation project.

As a result, the EPA has developed, within RCRA, a cleanup program. This program is intended for facilites which need an RCRA permit only to treat, store, or dispose of remediation wastes (remediation-only facilities). The Hazardous Remediation Waste Management Regulations were published in final form on 30 November 1998 (9a). The remediation waste management regulations created a new type of RCRA permit, the Remedial Action Plan (RAP) permit. A RAP is the mechanism under which on-site bioremediation of hazardous materials would be approved and permitted. Under 40 CFR §271.110, persons wishing to obtain a permit for remediation would be required to submit an application which provides:

> Sec. 270.110 (3)—A description of the processes you will use to treat, store, or dispose of this waste including technologies, handling systems, design and operating parameters you will use to treat hazardous remediation wastes before disposing of them according to the LDR standards of part 268 of this chapter, as applicable.

In order to perform an in situ remediation, it would be necessary to treat the soils to come into compliance with the Land Disposal Restrictions identified in 40 CFR §268.49.

Clearly, RCRA and CERCLA are key regulatory programs that impact the development of bioremediation (Box 3.1).

RCRA SUBTITLE D—SOLID WASTE PROGRAM

Subtitle D of RCRA covers solid wastes not covered under Subtitle C. RCRA defines solid waste as:

- garbage;
- refuse;
- sludges from waste treatment plants, water supply treatment plants, or pollution control facilities;
- nonhazardous industrial wastes; and
- other discarded materials, including solid, semisolid, liquid, or contained gaseous materials resulting from industrial, commercial, mining, agricultural, and community activities.

Solid wastes are regulated predominately by state and local governments. The major regulation under Subtitle D addresses how disposal facilities should be designed and operated.

In general, the provisions of Subtitle D, with one exception, are not of concern in the biotechnology field. The exception is for bioreactor landfills. A bioreactor landfill utilizes enhanced microbial processes to transform and degrade organic wastes in municipal waste landfills. The EPA is currently collecting information on the advantages and disadvantages of bioreactor landfills to identify specific bioreactor standards and recommend operating parameters. This could likely lead to new or modified regulations under Subtitle D.

RCRA SUBTITLE I—UNDERGROUND STORAGE TANK REGULATION

Possibly the largest source of widespread groundwater contamination in the United States is leaking underground storage tanks. There are approximately 700,000 federally regulated underground storage tanks in use that store petroleum or hazardous substances. The vast majority of these tanks are used to store petroleum products at retail establishments. The threat of groundwater contamination from these tanks, many of which were constructed of bare, unprotected steel, prompted Congress to pass a law which required the EPA to develop a comprehensive regulatory program for underground storage tanks holding petroleum or regulated hazardous substances. This law was enacted as part of RCRA. The underground storage tank provisions were incorporated as Subtitle I to RCRA.

The regulatory program set up by the EPA includes the following elements:

- design standards for new tanks;

BOX 3.1
RCRA, CERCLA, and Bioremediation

Many environmental regulations, including RCRA and CERCLA, have driven the development of bioremediation activities; these have been discussed by Timian and Connolly (17), who point to both the incentives and the disincentives provided by RCRA and CERCLA. They point out that since the passage of RCRA and CERCLA, industry has had an enormous incentive to properly treat and dispose of its hazardous wastes, but that even though RCRA and CERCLA require treatment of hazardous wastes and the cleanup of contaminated sites, these statutes also have standards which discourage the use of bioremediation. For example, they highlight that RCRA and CERCLA both require the use of the "best demonstrated available technology" for treatment and cleanup, which creates artificially high standards which cannot be reached with biological technologies. Nevertheless, the rapid growth of the bioremediation industry has been fueled by these regulations which have driven up the costs of traditional disposal methods.

In order to comply with RCRA and CERCLA regulations, a number of government organizations have implemented remediation programs, many of which employ bioremediation. For example, the U.S. Department of Energy (DOE) Savannah River Site (SRS) past disposal practices have resulted in soil and groundwater contamination. The U.S. EPA, South Carolina Department of Health and Environmental Control, and DOE are addressing these releases under an RCRA permit and CERCLA 120 Federal Facility Agreement. On 21 December 1989, SRS was included on the NPL. The inclusion created a need to integrate the established RFI program with CERCLA requirements to provide for a focused environmental program. In accordance with Section 120 of CERCLA 42 U.S.C. Section 9620, U.S. DOE has negotiated a Federal Facilities Agreement (FFA) with the U.S. EPA and the South Carolina Department of Health and Environmental Control, to coordinate remedial activities at SRS into one comprehensive strategy which fulfills these dual regulatory requirements. As illustrated in Box Fig. 3.1.1, many different approaches, including the use of bioremediation, are being carried out to meet the RCRA and CERCLA regulatory requirements.

The Jet Propulsion Laboratory (JPL) in Pasadena, Calif., is another federal site where activities have resulted in environmental contamination that must be remediated in accordance with CERCLA. The JPL site is organized into several operational units to deal with different conditions and risk. Operable Unit 2 (OU-2) covers all soil located beneath the JPL facility and above the groundwater table (also known as vadose zone soil). OU-1 includes all on-facility groundwater, and OU-3 includes all off-facility groundwater. The ROD included the identification of hazardous chemicals detected in the groundwater at the JPL facility; these included perchlorate and volatile organic compounds (VOCs) such as carbon tetrachloride, Freon 113, trichloroethene, and dichloroethene. The full-scale groundwater remedy for OU-3 (off-facility) groundwater may involve pumping groundwater from a network of wells above ground and into a treatment system. The primary method of treatment under consideration for the VOCs in groundwater is granular activated carbon. This process is already being used at municipal and private wells. The methods under consideration for treating perchlorate at off-facility wells include three types of ion exchange. Several options are being considered for OU-1 (on-facility) groundwater. The remedy may involve pumping groundwater from a network of wells above ground and into a treatment system, or it could involve treatment of the groundwater in place. Above-ground treatment options in-

BOX 3.1 continues

BOX 3.1
RCRA, CERCLA, and Bioremediation *(continued)*

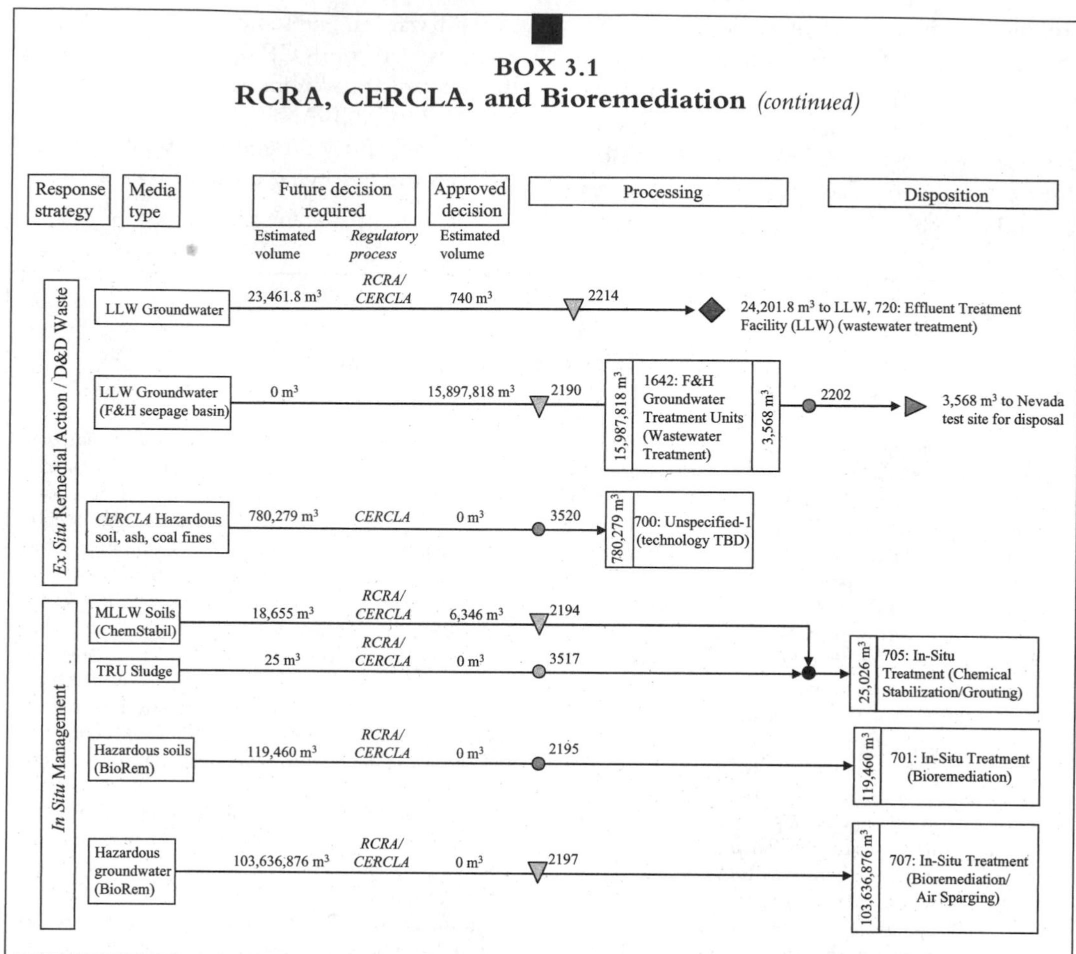

BOX FIGURE 3.1.1 Some remedial technologies at the U.S. DOE SRS. It is noteworthy that the bioremediation options at this site are predominantly in situ. Although limited in number compared to the total remedial technologies being considered, the bioremediation options are to treat huge volumes.

clude ion exchange or biological treatment. The biological treatment method uses microorganisms to breakdown the perchlorate into chloride and oxygen. Another option is treatment of the groundwater in place through nutrient and electron donor addition to promote the breakdown of perchlorate and VOCs by microbes. The National Aeronautics and Space Administration (NASA) has completed several pilot tests to confirm that the above treatment technologies will work at the JPL site. NASA is currently working on testing enhanced bioremediation in groundwater at the NASA JPL.

- upgrade requirements for existing tanks;
- notification of the installation of new tanks to the state or local agency or department designated to receive such notice;
- installation of new controls and operation and maintenance requirements;
- installation of leak detection systems;
- release reporting requirements; and
- release response and corrective action plans.

In the event of a release of petroleum or hazardous substances from an underground tank, the owner and/or operator of the tank must take the necessary steps to correct the problem. The following steps are required in the event of a release:

- 24-h notification of release,
- initial abatement measures,
- initial site characterization,
- free product removal,
- soil and groundwater investigation, and
- preparation of a corrective action plan and implementation of the approved plan.

In certain circumstances, the use of bioremediation technology could be appropriate as the selected corrective action. This would have be evaluated in light of the site-specific conditions in place.

OPA and National Contingency Plan (NCP). The Oil Pollution Act of 1990 (OPA) (33 U.S.C. §§2701 et seq. [2000]) amended the FWPCA to provide enhanced ability to prevent and respond to catastrophic oil spills. This statute was passed in response to the 1989 *Exxon Valdez* spill, which spurred both legislation aimed at preventing future spills and research into ways to remediate spills. One of the side effects of the *Exxon Valdez* accident is that it allowed the U.S. EPA and researchers from other organizations to perform a large-scale test of in situ bioremediation of oil. In that case, the addition of fertilizer was shown to significantly increase the rate of oil biodegradation (1).

The OPA contained provisions that expanded the federal government's ability to respond to large-scale oil spills and created a trust fund to finance response efforts. The OPA also modified the NCP (40 CFR §300 [2003]). Subpart J of the NCP (40 CFR §§300.900 to 300.920) contains the provisions for use of dispersants and other chemicals in response to a spill. The list of compounds which are considered for possible use include "dispersants, surface washing agents, surface collecting agents, bioremediation agents, or miscellaneous oil spill control agents." In order for products, including bioremediation agents, to be utilized in a spill response, the products must be authorized and included on the NCP Product Schedule. For products to become authorized and included on the list, the manufacturer must submit an application to the EPA in accordance with 40 CFR §300.920.

CAA

The Clean Air Act (CAA) (42 U.S.C. §§7401 et seq. [2000]) is the comprehensive federal law that regulates air emissions from area, stationary, and mobile sources. The Act has various aspects that have been significantly modified since it was first promulgated in 1970, with the most significant changes occurring as a result of the 1990 amendments to the CAA. A detailed description of the CAA and its requirements is beyond the scope of this chapter. There are certain aspects of the Act which could impact bioremediation activities and which should be highlighted.

The current version of the CAA is composed of 11 titles:

Title I, Attainment and Maintenance of National Ambient Air Quality Standards;

Title II, Mobile Sources;

Title III, Hazardous Air Pollutants;

Title IV, Acid Deposition Control;

Title V, Permits;

Title VI, Stratospheric Ozone Protection;

Title VII, Enforcement;

Title VIII, Miscellaneous;

Title IX, Clean Air Research;

Title X, Disadvantaged Business Concerns; and

Title XI, Clean Air Employment Transition Assistance

Bioremediation activities under the CAA can be either a source of emissions or a treatment technology utilized to control emissions from an emissions source.

TSCA

The Toxics Substances Control Act (TSCA) (15 U.S.C. §§2601 et seq. [2000]) is the only federal law that impacts bioremediation activities independently from the proposed use of the bioremediation activity. TSCA was enacted by Congress in 1976 in order to create a system to track industrial chemicals used in or imported into the United States. Chemicals that potentially pose an unreasonable environmental or human health risk may be banned from manufacturing or importation. Once-common chemicals that are currently banned under TSCA include asbestos compounds and polychlorinated biphenyl oils.

In 1984, the U.S. EPA issued a proposed policy statement that defined "new chemical substances" under TSCA to include living organisms and microorganisms (9b). This policy statement was officially adopted in 1986. Over the next 11 years, the EPA went through a rulemaking process to formally implement the policies and issued the final rule regarding approval of microorganisms in 1997 (9c). Under the Microbial Products of Biotechnology Rule, the EPA has set up a screening program for "new" microorganisms used commercially for such purposes as production of industrial enzymes and other specialty chemicals, agricultural practices (e.g., biofertilizers), and breakdown of chemical pollutants in the environment. New microorganisms are those microorganisms formed by combining genetic material from organisms in different genera (intergeneric). Microorganisms that are not intergeneric would not be considered new and would not be subject to TSA requirements.

Persons intending to use intergeneric microorganisms for commercial purposes in the United States must submit a Microbial Commercial Activity Notice (MCAN) to the EPA at least 90 days before such use. Within that 90-day period, the EPA is to review the submission in order to determine whether the microorganism presents an unreasonable risk to human health or the environment. Upon approval, or upon expiration of the MCAN review period, the submitter may begin manufacture or import of the microorganism. Expiration of the review period without approval does not constitute approval or certification of the new microorganism and does not mean that the EPA may not take regulatory action against the microorganism in the future.

The rule also addresses intergeneric microorganisms used in research and development for commercial purposes. It creates a mechanism for reporting on testing of new microorganisms in the environment called a TSCA Experimental Release Application (TERA). A TERA would be submitted to the EPA at least 60 days prior to initiation of such field trials.

The rule contains certain exemptions, including those for research and development activities and test marketing and general exemptions for certain specific microorganisms.

State Regulations

It is not possible to identify all the state regulations that could potentially impact a bioremediation project. The fact that a federal environmental program is in place does not limit the ability of a state to impose additional and more restrictive requirements. As an example, under the CWA, it is specifically stated that "Except as expressly provided in this chapter, nothing in this chapter shall (1) preclude or deny the right of any State or political subdivision thereof or interstate agency to adopt or enforce (A) any standard or limitation respecting discharges of pollutants, or (B) any requirement respecting control or abatement of pollution . . ." (33 U.S.C. 1370 [2000]). The only limitation is that the state requirements can be no less stringent that those enacted in the CWA.

In addition to retaining the right to impose additional or more stringent regulations in those areas where the federal government has enacted legislation, the states are free to impose regulation in other environmental areas. Most states have a variety of environmental regulations that either predate the federal regulation in that area or were enacted in order to address a problem or concern unique to the citizens of that state. These regulations may or may not impact a proposed bioremediation activity, but each project must be evaluated to ensure that it complies with all state regulations, in addition to the federal regulations.

Regulatory and Liability Concerns Associated with Bioremediation Activities

The use of bioremediation treatment technologies in place of more conventional physical or chemical treatment technologies raises significant regulatory and compliance issues, depending upon the application selected. There are also significant risk and liability issues associated with use of this technology that might not be present with the use of more conventional technologies. It is important to be aware of this during the evaluation phase of any project.

Most environmental regulations have compliance standards. For example, discharges to the waters of the United States are required to comply with effluent limitations based upon either water quality or technology-based standards. Remediation activities under CERCLA will be based upon either compliance with standards, such as the maximum contaminant levels, or compliance with some risk standard based upon harm. If the project or system is not able to meet the limitation or standard, the facility will either be in violation or, in the case of a remediation, will not be approved as a remediation mechanism.

Because of the compliance requirements, a treatment is typically chosen to ensure, with a high degree of confidence, that the technology will meet the necessary limit or standard. Selection of a process that cannot provide that degree of confidence may increase the costs and liability to the company. That is why many companies typically select the option that provides the highest degree of certainty and the lowest degree of future risk, even if the initial costs may be higher.

As an example, suppose a facility has an area of contaminated soil. The company could choose to excavate the soil and send it to an approved landfill. As an alternative, the company could attempt an in situ remediation. Excavation and disposal would remove all material from the site and essentially guarantee no future liability issues. The results from a bioremediation project are not as clear-cut. There are a variety of potential outcomes. The treatment may not work, and the material would still be on-site. The treatment might work but might not reduce pollutant levels enough to end liability, or the biological degradation products might create a secondary source of liability. If the main goal of the project is to ensure an end to all future liability for that contamination, a company might select excavation and disposal even if the up-front costs are significantly less for bioremediation.

In many cases, the decision to implement a bioremediation project will be more of a corporate decision than a technical or legal decision. The affected party will have to make a determination among its possible options, based upon the best available information regarding expected effectiveness, cost and benefits, regulatory framework, and politics and public opinion, whether to implement a conventional treatment-remediation technology or a more innovative technology, such as bioremediation.

REGULATORY DRIVERS AND BARRIERS TO IMPLEMENTATION OF BIOREMEDIATION IN THE UNITED KINGDOM

In this section, we review the regulatory framework in the United Kingdom for assessment and remediation of contamination and controls on such activity. An exhaustive review of the entire legal framework is not presented; rather, an overview of key elements and their impact on the implementation of innovative treatment technologies such as bioremediation is given. The full extent of the number of sites affected by contamination throughout the United Kingdom is difficult to estimate. An indicative figure of as many as 100,000 sites has been given for England and Wales, with between 5 and 20% of those having an impact on human health or the wider environment (6).

The United Kingdom regulatory framework is complex, being driven by European Commission Directives. These are transposed into legislation either as primary legislation as an Act of Parliament or through secondary legislation as regulations. Although differences in implementation exist between Scotland, England, Wales, and Northern Ireland, there is generally uniformity and coordination throughout the United Kingdom. The key agencies involved are the Scottish Environmental Protection Agency (SEPA), the Environmental Agency (EA) in England and Wales, and the Environment and Heritage Service (EHS) in Northern Ireland. These agencies cooperate at the regional and local scale with local authorities and district councils.

Legislation may be considered to be a driver for the treatment of contaminated land and, as more sustainable solutions are increasingly sought, as a driver for the use of "innovative" treatment technologies such as bioremediation. A typical example of this that is often presented is the perceived move away from excavation and disposal of contaminated soils to landfill (otherwise known as "dig and dump") towards on-site or off-site treatment. However, in a recent survey of remediation undertaken between 1996 and 1999 in England and Wales, bioremediation was used at less than 3% of sites compared to 76% for excavation and disposal to landfill (15). Conversely, the licensing and control of remediation activity may be considered to be a barrier to more widespread implementation of innovative technologies.

During the 1990s, the demand for housing and recognition of the need to preserve the countryside and protect greenfield sites from development became a significant driver of the remediation market. Recent United Kingdom government policy aims to achieve the building of 60% of all new housing in England on previously developed (brownfield) land or reuse of existing buildings (5). Planning Policy Guidance Note 3 (12) highlights the fact that the 60% target is a national one and regions with significant amounts of brownfield land should aspire to higher rates of reuse.

The government-established Urban Task Force examined the broader issues of urban regeneration, including redevelopment of brownfield sites, and recommended a target of bringing all contaminated land back into reuse by 2030 (4). The same group also recommended simplification of the permit system for undertaking remediation. An increased awareness of potential environmental liabilities may also have contributed to site owners being proactive with respect to remediation, but it is likely that the economics of matching the increasing demand for housing on brownfield sites has been uppermost.

In the United Kingdom, bioremediation is still considered an innovative technology, and a lack of confidence from stakeholders and site owners has been viewed as a barrier to more widespread implementation. Nevertheless, despite the dominance of dig-and-dump, recent predictions have estimated that the bioremediation market in the United Kingdom will grow by 36% from £36 million (U.S. $66 million) to £49 million ($89 million) per annum and will be one of the most rapidly growing remediation technologies during the period 2002 to 2006 (11).

Part IIA—Contaminated Land Regimen

In practice, contaminated land is most often handled under planning in the context of redevelopment of brownfield sites. However, the key legislation for dealing with historically contaminated sites is Section 57 of the Environment Act 1995, which inserted new elements to Part IIA of the Environmental Protection Act 1990. Implemented in 2000, and generally known as Part IIA, the regimen requires a risk-based approach to the assessment and remediation of contaminated sites based on their current use. Part IIA is implemented through regulations and statutory guidance where the "polluter pays" principle applies. The polluter-pays principle is also being enshrined in the forthcoming Environmental Liability Directive from the European Commission, anticipated to come into force in member states in 2007.

Part IIA also adopts a suitable-for-use approach, requiring that risks be dealt with on an individual-site basis, and therefore no uniform standards define what is or is not classed as contaminated land or what constitutes acceptable levels of contamination. One of the key provisions of Part IIA is a statutory definition of contaminated land:

> any land which appears to the local authority in whose area it is situated to be in such a condition, by reason of substances in, on or under the land that:
> - significant harm is being caused or there is a significant possibility of such harm being caused; or
> - pollution of controlled waters is being, or is likely to be, caused.

Controlled waters remain those defined in Secton 30A of the Control of Pollution Act 1974. These include territorial and coastal waters, rivers, lochs, lakes, ponds, and groundwater.

Another key provision of Part IIA was the requirement for local authorities to develop and implement an inspection strategy to identify contaminated land in their areas. As a result of this investigation, the authorities then have a requirement to ensure that remediation takes place. This is undertaken as a minimum to the effect that the site no longer meets the statutory definition of contaminated land. An enforcement notice (remediation notice) can be served if required, although voluntary remediation is encouraged. Under certain circumstances, the management of "special sites" requires consultation with and/or transfer to control of the key regulator SEPA, EA, or EHS. Sites designated as special cover a variety of sensitive situations, including those where pollution of controlled waters is occurring, nuclear and military sites, and land contaminated with waste acid tars.

From the definition of contaminated land under the Part IIA regimen, land that is contaminated may not meet the statutory definition. Before a site can be classified as contaminated land, the presence of a significant pollutant linkage must be demonstrated. Such a linkage must show that the contamination or source can migrate via a pathway and reach a receptor. Unless the source-pathway-receptor linkage is made, the land should not be identified as contaminated land. Receptors include humans; certain ecological systems (or living organisms forming part of such systems) that are protected, e.g., nature reserves; and also property, including crops, timber, produce, livestock, and buildings.

Historically, in the United Kingdom, a series of trigger values produced in 1983 and later revised in 1987 (10) gave threshold and action values for 18 organic and inorganic contaminants in soils, depending on the planned use of the site. The ethos was that threshold and action levels represented acceptable and unacceptable risks, respectively. However, action values were not produced for the majority of metals, and threshold values often became synonymous with remedial targets.

The ICRCL Guidance Note 59/83 was formally withdrawn in 2002, and the Department for Environment, Food and Rural Affairs (DEFRA), EA, and SEPA have an ongoing program of work to develop a scientific framework for assessing risks to human health from contaminated sites under United Kingdom conditions. The framework is based on toxicological criteria establishing unacceptable levels of human intake for a contaminant derived from soil and an estimation of the human exposure to soil contamination through various routes based on generic land use.

The first Contaminated Land Exposure Assessment (CLEA) model was published in 2002. CLEA is used to derive soil guideline values (SGVs) for the typical land reuse scenarios of residential uses (with and without vegetable growing), allotments, and commercial or industrial uses for a number of contaminants. Like the former Interdepartmental Committee on the Redevelopment of Contaminated Land levels, the SGVs are not remedial targets but intervention values to indicate when soil concentrations above certain levels may present an unacceptable risk to the health of site users and that further investigation and/or remediation is required.

Note that SGVs will never be available for all potential contaminants and that they are generic guidelines. As such, they do not replace the development of site-specific risk assessments incorporating a conceptual model or models identifying sources, pathways, and receptors as discussed in chapter 2. CLEA or another risk-based methodology such as risk-based corrective action would then be used to undertake such assessments. The full CLEA package includes Contaminated Land Reports 7 to 10, which give information on the development of SGVs and other technical aspects of the CLEA model. At the time of writing, these are available only from the EA website (http://www.environment-agency.gov.uk/).

As an instrument to deal with historically contaminated sites meeting the strict criteria for the classification of contaminated land, Part IIA represents a pragmatic approach to tackling the legacy of environmental pollution in the United Kingdom. Part IIA and its regulations and statutory guidance are intended to complement other legislation and the methods of site-specific risk-based assessment used voluntarily. An issue of concern for those undertaking remediation of sites as a voluntary action or as part of redevelopment covered by the planning system is to ensure that a site which may not have been designated as contaminated land under Part IIA cannot become so at a later date. This is possible where a change of use occurs or a new pathway is created.

Planning

The planning system in the United Kingdom controls (re)development and land use. As such, it is the main mechanism controlling reuse of brownfield sites. Whereas the Part IIA regimen applies to current land use, almost all future uses (including sites identified under Part IIA as contaminated land) will be considered under the planning system. Although separate guidance is issued in Scotland, England, Wales, and Northern Ireland, the system may be considered relatively uniform in the main. At the time of writing, new technical advice, "Development on Land Affected by Contamination," is being prepared to replace Planning Policy Guidance Note 23 ("Planning and Pollution Control" [1994]) and will apply to England. Similar updates have already taken place in Scotland, Wales, and Northern Ireland to take into account the implementation of Part IIA.

Contamination, be it actual or potential, is a material consideration in the planning process, and developers are responsible for investigating whether the proposed development, including its occupants, will be affected by contamination or cause contamination to affect neighboring properties or receptors such as water bodies. Local authorities may have identified the land being developed as contaminated land while investigating their area. Otherwise, it is incumbent on the developer to seek advice whether there is a reason to suspect contamination by virtue of former industrial use. The planning section of the local authority is unlikely to request intrusive site investigations or a detailed desk study unless there are grounds to suspect contamination. In keeping with the diverse nature of contamination scenarios, sites are assessed on a case-by-case basis.

Where contamination is identified, the developer will be required to come up with a remediation plan to remove the unacceptable risk. Essentially, these are the same risks as given under Part IIA, with a key difference being that the scope of the control is not limited to "significant" risk or harm. Ultimately, the goal of the remediation plan is the removal of unacceptable risk. One important aspect of the planning process is that it is highly pragmatic and does not try to predict site usage beyond the proposed use and impose stricter remedial targets based on a presupposed receptor or pathway in the future.

The planning system in the United Kingdom therefore has a key role in ensuring that developments are given approval only if due regard has been given to the environmental and, especially, human health risks arising from the change in land use and appropriate remedial measures have been or will be undertaken. By virtue of the numbers of brownfield sites undergoing redevelopment and therefore being considered under the planning system, one may expect there to be an impact on the adoption of innovative technologies. However, the planning system, while not presenting any barriers, does not in itself represent a driver in the same way as Part IIA. Whether this has any impact on the application is debatable, but the likelihood is that it does not. Bioremediation is not a technology that is lacking in benefits and unduly burdened by limitations.

As with so many considerations in contaminated land assessment and remediation, site-specific considerations, especially the nature of contamination as well as economic considerations, will often dictate the potential applicability of one technique over another. For contamination by petroleum hydrocarbons, where bioremediation would be expected to feature as a viable treatment option, for example, at a former petrol station, the key competitor for bioremediation in the United Kingdom has long been excavation and disposal to landfill.

The Landfill Directive

For practitioners and stakeholders involved in the remediation of contaminated land and for the waste management industry in general, the Landfill Directive (99/31/EC) will have a profound impact. The 20 Articles and 3 Annexes apply to all countries of the European Union and were transposed into United Kingdom law in 2001. Key elements of the Landfill Directive are as follows.

- All landfill sites are classified into one of three categories: "hazardous," "non-hazardous," and "inert."
- Higher engineering and management standards will be applied.
- Wastes must be treated prior to landfilling.
- Codisposal of hazardous and nonhazardous wastes is banned.
- Disposal of liquid wastes, certain hazardous and clinical wastes, and tires is banned.
- The amount of biodegradable waste going to landfill will be progressively reduced.

Producers of waste must code materials on the basis of the European Waste Catalogue (EWC 2002), which is subject to periodic review and amendment by the European Commission. Contaminated soils are included in the EWC under code 17 for "Construction and Demolition Wastes":

17 05 03*—soil and stones containing dangerous substances

17 05 04—soil and stones other than those mentioned in 17 05 03

The first, marked with an asterisk, is termed a "mirror entry." These are considered hazardous waste only if dangerous substances are present above threshold concentrations. Certain classifications are "absolute entries," and these are considered to be hazardous waste regardless of any threshold concentrations (7). Classification of soils and other materials from contaminated sites will require "basic characterization" testing to determine their properties. Soils are assessed against hazardous properties H1 to H14 taken from the Hazardous Waste Directive (91/689/EC). These include harmful, toxic, carcinogenic, and ecotoxic properties. Total concentrations of contaminants consisting of worst-case compounds are compared against thresholds for the various risk phrases, e.g., harmful if swallowed. On the basis of this initial assessment, soils can be classified as non-hazardous (including inert) or hazardous for the purposes of disposal to landfill and will in most cases require treatment prior to disposal.

Treatment applies to all wastes with few exceptions and is defined in Article 2 as "the physical, thermal, chemical or biological processes, including sorting, that change the characteristics of the waste in order to reduce its volume or hazardous nature, or facilitate its handling or enhance recovery." Exceptions include inert waste where treatment is not technically feasible or other wastes where the quantity or hazards to human health or the environment would not be reduced by treatment. Waste acceptance criteria, which include a variety of leaching limits and limits on levels of total organic carbon, must then be met. Testing requirements vary according to landfill type and at the time of writing have been published for inert and hazardous wastes. A third category known as stable nonreactive hazardous wastes applies where nonhazardous landfills have specially constructed cells able to receive this type of waste. Note that dilution of contaminated soil with uncontaminated material solely to meet waste acceptance criteria is precluded by Article 5(4) of the Landfill Directive.

From even so brief an overview, it is apparent that the Landfill Directive will have profound implications for the contaminated land industry in the United Kingdom and throughout the European Union. Contaminated sites undergoing redevelopment in the United Kingdom and Europe are often small, for example, the former petrol station being developed for housing. The limitations of time and space on-site and availability of relatively low-cost disposal to widely available licensed landfill sites have historically made this the most favorable option. The requirement for treatment of most wastes, including contaminated soils, prior to disposal coupled with increasing landfill fees and haulage costs will undoubtedly drive the market for both in situ and ex situ treatments, including bioremediation.

More detailed site investigations are likely to be employed, ensuring a greater understanding of the extent and degree of contamination on-site to allow as full a range of remediation options as possible to be considered. The most favorable outcome will be to treat and reuse soils on-site where possible and avoid disposal even where treated soils meet inert-waste criteria. Although it is difficult to predict how the contaminated land community will respond to the impact of the directive, what is certain is that unless alternative technologies such as bioremediation become more widely accepted and commercially viable, brownfield regeneration activity may be suppressed.

The WFD

The Water Framework Directive (WFD) (2000/60/EC) is a wide-ranging instrument that integrates the management of water in terms of sustainable use, protection, and improvement as well as reducing the effects of droughts and flooding. The WFD adopts the river basin as the unit for management including rivers, lakes, estuaries, coastal waters, and groundwater.

The largest impact on the use of bioremediation from the WFD will be with respect to groundwater. The WFD builds on the earlier 80/68/EEC Directive on protection of groundwater against pollution caused by certain dangerous substances. This aimed to prevent direct discharge of high-priority pollutants (list I) and to control the discharge of other pollutants (list II). Article 22(2) of the WFD sets out to repeal the earlier 80/68/EEC Directive in 2013. Altogether, seven directives will be repealed by the WFD, representing a significant rationalization of water legislation by the European Community.

Rather than producing lists of acceptable concentrations for a wide range of pollutants, there is a move towards a presumption that groundwater should not be polluted at all. A combination of preventing discharges to groundwater and monitoring to detect changes in chemical status is being applied. There is also a requirement to enhance and restore the chemical status of aquifers and to reverse any "significant and upward trend" in concentrations of pollutants in groundwater. Quantitative status is also considered, i.e., groundwater levels and the balance between any abstraction and recharge. There are only two classes of groundwater status, either good or poor.

There is a long timescale for implementation of the WFD, and improvements in environmental objectives in terms of achievement of good groundwater quantitative and chemical status are not expected until 2015. Where priority substances are present in contaminated soils, there will be an additional demand on any remediation strategy to achieve zero concentration in groundwater. The WFD is likely to be a driver for remediation overall, especially to meet the groundwater requirements. Long-term bioremediation processes such as enhanced monitored natural attenuation of groundwater are likely to be among the only cost-effective solutions to achieve gradual improvements in groundwater quality at many sites.

PPC

The Pollution Prevention and Control (PPC) regimen implements the 96/61/EC Directive on Integrated Pollution Prevention and Control (IPPC). The regimen is aimed at reducing the environmental impact of industrial activities by a system of permitting, controlling emissions to air, land, and water (including discharge to sewers). The regulators (SEPA, EA, and EHS) control processes where a greater potential to cause pollution exists; otherwise, local authorities manage the system.

Underlying principles for the environmental assessment of industrial sites include the use of best available techniques to prevent pollution, conservation of energy, minimization of waste, and limitation of environmental impact of accidents (3). Of most relevance to the (bio)-remediation industry is the strict approach to monitoring and controlling any potential deterioration in the environmental quality of the site.

Operators are required to undertake a baseline assessment and produce a site report as part of a permit application. The framework for the site report follows the source-pathway-receptor model discussed for the Part IIA regimen. This raises the potential for identification of contaminated land under the Part IIA regimen and may well have been carried out as part of the planning process. If, during operation, there is a breach of the conditions of the PPC permit resulting in pollution, the regulator can issue an enforcement notice to remedy the situation. When an operator ceases the permitted work, the operator must apply to the regulator to surrender the permit. This application must include a site report that identifies any changes from the original. The regulator must be satisfied that no deterioration has taken place, that the site has been returned to a "satisfactory state," and that appropriate steps have been taken to avoid any pollution risk. Unlike the suitable-for-use approach described for the Part IIA regimen, "restoration under IPPC is not constrained by the future use of the land" (3), in keeping with the preventive nature of the regime.

The scale of the PPC regimen is new, and the scale of its impact as a driver for bioremediation is difficult to predict. However, it may be assumed that along with other legislative drivers requiring remediation of sites and the profound implications of the Landfill Directive, increased opportunities to apply bioremediation are likely. The use of in situ techniques may be favored where avoiding disruption to site infrastructure is paramount.

Control of Remediation Activities

It cannot be assumed that remediation can bring only a positive outcome and both in situ and ex situ remedial treatments may result in a negative environmental impact. Almost all site remediation activities in the United Kingdom are subject to control by the environmental regulators (SEPA, EA, and EHS) with respect to waste management, water and air pollution, and health and safety. The current controls were introduced with Part IIA of the Environmental Protection Act 1990 with the key regulations included in the Waste Management Licensing Regulations 1994. A waste management license is required for (bio)-remediation, as it is considered a waste treatment activity. Site licenses apply to the actual location, whereas mobile plant licenses apply to the process, not the site, and allow a plant to be moved from one location to another. Each new location requires a site-specific working plan ensuring that environmental risk can be controlled. Control of remedial activity has in the past few years been the subject of debate and lobbying by industry bodies such as the Environmental Industries Commission to increase the flexibility of licensing. The timescale required to obtain the necessary license may be viewed as being out of step with commercial expectations from developers working under tight deadlines during brownfield redevelopment projects.

In their report *The Remediation Permit—Towards a Single Regeneration License*, the Remediation Permit Working Group of the Urban Task Force concluded that the current system imposes unjustified costs and discourages redevelopment of brownfield sites, as well as failing to provide sufficient incentive to alternative technologies. The group recommended a series of improved guidelines and regulatory procedures offering more exemptions to streamline the timescales and costs of licensing and monitoring remediation activity. At the time of writing, DEFRA is considering responses from a recent consultation. Undoubtedly, the control of (bio)remediation activity has been recognized by government, regulators, and the remediation industry as being less than optimal. There is still much work to be done in the United Kingdom to implement recommendations from review and consultations and provide a system of control that can operate in partnership with the legislative drivers.

Regulatory Oversight of Environmental Releases of Genetically Modified Microorganisms

The Organization for Economic Cooperation and Development (OECD) has had an active role in developing guidance for the safe use of recombinant DNA technology, identifying scientific principles and criteria for the safe use of genetically modified organisms (GMOs) in industry, agriculture, and the environment (13, 14, 16). The OECD focused on the development of a rational and flexible approach to the evaluation of safety that could be used throughout the world. A landmark report by the OECD entitled *Recombinant DNA Safety Considerations*, now known as the "Blue Book," set out the first international scientific principles and criteria for the safe handling of GMOs outside of contained laboratory conditions (13). Its criteria and principles have been adopted throughout Europe for good large-scale industrial practices. The OECD continues to issue guidance for the safe use of recombinant DNA technology.

With regard to deliberate environmental releases, such as those that would be used for bioremediation applications, the OECD held that it is important to evaluate recombinant DNA organisms for potential risk, prior to applications in the environment. However, because the OECD was unable to develop general international guidelines governing deliberate releases of genetically modified microorganisms into the environment, it recommended that independent reviews of potential risks be conducted on a case-by-case basis prior to any field application. The report defined case-by-case evaluation as a review of the proposed application against assessment criteria which were relevant to the particular proposal, e.g., the properties of the organism proposed for release and its probable environmental fate. The report held that development of organisms for environmental applications should be conducted in a stepwise fashion, moving, where appropriate, from the laboratory to the growth chamber, to limited field testing, and finally to large-scale field testing.

In 1992, the OECD issued a follow-up report that dealt specifically with the safety of small-scale field tests with GMOs (14). This report laid out a set of principles, called good development principles, for designing safe experiments at the stage of basic or initial field research. These good development principles set out the characteristics of microorganisms to be considered to include dispersal, survival, and multiplication; interactions with other species and/or biological systems; potential for gene transfer; and the mode of action, persistence, and degradation of any newly acquired toxic compound. They also set out characteristics of the research site to include important ecological and/or environmental considerations relative to safety in the specific geographical location, climatic conditions, size, and an appropriate geographical location in relation to proximity to specific biota that could be affected.

The OECD reports have guided the regulatory frameworks employed throughout Europe. The European Union adopted specific legislation governing the deliberate release of genetically engineered microorganisms into the environment that requires case-by-case evaluation (Europa, 24 July 2001 Press Release; http://europa.eu.int/rapid). The main legislation which authorizes experimental releases of GMOs in the European Community is Directive 90/220/EEC. European Directive 90/220/EEC, which governs the release and marketing of GMOs in the European Union, recognizes two broad categories of release of GMOs into the environment: part B, releases for small-scale research trials; and part C, releases for commercial purposes (2). An updated Directive 2001/18/EC on the deliberate release of GMOs was adopted by the European Parliament and the Council of Ministers in 2001 and entered into force in 2002. Directive 90/220/EEC put in place a step-by step approval process on a case-by-case assessment of the risks to human health, animal health, and the environment before any GMO or product consisting of or containing GMOs can be released into the environment (8).

The United Kingdom has also adopted parallel regulations that involve public consultations as part of the implementation strategy. In Great Britain, Directive 90/220 has been implemented by Part VI of the Environmental Protection Act 1990 and the Genetically Modified Organisms (Deliberate Release) Regulations 1992, as amended in 1995 and 1997, and in Northern Ireland there is separate but equivalent legislation. The release or marketing of GMOs cannot take place without the explicit consent of the regulatory authorities (2). In the United Kingdom, DEFRA and the devolved administrations are responsible for implementing the Directive in England, Scotland, Wales, and Northern Ireland. In the United Kingdom, the Advisory Committee on Releases to the Environment (ACRE [http://www.defra.gov.uk/environment/acre/]) is an independent advisory committee composed of leading scientists that functions to give statutory advice to ministers and local administrations on the risks to human health and the environment from the release and marketing of GMOs. The entire regulatory process is underpinned by a detailed environmental risk assessment, prepared by the applicant, which examines and evaluates any possible harmful consequences of releasing a particular GMO (2). Government scientists in the Joint Regulatory Authority based at DEFRA review this assessment and seek the advice of an expert committee. In assessing applications, every possible precaution is taken to ensure that human health and the environment are protected. Only if the risks are considered to be very low will the release be allowed to proceed. To gain approval for a research trial, the application has to include information on the nature of the GMO, how it has been modified, the precise nature of the research program proposed, where it will be released, and how the release will be monitored. The applicant must also supply information necessary for evaluating foreseeable risks, whether they are immediate or delayed, from the release of the GMO. The application dossier is reviewed by scientists in the Joint Regulatory Authority and is forwarded to experts in other government departments, devolved administrations, and statutory conservation bodies. All applications are scrutinized by ACRE, an independent scientific committee whose members include leading academic scientists.

CONCLUDING REMARKS

The regulatory framework in the United States, United Kingdom, and elsewhere is complex and evolving with interactions between different regulations and regimes. When contaminated sites undergo remediation, there is a clear requirement to protect the environment while ensuring that risks to human health and the environment are minimized. Permitting, licensing, and regulation have a crucial role in controlling the remediation process. Often this leads to the selection of traditional remediation methods, such as removal of contaminated soil to landfills. However, a balance is required to ensure that brownfield land is developed preferentially over virgin greenfield sites in as sustainable a manner as possible. It is likely that some conflict between a desire for redevelopment of brownfield sites and protection of the environment during remediation activities will always exist or be perceived to exist by those subject to control. Opportunities for bioremediation in its widest context, including constructed wetlands, may find increasing application in dealing with surface runoff from roads and new developments as sustainable urban drainage requirements and growing awareness of diffuse pollution come to the fore.

Overall, the redevelopment of brownfield sites may prove to be the biggest driver of remediation and therefore bioremediation activity. In the United Kingdom, the Landfill Directive will have widespread implications for the remediation industry with the requirement for treatment before disposal. Bioremediation of soils or waters, whether in situ or ex situ, is clearly not applicable to every site or for all contaminants. However, if this is to be the end of "dig and dump," at least for the bulk of contaminated soils, the permitting of bioremediation processes must be flexible enough to ensure broader application where it can offer benefits in terms of sustainability and cost-effectiveness. The use of biotechnology can help implement regulatory goals. To do so, it must meet the requirements of the regulatory framework and the oversight at the national, state, and local levels. Despite the complexities of the regulatory frameworks, a knowledge of the basic principles of the legal requirements is essential for understanding how and when bioremediation can be applied for the restoration of contaminated soils and waters.

REFERENCES

1. **Bragg, J. R., R. C. Prince, E. J. Harner, and R. M. Atlas.** 1994. Effectiveness of bioremediation for the *Exxon Valdez* oil spill. *Nature* **368:** 413–418.
2. **Department for Environment, Food and Rural Affairs.** 2001. Genetically modified organisms: the regulatory process. http://www.defra.gov.uk/environment/gm/.
3. **Department for Environment, Food and Rural Affairs.** 2004. *Integrated Pollution Prevention and Control—A Practical Guide*, 3rd ed. The Stationery Office Ltd., London, United Kingdom.
4. **Department of Environment, Transport and Regions.** 1999. *Towards an Urban Renaissance.* The Stationery Office Ltd., London, United Kingdom.
5. **Department of Environment, Transport and Regions.** 2000. *Our Towns and Cities: The Future Delivering on Urban Renaissance.* The Stationery Office Ltd., London, United Kingdom.
6. **Environment Agency.** 2000. *The State of the Environment of England & Wales: The Land.* The Stationery Office Ltd., London, United Kingdom.
7. **Environment Agency.** 2003. *Interpretation of the Definition and Classification of Hazardous Waste Technical Guidance WM2.* Environment Agency, London, United Kingdom. http://www.environment-agency.gov.uk/business.
8. **European Council.** 2001. Directive 90/220/EEC, Directive 2001/18/Ec of the European Parliament and of the Council of 12 March 2001 on the deliberate release into the environment of genetically modified organisms and repealing. *Off. J. Eur. Communities.* http://europa.eu.int/eur-lex.
9. **Federal Register.** 2004. Civil monetary penalty inflation adjustment rule. Federal regulation 7121. *Fed. Regist.* **69**.

9a. **Federal Register.** 1998. Hazardous remediation waste management requirements (HWIR-media). *Fed. Regist.* **63:** 65874–65947.

9b. **Federal Register.** 1984. Proposed policy regarding certain microbial products. *Fed. Regist.* **49:** 50886.

9c. **Federal Register.** 1997. Microbial products of biotechnology: final regulation under the Toxic Substances Control Act, final rule. *Fed. Regist.* **62:** 1710.

10. **Interdepartmental Committee on the Redevelopment of Contaminated Land.** 1987. *Guidance on the Assessment and Redevelopment of Contaminated Land.* ICRCL 59/83, 2nd ed. Department of the Environment, London, United Kingdom.
11. **MSI.** 2002. *Data Report: Contaminated Land Treatment: UK.* MSI Marketing Research for Industry Ltd., Chester, United Kingdom.
12. **Office of the Deputy Prime Minister.** 2000. *The Government's Response to the Environment, Transport and Regional Affairs: Seventeenth Report—Housing PPG3.* Cm 4667. Office of the Deputy Prime Minister, London, United Kingdom.

13. **Organization for Economic Development and Cooperation.** 1986. *Recombinant DNA Safety Considerations*. Organization for Economic Development and Cooperation, Paris, France. http://www.oecd.org.
14. **Organization for Economic Development and Cooperation.** 1992. *Safety Considerations for Biotechnology*. Organization for Economic Development and Cooperation, Paris, France. http://www.oecd.org.
15. **Rivett, M. O., J. Petts, B. Butler, and I. Martin.** 2002. Remediation of contaminated land and groundwater: experience in England and Wales. *J. Environ. Manag.* **65:**251–268.
16. **Teso, B.** 1992. International harmonization of safety principles for biotechnology. *In Proceedings of the 2nd International Symposium on The Biosafety Results of Field Tests of Genetically Modified Plants and Microorganisms*, 11 to 14 May, Goslar, Germany.
17. **Timian, S. J., and M. Connolly.** 1996. The regulation and development of bioremediation. *Risk Health Safety Environ.* **7:**279–290. http://www.piercelaw.edu.

MODELING BIOREMEDIATION OF CONTAMINATED GROUNDWATER

Henning Prommer and D. Andrew Barry

4

INTRODUCTION

Over the last two decades, mathematical modeling has become an important tool to assist in analyzing and understanding complex environmental systems. Wherever a multitude of processes, of either a physical, chemical, or biological nature, interact with each other, mathematical modeling provides a rational framework to formulate and integrate knowledge that has been otherwise derived from (i) theoretical work, (ii) fundamental (e.g., laboratory) investigations, and (iii) site-specific experimental work. In the case of subsurface systems, data acquisition is typically very expensive, especially in the field, so data sets are usually sparse. Thus, validation of complex models can be difficult. At the same time, it is the lack of spatially and temporally dense information and the need to fill the gaps between measured data that provide an important driving force for integrated modeling.

In situ processes making use of bioremediation are prime candidates for an integrated modeling approach. Whether microbial activity is responsible for direct breakdown of organic contaminants (such as dissolved petroleum products) or whether it is employed more indirectly to alter geochemical conditions (such that metal precipitation, for example, occurs), it is evident that predictions of the combined biogeochemical-hydrodynamic system become very difficult if isolated aspects of the total problem are considered separately. The purpose of this chapter is to show how such processes can be dealt within a single comprehensive yet realistic framework.

We have previously reviewed modeling of the fate of oxidizable organic contaminants in groundwater (9) and the physical and reactive processes during biodegradation of hydrocarbons in groundwater (50). Here, we provide an introduction and overview of the mathematical/mechanistic descriptions of the important processes governing bioremediation, considering the critical factors of microbial processes (growth and decay of bacteria) and physical processes (advection and dispersion) as they relate to the applicability of bioremediation to the removal of organic pollutants from contaminated groundwater.

Henning Prommer, Department of Earth Sciences, Faculty of Geosciences, University of Utrecht, P.O. Box 80021, 3508 TA Utrecht, The Netherlands, and CSIRO Land and Water, Private Bag No. 5, Wembley WA 6913, Australia. *D. Andrew Barry*, Contaminated Land Assessment and Remediation Research Centre, Institute for Infrastructure and Environment, School of Civil Engineering and Electronics, University of Edinburgh, Edinburgh EH9 3JL, Scotland, United Kingdom.

Bioremediation: Applied Microbial Solutions for Real-World Environmental Cleanup
Edited by Ronald M. Atlas and Jim C. Philp © 2005 ASM Press, Washington, D.C.

ROLE OF MODELING BIOREMEDIATION PROCESSES

Both passive and enhanced (active) in situ bioremediations are cost-effective contaminant cleanup methods compared to other methods such as landfill disposal or incineration (see Table 1.3). However, traditional methods are still, in many cases, the preferred remediation option despite the apparent economic benefits of bioremediation (29). Certainly, one important reason for this is the lack of reliable a priori predictions concerning the feasibility, duration, and cost of bioremediation. Site managers dealing with soil or aquifer contamination are faced with such questions as:

- To what extent will environmentally important receptors down gradient of the source zone be impacted by a contaminant?
- What are the expected average and maximum concentration levels?
- What are the timescales for cleanup to below given limits for different remediation schemes?
- What is the optimal design (in a multiobjective environment) of a particular (active/passive) remediation scheme?
- What is the sensitivity of, say, the duration of the remediation process to changes in physical or biogeochemical conditions?
- What is the probability of failure of the proposed remediation scheme?

The role of modeling in a remediation investigation is more than that of gaining an increased quantitative understanding of the biological system. The process of developing a clearly formulated conceptual model at the beginning of the modeling process (or its ongoing revision) often leads ultimately to a more rigorous scientific understanding for model developers and users. An additional and important benefit of modeling is, of course, its predictive capability. However, such predictions do bear, to a variable degree, uncertainty that originates from the following:

- incomplete hydrogeological and hydrogeochemical site characterization,
- incomplete process understanding, and
- parameter ambiguity due to spatial and/or temporal scaling issues.

Predictive modeling must therefore be viewed with appropriate caution. Given that the modeling framework is sufficiently comprehensive, issues such as parameter uncertainty can be implemented into the framework and directly quantified. The approach also contributes to the broader question of probabilistic risk assessment, which often guides engineering design of remediation schemes.

In general, modeling provides the best means to incorporate observed data into a systematic site investigation or, where data are lacking, to investigate quickly a suite of scenarios that assist in gaining a better understanding of factors dominating the duration and effectiveness of site cleanup. Below, an overview of some of the mathematical descriptions and modeling approaches involved in the simulation of bioremediation problems is provided. For the sake of clarity, we will focus initially on the quantitative description of biochemical processes (as a function of time) in batch-type systems. The following discussion of batch-type modeling uses simple examples to discuss some of the common concepts needed to model microbially mediated biodegradation reactions (9). This discussion of biotic processes will be succeeded by an introduction to some of the fundamentals of modeling physical transport before we move on to biogeochemical transport, in which these biotic and physical processes are combined.

MODELING BIOTIC REACTIVE PROCESSES

In order to be able to design active bioremediation systems and to understand passive bioremediation (natural attenuation), mechanistic descriptions that quantify microbial activity are needed. Since both the rate of microbial growth and the rate of contaminant utilization are highly dependent on the amount of biomass available to catalyze the reactions (38), such models must have the capability to predict both transient and spatial variations in biomass. To model the production of biomass and the related consumption and production of other chemicals, the key steps are (i) to determine the stoichiometry of the biodegradation reactions and, in order to describe the temporal variations, (ii) to formulate the rate expressions for the reaction kinetics. Below, we describe modeling of biodegradation under oxidizing conditions. (Note: for simplicity we do not describe here biodegradation reactions in which the contaminants act as electron acceptors.)

Biodegradation Reactions for Oxidizable Organic Contaminants

The electron flow in microbially mediated redox reactions that mineralize, e.g., petroleum hydrocarbons might be simplified into two (sub)steps. In the first step, which is the oxidation of an organic substrate, electrons are transferred to electron carriers such as NADH (14, 53). For example, the complete mineralization of toluene coupled to the reduction of NAD^+ can be written as

$$C_7H_8 + 36\ NAD^+ + 21\ H_2O \rightarrow 36\ NADH + 36\ H^+ + 7\ H_2CO_3 \quad (1)$$

In this step, electrons are gained, which can now be further transferred (14) to extracellular electron acceptors (for example, oxygen or nitrate) or, alternatively, used for the formation of additional biomass. The electron transfer to extracellular electron acceptors is referred to as respiration and involves the (re)oxidation of NADH. The appropriate reactions for the major electron accepting processes are listed in Table 4.1. In the second step, the free energy gained in these reactions can then be stored as ATP and, together with NADH, reinvested to generate new biomass. If a simplified chemical composition for biomass ($C_5H_7O_2N$) is assumed, the reaction, which diverts the electron flow towards the generation of new biomass, is described by (53):

$$10\ NADH + 5\ H_2CO_3 + NH_4^+ + 9\ H^+ + \text{free energy} \rightarrow C_5H_7O_2N + 10\ NAD^+ + 13\ H_2O \quad (2)$$

The efficiency of the microorganisms determines the fraction of electrons (gained in the oxidation step described by equation 1) that is diverted either to biomass generation according to equation 2 or towards the electron-accepting step according to the reactions listed in Table 4.1. The lower the efficiency of the bacteria, the lower the fraction of organic carbon that is incorporated into biomass and the higher the fraction that is converted to carbon dioxide. Efficiencies can vary over a wide range, depending on both organic substrates and electron acceptors, as well as the ambient conditions, such as temperature. For the reactions which yield the most energy, i.e., oxygen and nitrate reduction, efficiencies can be as high as 50 to 70%. On the other hand, they can be as low as 5%, as found for CO_2 fixation (53). Edwards et al. (22), for example, used ^{14}C-labeled substrate to investigate the degradation of toluene and xylene under sulfate-reducing conditions and found that only 10% of the organic carbon was converted to cell material, while the

TABLE 4.1 Electron-accepting processes

Electron-accepting reaction
$O_2 + 2\ NADH + 2\ H^+ \rightarrow 2\ H_2O + 2\ NAD^+$ + free energy
$2\ NO_3^- + 5\ NADH + 7\ H^+ \rightarrow N_2 + 6\ H_2O + 5\ NAD^+$ + free energy
$SO_4^{2-} + 4\ NADH + 5\ H^+ \rightarrow HS^- + 4\ H_2O + 4\ NAD^+$ + free energy
$FeOOH + NADH + 3\ H^+ \rightarrow Fe^{2+} + 2\ H_2O + NAD^+$ + free energy
$MnO_2 + 2\ NADH + 4\ H^+ \rightarrow Mn^{2+} + 2\ H_2O + 2\ NAD^+$ + free energy
$CO_2(g) + 8\ H^+ + 8\ NADH \rightarrow CH_4(g) + 2\ H_2O + 8\ NAD^+$ + free energy

rest was used as an energy source and converted to carbon dioxide. In addition to an experimentally based determination of microbial efficiency, it might also be determined from thermodynamic considerations by the method proposed by Van Briesen and Rittmann (59). On the basis of a known or estimated efficiency, the redox reactions can then be balanced. For example, in the above-mentioned case (toluene mineralization under aerobic conditions) an efficiency estimate of 60% leads to the following reaction, in which 40% of the organic carbon (C_7H_8) is converted to inorganic carbon (HCO_3^-) and 60% is converted to biomass ($C_5H_7O_2N$):

$$C_7H_8 + 0.84\ NH_4^+ + 4.8\ O_2 + 0.48\ H_2O \rightarrow 2.8\ HCO_3^- + 0.84\ C_5H_7O_2N + 3.64\ H^+ \quad (3)$$

For the oxidation/mineralization of toluene under sulfate-reducing conditions, assuming a lower (10%) efficiency, balancing leads to

$$C_7H_8 + 0.14\ NH_4^+ + 4.15\ SO_4^{2-} + 2.58\ H_2O + 1.86\ H^+ \rightarrow 6.30\ HCO_3^- + 0.14\ C_5H_7O_2N + 4.15\ H_2S \quad (4)$$

Under the same assumptions, but for a different reducible, organic compound (benzene), one obtains

$$C_6H_6 + 0.12\ NH_4^+ + 3.45\ SO_4^{2-} + 2.64\ H_2O + 1.38\ H^+ \rightarrow 5.4\ HCO_3^- + 0.12\ C_5H_7O_2N + 3.45\ H_2S \quad (5)$$

whereas the same exercise for ethylbenzene and xylene(s) leads to

$$C_8H_{10} + 0.16\ NH_4^+ + 2.52\ H_2O + 2.34\ H^+ \rightarrow 7.2\ HCO_3^- + 0.16\ C_5H_7O_2N + 4.85\ H_2S \quad (6)$$

Reactions 3 to 6 and their stoichiometry reflect the conditions for microbial growth. If cell growth is neglected, the stoichiometry of the mineralization reaction simplifies, for example, from reaction 4 to

$$C_7H_8 + \mathbf{4.5}\ SO_4^{2-} + \mathbf{3}\ H_2O + \mathbf{2}\ H^+ \rightarrow \mathbf{7}\ HCO_3^- + \mathbf{4.5}\ H_2S \quad (7)$$

In this reaction, there is obviously more sulfate consumed than in reaction 4, which considers microbial growth (4.5 versus 4.15 mol of sulfate per mol of toluene degraded). This apparent discrepancy can be accounted for by noting the concentrations and valence of the redox-sensitive reactants before and after the reaction. For the complete mineralization of toluene, electron balance requires (50)

$$Y_{C_7H_8} ov_{C_7H_8} + Y_{SO_4\text{-}2} ov_{SO_4^{2-}} = Y_{HCO_3^-}\ ov_{HCO_3^-} + Y_{H_2S} ov_{H_2S} \quad (8)$$

where $Y_{C_7H_8}$, $Y_{SO_4^{2-}}$, $Y_{HCO_3^-}$, and Y_{H_2S} are the stoichiometric factors of toluene, sulfate, bicarbonate, and hydrogen sulfide, respectively, and $ov_{C_7H_8}$, $ov_{SO_4^{2-}}$, $ov_{HCO_3^-}$, and ov_{H_2S} are the appropriate valences of these species. Maintaining mass balance for sulfur and carbon requires that $Y_{SO_4^{2-}} = Y_{H_2S}$ and $Y_{HCO_3^-} = 7$, respectively. Consideration of microbial growth mandates that ammonium and bacteria are included in the electron balance:

$$Y_{C_7H_8} ov_{C_7H_8} + Y_{SO_4^{2-}}\ ov_{SO_4^{2-}} + Y_{NH_4^+} ov_{NH_4^+} = Y_{HCO_3^-}\ ov_{HCO_3^-} + Y_{C_5H_7O_2N}\ ov Y_{C_5H_7O_2N} + Y_{H_2S}\ ov_{H_2S} \quad (9)$$

where $Y_{NH_4^+}$ and $Y_{C_5H_7O_2N}$ are the stoichiometric factors for ammonium and bacteria, respectively, and $ov_{NH_4^+}$ and $ov_{C_5H_7O_2N}$ are the corresponding valences (50). However, nitrogen balance requires that

$$Y_{NH_4^+} = Y_{C_5H_7O_2N} \quad (10)$$

At the same time, bacteria and ammonium have the same valence, in which case it turns out that the respective terms are equivalent. Thus, the difference in sulfate consumption between the case described by reaction 4 and the corresponding case described by reaction 7 can be calculated from (50):

$$\Delta Y_{SO_4^{2-}} = \frac{(7-6.3)ov_{HCO_3^-}}{ov_{SO_4^{2-}} - ov_{H_2S}} = \frac{0.7 \times 4}{6-(-2)} = 0.35 \quad \textbf{(11)}$$

As we will see, the electron acceptor "saving" that occurs during this step might subsequently be consumed during microbial decay. The reaction for the decay of bacteria (under sulfate-reducing conditions) and the corresponding consumption of electron acceptors is described by (50):

$$C_5H_7O_2N + \mathbf{3}\ H_2O + \mathbf{2.5}\ SO_4^{2-} \rightarrow \mathbf{5}\ HCO_3^- + \mathbf{2.5}\ HS^- + \mathbf{1.5}\ H^+ + NH_4^+ \quad (12)$$

It follows from the stoichiometry of the reaction that the decay of 1 mol of bacteria ($C_5H_7O_2N$) requires 2.5 mol of sulfate. Consequently, the decay of the 0.14 mol of bacteria produced during growth (see reaction 4) consumes 0.14 × 2.5 = 0.35 mol of sulfate. This amount equals the difference in sulfate consumption between reactions 4 and 7. Accordingly, where microbial concentrations become steady state or quasi-steady state, i.e., where microbial growth occurs at the same rate as bacterial decay, the stoichiometry from reaction 7 applies. Depending, for example, on whether the assumption of a quasi-steady-state microbial concentration is a sufficiently good approximation for a given problem, kinetic reaction models of different complexities can be built on the basis of the above stoichiometric relationships.

Rate expressions for the mineralization reactions typically used in reactive transport models for biodegradation are based on three conceptual reaction models. In the first and simplest model (13), the biodegradation reaction is assumed to be instantaneous. This means that the reaction is limited only through the availability of either the electron acceptor or the contaminant (electron donor). The reaction is assumed to proceed at an infinite rate until either of these is depleted. A widely used biodegradation model that incorporates this approach for aerobic (only) biodegradation of hydrocarbons is BIOPLUME (13, 51), and an analytical solution for this problem was provided by Ham et al. (30). Note that in this approach, bacterial concentrations are not modeled explicitly. This assumption is, however, not necessarily suitable for simulating slowly degrading hydrocarbon contaminants. An improved description of the reaction kinetics, using a dual-substrate Monod kinetics formulation, was proposed previously (52). In this description, the biomass concentration is assumed to remain constant with time (although it can vary spatially) throughout the subsurface (i.e., model domain). The modeling approach described by Lu et al. (39) also incorporates this assumption but provides a framework that allows the simulation of multiple electron-accepting processes. In many cases, however, it is obvious that the dynamics of biomass growth will play an important role in biodegradation. Hence, in the most realistic description of the bioremediation process, the contaminant degradation kinetics are typically described by using Monod kinetics, and the biomass concentration is allowed to change as a function of space and time as the biomass grows and decays.

Microbial Growth and Decay

In the macroscopic mathematical descriptions of microbial growth dynamics aimed at governing laboratory- or field-scale processes, many of the complex interdependencies that are known at the microscopic scale are commonly neglected and described by empirical formulations based on the classical works of Michaelis and Menten (41) and Monod (42). The expression describing a specific bacterial growth rate, v_{sp}, observed in many batch experiments is:

$$v_{sp} = v_{max} \frac{C_{org}}{K_{org} + C_{org}} \quad (13)$$

where C_{org} is the concentration of the organic substrate, v_{max} is an asymptotic maximum specific uptake rate, and K_{org} is the half-saturation constant, the substrate concentration at which the actual uptake rate equals $v_{max}/2$. Based on equation 13, the total uptake rate v_m considers the dependency of the change of microbial mass on the actual microbial concentration X itself:

$$v_m = v_{max} \frac{C_{org}}{K_{org} + C_{org}} X \quad (14)$$

An additional (potential) growth limitation by electron acceptor availability might be incorporated into equation 14, leading to

$$v_m = v_{max} \frac{C_{org}}{K_{org} + C_{org}} \frac{C_{ea}}{K_{ea} + C_{ea}} X \quad (15)$$

where C_{ea} is the electron acceptor concentration and K_{ea} is the appropriate half-saturation constant (3,4). The complete mass balance equation for the microbial mass, X, describing the change of microbial concentration as a function of time, also needs to include microbial decay:

$$\frac{\partial X}{\partial t} = \frac{\partial X_{growth}}{\partial t} + \frac{\partial X_{decay}}{\partial t} \quad (16)$$

with

$$\frac{\partial X_{growth}}{\partial t} = v_{max} Y_x \frac{C_{org}}{K_{org} + C_{org}} \frac{C_{ea}}{K_{ea} + C_{ea}} X \quad (17)$$

and

$$\frac{\partial X_{decay}}{\partial t} = -v_{dec} X \quad (18)$$

where v_{dec} is a decay rate constant and Y_x is a stoichiometric factor. During growth ($v_m > 0$), both organic substrate and electron acceptors are consumed at rates that are proportional to v_m. Thus, for a known reaction stoichiometry, the actual rates can be easily determined. For example, in the previously discussed case of toluene degradation under sulfate-reducing conditions (equation 4), the complete mineralization of 1 mol of toluene consumes 4.15 mol of sulfate (during growth) and yields 0.14 mol of sulfate-reducing bacteria (SRB), thus:

$$\frac{\partial X_{growth}}{\partial t} = 0.14 v_m \quad (19)$$

and

$$\frac{\partial C_{sulf}}{\partial t} = -4.15 v_m \quad (20)$$

where C_{sulf} is the concentration of sulfate. As shown in Fig. 4.1, for given initial concentrations (at time $t = 0$ [see Table 4.2]), we can now compute the temporal development of toluene and sulfate and of the microbial mass in a closed batch-type system (case 1a). This can be done, for example, with the geochemical model PHREEQC-2 (45) by using its capability to compute arbitrary user-defined kinetic reactions. Notable in the upper plot is the lag period of several days before the degradation affects the aqueous concentrations of toluene and sulfate. Its length depends largely on the initial bacterial concentration and on v_{max}, the maximum uptake rate (values given in Tables 4.2 and 4.3). The removal of the initial toluene mass (0.1 mmol) is reached after 14 days, at a time when approximately 0.415 mmol of sulfate is depleted. The microbial (net) growth then stops immediately ($v_m = 0$) and the microbial mass is subsequently changing at the rate given by equation 18, thereby consuming the remaining 0.035 mmol of sulfate, as discussed previously (Fig. 4.1, case 1a). By comparison, Fig. 4.1 shows also the temporal development of the sulfate concentration if the effects of the different valence states of bacteria (compared to the end product CO_2) and the related geochemical changes are not considered (Fig. 4.1, case 1b). In the latter case, all sulfate is consumed during bacterial growth and none

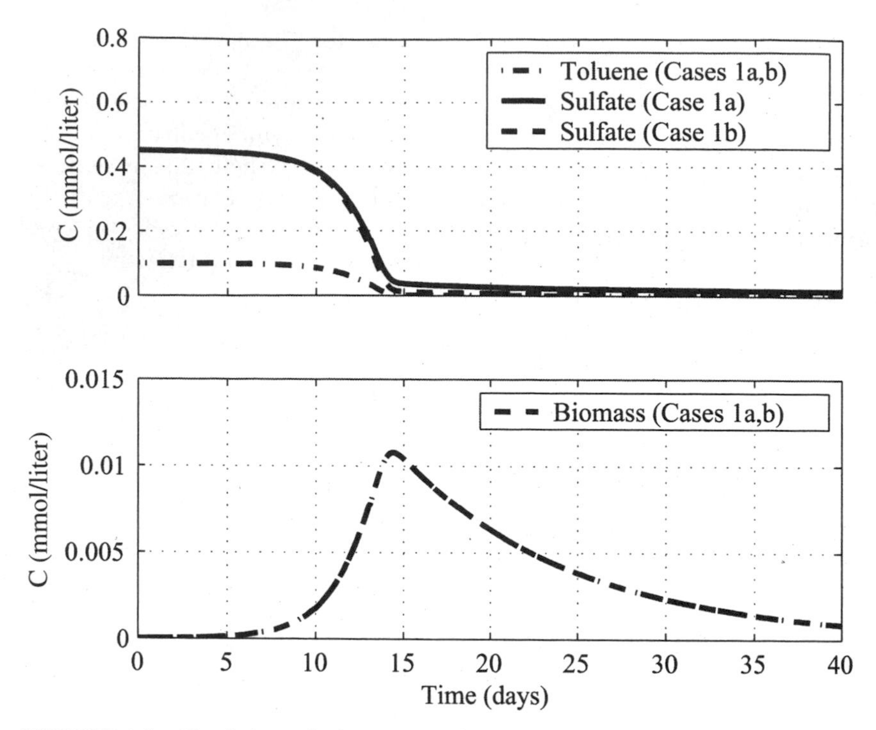

FIGURE 4.1 Simulation of toluene mineralization by SRB in a closed batch system.

TABLE 4.2 Initial concentrations of aqueous components and minerals in the batch-type biodegradation simulations

Aqueous component	Initial concn (mol/liter)				
	Case 1	Case 2	Case 3	Case 4	Case 5
Benzene	—[b]	4.0×10^{-4}	—	—	—
Toluene	1.0×10^{-4}	3.0×10^{-4}	3.0×10^{-4}	3.0×10^{-4}	2.0×10^{-4}
Ethylbenzene	—	1.0×10^{-4}	—	—	—
Xylene	—	2.5×10^{-4}	—	—	—
O(0)	—	—	6.25×10^{-4}	—	6.25×10^{-4}
N[a]	1.0×10^{-4}	1.0×10^{-4}	5.0×10^{-4}	1.0×10^{-4}	5.0×10^{-4}
S(6)	4.5×10^{-4}	5.0×10^{-3}	8.0×10^{-4}	1.0×10^{-3}	—
C(4)	—	—	—	—	1.12×10^{-3}
Ca	—	—	—	—	3.32×10^{-3}
Na	—	—	—	—	1.00×10^{-3}
Cl	—	—	—	—	6.17×10^{-3}
Aerobes	—	—	1.0×10^{-8}	—	—
Denitrifiers			1.0×10^{-8}		
SRB	1.0×10^{-8}	1.0×10^{-8}	1.0×10^{-8}	1.0×10^{-8}	—
Calcite	—	—	—	—	1.0×10^{-3}
$Fe(OH)_3$(a)	—	—	—	—	4.0×10^{-3}
Siderite	—	—	—	—	0

[a] Total nitrogen, except for cases 3 and 5, where it is defined as nitrate/N (5).
[b] —, not considered in simulation.

TABLE 4.3 Parameters used in the batch-type biodegradation modeling (cases 1 to 5)

Aqueous component	Value for:				
	Case 1	Case 2	Case 3	Case 4	Case 5
K_{ox} (mol liter^{-1})	—[a]		1.0×10^{-5}	—	1.0×10^{-5}
K_{nit} (mol liter^{-1})	—	—	—	—	1.0×10^{-5}
K_{sulf} (mol liter^{-1})	1.0×10^{-5}	1.0×10^{-5}	1.0×10^{-5}	1.0×10^{-5}	—
$K_{Fe(OH)_3}$ (mol liter^{-1})	—	—	—	—	1.0×10^{-5}
K_N (mol liter^{-1}	—	—	2.0×10^{-6}	—	—
$K_{benzene}$ (mol liter^{-1}	—	1.0×10^{-5}	—	—	—
$K_{toluene}$ (mol liter^{-1})	1.0×10^{-5}	1.0×10^{-5}	1.0×10^{-5}	1.0×10^{-5}	—
$K_{ethylbenzene}$ (mol liter^{-1})	—	1.0×10^{-5}	—	—	—
K_{xylene} (mol liter^{-1})	—	1.0×10^{-5}	—	—	—
$K_{inhib,\ ox}$ (mol liter^{-1})	—	—	5.0×10^{-6}	—	—
$K_{inhib,\ bio}$ (mol liter^{-1})				1.0×10^{-4} 2.5×10^{-5} 1.0×10^{-5} 2.5×10^{-6}	—
$\nu_{max,\ sulf,\ benzene}$ (day^{-1})	—	0.1	—	—	—
$\nu_{max,\ herob,\ toluene}$ (day^{-1})	—	—	10.0	—	—
$\nu_{max,\ sulf,\ toluene}$ (day^{-1})	5.0	5.0	5.0	5.0	—
$\nu_{max,\ sulf,\ ethylbenzene}$ (day^{-1})	—	0.5	—	—	—
$\nu_{max,\ sulf,\ xylene}$ (day^{-1})	—	1.0	—	—	—
ν_{decay} (day^{-1})	0.1	0.1	0.1	0.1	0.1
r_{ox} (day^{-1})	—	—	—	—	4.32×10^{-5}
r_{nit} (day^{-1})	—	—	—	—	8.64×10^{-6}
$r_{Fe(OH)_3}$ (day^{-1})	—	—	—	—	8.64×10^{-7}

[a] —, not considered in simulation.

is consumed during bacterial decay. Note that the differences between the two cases become more pronounced with increasing bacterial efficiency (here 10%), as shown for example by Barry et al. (9) for simulations involving dissolved organic carbon, CH_2O, and oxygen.

Of course, the formulations for microbial growth above apply only to the uptake of a single substrate, whereas contamination often involves numerous (organic) compounds. The (possibly simultaneous) uptake of these substrates can be incorporated into the modeling approach described above. The model of Kindred and Celia (35), also used in other studies (24, 48, 55), states that the growth of the degrading microbial community is simply the sum of the growth rates arising from degradation of individual organic contaminants:

$$\frac{\partial X}{\partial t} = \left[\left(\sum_{n=1,\, n_{org}} \frac{\partial X_n}{\partial t} \right) - \nu_{dec} \right] X \qquad \textbf{(21)}$$

where, in analogy to equation 17, each of the growth terms $\partial X_n / \partial t$ can be derived from

$$\frac{\partial X_n}{\partial t} = \nu^{n}_{max} Y_x \frac{C_{org,\, n}}{K_{org,\, n} + C_{org,\, n}} \frac{C_{ea}}{K_{ea} + C_{ea}} \qquad \textbf{(22)}$$

The uptake rates ν^{n}_{max} can differ between different substrates. In this way, it is possible to model varying degradation rates of different electron donors. This can be necessary when, for example, benzene degrades more slowly than the other compounds, as reported by Davis et al. (19). The case of a simultaneous uptake of BTEX constituents (benzene, toluene, ethylbenzene, xylenes) by one microbial group is demonstrated by the next simulation example (case 2). Different initial amounts of organic compounds were assumed to be present (Table 4.2) and the initial sulfate mass was increased compared to that of the first example (thus, still present in excess). As can be seen in Fig. 4.2, for the chosen parameters

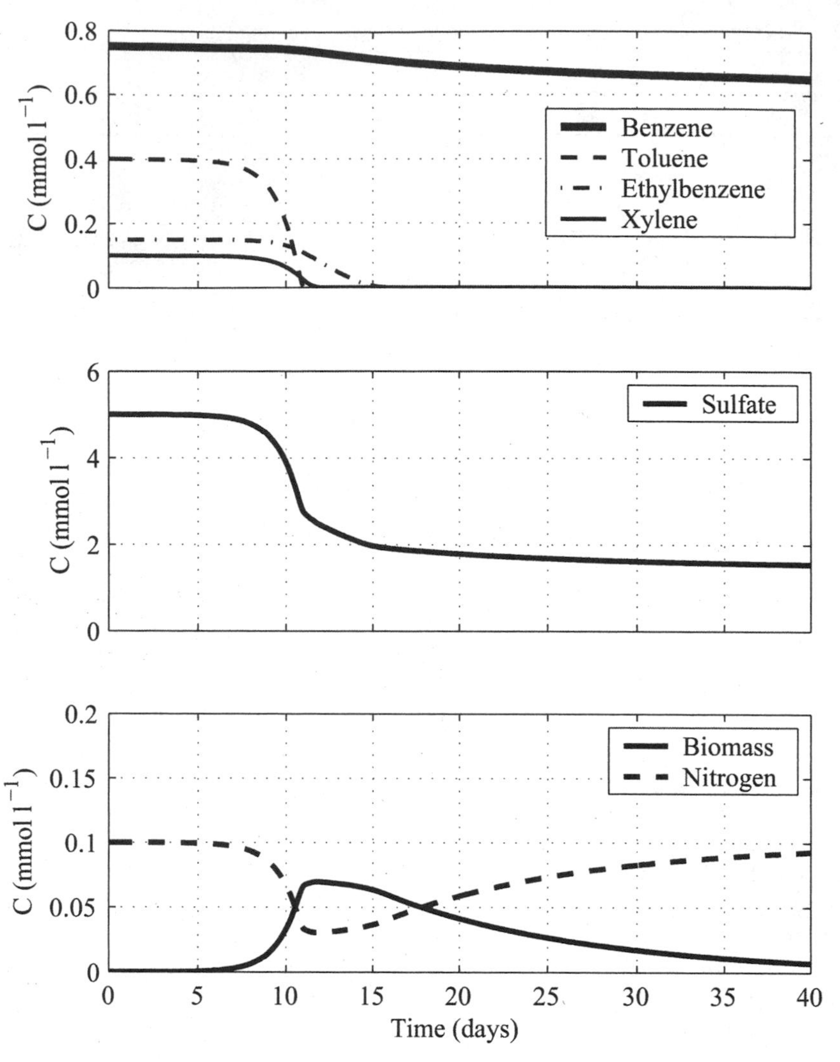

FIGURE 4.2 Simulation of BTEX mineralization by SRB in a closed batch system (after the work of Barry et al. [9]).

(Table 4.3), toluene and ethylbenzene degrade first. They are followed by xylene, while degradation of benzene takes longest. The rate of sulfate consumption decreases once toluene and ethylbenzene are depleted and slows down further as soon as only benzene is left to degrade. Figure 4.2 shows also the temporal development of the nitrogen concentration within the aqueous phase. During bacterial growth, nitrogen is removed from the aqueous phase and incorporated into biomass. Once the bacterial decay rate exceeds the growth rate, the nitrogen released to the aqueous phase eventually returns the nitrogen concentration to its initial value.

Nutrient Limitation

While in many cases bacterial activity is limited by substrate or electron acceptor availability, growth might also be limited by nutrients such as nitrogen or phosphorus. The most common mathematical model to consider this effect is simply to add an additional Monod term to equation 17. For the case of a potential growth limitation by nitrogen, this gives

$$\frac{\partial X_{\text{growth}}}{\partial t} = \nu_{\max} Y_x \frac{C_{\text{org}}}{K_{\text{org}} + C_{\text{org}}} \frac{C_{\text{ea}}}{K_{\text{ea}} + C_{\text{ea}}} \frac{C_n}{K_n + C_n} X \quad (23)$$

where K_n is a half-saturation constant for nitrogen and C_n is the concentration of nitrogen in the aqueous phase.

Multiple Bacterial Groups and Growth Inhibition

So far, we have assumed that the contaminant degradation is dominated by a single biogeochemical process in which one particular bacterial group is largely responsible for the contaminant breakdown. This assumption might indeed be valid for some cases of active bioremediation schemes in which electron acceptors are actively replenished. However, in most instances where contaminants degrade naturally, i.e., without any intervention, it is very likely that more than one bacterial group will be involved in the degradation process as electron acceptors become locally depleted. When this is the case, bacterial groups that rely on less thermodynamically favorable electron acceptors might take over and proceed with the contaminant breakdown. Eventually this leads in most cases to the formation of distinct redox zones whereby the dominating bacterial group corresponds to the most thermodynamically favorable electron acceptor within each zone. In numerical models, the sequential use of electron acceptors can be achieved through the introduction of additional Monod-type inhibition terms. For example, the inhibited growth of sulfate-reducing bacteria in the presence of oxygen and nitrate can be expressed by

$$\frac{\partial X_{\text{sulf}}}{\partial t} = \nu_{\text{sulf}}^{\max} Y_x I_{\text{inh,ox}} I_{\text{inh,nit}} \frac{C_{\text{org}}}{K_{\text{org}} + C_{\text{org}}} \frac{C_{\text{sulf}}}{K_{\text{sulf}} + C_{\text{sulf}}} X_{\text{sulf}} \quad (24)$$

and using for the inhibition terms:

$$I_{\text{inh, ox}} = \frac{K_{\text{inh, ox}}}{K_{\text{inh, ox}} + C_{\text{ox}}} \quad (25)$$

and

$$I_{\text{inh, nit}} = \frac{K_{\text{inh, nit}}}{K_{\text{inh, nit}} + C_{\text{nit}}} \quad (26)$$

where C_{ox} and C_{nit} are the concentrations of oxygen and nitrate, respectively, and K_{org} and K_{sulf} are half-saturation constants. $K_{\text{inh, ox}}$ and $K_{\text{inh, nit}}$ are inhibition constants that need to be much smaller than typical nitrate or oxygen concentrations under ambient conditions. The Monod-type inhibition term $I_{\text{inh, ox}}$ will then remain ≈ 0 as long as oxygen is present in significant amounts but reaches its maximum value of ≈ 1 as soon as oxygen is depleted. Growth of SRB remains inhibited after $I_{\text{inh, ox}}$ becomes ≈ 1, because $I_{\text{inh, nit}}$ is still ≈ 0 as long as nitrate is present. Once nitrate concentrations become very low, $I_{\text{inh, nit}}$ also approaches ≈ 1 and growth of SRB can start to increase to rates that will affect the concentration of the organic substrate (no more growth inhibition). Mathematically, the form of the inhibition terms resembles that used in model approaches that do not explicitly consider bacterial growth and decay for the computation of the oxidation rates of the organic compounds (39). Van Cappellen et al. (60) have used inhibition terms of this form for the simulation of the oxidation of organic matter in aquatic sediments. Most comprehensive biodegradation models incorporate multiple inhibition terms into the growth equation(s). Following equation 21 and assuming that microbial groups are distinguished by their capacity of using a particular electron acceptor, a generalized formulation for microbial growth is then

$$\frac{\partial X}{\partial t} = \left\{ \left[\prod_{i=1,\, n_{\text{inh}}} I_{\text{inh},\, i} \left(\sum_{n=1,\, n_{\text{org}}} \frac{\partial X_n}{\partial t} \right) \right] - \nu_{\text{dec}} \right\} X \quad (27)$$

where $I_{\text{inh},\,i}$ is an inhibition term similar to that in equation 25 for each of the n_{inh} more favorable electron acceptors. An illustrative example for the sequential use of electron acceptors is given by the third simulation example (case 3). It involves the sequential consumption of oxygen, nitrate, and sulfate during oxidation of toluene. For the initial conditions and parameters chosen in this example (Table 4.2), full mineralization of toluene occurs during the course of the numerical experiment, as can be seen in Fig. 4.3. However, it can also be seen that the degradation of toluene occurs in three distinct phases, i.e., under aerobic, denitrifying, and finally sulfate-reducing conditions. During the first phase, the oxygen that is initially present in the system is consumed rapidly, subsequent to an initial lag period during which toluene concentrations remain essentially unchanged. The lag period results from the very small initial microbial concentrations of the aerobic degraders (aerobes). Once oxygen is depleted, the oxidation of toluene then continues after a further lag period. It involves the growth of denitrifying bacteria, which stops once nitrate is also depleted. In the final stage, the toluene removal is completed by the activity of the SRB.

Some conceptual models for biologically mediated degradation reactions suggest that with increasing biomass concentrations and thus increasing biofilm thickness, degradation rates might become limited by the supply of reactants. In order to avoid solving the diffusional transport of reactants at a microscopic (i.e., within-biofilm) level, Kindred and Celia (35) have proposed use of a macroscopic formulation to account for the rate limitation resulting from excessive biomass accumulation. The mass balance equation for bacteria becomes

$$\frac{\partial X}{\partial t} = (v_{\text{max}} Y_x I_{\text{bio}} \frac{C_{\text{org}}}{K_{\text{org}} + C_{\text{org}}} \frac{C_{\text{ea}}}{K_{\text{ea}} + C_{\text{ea}}} - v_{\text{dec}})X \tag{28}$$

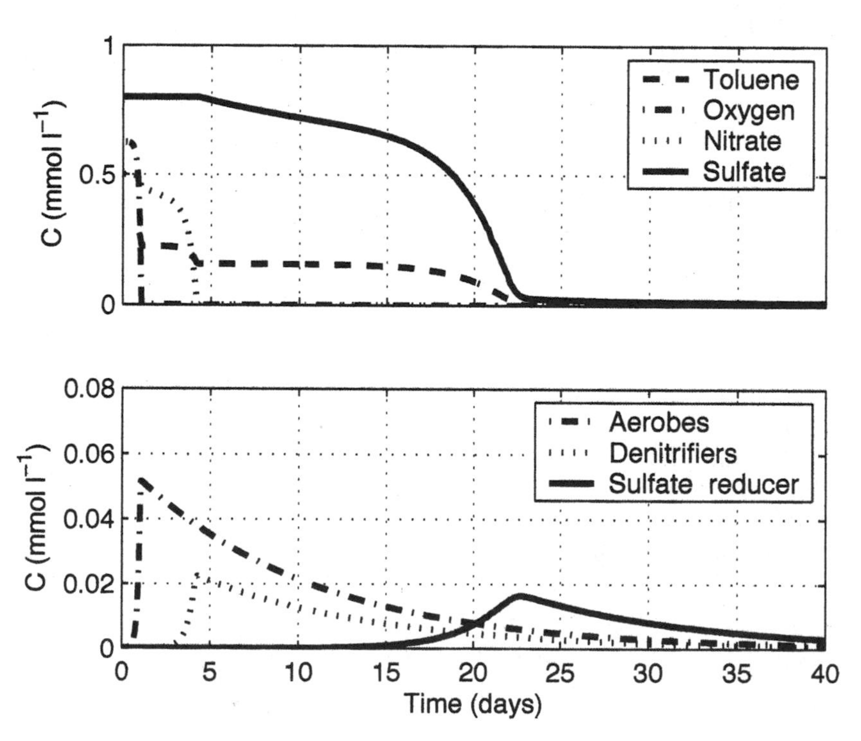

FIGURE 4.3 Simulation of toluene mineralization by sequential use of oxygen, nitrate, and sulfate in a closed batch system (after the work of Barry et al. [9]).

with: $$I_{bio} = \frac{K_{bio}}{K_{bio} + X} \quad (29)$$

where K_{bio} is an inhibiting biomass concentration. As Kindred and Celia (35) pointed out, when a microbial concentration becomes much larger than K_{bio}, the growth term will become similar to the basic Michaelis-Menten expression. Conceptually, this represents a situation where the real biofilm thickness becomes irrelevant since the metabolic activity occurs predominantly on the upper layers of a biofilm that are more exposed to the nutrient-bearing aqueous phase. A mathematically equivalent inhibition term aimed at suppressing excessive microbial growth was proposed by Zysset et al. (69):

$$I_{bio} = \frac{\theta_{max} - X}{\theta_{max}} \quad (30)$$

where θ_{max} is a maximum bacterial concentration.

The sensitivity of modeling results to the selected magnitude of K_{bio} is illustrated by case 4 (Fig. 4.4). It can be seen that if $K_{bio} = 1 \times 10^{-4}$ mol liter^{-1}, the simulated evolution of the biomass concentrations is almost similar to

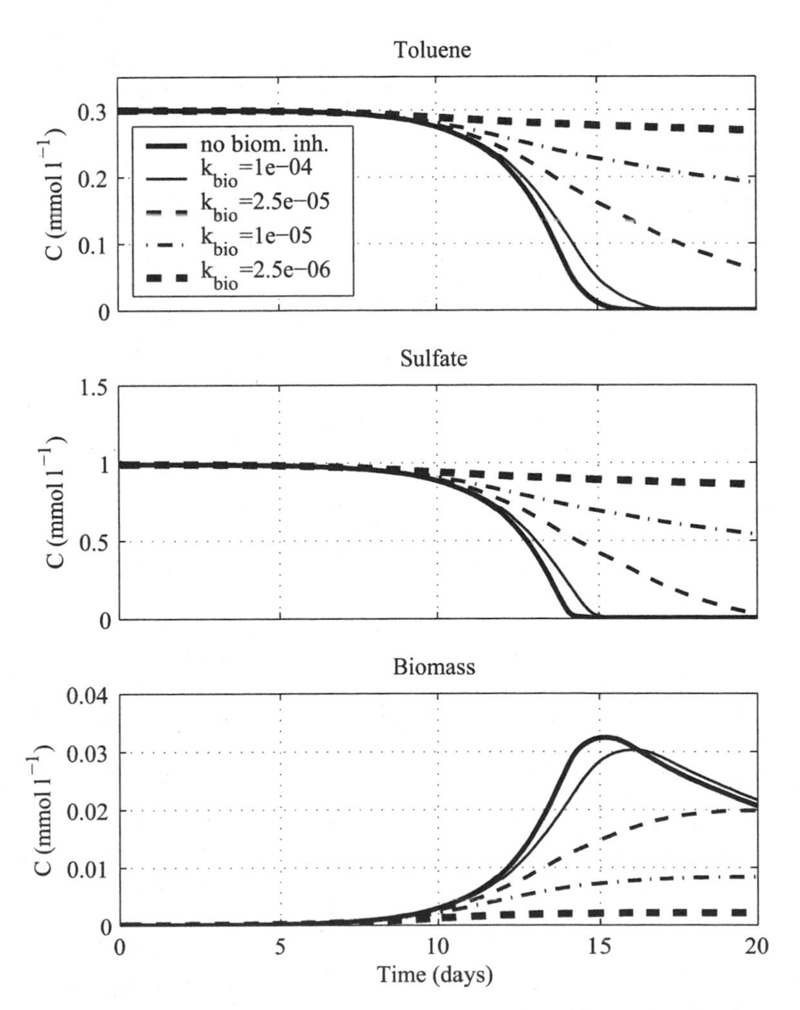

FIGURE 4.4 Simulation of toluene mineralization by sulfate with and without growth inhibition.

that of the uninhibited case. In contrast, for the example simulation that uses the smallest K_{bio} (2.5×10^{-6} mol liter^{-1}), the microbial growth is strongly inhibited, resulting in a much lower contaminant removal rate.

Modeling Geochemical Changes during Biodegradation

In many cases, the reactions that are critical in determining the rates of removal of contaminants from groundwater involve the production and consumption of protons and other reactants, which subsequently can trigger the precipitation or dissolution of minerals, ion exchange, or surface complexation reactions (21). For example, the oxidation of BTEX compounds not only consumes electron acceptors, but also causes changes in alkalinity and pH and produces reduced forms of electron acceptors such as sulfide or ferrous iron. The latter species may react further and precipitate as minerals such as siderite ($FeCO_3$), iron sulfide (FeS), or pyrite (FeS_2). For simplicity, those effects are often neglected in modeling studies.

There are situations, however, in which the simulation of only the primary reactions provides insufficient information or delivers erroneous results. Where reaction rates are dependent on the pH of the groundwater, the simulation of the hydrogeochemistry and thus of secondary reactions might be required, which itself is changing during biodegradation. When monitored natural attenuation is considered for use as a remediation technique, understanding and quantifying the detailed geochemical footprint of degradation reactions can be an important component for the successful verification of the attenuation processes (64). Given the increased use of monitored natural attenuation for treatment of BTEX-contaminated groundwater, the inclusion of the geochemical changes that occur during the bioremediation process can be essential for accurate modeling of the fate of the contaminants.

The next example that we will discuss here (case 5) provides a relatively simple illustration of a situation where geochemical interactions occur in response to the oxidation of organic compounds. As above, we use a batch-type model to look at the fate of the reaction products. Toluene mineralization is taken again as the primary reaction, which now also acts as the driving force for secondary reactions. As noted by Brun and Engesgaard (15), such problems can be modeled by a two-step method, or so-called partial equilibrium approach. This approach splits the redox reaction to be modeled into two separate steps: (i) the electron-donating oxidation step and (ii) the electron-accepting step, as discussed in earlier sections of this chapter. According to Postma and Jakobsen (47), the assumption that the first step is rate limiting is made and thus the second step can be simply modeled as an equilibrium reaction. For illustration purposes, we use, in contrast to the previous simulation examples, an approach in which microbial growth and decay are not included in the model. Instead, we use a simpler rate equation for the time-dependent degradation of toluene, based on an additive Monod expression:

$$\frac{\partial C_{org}}{\partial t} = -\sum_{i=1,\, n_{ea}} r_{ea,\, i} \frac{C_{ea,\, i}}{K_{ea,\, i} + C_{ea,\, i}} \qquad \textbf{(31)}$$

where C_{org} is here the toluene concentration, $C_{ea,\, i}$ is the concentration of the *i*th electron acceptor, n_{ea} is the number of electron acceptors, and $r_{ea,\, i}$ and $K_{ea,\, i}$ are the reaction rate and the half-saturation constants corresponding to the *i*th electron acceptor, respectively. The form of equation 31 causes the oxidation rate of toluene to decrease successively as more and more oxidation capacity gets depleted, i.e., as redox conditions become more and more reducing. In the simulation, we assume that oxygen, nitrate, and amorphous iron oxide [$Fe(OH)_3$] are initially present as electron acceptors. The parameter values and initial concentrations used for this numerical experiment are shown in Tables 4.2 and 4.3, while the simulation results are shown in Fig. 4.5. As microbial growth and decay are not simulated, toluene degradation begins with-

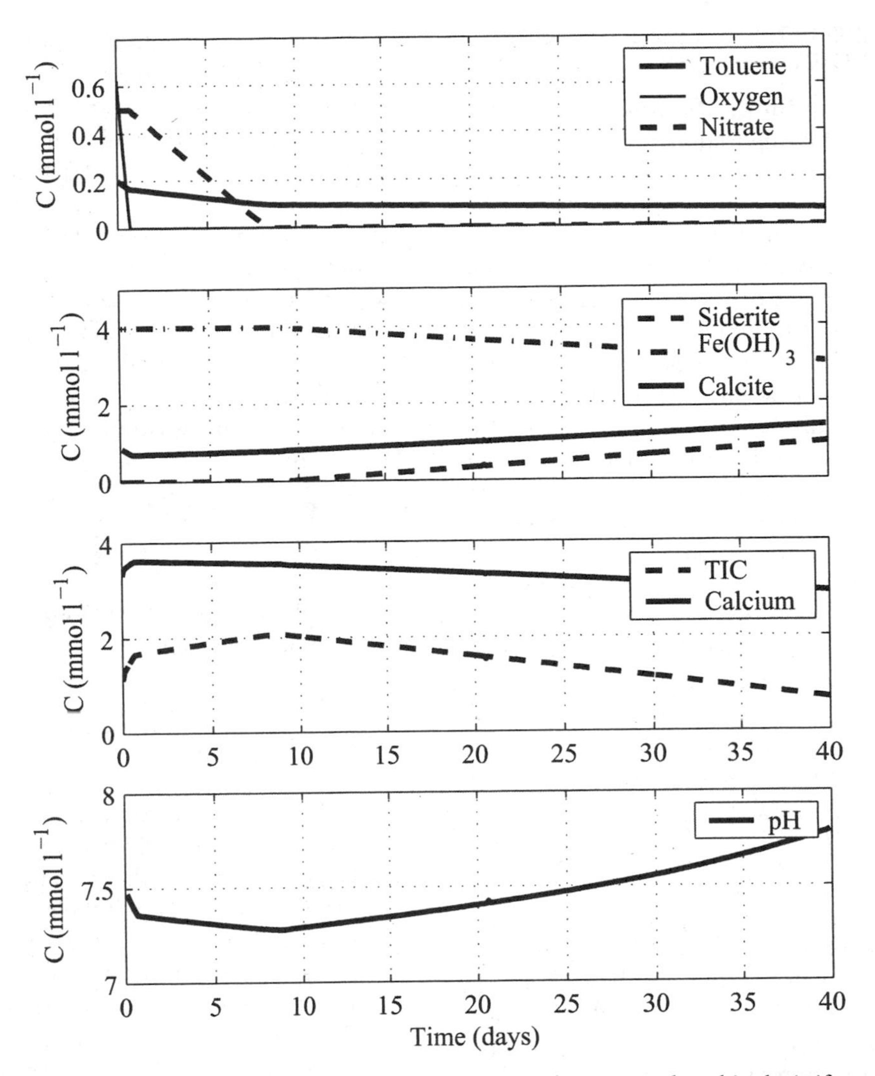

FIGURE 4.5 Simulation of toluene mineralization under sequential aerobic, denitrifying, and iron-reducing conditions in a closed batch system. TIC, total inorganic carbon.

out any lag period immediately after the start of the simulation. The degradation rate is highest at the beginning while oxygen is still present but decreases constantly over time. Once nitrate is depleted (after 8 days), the simulated $Fe(OH)_3$ concentration also starts to decrease. The simulation results in Fig. 4.5 show also that the total inorganic carbon concentration increases during the phase at which oxygen and nitrate act as electron acceptors. However, once iron reduction has started, the total inorganic carbon concentration starts to decrease again as the solution becomes (or, more precisely, would become) oversaturated with respect to siderite ($FeCO_3$), which therefore precipitates. Figure 4.5 also depicts pH changes during the three phases of the numerical experiment. The changes would even be more pronounced if calcite ($CaCO_3$) did not buffer the pH changes. The simulation results indicate therefore that even the (dissolved) calcium concentrations are ultimately affected by toluene oxidation.

MODELING FLOW AND PHYSICAL TRANSPORT

Nonreactive Single-Species Transport

The successful application of bioremediation techniques to contaminated groundwater is based on the availability of all essential ingredients required for the microbial activity to occur. In many groundwater systems, the rate of delivery of one or more of the reactants that take part in the transformation limits the progress of contaminant destruction. Clearly, groundwater flow and the resulting mixing caused by hydrodynamic dispersion play a key role for remediation success. Detailed understanding and quantification of the natural groundwater dynamics and/or the dynamics induced by a remediation scheme are essential for optimizing engineered schemes or predicting natural attenuation.

The macroscopic mass balance for the transport of a single, dissolved, nonreactive chemical species within a three-dimensional domain is described mathematically by the partial differential equation (10, 11, 67)

$$\frac{\partial C}{\partial t} = \frac{\partial}{\partial x_i}(D_{ij} \frac{\partial C}{\partial x_j}) - \frac{\partial}{\partial x_i}(v_i C) + \frac{q_s}{\theta} C_q \quad (32)$$

where C [ML^{-3}] is the aqueous concentration of the chemical species, v_i [LT^{-1}] is the pore water velocity in direction x_i [L], D_{ij} [L^2T^{-1}] is the hydrodynamic dispersion coefficient tensor (summation convention assumed), θ is the porosity of the porous medium, q_s [$L^3L^{-3}T^{-1}$] is the volumetric flux rate per unit volume of water representing external sources and sinks, and C_q [ML^{-3}] is the concentration of the species within this flux if q_s is positive (injection), otherwise $C_q = C$. In equation 32, the chemical species is subject to advective and dispersive transport by the flowing fluid.

The advection-dispersion equation results from averaging microscopic processes occurring at the pore scale within a representative elementary volume, leading to a continuum model at the macroscopic level (10). The advection term describes the transport of a dissolved species at the same mean velocity as the groundwater, which is the dominating physical process in most field-scale contamination problems within the saturated groundwater zone. The dispersion term represents two processes, mechanical dispersion and molecular diffusion. Mechanical dispersion results from the fluctuation of the (microscopic) streamlines in space with respect to the mean flow direction and inhomogeneous conductivities within the representative elementary volume. Molecular diffusion is caused by the random movement of the molecules in a fluid. It is usually negligible compared to mechanical dispersion (10). The (macroscopic) pore velocity v_i in equation 32 is derived from Darcy's law

$$v_i = -\frac{K_{ij}}{\theta}\frac{\partial h}{\partial x_j} \quad (33)$$

and the three-dimensional flow equation for saturated groundwater

$$\frac{\partial}{\partial x_i}\left(K_{ij}\frac{\partial h}{\partial x_j}\right) + q_s = S_s\frac{\partial h}{\partial t} \quad (34)$$

where K_{ij} is the hydraulic conductivity tensor and h [L] is the hydraulic head. Note that the off-diagonal entries of the hydraulic conductivity tensor become zero if the principal components are aligned with the x, y, and z axes of the flow domain. Analytical solutions for equations 32 to 34 exist only for relatively simple cases. Thus, for solving more complicated cases, e.g., involving heterogeneous aquifers, transient boundary conditions, etc., numerical techniques such as the finite difference and finite element methods (11, 34, 46, 62) are required. Groundwater flow models such as MODFLOW (40), HST3D (37), FEMWATER (65), and FEFLOW (20) that incorporate information on the hydrological and hydrogeological properties of a site will typically form the basis for subsequent contaminant transport simulations. However, in many cases, a proper groundwater flow model itself can already provide useful information, e.g., to estimate how fast the edge of a contamination front would migrate in a nonreactive case, i.e., if no biodegradation or sorption occurred.

Modeling packages such as PMPATH (16) allow predictions of the contaminant flow path and travel times of nonreactive contaminants.

The first step in building such a site-specific groundwater flow model is, based on a preliminary site characterization, the development of a conceptual hydrological and hydrogeological model. At this stage, all available geologic and hydrographic information is collated and analyzed. The conceptual model formulates the following qualitatively:

- the general groundwater flow direction;
- (natural) boundaries that might be used as boundaries in the numerical model, such as (subsurface) catchment boundaries and (dividing) streamlines;
- which stratigraphic layer(s) is more and which one is less or much less permeable, and which layer(s) is suitable to form a boundary in the numerical model; and
- the fluxes into and out of a chosen (model) domain and how to determine or estimate these fluxes quantitatively.

The so-constructed conceptual model might be translated into a simple numerical model, pointing to data gaps and thus assisting in the design of a refined characterization program and monitoring network. Details of this procedure can be found in many groundwater hydrology-specific textbooks such as those by Freeze and Cherry (28) or Fetter (26). Modeling texts such as those by Anderson and Woessner (2), Chiang and Kinzelbach (16), and Zheng and Bennett (68) are available also.

Reactive Single-Species Transport

The chemical species transported by groundwater as described by equation 32 will, in most cases, either interact with the aquifer material (e.g., sorption, ion exchange, or precipitation-dissolution) or react with other chemicals in the aqueous phase (e.g., acid-base reactions or redox reactions) or both. In a very general, unspecified form this can be expressed mathematically by including an additional reaction term, R_{chem}, in the above-mentioned transport equation, leading to

$$\frac{\partial C}{\partial t} = \frac{\partial}{\partial x_i}\left(D_{ij}\frac{\partial C}{\partial x_j}\right) - \frac{\partial}{\partial x_i}(v_i C) + \frac{q_s}{\theta}C_q + R_{\text{chem}} \tag{35}$$

This reaction term is typically a complicated function of the concentrations of many other chemical species and/or of bacterial concentrations, for example. Limited by the availability of appropriate computational resources, these interactions between multiple groundwater and soil constituents were, until recently, neglected in most modeling applications. However, with the steady disappearance of this constraint, models of increasing complexity have been developed.

COUPLED PHYSICAL TRANSPORT AND REACTIVE PROCESSES

We now consider the case where physical transport and reactive processes occur simultaneously. To solve this type of problem, a range of numerical schemes exists (e.g., see reference 56). We focus here, however, on only one particular method, the split-operator technique (6–8, 33, 66), which is (i) simple to understand, (ii) relatively easy to implement on parallel computers, and (iii) with few exceptions (e.g., OS3D/GIMRT software from Pacific Northwest National Laboratory, Richland, Wash.) has become the standard method for solving such combined physicochemical problems. The operator-splitting technique involves separating the processes (e.g., flow, transport of individual chemical components or species, chemical reactions, and microbial activity) within the numerical model and solving each submodel independently. Different implementations of this method exist, resulting in different degrees of accuracy and computational burden. In the most simple and commonly applied variation, equation 32 is initially solved separately for each transported chemical for each time step of length Δt before, in a second step, the concentration changes due to chemical reactions during Δt are determined. The latter computation is carried out for each grid cell (in a spatially discretized domain) independently from other grid cells, thereby simplifying the computational burden significantly compared with that of the fully coupled case. The rates at which the reactions proceed in relation to the velocity at which the chemicals are transported provide the criteria for choosing an appropriate solution technique to accurately describe the reactive processes. Under the assumption that all reactive processes proceed rapidly in comparison to groundwater flow and transport, known as the local equilibrium assumption, equilibrium thermodynamics can be used to compute the concentration changes. Note that within this (equilibration) step, each grid cell is treated as a closed system. That is, the total concentration of each chemical component, consisting of the sum of aqueous, sorbed, precipitated and gas-phase concentrations, does not change, although the distribution within these phases and the concentration of (complexed) species will vary. Batch-type geochemical equilibrium packages such as PHREEQE/PHREEQC (43, 44) or MINTEQA2 (1) can be used for this step. Engesgaard and Kipp (23) and Walter et al. (61) presented examples of models employing this technique. More recently, the split-operator technique has been applied to model sets of multiple, exclusively kinetically reacting species (e.g., see reference 18) or for problems where both equilibrium and kinetic reactions occur. The PHREEQC-2 model (45), which we have already used for the batch-type examples, can be used for a modeling approach of the latter class. Coupled to a transport model via operator splitting, it is capable of determining the reaction term R_{chem} within equation 35 for a wide variety of sets of mixed equilibrium and kinetic reactions (15, 49, 50; also PHAST [http://www.brr.cr.usgs.gov/projects/GWC_coupled/phast/] from the U.S. Geological Survey). Below, we will explore some of the major aspects and typical applications related to bioremediation.

MODELING OF NATURAL ATTENUATION PROCESSES

More than in other areas of bioremediation, numerical modeling is applied to assist the assessment of site-specific risks when natural attenuation is the preferred remediation option. In many cases it can be expected that the dissolution from free product, i.e., non-aqueous-phase liquids (NAPLs), in the vicinity of the saturated groundwater zone will provide a long-term contamination source for the groundwater, perhaps lasting decades or even hundreds of years. Under most circumstances that involve NAPL dissolution, groundwater contaminant plumes of a more or less stable length will form during a relatively short timescale (30). Stable plumes occur under conditions where the total mass dissolved from the free product (per unit of time) equals the rate of total contaminant mass destruction by biodegradation. In the case of oxidizable organic compounds such as petroleum hydrocarbons (or landfill leachates), typically one or more electron acceptors will be consumed so that distinct redox zones form downstream of the contaminant source (5, 12). They are the result of the sequential use of electron acceptors, with the most reduced chemical environment being the plume center in proximity of the source region. The key factors that are largely responsible for the length of steady-state plumes are:

- groundwater flow velocity,
- rate of organic compound dissolution from free product or NAPL in the source zone,
- oxidation capacity provided by the background or uncontaminated water, and
- mixing of electron acceptors through transverse hydrodynamic dispersion.

In the following section, we will discuss some of the important modeling aspects of these governing factors.

Modeling the Contamination Source

In most cases where numerical modeling is applied to field-scale contamination problems, the extent and the location of the source of the contamination are not well characterized. In order to reduce the uncertainty associated with the lack of detailed information, the migration of the contaminant might be simulated as a multiphase problem involving water, air, and oil (25). However, in many instances, in particular when the oil phase is denser than water, the source location and extension must be largely guessed. With the known or estimated location of the contamination source, the (simultaneous) mass transfer of multiple organic compounds from the NAPL phase to the water phase and the (dissolved) concentrations of the organic compounds can be approximated. For the n_{org} organic compounds that are present in the NAPL phase, nonequilibrium mass transfer, r_{dis}, from a NAPL pool is expressed as

$$\frac{\partial C_i^{dis}}{\partial t} = \varpi_i \left(C_i^{sat,\,mc} - C_i \right) \quad (36)$$

where ϖ_i [T^{-1}] is a rate transfer coefficient approaching infinity for equilibrium dissolution, C_i [ML^{-3}] is the aqueous species concentration of the ith organic compound, and $C_i^{sat,\,mc}$ [ML^{-3}] is the multicomponent solubility of the ith organic compound. The multicomponent solubility is calculated according to Raoult's law:

$$C_i^{sat,\,mc} = C_i^{sat} \gamma_i m_i \quad (37)$$

where C_i^{sat} [ML^{-3}] is the single-species solubility, γ_i is the activity coefficient of the ith organic compound (typically for simplicity assumed to be unity), and m_i is the mole fraction of the ith organic compound. The mole fraction is defined as

$$m_i = \frac{n_i}{n_{tot}} \quad (38)$$

where n_i is the molar concentration of compound i in the NAPL and n_{tot} is the total molar concentration of all organic compounds in the NAPL. From equations 36 to 38 it becomes clear that the near-source concentrations might change with time due to the temporal change in the molar fraction. When the dissolution process is kinetically controlled, concentrations will also depend on the groundwater flow velocity.

As a first example for a coupled transport and reaction simulation, we look at a simple, one-dimensional transport problem where uncontaminated, anaerobic, sulfate-rich ground water is passing a 4-m-long aquifer zone (located between 5 and 9 m from the left influent boundary) contaminated with residual NAPL blobs. The groundwater flow rate is constant, i.e., pore velocity of 0.75 m day^{-1} (case 6; initial concentrations are given in Table 4.4). We assume that the residual NAPL consists only of the four BTEX compounds as, in practice, they are the most soluble gasoline constituents. As before, we assume that sulfate and dissolved BTEX compounds act as electron acceptors and electron donors, respectively. Only the activity of SRB attached to the aquifer matrix is simulated; i.e., any detachment-attachment and transport processes of SRB are considered negligible. As the simulation results for this simple system show (Fig. 4.6), there is an increase of BTEX concentra-

TABLE 4.4 Initial concentrations of NAPLs, aqueous components, and microbes in the uncontaminated aquifer (cases 6 and 7)

Component or bacteria	Concn (mol/liter)
Inorganic aqueous components[a]	
O(0)[b]	0.0
S(VI)	2.50×10^{-3}
S(−II)	0.0
Fe(II)	5.78×10^{-5}
Fe(III)	1.60×10^{-12}
C(IV)	7.95×10^{-3}
Ca	5.00×10^{-3}
Mg	5.66×10^{-4}
Na	5.24×10^{-3}
K	2.59×10^{-4}
Cl	6.44×10^{-3}
Organic compound or bacteria	
Non-NAPL	
Benzene	0.0
Toluene	0.0
Ethylbenzene	0.0
Xylene	0.0
NAPL	
Benzene	2.0
Toluene	5.0
Ethylbenzene	2.0
Xylene	1.0
SRB	1.0×10^{-8}

[a] pH, 6.57; pe (measure of electron activity in aqueous solution), −2.08.
[b] Values in parentheses indicate valence state.

tions within the NAPL-containing zone, with the maximum concentrations varying significantly due to the different mole fractions of the compounds in the NAPL mix and due to their different (single-component) aqueous solubilities (C^{sat}). The BTEX compounds are undergoing biodegradation within and downstream of the source zone as long as sulfate is available, so that toluene, ethylbenzene, and xylene are completely removed, but not benzene. Since no other degradation processes that involve less favorable electron acceptors such as iron reduction or methanogenesis are modeled here, no further contaminant destruction occurs once sulfate is depleted.

Role of Dispersive Mixing

Under such circumstances, transverse dispersion becomes the mechanism that controls the fate of the remaining, undegraded portion of the

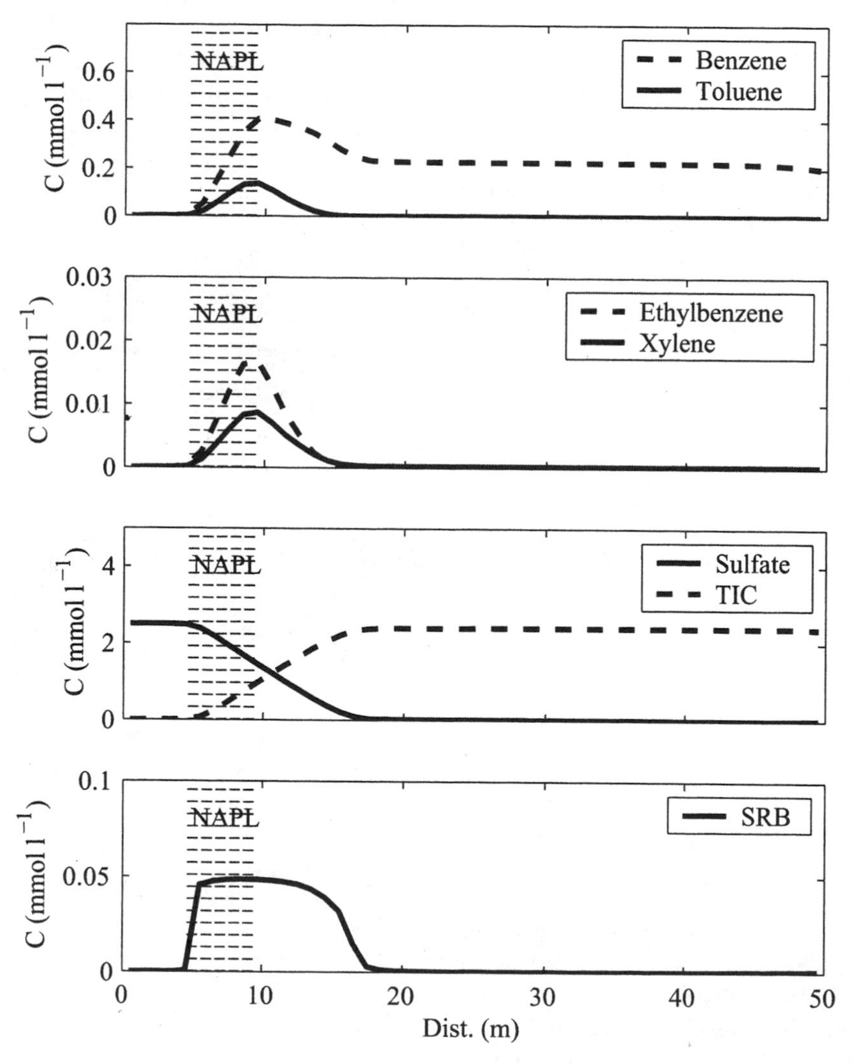

FIGURE 4.6 Simulation of BTEX compound dissolution, transport, and mineralization under sulfate-reducing conditions in a one-dimensional domain. The NAPL source zone is located between 5 and 9 m from the influent end (left). TIC, total inorganic carbon.

mass released from the NAPL source. Of course, the one-dimensional model is unable to capture this, at least in a direct manner. Thus, a multidimensional transport simulation is necessary. Using the same "chemical" system, i.e., the same initial or inflow concentrations (Table 4.4) and parameters, and a groundwater flow rate similar to that of the previous case, the next case (case 7), a two-dimensional simulation, demonstrates the effect of (horizontal) transverse dispersion (Fig. 4.7). A microbial "fringe" develops downstream of the source zone at locations where degradable organic substances and electron acceptor are present simultaneously. A major problem in terms of modeling such cases arises from the fact that the apparent transverse dispersivity of nonreacting solutes does not necessarily provide a good estimate for bioreactive transport, as was pointed out, e.g., by Cirpka et al. (17). Using a streamline-oriented grid, they investigated the two-dimensional physical transport in a heterogeneous aquifer coupled to biodegradation reactions of a single-substrate, single-electron-acceptor system. They demonstrated that employing a Fickian macrodispersion model and using transverse dispersivities (deduced from the second spatial concentration moment transverse to flow direction) will significantly overestimate local-scale mixing and thus biodegradation rates. When physical transport is dominated by advection, i.e., hydrodynamic dispersion is small compared to advective

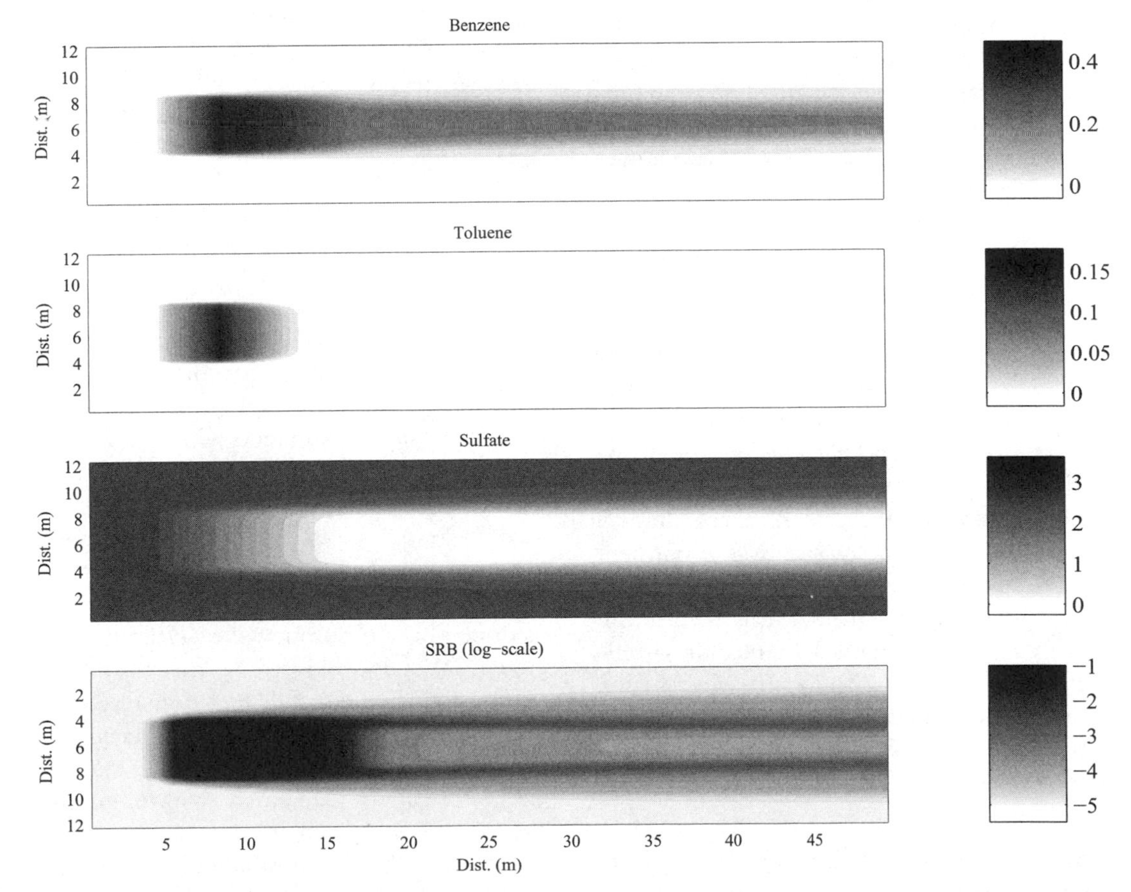

FIGURE 4.7 Simulation of BTEX compound dissolution, transport, and mineralization under sulfate-reducing conditions in a two-dimensional domain. The NAPL source zone is located between 5 and 9 m from the influent end (left) and has a width of 6 m. Dist., distance.

transport, these fringes might be very thin, resulting in steep chemical gradients. Resolution of the physical transport mechanisms in local regions without introducing large numerical errors remains a challenging and computationally expensive task, in particular for heterogeneous aquifers.

FIELD-SCALE APPLICATION

A numerical study was carried out for a BTEX-contaminated field site for which a wealth of detailed information of hydrochemical and microbial parameters has been intensively recorded. The field site is located in Perth, Western Australia (e.g., see reference 19), where toluene, ethylbenzene, and xylenes are mineralized under exclusively sulfate-reducing conditions. The modeling study is described in full detail elsewhere (50). Vertical concentration profiles obtained from multilevel sampling devices showed only limited vertical spreading of the long, thin plumes, indicating that the porous medium is fairly homogeneous and that transverse dispersion is rather small. As part of a more detailed modeling study, the flow, transport, and mixing processes and how they affect the site-specific degradation behavior were initially studied in a vertical cross-sectional model along the (flow) path of the contaminants, reported here as case 8. The specific scope of the numerical study was to further understand the site-specific role of the additional transverse dispersion, caused by the transient flow field, on reactive processes and, in particular, to determine whether this mechanism is likely to enhance natural attenuation significantly. In order to quantify the effect of the transient flow field, comparisons are made with transport simulations based on a steady-state condition.

Nonreactive-Single-Species Transport

The schematic vertical cross-sectional model was developed such that it approximately reproduces the hydro(geo)logical properties of the field site (19, 27, 54, 57). For example, recharge and inflow at the model upstream boundary were chosen so that the model would schematically represent the typical Perth climate of a 5-month rainy period (recharge of 1.5 mm day^{-1}) and a 7-month dry period without recharge. With a hydraulic conductivity of 22 m day^{-1}, the transient-flow model of the unconfined aquifer does indeed mimic the observed annual pattern of the piezometric heads. Water table fluctuations of almost 2 m occur at the influent boundary in response to the seasonally varying recharge. In the steady-state model, which was used for comparison, the recharge and boundary inflows were averaged over one annual cycle. For both the transient and the steady-state simulations, the groundwater flow field is characterized by increasing flow velocities towards the effluent boundary. This effect can be attributed to the nonsloping aquifer bottom, which, together with the falling water table (in flow direction), leads to a notable reduction of the effective thickness of the aquifer and consequently higher flow velocities. The computation of the first and second spatial moments can be used to indicate the location of the plume center and to quantify the spreading that the plume has undergone. The plots in Fig. 4.8 show the computed spatial moments along the flow direction. Snapshots for the transient plume behavior from different stages during the annual cycle are plotted in comparison with the corresponding results from the steady-state model. It can be seen that independently of dispersion and sorption the vertically integrated mass decreases along the flow path despite the absence of degradation reactions. Of course, in reality no mass has been destroyed. The example shows, rather, that the interpretation of measured concentrations, even when integrated in one or two spatial directions, can be misleading if no flux weighting is applied. In the absence of flux weighting, a proper conservation of mass is not warranted and the disappearance of the

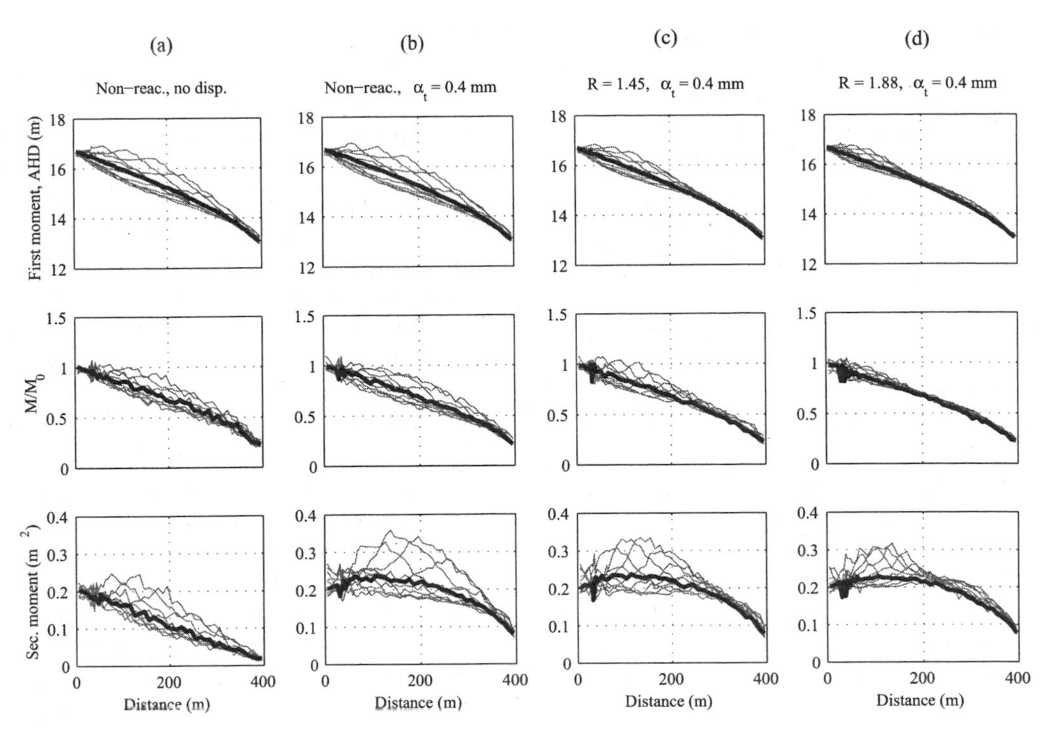

FIGURE 4.8 First-moment, vertically integrated mass and second moment for advective, nonreactive transport (a), advective, dispersive, nonreactive transport (b), and advective, dispersive transport with linear equilibrium sorption (retardation factors R of 1.45 and 1.88, respectively) (c and d). The solution for steady-state flow (thick lines) and solutions from the transient simulations (thin lines) are indicated. Reprinted from reference 50 with permission from Elsevier.

vertically integrated concentrations could be interpreted as biodegradation.

The effect of sorption reactions on the discussed plume characteristics was investigated by comparing "no-sorption" simulations with two scenarios that included linear equilibrium sorption reactions. In the first scenario, the sorption corresponds to a retardation factor of $R = 1.45$ and, in the second, to $R = 1.88$. The simulation results shown in Fig. 4.8 demonstrate that the inclusion of the sorption reactions exceeds a "buffering" effect on the vertical movement of the plume that results from seasonal hydraulic changes. This is indicated by the smaller variability of the computed first spatial moment (vertical position of plume center) and the computed second spatial moment (plume spreading). The results also show that for the transverse dispersivity used in these simulations (0.4 mm), the accelerating flow has a converging effect for the plume that is even stronger than the spreading caused by hydrodynamic dispersion. More details are given elsewhere (50).

Reactive Multicomponent Transport

Based on the previous simulations, in which degradation reactions were ignored, the influence of the transient flow field on reactive processes was studied in a second step. As discussed earlier in this chapter, biodegradation of oxidizable contaminants (in this case, BTEX compounds) might occur mainly at contaminant plume fringes where organic substrates and electron acceptors (in the present case sulfate) are simultaneously available. Thus, it might be intuitively expected that the vertical movement of the plume caused by the seasonal water table fluctuations would facilitate increased mixing of reactants and that in response the rates of total mass removal by biodegradation would increase. This increased mixing of reactants might be expected when electron donor(s) and electron acceptor(s) have different sorption characteristics. The reactive multicomponent transport model PHT3D (49) was used to investigate the magnitude of this effect by comparing the results from reactive transport simulations that use a transient flow field with the corresponding results from simulations based on a steady-state flow field.

Previous studies (19, 48) identified sulfate reduction as the dominant mechanism for the attenuation of BTEX compounds. Oxygen was observed only in low concentrations in the proximity of the water table during times of recharge. There it is consumed rapidly, either directly by organic matter or by groundwater constituents and/or minerals that have been reduced previously. Nitrate was generally not found in this portion of the aquifer. The water composition used for the numerical study, which was assumed to be representative for the uncontaminated portion of the aquifer, was based on a water sample taken upstream of the contaminated area. The recharge water was assumed to have an identical composition, except that the solution had been equilibrated with an additional 3.2 mg of oxygen liter^{-1}. Two minerals (pyrite and siderite) that had been identified as reaction products in a soil core taken within the contaminated zone (48) were also included in the simulation. SRB and toluene (other organic compounds were excluded) were treated as kinetic species for which rate expressions were formulated on the basis of the concepts discussed earlier in this chapter.

Contamination Source

The investigation of core material from the contamination source zone showed that residual NAPLs were distributed over a 1- to 1.5-m-thick zone below the water table (19). The measurements of dissolved concentrations from a nearby multilevel sampling device indicated that the location and vertical extent of the zone from which elevated concentrations of BTEX compounds dissolve would largely follow the

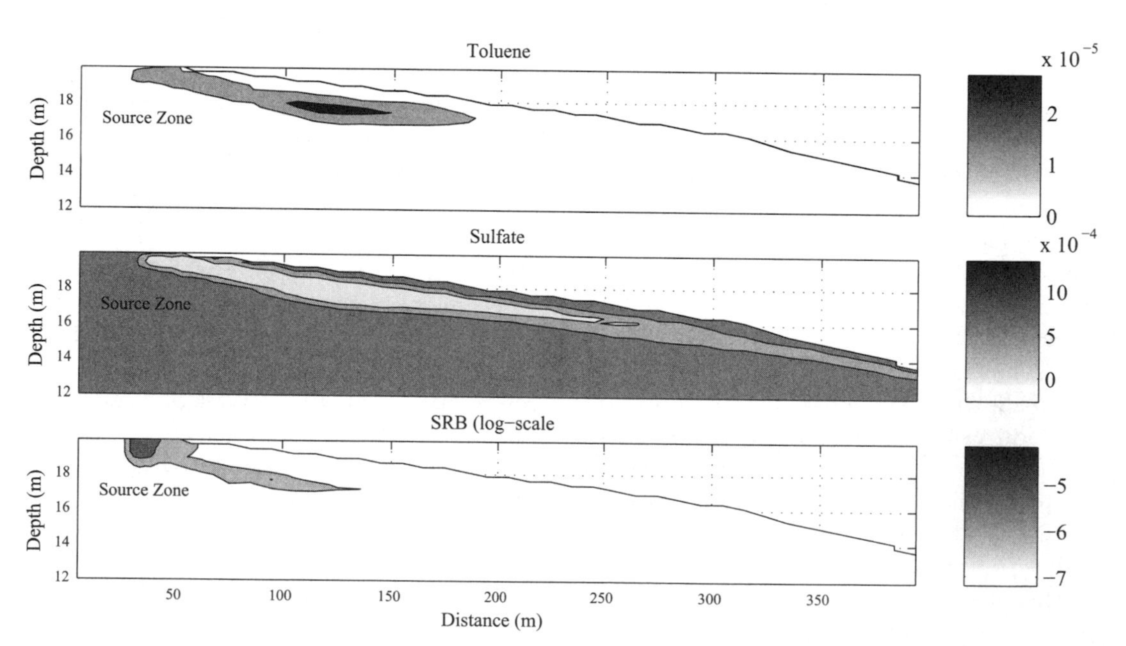

FIGURE 4.9 Contour plots for toluene, sulfate, and SRB for a snapshot from the transient simulations (concentrations are given in moles per liter) (after the work of Prommer et al. [50]).

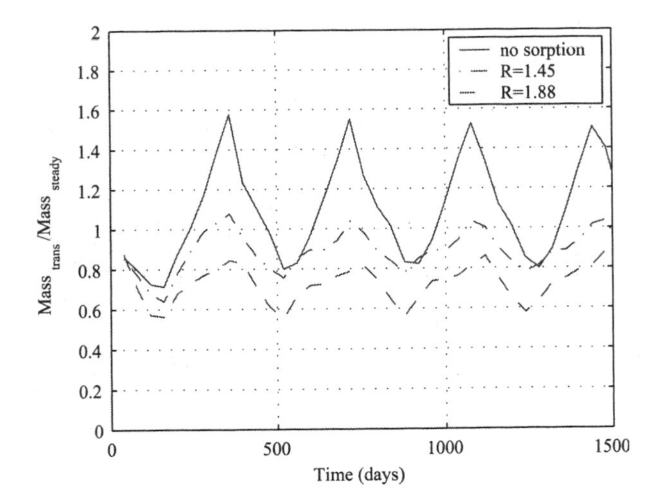

FIGURE 4.10 Simulated total mass (concentrations integrated over the model domain) of toluene during the transient (trans) simulations. Reprinted from reference 50 with permission from Elsevier.

movement of the water table. However, the lower portion of this NAPL dissolution zone appeared to move somewhat more slowly than the water table itself at times of a rising water table. The latter effect was not considered in the numerical model study. Instead, the vertically moving NAPL dissolution zone was modeled to occur from a defined, constant thickness below the moving water table.

Steady-State versus Transient Simulations

The reactive transport simulations reproduce the observed thin toluene plumes and their approximate length, which is limited by the mass removal through biodegradation. The results show the observed limited vertical extent of the sulfate-depleted zone, confirming that the transverse dispersivity at this site is small. The simulations illustrate that the activity of the bacteria, i.e., their highest concentrations, occurs mainly in the vicinity of the contamination source and, further downstream, at the plume fringes where the presence of sulfate and toluene overlaps. The seasonally changing flow and, in particular, the "moving" contaminant source lead to pronounced seasonal variability of contaminant dissolution and its flow path and biogeochemical reaction rates. A snapshot from those transient simulations is shown in Fig. 4.9 for toluene, sulfate, and SRB concentrations. The local (integrated) mass of dissolved toluene in the model varies between 60 and 160% of the corresponding steady-state mass, as can be seen in Fig. 4.10. The plot also identifies the effect of toluene sorption, which reduces the seasonal variability of the total mass. It can be seen that on average, the total mass decreases with increasing

TABLE 4.5 Initial concentrations of aqueous components, minerals, and microbes in the uncontaminated aquifer used in the field-scale natural attenuation simulation (case 8)

Component or bacteria	Concn (mol/liter)
Aqueous components[a]	
O(0)[b]	0.0
S(VI)	7.84×10^{-4}
S(−II)	0.0
Fe(II)	6.65×10^{-5}
Fe(III)	2.14×10^{-12}
C(IV)	7.95×10^{-3}
Ca	8.41×10^{-4}
Cl	6.44×10^{-3}
Mg	5.66×10^{-4}
Na	5.24×10^{-3}
K	2.59×10^{-4}
Minerals or bacteria	
Pyrite	0
Siderite	0
SRB	1.00×10^{-8}

[a] pH, 5.06; pe, 1.21.
[b] Values in parentheses indicate valence state.

sorption. This shows that the modeled toluene sorption indeed causes a notable chromatographic effect by slowing down the vertical movement of the toluene plume relative to the somewhat faster vertical movement of the sulfate-depleted zone. The effect leads to enhanced mixing between toluene and sulfate-containing water. If toluene sorption is excluded, the overall influence of the seasonally changing flow on mass removal by biodegradation remains rather small.

MODELING OF ENHANCED BIOREMEDIATION

As stated above, very often the supply of electron acceptors is the critical factor for natural attenuation, i.e., the length of a plume. In such cases, the addition of electron acceptors, in many cases together with other potentially growth-limiting substances, might be used as a means to enhance and accelerate naturally occurring degradation processes. However, the choice among electron acceptors is not necessarily straightforward, as discussed, e.g., by Kinzelbach et al. (36). Oxygen has the advantage of having the highest molecular energy yield. Unfortunately, its low solubility in water is a limiting factor for the transfer of oxidation capacity to an aquifer. Sulfate has a higher solubility but is accompanied by a sometimes undesired hydrogen sulfide production. Furthermore, sulfate reduction might not be able to degrade efficiently more recalcitrant oxidizable organic substances. In terms of modeling enhanced remediation processes, the two major challenges are:

- choice of reactive processes to be included in a simulation, and
- accurate representation of the local-scale mixing processes that mix the injected water with the contaminated water.

TABLE 4.6 Initial concentrations of aqueous components, minerals, and bacteria of the initially contaminated aquifer and of the oxygenated water that is injected to enhance remediation

Aqueous component, mineral, or bacterial group	Concn (mol/liter) in case 9a (primary reactions only)		Concn (mol/liter) in case 9b (primary and secondary reactions)[a]	
	$C_{initial}$	C_{inflow}	$C_{initial}$	C_{inflow}
Toluene	3×10^{-5}	0	3×10^{-5}	0
N(5)	—[b]	—	0.0	5.0×10^{-4}
N(3)	—	—	0.0	0.0
N(0)	—	—	5.0×10^{-4}	0.0
N(−3)	—	—	1.10×10^{-6}	0.0
O(0)	0.0	6.32×10^{-4}	0.0	6.32×10^{-4}
S(VI)	—	—	5.26×10^{-4}	5.0×10^{-4}
S(-II)	—	—	5.07×10^{-11}	0.0
Fe(II)	—	—	2.58×10^{-6}	0.0
Fe(III)	—	—	3.48×10^{-13}	1.79×10^{-12}
C(IV)	—	—	3.19×10^{-3}	1.70×10^{-4}
Ca	—	—	1.90×10^{-3}	1.43×10^{-3}
Na	—	—	4.35×10^{-4}	4.35×10^{-4}
Cl	—	—	2.82×10^{-4}	2.82×10^{-4}
Pyrite	—	—	8.00×10^{-3}	—
Goethite	—	—	0	—
Aerobes[c]	1.0×10^{-8}	—	1.0×10^{-8}	—

[a] pH, 7.32; pe (measure of electron activity in an aqueous solution), −3.16.
[b] —, not considered in simulation.
[c] Concentrations given in number of organisms per milliliter.

In order to illustrate the first point, i.e., the importance of choosing the correct conceptual model, we set up a simple example of enhanced remediation via pulsed injection of oxygenated water into an anaerobic, heterogeneous, initially toluene-contaminated aquifer (case 9) and compare the results for two sets of reactions. In the first case, we consider only the (primary) biodegradation reaction, here the degradation of toluene under aerobic conditions. To model this, the reactions include only toluene, oxygen, and aerobic degraders. In contrast, in the second modeling approach, we additionally describe the composition of the groundwater and of major minerals in more detail and thus define the oxidation state of the aquifer (Tables 4.5 and 4.6). In this example, we include the presence of reduced aqueous species and of pyrite (FeS_2). As a consequence, oxygen that is injected to enhance the breakdown of the organic contaminants is partially diverted to other oxygen-consuming processes, i.e., the reoxidation of pyrite and of ammonia. As shown, e.g., by the biogeochemical modeling exercise of Thullner and Schäfer (58), such a diversion to minerals might consume a large fraction of the oxidation capacity that is added by the remediation scheme. In this case, it was estimated that only 2% of the oxygen mass added was consumed by the contaminants, whereas 56% of the oxygen injected was diverted to pyrite oxidation. Figure 4.11 shows a comparison of these two reaction

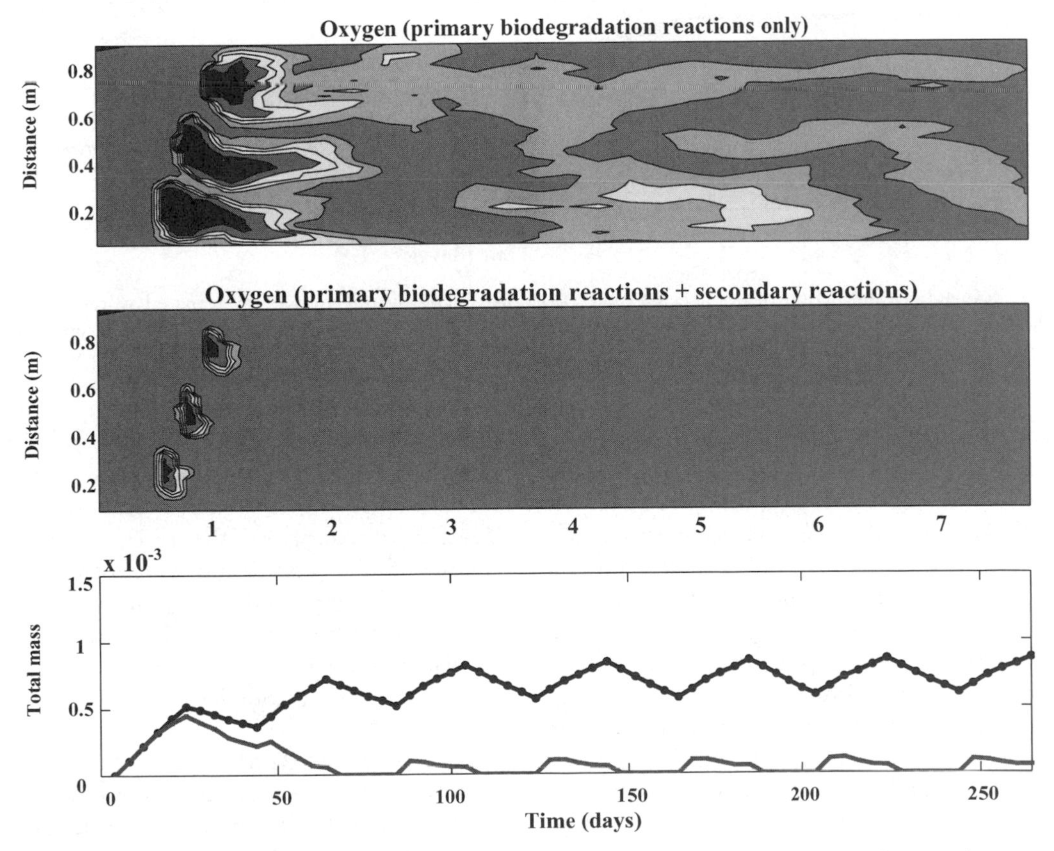

FIGURE 4.11 Simulation of enhanced remediation of toluene by injection of oxygenated water, showing oxygen concentration distribution when primary reactions only (above) and primary and secondary oxygen-consuming reactions (middle) are considered.

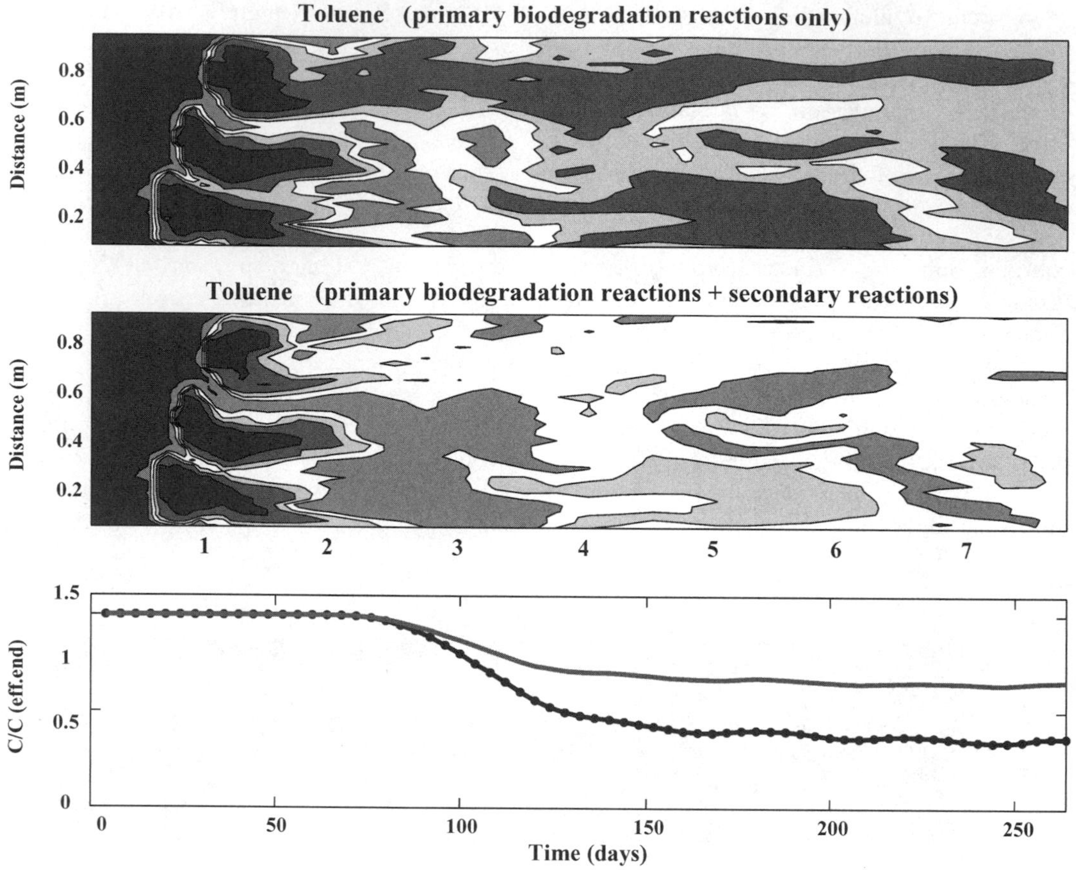

FIGURE 4.12 Simulation of enhanced remediation of toluene by injection of oxygenated water, showing toluene concentration distribution when only primary reactions (top) and both primary and secondary reactions (middle) are considered.

models. The figure shows the oxygen concentration 200 days after the beginning of the remediation and the temporal development of the integrated oxygen concentration, i.e., the total mass of oxygen in the model domain versus time. The corresponding toluene concentrations are shown in Fig. 4.12, where the large difference in the amount of toluene mass that is degraded within the model boundaries is apparent. Confirming the findings of our simplified case, Eckert and Appelo (21) reported the results from a BTEX-contaminated site in Germany where KNO_3 injection was used for enhanced remediation. The associated geochemical transport model study also highlights that a significant portion of the injected oxidation capacity was consumed by sulfide minerals. The fraction of the injected oxidants that is lost to untargeted reactions depends of course on the reaction kinetics (rates) of the competing reactions. Some knowledge regarding the reactivity of reduced sediments has been developed in studies such as that presented by Hartog et al. (31, 32), but it is far from complete.

CONCLUDING REMARKS

In this chapter, we have introduced some of the key concepts that might play a role for the use of numerical modeling in the context of bioremediation. Emphasis was put on reviewing how the underlying chemical and physical processes are translated into a coupled framework that simultaneously deals with both aspects. The theory and examples that we discussed here were confined to problems in the water-saturated zone, where biodegradation modeling probably finds its major applications. For simplicity, we have also focused the discussions on one particular contaminant class (oxidizable organic compounds). However, topics where modeling of transport in combination with microbially mediated reactions has also been applied include, for example:

- biodegradation modeling in the unsaturated zone,
- biodegradation of NAPLs in combination with multiphase transport,
- in situ metal precipitation by injection of nontoxic degradable organic substances, and
- biodegradation of chlorinated hydrocarbons.

All of these problems have in common the fact that they integrate interdisciplinary knowledge developed by microbiologists, geochemists, hydrogeologists, mathematicians, and engineers. Of course, a key factor for the successful development and application of such biodegradation models is always to find the necessary balance of process detail for each of the physical, chemical, or microbial processes that are considered. Therefore, it is important to have toolbox-like models at hand, so that additional process detail can be quickly incorporated (or redundant details can be eliminated) and to allow a quick adjustment of simulation tools to evolutionary changes of conceptual models. In all cases, the right balance between the complexity of models and the data that are available to underpin its use must also be found.

REFERENCES

1. **Allison, J. D., D. S. Brown, and K. J. Novo-Gradac.** 1991. *MINTEQA2/PRODEF2, A Geochemical Assessment Model for Environmental Systems: Version 3.0 User's Manual.* Technical report EPA/600/3–91/021. U.S. Environmental Protection Agency, Athens, Ga.
2. **Anderson, M. P., and W. W. Woessner.** 1992. *Applied Groundwater Modeling Simulation of Flow and Advective Transport.* Academic Press, San Diego, Calif.
3. **Bae, W., and B. E. Rittmann.** 1996. A structured model of dual-limitation kinetics. *Biotechnol. Bioeng.* **49:**683–689.
4. **Bae, W., and B. E. Rittmann.** 1996. Responses of intracellular cofactors to single and dual limitation. *Biotechnol. Bioeng.* **49:**690–699.
5. **Baedecker, M. J., I. M. Cozarelli, D. I. Siegel, P. C. Bennett, and R. P. Eganhouse.** 1993. Crude oil in a shallow sand and gravel aquifer. 3. Biogeochemical reactions and mass balance modeling. *Appl. Geochem.* **8:**569–586.
6. **Barry, D. A., K. Bajracharya, and C. T. Miller.** 1996. Alternative split-operator approach for solving chemical reaction/groundwater transport models. *Adv. Water Resouces* **19:**261–275.
7. **Barry, D. A., C. T. Miller, and P. J. Culligan-Hensley.** 1996. Temporal discretisation errors in non-iterative split-operator approaches to solving chemical reaction/groundwater transport models. *J. Contam. Hydrol.* **22:**1–17.
8. **Barry, D. A., C. T. Miller, P. J. Culligan, and K. Bajracharya.** 1997. Analysis of split operator methods for nonlinear and multispecies groundwater chemical transport models. *Math. Comput. Sim.* **43:**331–341.
9. **Barry, D. A., H. Prommer, C. T. Miller, P. Engesgaard, and C. Zheng.** 2002. Modelling the fate of oxidisable organic contaminants in groundwater. *Adv. Water Resources* **25:** 945–983.
10. **Bear, J.** 1972. *Dynamics of Fluids in Porous Media.* Dover Publications Inc., New York, N. Y.
11. **Bear, J., and A. Verruijt.** 1987. *Modeling Groundwater Flow and Pollution.* D. Reidel Publishing Co., Dordrecht, The Netherlands.
12. **Bjerg, P. L., K. Rügge, J. K. Pedersen, and T. H. Christensen.** 1995. Distribution of redox sensitive groundwater quality parameters downgradient of a landfill (Grindsted, Denmark). *Environ. Sci. Technol.* **29:**1387–1394.
13. **Borden, R. C., P. B. Bedient, M. D. Lee, C. H. Ward, and J. T. Wilson.** 1986. Transport of dissolved hydrocarbons influenced by oxygen-limited biodegradation. 2. Field application. *Water Resource Res.* **22:**1973–1982.
14. **Brock, T. D., M. T. Madigan, J. M. Martinko, and J. Parker.** 1994. *Biology of Microorganisms.* Prentice-Hall, Englewood Cliffs, N.J.
15. **Brun, A., and P. Engesgaard.** 2002. Modelling of transport and biogeochemical processes in pollution plumes: literature review and model development. *J. Hydrol.* **256:**211–227.

16. **Chiang, W.-H., and W. Kinzelbach.** 2000. *3D Groundwater Modeling with PMWIN: A Simulation System for Modeling Groundwater Flow and Pollution.* Springer-Verlag, Berlin, Germany.
17. **Cirpka, O. A., E. O. Frind, and R. Helmig.** 1999. Numerical simulation of biodegradation controlled by transverse mixing. *J. Contam. Hydrol.* **40:**159–182.
18. **Clement, T. P.** 1997. *RT3D—A Modular Computer Code for Simulating Reactive Multi-Species Transport in 3-Dimensional Groundwater Systems.* PNNL-SA-28967. Battelle Pacific Northwest National Laboratory, Richland, Wash.
19. **Davis, G. B., C. Barber, T. R. Power, J. Thierrin, B. M. Patterson, J. L. Rayner, and W. Qinglong.** 1999. The variability and intrinsic remediation of a BTEX plume in anaerobic sulfate-rich groundwater. *J. Contam. Hydrol.* **36:** 265–290.
20. **Diersch, H.-J. G.** 1997. *Interactive, Graphics-Based Finite-Element Simulation System FEFLOW for Modelling Groundwater Flow. Contaminant Mass and Heat Transport Processes.* User's manual version 4.6. WASY. Institute for Water Resources Planning and System Research Ltd., Berlin, Germany.
21. **Eckert, P., and C. A. J. Appelo.** 2002. Hydrogeochemical modeling of enhanced benzene, toluene, ethylbenzene, xylene (BTEX) remediation with nitrate. *Water Resource Res.* **38:**VI–VII.
22. **Edwards, E. A., L. E. Wills, M. Reinhard, and D. Grbic-Galic.** 1992. Anaerobic degradation of toluene and xylene by aquifer microorganisms under sulfate-reducing conditions. *Appl. Environ. Microbiol.* **58:**794–800.
23. **Engesgaard, P., and K. L. Kipp.** 1992. A geochemical transport model for redox-controlled movement of mineral fronts in groundwater flow systems: a case of nitrate removal by oxidation of pyrite. *Water Resource Res.* **28:**2829–2843.
24. **Essaid, H. I., B. A. Bekins, E. M. Godsy, E. Warren, M. J. Baedecker, and I. M. Cozzarelli.** 1995. Simulation of aerobic and anaerobic biodegradation processes at a crude-oil spill site. *Water Resource Res.* **31:**3309–3327.
25. **Essaid, H. I., W. N. Herkelrath, and K. M. Hess.** 1993. Simulation of fluid distributions observed at a crude-oil spill site incorporating hysteresis, oil entrapment, and spatial variability of hydraulic properties. *Water Resource Res.* **29:** 1753–1770.
26. **Fetter, C. W.** 1999. *Contaminant Hydrogeology,* 2nd ed. Prentice-Hall, Englewood Cliffs, N. J.
27. **Franzmann, P. D., L. R. Zappia, T. R. Power, G. B. Davis, and B. M. Patterson.** 1999. Microbial mineralisation of benzene and characterisation of microbial biomass in soil above hydrocarbon contaminated groundwater. *FEMS Microbiol. Ecol.* **30:**67–76.
28. **Freeze, R. A., and J. A. Cherry.** 1979. *Groundwater.* Prentice-Hall, Englewood Cliffs, N. J.
29. **Griffiths, S. K., R. H. Nilson, and R. W. Bradshaw.** 1997. *In situ* bioremediation: a network model of diffusion and flow in granular porous media. SAND97-8250. Sandia National Laboratories, Albuquerque, N. Mex. http://infoserve.sandia.gov.
30. **Ham, R. J. Schotting, H. Prommer, and G. B. Davis.** 2004. Effects of hydrodynamic dispersion on plume lengths for instantaneous bimolecular reactions. *Adv. Water Resource* **27:**803–813.
31. **Hartog, N., J. Griffioen, and C. H. Van Der Weijden.** 2002. Distribution and reactivity of O_2-reducing components in sediments from a layered aquifer. *Environ. Sci. Technol.* **36:**2436–2442.
32. **Hartog, N., P. F. van Bergen, J. W. de Leeuw, and J. Griffioen.** 2004. Reactivity of organic matter in aquifer sediments: geological and geochemical controls. *Geochim. Cosmochim. Acta* **68:**1281–1292.
33. **Herzer, J., and W. Kinzelbach.** 1989. Coupling of transport and chemical processes in numerical transport models. *Geoderma* **44:**115–127.
34. **Istok, J. D.** 1989. Groundwater modeling by the finite element method. Water Resources Monograph no. 13. American Geophysical Union, Washington, D.C.
35. **Kindred, J. S., and M. A. Celia.** 1989. Contaminant transport and biodegradation. 2. Conceptual model and test simulation. *Water Resource Res.* **25:**1149–1160.
36. **Kinzelbach, W., W. Schäfer, and J. Herzer.** 1991. Modeling of natural and enhanced denitrification processes in groundwater. *Water Resource Res.* **27:**1123–1135.
37. **Kipp, K. L., Jr.** 1987. *HST3D—A Computer Code for Simulation of Heat and Solute Transport in 3D Ground-Water Flow Systems.* U.S. Geological Survey Water-Resources Investigations report 86-4095. U.S. Geological Survey, Reston, Va.
38. **Lawrence, J. R., and M. J. Hendry.** 1996. Transport of bacteria through geologic media. *Can. J. Microbiol.* **42:**410–422.
39. **Lu, G., T. C. Clement, C. Zheng, and T. H. Wiedemeier.** 1999. Natural attenuation of BTEX compounds: model development and field-scale application. *Ground Water* **37:**707–717.
40. **McDonald, J. M., and A. W. Harbaugh.** 1988. *A Modular 3D Finite-Difference Ground-Water Flow Model.* Techniques of Water-Resources Investigations report TWI 06-A1. U.S. Geological Survey, Reston, Va.
41. **Michaelis, L., and M. L. Menten.** 1913. Die Kinetik der Invertinwerkung. *Biochem. Z.* **49:** 333–369.

42. **Monod, J.** 1949. The growth of bacterial cultures. *Annu. Rev. Microbiol.* **3:**371–394.
43. **Parkhurst, D. L., D. C. Thorstenson, and L. N. Plummer.** 1980. *PHREEQE—A Computer Program for Geochemical Calculations.* U.S. Geological Survey. Water-Resources Investigations report 80-96. U.S. Geological Survey, Reston, Va.
44. **Parkhurst, D. L.** 1995. *User's Guide to PHREEQC—a Computer Program for Speciation, Reaction-Path, Advective-Transport, and Inverse Geochemical Calculations.* U.S. Geological Survey Water-Resources Investigations report 4227. U.S. Geological Survey, Reston, Va.
45. **Parkhurst, D. L., and C. A. J. Appelo.** 1999. *User's Guide to PHREEQC—A Computer Program for Speciation, Reaction-Path, 1D-transport, and Inverse Geochemical Calculations.* U.S. Geological Survey Water-Resources Investigations report 99-4259. U.S. Geological Survey, Reston, Va.
46. **Pinder, G. F., and W. G. Gray.** 1977. *Finite Element Simulation in Surface and Subsurface Hydrology.* Academic Press, London, U.K.
47. **Postma, D., and R. Jakobsen.** 1996. Redox zonation: equilibrium constraints on the Fe(III)/SO_4-reduction interface. *Geochim. Cosmochim. Acta* **60:**3169–3175.
48. **Prommer, H., G. B. Davis, and D. A. Barry.** 1999. Geochemical changes during biodegradation of petroleum hydrocarbons: field investigations and biogeochemical modeling. *Org. Geochem.* **30:**423–435.
49. **Prommer, H., D. A. Barry, and C. Zheng.** 2003. MODFLOW/MT3DMS-based reactive multicomponent transport modeling. *Ground Water* **41:**247–257.
50. **Prommer, H., D. A. Barry, and G. B. Davis.** 2002. Influence of transient groundwater flow on physical and reactive processes during biodegradation of a hydrocarbon plume. *J. Contam. Hydrol.* **59:**113–132.
51. **Rifai, H. S., P. B. Bedient, R. C. Borden, and J. F. Haasbeek.** 1987. *BIOPLUME II: Computer Model of Two-Dimensional Contaminant Transport under the Influence of Oxygen Limited Biodegradation in Ground Water Users' Manual.* Rice University, Houston, Tex.
52. **Rifai, S. H., and P. B. Bedient.** 1990. Comparison of biodegradation kinetics with an instantaneous reaction model for groundwater. *Water Resources Res.* **26:**637–645.
53. **Rittmann, B. E., and J. M. van Briesen.** 1996. Microbiological processes in reactive modeling, p. 311–334. *In* P. C. Lichtner, C. I. Steefel, and E. H. Oelkers (ed.), *Reactive Transport in Porous Media. Reviews in Mineralogy*, vol. 34. Mineralogical Society of America, Washington, D.C.
54. **Robertson, W. J., P. D. Franzmann, and B. J. Mee.** 2000. Spore-forming, Desulfosporosinus-like sulphate-reducing bacteria from a shallow aquifer contaminated with gasoline. *J. Appl. Microbiol.* **88:**248–259.
55. **Schäfer, D., W. Schäfer, and W. Kinzelbach.** 1998. Simulation of processes related to biodegradation of aquifers. 2. Structure of the 3D transport model. *J. Contam. Hydrol.* **31:**167–186.
56. **Steefel, C. I., and K. T. B. MacQuarrie.** 1996. Approaches to modeling of reactive transport in porous media, p. 83–129. *In* P. C. Lichtner, C. I. Steefel, and E. H. Oelkers (ed.), *Reactive Transport in Porous Media. Reviews in Mineralogy*, vol. 34. Mineralogical Society of America, Washington, D.C.
57. **Thierrin, J., G. B. Davis, and C. Barber.** 1995. A ground water tracer test with deuterated compounds for monitoring in situ biodegradation and retardation of aromatic hydrocarbons. *Ground Water* **33:**469–475.
58. **Thullner, M., and W. Schäfer.** 1999. Modeling of a field experiment on bioremediation of chlorobenzenes in groundwater. *Bioremed. J.* **3:**247–267.
59. **Van Briesen, J. M., and B. E. Rittmann.** 2000. Mathematical description of microbial reactions involving intermediates. *Biotechnol. Bioeng.* **67:**35–52.
60. **Van Cappellen, P., J.-F. Gaillard, and C. Rabouille.** 1993. Biogeochemical transformations in sediments: kinetic models of early diagenesis, p. 401–447. *In* R. Wollast, F. T. Mackenzie, and L. Chou (ed.), *Interactions of C, N, P and S—Biogeochemical Cycles and Global Change.* Springer-Verlag, Berlin, Germany.
61. **Walter, A. L., E. O. Frind, D. W. Blowes, C. J. Ptacek, and J. W. Molson.** 1994. Modeling of multi-component reactive transport in groundwater. 1. Model development and evaluation. *Water Resource Res.* **30:**3137–3148.
62. **Wang, H., and M. Anderson.** 1982. *Introduction to Groundwater Modeling: Finite Difference and Finite Element Methods.* W. H. Freeman and Co., New York, N.Y.
63. **Watson, I. A., S. E. Oswald, K. U. Mayer, Y. Wu, and S. A. Banwart.** 2003. Modeling kinetic processes controlling hydrogen and acetate concentrations in an aquifer-derived microcosm. *Environ. Sci. Technol.* **37:**3910–3919.
64. **Wiedemeier, T. H., J. T. Wilson, D. H. Kampbell, R. N. Miller, and J. E. Hansen.** 1995. *Technical Protocol for Implementing Intrinsic Remediation with Long-Term Monitoring for Natural Attenuation of Fuel Contamination Dissolved in Groundwater.* Technology Transfer Divisions 1 & 2, Air Force Center for Technical Excellence. Brooks Air Force Base, San Antonio, Tex.
65. **Yeh, G. T., S. Sharp-Hansen, B. Lester, R.**

Strobl, and J. Scarbrough. 1992. *Three-Dimensional Finite Element Model of Water Flow through Saturated-Unsaturated Media (3DFEMWATER)/ Three-Dimensional Lagrangian-Eulerian Finite Element Model of Waste Transport through Saturated-Unsaturated Media (3DLEWASTE): Numerical Codes for Delineating Wellhead Protection Areas in Agricultural Regions Based on the Assimilative Capacity Criterion.* EPA report/600/R-92/223. Environmental Protection Agency, Athens, Ga.

66. **Yeh, G. T., and V. S. Tripathi.** 1989. A critical evaluation of recent developments in hydrogeochemical transport models of reactive multichemical components. *Water Resource Res.* **25:** 93–108.

67. **Zheng, C., and P. P. Wang.** 1999. *MT3DMS: a Modular Three-Dimensional Multispecies Transport Model for Simulation of Advection, Dispersion and Chemical Reactions of Contaminants in Groundwater Systems.* Documentation and user's guide, contract report SERDP-99-1. U.S. Army Engineer Research and Development Center, Vicksburg, Miss.

68. **Zheng, C., and Bennett.** 2002. *Applied Contaminant Transport Modeling,* 2nd ed. Wiley, New York, N.Y.

69. **Zysset, A., F. Stauffer, and T. Dracos.** 1994. Modeling reactive groundwater transport governed by biodegradation. *Water Resources Res.* **30:** 2435–2448.

BIOREMEDIATION OF CONTAMINATED SOILS AND AQUIFERS

Jim C. Philp and Ronald M. Atlas

5

APPROACHES TO SOIL AND AQUIFER BIOREMEDIATION

Numerous industries have viewed soil as a waste disposal site, dumping vast quantities of hazardous materials—oil and oily sludges from refineries; solvents and a variety of chlorinated hydrocarbons from the cleaning, printing, and many other industries; radionuclides from nuclear power plants and nuclear weapon facilities; coal tars from town gasification plants; creosote from wood treatment facilities; pesticides from agriculture; and so forth, ad infinitum. Large quantities of hazardous substances carelessly disposed of in the environment are creating enormous pollution problems in soils and waters around the world. Some are so heavily contaminated that they are designated Superfund sites.

The contamination that has resulted from the intentional spillage and burying of hazardous materials, accidental spills, and migration of hazardous substances from spillages that have occurred elsewhere has rendered many soils and waters dangerous for human health and harmful to the ecology of the area. Elevated rates of human cancer, dead birds, and landscapes devoid of plants are the unfortunate results of uncontrolled human activities. The environmental consequences and the human health threats of industrial pollution led to public demands for environmental cleanup and restoration and a cadre of government regulations and legal actions. The result has been a search for remediation technologies that can be actively applied to soil and aquifer cleanup.

Most of the contaminants that commonly cause concern originate above ground or in surface soils where pollutants are spilled or buried. Soil contaminants become mixed with the naturally occurring soil. Many of the contaminants in the soil become physically or chemically attached to soil particles or become trapped in the small spaces between soil particles. Over time, many of these contaminants, however, begin to migrate into nearby waterways and into the underlying groundwater.

Soil permeability and pores and fractures in the rock are critical factors controlling the movement of water from the surface into the underlying layers. If water can move rapidly through the rock material pores and fractures, an aquifer forms. The water that accumulates in an aquifer can be contaminated when surface

Jim C. Philp, Department of Biological Sciences, Napier University, Merchiston Campus, 10 Colinton Road, Edinburgh EH10 5DT, Scotland, United Kingdom. *Ronald M. Atlas*, Graduate School, University of Louisville, Louisville, KY 40292.

Bioremediation: Applied Microbial Solutions for Real-World Environmental Cleanup
Edited by Ronald M. Atlas and Jim C. Philp

water, which recharges an aquifer, is polluted or when hazardous substances soak through the soil into the groundwater. Groundwater begins to accumulate within the unsaturated zone of soil, a layer that contains air and water filling the pores. Below this layer lies the saturated zone, in which all the pores and rock fractures are filled with water. The top of the saturated zone is referred to as the water table. Soil overlying the water table provides the primary protection against groundwater pollution. Some potential pollutants nevertheless reach the groundwater.

The potential vulnerability of an aquifer to groundwater contamination is in large part a function of the susceptibility of its recharge area to infiltration. Areas that are replenished at a high rate are generally more vulnerable to pollution than those replenished at a lower rate. Unconfined aquifers that do not have a cover of dense material are susceptible to contamination. Bedrock areas with large fractures are also susceptible by providing pathways for the contaminants. Confined, deep aquifers tend to be better protected with a dense layer of clay material.

There are three general approaches to cleaning up contaminated soil: (i) soil can be excavated from the ground and be either treated or disposed of (ex situ treatment), (ii) soil can be left in the ground and treated in place (in situ treatment), or (iii) soil can be left in the ground and contained to prevent the contamination from becoming more widespread and reaching plants, animals, or humans (containment and intrinsic remediation). Containment of soil in place is often done by placing a large plastic cover or concrete barrier over the contaminated soil to prevent direct contact and keep rainwater from seeping into the soil and spreading the contamination. In situ and ex situ treatment approaches can include flushing contaminants out of the soil by using water, chemical solvents, or air; destroying the contaminants by incineration; encouraging natural organisms in the soil to break them down; or adding material to the soil to encapsulate the contaminants and to prevent them from spreading.

Bioremediation is one of the technologies that can be applied by each of these general approaches. In some cases, one relies upon the intrinsic biodegradative capabilities of the indigenous microbial communities and monitors the movement and progressive slow decline in contaminant concentrations. This monitored natural attenuation (MNA) approach is valuable when there are no acute threats to human health and where the impact is not spreading rapidly. In other cases, active remediation is needed to curtail the impact. Depending upon the nature of the problem, it may be necessary to excavate the contaminated soil and move it to a site for its safe disposal or treatment. This ex situ approach to bioremediation is analogous to what is done for traditional sewage and solid waste treatment. Bioremediation may also be conducted in situ, for example, by bioventing, in which air is used to move the contaminants from the groundwater into a phase where evaporation and biodegradation can occur simultaneously. Depending upon the nature of the pollutant, there may already be sufficient populations of microorganisms to degrade the contaminant—in which case stimulation of those microbial populations by environmental medications (e.g., addition of fertilizer or oxygen) may be all that is required. In other situations, it may be beneficial to consider augmenting the indigenous microbial populations by seeding with specialized microorganisms, including possibly with genetically modified microorganisms. In this chapter, we review the various bioremediation technologies and the situations to which they are applicable.

Because groundwater is a hidden resource, it is too easy to forget that its misuse is a hidden problem. For the purpose of context, the global importance of groundwater needs to be emphasized. As many as 2 billion people rely directly on aquifers for drinking water, and 40% of the world's food is produced by irrigated agriculture that relies largely on groundwater (220). At least 12 megacities (populations of over 10 million) could not function without groundwater, and typically at least 25% of the water for these cities comes from aquifers.

China alone has over 500 cities, and two-thirds of the water for them comes from aquifers.

Despite this importance, the number of instances of groundwater contamination due to accidental spills or unsatisfactory disposal is beyond counting. Up to a certain point, natural processes, especially biodegradation, can attenuate contamination. In this regard, the biological active zone is the vadose (unsaturated) zone, where attenuation rates are highest. Contaminant removal continues in the saturated zone but usually at much lower rates, and migration of contaminants to the saturated zone can have the effect of dispersion of the contaminants. While bringing about dilution, the latter process often cannot be relied upon for complete decontamination.

The easy availability of groundwater and its vast supply (95% of the freshwater on the planet, apart from the locked water of the polar ice caps, is groundwater) have been its undoing. A great deal of it lurks fairly close to the surface, but intrusive disturbance of the subsurface has very high potential for destroying its flow and distribution. Therefore, techniques for remediating contaminated groundwater should operate by minimal disturbance. The sheer scale of the problem dictates that the remediation technologies should be as inexpensive as possible, and it is for that reason that bioremediation is often considered.

EX SITU BIOREMEDIATION TECHNOLOGIES

In many cases, it is necessary to move the contaminated soil or groundwater to a site where a suitable treatment system can be engineered. Contaminated soil may be excavated and moved to landfills; to thermal treatment systems, e.g., incinerators; or to a variety of bioremediation systems including biopiles, windrows for composting, landfarms, and soil slurry reactors. All have their merits. The choice of which technology to use often is driven by the required performance criteria, i.e., the nature of the contamination and the levels of cleanliness that must be achieved and the cost of remediation, including the cost of transporting the contaminated soil or water from the site of contamination to the site of treatment. Bioremediation technologies often can be performed at or very near the site of contamination, reducing the cost of transporting contaminated materials and thereby making the economics of bioremediation more favorable than the other physical disposal and treatment technologies. If the economic and technical analysis favors bioremediation and if a risk-based remedial design has concluded that ex situ treatment is the optimal approach, there are a variety of technologies available which can be considered.

Two technologies—biopiles and windrow composting—currently dominate the ex situ bioremediation market for treatment of contaminated soils. Both are aerobic processes in which the soil is excavated and heaped into a defined space for treatment. In composting, an organic material is added so that microorganisms generate heat through their metabolism, often causing the temperature to rise to at least 60°C (150). Except for the addition of organic material to support heat generation, biopiles and composting are essentially identical processes. Both technologies involve preparation of the contaminated soil to favor aerobic microbial metabolism of the contaminants; this may involve the addition of bulking agents, fertilizers, and water. Composting windrows for contaminated land bioremediation are aerated by periodic turning of the windrows with a modified windrow turner. In a biopile, aeration is accomplished through a network of slotted plastic pipes, either passively or by forced aeration. If space at a site is a constraint, then biopiles would be favored since a compost system requires sufficient space between windrows for access for the turning equipment. Biopiles can be formed with much larger volumes of soil than can be achieved for windrows, since the height of the windrow is limited by the size of the machinery used to turn it.

Biopiles and windrows have been used successfully at pilot scale and full scale for the bioremediation of a wide range of contaminants. While the majority of applications have been at petroleum hydrocarbon-contaminated sites (223, 339), they have also been used for manufactured gas plant sites (274), pharmaceutical wastes (118), chlorophenols (167), creosote (containing high concentrations of polynuclear aromatic hydrocarbons [PAHs]) (22, 65), pesticides (206), polychlorinated biphenyls (PCBs) (205), and nitroaromatics (46). Box 5.1 presents an example of a large-scale application in which biopiles were successfully employed as part of a large-scale bioremediation effort in Italy.

An emerging market for ex situ bioremediation by biopiles and windrows is cold-climate cleanup of fuel and oil spills. Crude oil is generally more persistent in Arctic tundra than in other regions, and a number of oil fields are situated in cold regions, e.g., in Alaska and Siberia. The feasibility of Arctic tundra ex situ bioremediation has been demonstrated (212), and the need for temperature, nutrient, and moisture control makes ex situ bioremediation the likely technology of choice in the Antarctic (7). These technologies appear to be applicable to arid desert contaminated sands also (31).

The third generic choice of ex situ technology is landfarming, which is more of a niche choice that has been used in the oil industry, from such diverse climates as Canada to Bolivia (371) to Saudi Arabia (126, 127), for decades for the treatment of refinery residues. It is a shallow treatment that is land-intensive, which makes it unpopular for most applications, particularly inner-city brownfield sites, typified by former gas stations and sites with leaking underground storage tanks (USTs).

The other generic technology is the use of slurry bioreactors, which offer quite different advantages over the other technologies, principally in the level of process control that is possible in a bioreactor and flexibility of operation. Combined slurry- and soil-phase approaches for bioremediation have also been investigated (224). In the following sections, we will discuss each of these approaches.

BOX 5.1
Biopile Treatment of Oil-Contaminated Soils in Italy

As a result of the Trecate 24 blowout that occurred in the course of drilling an oil exploration well in northern Italy, a very large area of highly productive agricultural land was contaminated. Biopiling and landfarming were successfully employed to provide an effective, integrated approach to petroleum hydrocarbon remediation. The unrecoverable crude oil impacted approximately 1,500 ha of superficial soils that were used primarily to cultivate rice, the vadose zone, and the underlying aquifer. To return the fields to production, the well's owner initiated a $45 million bioremediation project; bioremediation was used because of the need to remediate the valuable rice production soils without altering their pedelogic and rice production capacity (262).

Highly impacted soils (~5,000 to >100,000 mg of total petroleum hydrocarbon kg^{-1}) (approximately 12.5 ha with a volume of 26,000 m^3) were treated by using two biopiles, each approximately 135 m long and 50 m wide. The biopiles had an air injection system to provide oxygen to support aerobic hydrocarbon biodegradation, a moisture and nutrient delivery system to provide water and nutrients, a heat trace system to optimize temperature to ensure maximum biological activity, and a system of probes and sensors to monitor internal process control parameters. Biopile hydrocarbon biodegradation rates as high as 120 mg kg^{-1} day^{-1} were achieved.

Biopiles

Biopiles range in size from sub-cubic meters for pilot-scale investigations to 10,000 m^3 at full scale (163). Design is critical for achieving optimal performance, and remediation contractors have their own proprietary designs and engineering specifications for biopiles. In this discussion, design and engineering aspects will be considered in general terms with reference to real examples. The discussion will focus on the application of temporary biopiles, which are widely used around the world, as opposed to permanent installations, which have achieved success in the United States but are far less popular in Europe. The design and engineering considerations are similar but differ significantly with respect to the construction of the base upon which the contaminated soil is piled. Permanent installations usually have a concrete base, whereas temporary biopiles make use of different types of plastic liners.

Effective biopiles have been constructed in a large variety of shapes and sizes. There are no guidelines or limits for height or width of biopiles, but it is wise to build them such that the maximum reach of the front-end loader being used to form the piles is not exceeded. If this guidance is not heeded, then the front-end loader will inevitably run over the previous lift when adding the subsequent lift, thus compacting the soil and undoing the careful work previously done in soil preparation. In the early stages of construction, this might also destroy the piping runs at the lower levels in the biopile.

It is important that the shape and size of the pile should be considered as a means to creating conditions of even aeration throughout the pile. Sides that do not have a high slope might lead to overaeration of peripheral soil relative to soil in the core. This can also lead to "wasted" aeration, in that air is lost from the sides, which might have the added consequence of removing volatile contaminants with it, creating unwanted volatile organic compound (VOC) emissions and aerosols, and cause preferential drying of the peripheral soil if the pile is not covered. In practical terms, shallow sloping sides also waste space and complicate maneuvering with a front-end loader.

Pile height is again governed by aeration. The very simplest biopiles have no forced aeration and rely largely on temperature gradient-driven convection to create airflow through the slotted pipes. This limits the size of the piles considerably. Forced aeration, either by blower or vacuum pump, relieves this restriction. Experience has shown that a single piping layer close to the base of the pile is sufficient for piles up to 3 m in height (349). Electing to build pipes higher than this would necessitate adding a second layer of pipes, which greatly complicates the construction.

The width and length of the biopile are determined by the total volume of soil to be treated (after amendments) and the amount of space available on-site. For large sites, it is common practice to build multiple biopiles. They may all be identical or may be treating soils that have been identified as more or less contaminated. The equation for the volume of a biopile based on the above geometry is

$$V = 1/6h(B_1 + 4M + B_2)$$

Figure 5.1 shows how to derive the volume of a biopile from some known dimensions. Alternatively, given a volume of contaminated soil, the equation can be used to calculate the possible dimensions of a biopile relating to the amount of space available at a particular site. By assuming a contaminated soil density of 1.5 tons m^{-3}, sizing of a site for a known tonnage can be done, although greater accuracy would be achieved by measuring the density of the contaminated soil.

Biopiles have a variety of space requirements so that the site must be considerably larger than the biopile itself. A soil storage space is required to stockpile soil before processing. The size depends on the total volume of soil to be processed and the coordination with soil processing. It is quite feasible on a large site that a relatively small stockpiling area can be used since stockpiling, soil processing, and biopile construction can proceed at the same time. The majority of the soil processing space is required for stone removal, soil sieving, and shredding. A tank and manifold system may be required for leachate treatment. Another tank for water and/or nutrient supply may be needed. A secure container can be used to house blowers, with diesel generators kept close by outside. The container can also house spare parts, sampling equipment, and other sundry equipment. All of these have to be arranged logically on a site. At an early stage of site design, the turning circle of mobile plant, such as front-end loaders, tractors and trailers, and telescoping fork trucks, should be considered, especially in the critical areas of soil storage and processing. These should be located close to the biopile areas to maximize the efficiency of use of soil processing equipment. It is desirable to have the access road as short as possible and close to the soil processing area.

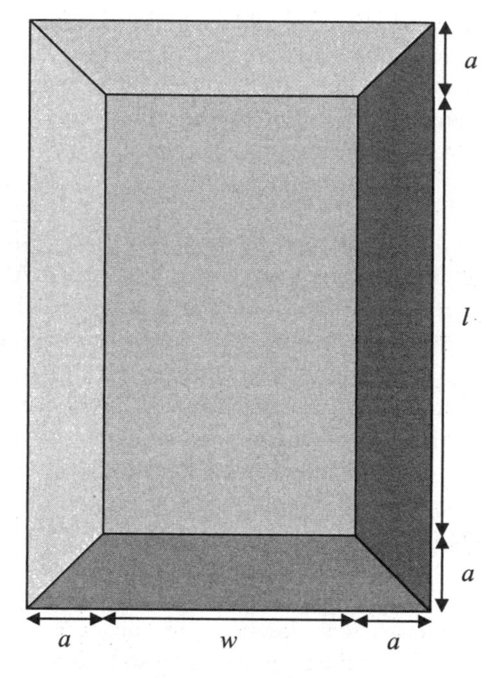

FIGURE 5.1 Calculation of volume of a biopile. After the work of von Fahnestock et al. (349). $V = 1/6h(B1 + 4M + B2)$, where V is the volume of the pile (in cubic meters), h is pile height (meters), $B1$ is the area of the lower base (square meters), $B2$ is the area of the upper base (square meters), and M is the area of the biopile midsection (square meters).

$$B1 = (l + 2a)(w + 2a) + lw + 2aw + 2al + 4a^2.$$
$$B2 = lw.$$
$$M = [1 + 2(a/2)] \times [w + 2(a/2)] = lw + la + aw + a^2.$$
$$V = 1/6h(lw + 2aw + 2al + 4a^2 + 4lw + 4aw + 4al + 4a^2 + lw).$$
$$= 1/6h(6lw + 6aw + 6al + 8a^2).$$
$$= h(lw + aw + al + 1.33a^2).$$
$$\tan\theta = h/a \Rightarrow a = h/\tan\theta.$$

An angle θ of 50 to 60° gives an approximate slope (h/a) of 1.2 to 1.75.

Soil sieving and stone removal equipment can be brought in on hire for short periods; the awkward size and shape of this equipment necessitate that the access make it easy to maneuver.

All of the space at a biopile treatment facility needs to be secure from vandalism. Access to the site for workers should be through a single entry point, where clothes can be changed. The idea of a "clean" and "dirty" side at the entry point reinforces the idea that a biological facility is being entered. The housing for this can also be used to store other materials deemed necessary for health and safety and emergencies, e.g., respirators, disinfecting solutions, and trays. The site office should be located on the clean side, and its size and facilities depend on the size of the project and the number of staff assigned to it on a full-time basis.

As a very rough estimate, excluding the site offices and access road, a site treating 1,000 m^3 employing biopiles would occupy about 2,500 m^2. On larger sites, economies of scale may be achieved since multiple biopiles should not require multiple stockpile and processing areas, although sufficient thought has to be given to the spacing between the biopiles for access for mobile plant. It should also be borne in mind that each contaminated site is unique, and site design is often constrained by the overall site size and shape. For example, a complex 4-ha site in the United Kingdom treating 34,000 m^3 of soil contaminated from coal coking operations by biopiles requires two full-time staff members, the site engineer and site scientist. The site engineer is responsible for contractual work, liaison with the regulator and clients, and dealing with subcontractors, and the site scientist takes all samples, performs field tests, and does liaison with analytical laboratories. The biopile formation and site closure stages require more staff, but once a project is running it is not labor-intensive.

BIOPILE COMPONENTS

The essential components and features of a biopile are shown in Fig. 5.2 (163, 349). The stylized geometry of a biopile is a trapezoidal encapsulated soil pile, with appropriate amendments, sitting on a base liner system (96). The biopile design includes piping for aeration and optional other design components, which may include an irrigation system and cover.

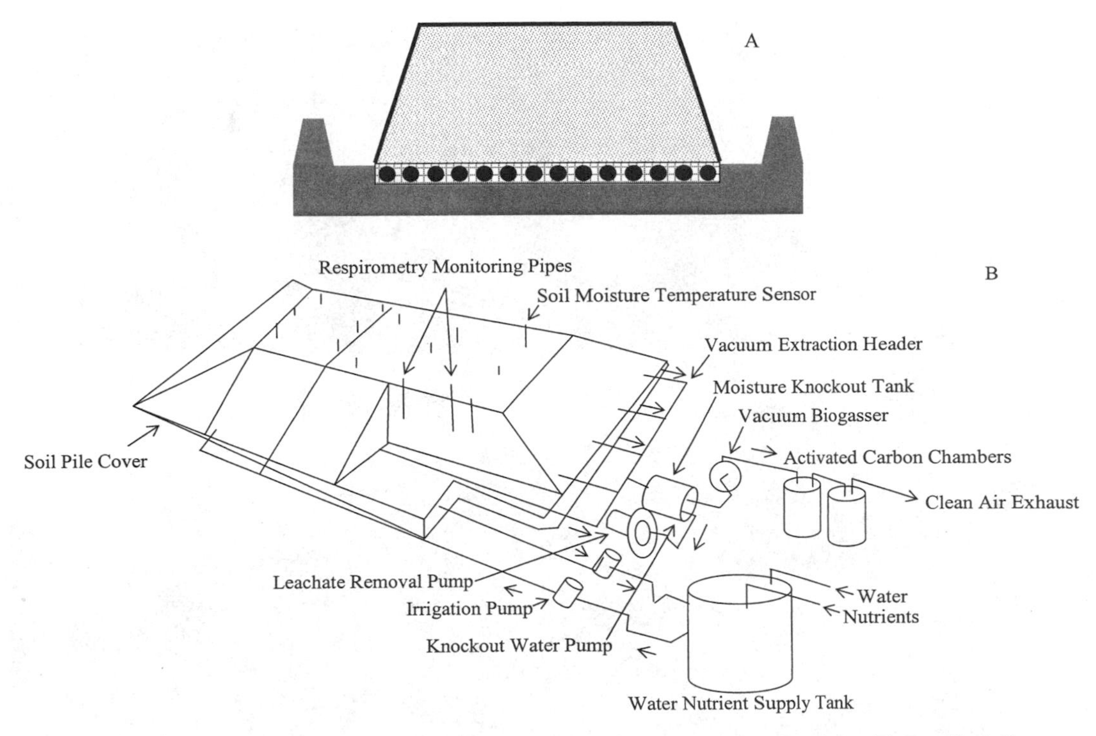

FIGURE 5.2 Elements of a biopile. After the work of Kodres (163). (A) Schematic. (B) Detailed diagram.

Biopile Base. The biopile base should be built on a relatively solid surface. At its most engineered, the base consists of a soil or clay foundation (up to 25 cm), an impermeable liner, and a bund to contain leachate (Fig. 5.3). The biopile base should have a slight slope of 1 or 2° to allow drainage of leachate to an appropriately sited leachate collection sump located at a corner of the biopile. The impermeable liner, usually clay or a synthetic material, is then placed over the base. Clay liners are not recommended for highly soluble contaminants such as phenol. Synthetic liners of a high-density polymer are recommended, with thicknesses from 40 to 80 mil (1 to 2 mm). High-density polyethylene (HDPE) with heat-welded seams is ideal for this purpose. Thinner liners should be capable of taking the weight of heavy, even-tracked plant without tearing. Clean soil can be compacted on top of the liner to further protect it.

Aeration System. Once a biopile base has been constructed, the aeration system is installed on top of the liner. Slotted plastic pipes of various thicknesses and materials are available as the main element of the aeration system within the pile. Polyvinyl chloride (PVC) slotted pipe (2- to 4-in. diameter) is a common choice (Fig. 5.4). The pipes are embedded within a highly permeable matrix, such as new wood chips or gravel, to act as an aeration manifold, whether operated with a blower or a vacuum pump. The depth of this layer is variable but obviously the pipes must be sufficiently covered to minimize short-circuiting from them.

The length of the biopile determines the length of each aeration pipe run. At the manifold header side, each aeration pipe run starts with a solid, not slotted, length of pipe of the same diameter as the slotted pipe. This is typically of the order of 3 m long, the actual length dictated by the distance to the aeration manifold. It should proceed about 3 m into the pile before it is joined to slotted pipe to ensure that short-circuiting does not occur right at the start

FIGURE 5.3 Photograph of a biopile base showing its preparation. Courtesy of WSP Remediation Ltd., Cardiff, United Kingdom.

of the pile. Next, the solid pipe is connected to the slotted pipe with a rubber or plastic connector. The length of the slotted pipe is dictated by the length of the biopile and, as at the start of the pipe run, should terminate some 3 m short of the far end of the biopile to prevent short-circuiting through the sloping edge. The slotted pipe is terminated with an end cap.

The start of the pipe run is connected to the header manifold by a gate valve of appropriate diameter. When several pipe runs are used to aerate a wide biopile, the gate valves are used to equalize the flow of air through each pipe run. The zone of influence of each pipe run is influenced by rate of airflow and also by soil porosity. Too-rapid airflow causes drying of the pile and may drive off VOCs creating an environmental concern in the vicinity of the pile. It also uses excessive electricity. By using relatively low-power blowers, this problem can be circumvented. To ensure even aeration within the pile, then, it is necessary to add several pipe runs parallel to each other. Once the biopile is built, it is very difficult to influence the porosity of the soil. As a general rule, parallel pipe runs are spaced about 2.5 to 3 m apart. Air velocity can be easily measured in each pipe run.

Connection of the header manifold to the blower then completes the system. A low-power centrifugal blower (Fig. 5.5) is sufficient for aerating large volumes of soil, and it is wise to install several small blowers of various powers (1 to 5 hp). Varying the rate of aeration over the duration of a project makes sense, as the need is greater at the start when there is a high level of contamination and microbial activity must be stimulated. Variable-speed blowers are therefore advantageous. It is not advisable to control aeration by on-off cycling, especially in the early stages, as even short periods with the blower off can lead to anaerobio-

FIGURE 5.4 Photograph of a biopile slotted pipe (plastic pipe, 4-in. diameter).

sis in regions of the pile, and this is very difficult to monitor. As an average figure, the blower should be capable of delivering about 0.14 m^3 of airflow per pipe run per min (130).

Depending upon the location of the project, operating in blower mode might have a high potential for stripping VOCs and creating an odor and health risk on and even off the site. Local legislation may then require containment measures that are rather expensive. In such circumstances, it is better to operate aeration by vacuum, as the captured air can be passed through a VOC removal system, typically granular activated carbon (GAC), and thus ameliorate the odor and risk. There has to be good justification for this, however, as operating in the extractive mode complicates the aeration setup. Pulling air through the pile will entrain condensate, and even leachate may be pulled through, so the vacuum pump must be preceded by water knockout and collection tanks. Granular activated carbon treatment of off-gases considerably increases the cost. It has been shown under laboratory conditions that VOC volatilization can be suppressed by the addition of activated carbon as a soil amendment (242). On a full-scale biopile, however, this would represent a large cost. If the extractive mode is deemed necessary, the pump should be capable of removing at least 15 pore volumes per day (131).

Passively aerated biopiles have also been used, in which a similar engineering system is used but without mechanical aeration. Airflow is created by differences in temperature between the soil and the outside air driving convective currents, facilitated by the slotted pipes, and by wind-induced pressure gradients. Naturally, this can achieve more limited aeration and might lead to uneven aeration as a result of a limited radius of influence as the air leaves the pipes and enters the soil mass. The central region of a pile would be particularly prone to local oxygen deficit (163). Wind-driven tur-

FIGURE 5.5 Photograph of a typical blower for full-scale biopile operations. Courtesy of WSP Remediation Ltd.

bines have been investigated to see whether improved aeration can be achieved without electrical energy (176). Results were inconclusive, in that improved airflow was demonstrated but without evidence of enhanced bioremediation.

Covers. Many bioremediation projects have been done without covering the biopiles, but covers offer some advantages. A primary advantage is that a cover prevents leachate formation and simplifies the biopile design accordingly. Another is that covered biopiles lose very little water. A typical ex situ bioremediation contract for petroleum hydrocarbon cleanup might last 3 or 4 months, during which time a covered biopile might lose only 1 to 2% of its initial water content. Thus, it would be sufficient to amend the water content during the construction phase only. If the alternative is to install an automated sprinkler or drip-type system to maintain water content within defined tolerances, then a cover is a much simpler engineering option. The moisture content, nevertheless, must be monitored since forced aeration, by either blower or suction, tends to remove moisture as the air entering the pile usually does so at less than 100% humidity.

The cover is often waterproof plastic sheeting, Visqueen, or a thin grade of HDPE liner, sufficiently thick to prevent tearing while being manually removed and refitted. If extractive aeration is being performed, completely impermeable liners may not allow sufficient air circulation. Framing systems have been tried but complicate the installation and are prone to water ponding and collapse. Alternatively, the pile can be covered with wood chips, which retain heat but allow air to flow.

Recently, fleece liners have been adopted from the composting of green wastes. Fleece liners are made from blown polypropylene.

They have properties similar to fleece garments. They resist moderate rainfall but are not completely impermeable to water and are also gas permeable. They allow some water to pass through to a biopile during operation, which will replace water lost through aeration. Gas permeability overcomes the problem with completely impermeable plastic covers mentioned above. Equipment is also available to mechanically remove and replace fleece liners (Fig. 5.6) from biopiles to cut down on labor requirements.

Irrigation Systems. If water has to be added, it is preferable to do it at the biopile formation stage if the batching procedures described above are being used. If an alternative water addition system needs to be employed, the preference is for a dripline irrigation system. Water flow from such a system can easily be monitored, the low rate of application prevents runoff, it can be set up to evenly irrigate a biopile, and it can be operated without supervision. However, it is inevitably a further engineering complication that is best done without if possible.

BIOPILE FORMATION

The standard way to form a biopile is to add all necessary amendments to the graded, sieved soil and then start lifting the soil onto the base and aeration system with a front-end loader. An alternative is to form the biopile with graded soil and any bulking agents and then add liquids at a later stage. The latter strategy carries risks in that it is difficult to achieve uniform distribution of materials. Usually, nitrogen and phosphorus are applied together. If a liquid source is sprayed onto the top of a biopile, the mobility of the nitrogen source will allow it to travel through the pile. However, phosphorus interacts with metals and other soil components and will be immobilized within a

FIGURE 5.6 Photograph of a fleece roller. A self-propelled windrow turner can be seen in the background. Courtesy of Shanks Waste Management Ltd., United Kingdom.

meter of the top of the biopile. Also, if the occasion dictates the use of inocula, then very quickly the microorganisms used will be immobilized and achieving uniform mixing by application after biopile formation is all but impossible. Many laboratory and field trials have shown that bacteria do not move appreciably through soil (10) as a result of physical filtration by small soil pores and adsorption to soil particles.

Therefore, the safest way to guarantee the maximum level of homogeneity of materials in a biopile is to mix everything just prior to, or even during, biopile formation. The standard practice is to hire in soil grading (screening) equipment to remove stones and grade the soil down to an average particle size, usually 30 to 50 mm (Fig. 5.7). Parallel bar screens remove large stones, and then trommel or vibrating screens are used to grade the soil down to a smaller size. Large amounts of clay will require soil shredding as the next step in the process.

Knowing the capacity of the loader being used to feed the screening equipment and the speed of the conveyor belt, it should be possible to spray liquid amendments at the screening stage with some accuracy. It should also be possible to add bulking agents to a known percentage (usually by volume). If the soil requires shredding, then the effort will have been wasted. After screening and shredding, various equipments are available for soil mixing, and this is the optimum stage for adding amendments. The Kuhn Knight Reel Auggie (Fig. 5.8) is typical of U.S. equipment for this purpose, although it is designed for compost mixing. Contaminated soil may be batched with the other liquid and solid amendments, and the mixing action allows efficient aeration. A large, slowly rotating reel drives materials forward to

FIGURE 5.7 Photograph of soil screening and addition of wood chips (the lighter material in the soil heap) and bioaugmentation culture (contained in the 1-m^3 Intermediate Bulk Container), which is being applied by spray at the top of the soil grader.

FIGURE 5.8 Photograph of a Knight Reel Auggie for soil mixing and distribution. Courtesy of Kuhn Knight.

two blending augers. The combined action of the augers mixes, aerates, and discharges the materials, and the discharge can be done directly to a forming biopile or compost windrow. This equipment can be truck mounted or tractor mounted and can be driven electrically or by self-contained diesel engine. As soil is batched, it is easier to control the volumes and thus the final concentrations of any amendments.

In the United Kingdom, hiring a soil screener is expensive, and its operation can force a reduction of activities on a small site. A relatively recent development has been the use of the ALLU bucket for the purpose of soil grading (Fig. 5.9). This equipment is able to crush stone, brick, and lightweight concrete while processing and aerating soil, and as the bucket size is known, amendments can be added accurately. It can be mounted on a variety of equipment, including front- and rear-end loaders and tracked and tire excavators. The high level of mobility means that mixing can be done directly onto a biopile or windrow, or it can be done directly onto a trailer. It has another convenient advantage in that the output tends to form soil piles in the shape required for windrows, so that subsequent manipulation to finish a biopile is easy.

Conventionally, a biopile is formed from back to front, the back being the end at which the aeration header is located. Common sense precautions to be taken are to make sure that previous lifts are not driven over, since this compacts the soil; driving over the aeration system will destroy it.

Windrow Composting

As described above, the main difference in materials used in windrow bioremediation and biopiles is that some organic, heat-generating material is added to the windrows. The objective is to increase the metabolic rate of indigenous hydrocarbon oxidizers. Most bacteria

FIGURE 5.9 Photograph of an ALLU bucket. It is fitted with rotating drums with blades, and also crushing bars, so that it can be used to grade soil containing materials such as brick. Courtesy of WSP Remediation Ltd.

grow over a range of approximately 40°C, whatever their optimum temperature for growth. Within the normal temperature range for growth of a bacterium, growth rate obeys the Arrhenius relationship between reaction rate and temperature. Increasing the temperature of a contaminated soil windrow from 20 to 30°C would be expected to more or less double the growth rate of hydrocarbon-oxidizing bacteria and therefore speed up the bioremediation process. Another effect of increasing temperature might be to increase the bioavailability of poorly water-soluble contaminants since increasing the temperature should increase solubility, as well as decreasing viscosity of oily contaminants. During composting of contaminated soil, the thermophilic stage is not reached, and the temperature does not exceed 45°C (282).

In this regard, bioremediation of contaminated land by windrow treatment is a very different operation from composting of organic wastes or sewage sludge. The main difference between contaminated soil composting and traditional composting is that the former lacks the concentration of organic materials of the latter. A direct consequence is that heat generation is normally lower in contaminated land composting. For traditional composting of waste materials, the high temperature generated in the procedure is essential: several ordinances stipulate that every part of the compost must reach a temperature of 65°C at a water content of at least 40% for 3 consecutive days (143), for the purpose of pathogen kill. For contaminated soils, the likelihood of the material containing large pathogen loads is much less, and also high temperatures are not suited to the metabolism of most hydrocarbon-oxidizing bacteria. In addition, such high temperatures would involve water loss at a level that would complicate windrowing operations.

One of the purposes of turning windrows is to dissipate heat, and the increased porosity achieved by adding bulking agents aids heat dissipation.

WINDROW COMPONENTS

A large variety of organic amendments has been used in composting bioremediation. Many are based on the application of manure, from either cows (22), pigs (368), or chickens (274). Sewage sludge is abundantly available globally, and it has been successfully used as an amendment in composting bioremediation (141). Virtually any putrescible material available in bulk can be used, such as vegetable wastes (22), spent mushroom compost (SMC) (95, 169), and even garden waste (118, 119, 205). The use of composting approaches to bioremediation of organic pollutants generally (282) and specifically the use of composting to treat PAHs (21) have been reviewed.

The use of SMC is an interesting case. SMC is the residual compost waste generated by the mushroom production industry. It is readily available, as mushroom production is the largest solid-state fermentation industry in the world; in the United Kingdom alone, there are some 400,000 to 500,000 tons of waste mushroom compost produced per annum (281). It consists of a combination of wheat straw, dried blood, horse or chicken manure, and ground chalk composted together. It is a good source of general nutrients (0.7% N, 0.3% P, 0.3% K plus a full range of trace elements), as well as a useful soil conditioner. A fascinating feature of SMC is that it may contain a relative abundance of extracellular ligninolytic fungal enzymes (169), which are relatively nonspecific in their substrate preference. Hence, they may assist in the biodegradation of aromatic molecules such as PAHs, giving SMC an additional role in composting bioremediation.

The construction of more-robust turning equipment, based on compost windrow turners but strong enough to turn large soil windrows, has allowed full-scale windrow bioremediation of contaminated land. Much of the discussion is similar to that for biopiling. For example, soil preparation is identical in the need for water, nutrients, and bulking agents, although the initial care in mixing need not be so rigorous since the whole point is that aeration is brought about by pile turning. The turning naturally mixes the contents of the windrow: the more often it is turned, the greater the homogeneity of materials, including moisture. Windrows are inherently simpler in design; therefore, this section will focus more on the differences between windrow composting and biopiling.

The natural shape of a windrow is more like a triangular prism than the characteristic trapezoidal, or incomplete pyramid, shape of the biopile. Windrows are suited to long, narrow sites, as they can be constructed to any length required (Fig. 5.10). Typically, windrows for bioremediation of contaminated soil would be of the order of 1.5 to 2 m high and perhaps 3 to 4 m wide. With the arrival of large, self-propelled windrow turners, windrows can now be constructed to greater heights and widths than before. Doing so necessitates more regular turning to prevent anaerobiosis, though.

Given that compost windrows can be much

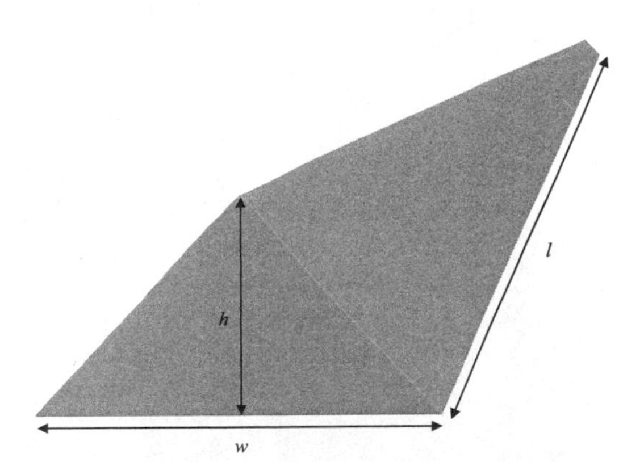

FIGURE 5.10 Calculating volume of a windrow. Volume of windrow = $1/2\ w \times h \times l$. If the mass of contaminated soil and its density are known, the total volume of material, including the bulk materials such as organic compost and bulking agents, can be calculated. Knowing the width of the site, and the space between windrows, the number of windrows of set width can be calculated.

longer than they are wide, the small volume of material at each end can be ignored in sizing. That is why the volume of a triangular prism is a realistic choice.

WINDROW FORMING AND TURNING

Comments similar to those for bed construction pertain to windrow composting and biopile construction, except that for windrows there is no added complication of installing piping runs for aeration. Mixing can be done during windrow formation. For example, if wood chips are to be used for bulking, and the soil-to-chips volume ratio has been calculated, then the wood chips can be laid out on the windrow bed along with, say, pellet NPK fertilizer, then the moist soil, and any additional heat-generating organic material, and the windrow turner will mix them together while forming the windrow.

A variety of equipments can be used to turn windrows. Turning can even be done with a small tractor with loader, depending on the scale of the operation. The quality of the mix depends greatly on the amount of time that the operator is willing to spend. The ALLU bucket can also be used in this function (Fig. 5.11). These options are relatively time-consuming compared to the use of dedicated windrow-turning equipment.

For medium-scale operations, a custom-made turner driven from a tractor power take-off (Fig. 5.12) is suitable for windrows around 1.25 m high and 4.35 m wide. Largest-scale operations require large, self-propelled windrow turners (Fig. 5.6, background), which can turn windrows around 2.5 m high and 5.5 m wide. Paradoxically, space is saved by using large self-propelled vehicles: the space between windrows is of the order of 1.25 to 3 m, whereas using a tractor-driven turner requires

FIGURE 5.11 Photograph of an ALLU bucket being used as a windrow turner. Courtesy of WSP Remediation Ltd. Its adaptability to different machinery and its low cost make it a flexible alternative to windrow turners, although slower.

FIGURE 5.12 Photograph of a tractor-driven windrow turner. Courtesy of Environmental Reclamation Services Ltd., Glasgow, Scotland, United Kingdom.

a space between windrows of 3 to 4.5 m, and self-propelled vehicles build higher windrows.

For an example of bioremediation using windrows, see Box 5.2.

In-Vessel Composting

At first sight, in-vessel composting, i.e., treatment within a bioreactor, offers some advantages, mostly relating to the higher degree of process control that can be applied, e.g., better temperature control, control of odors, and improved mechanical mixing. However, it is not a popular full-scale practice for at least two reasons. Very often the volumes of soil are too large to make this a cost-effective approach. Mobile in-vessel plant would be too small for mass application, and transporting large volumes of contaminated soils from the field to large fixed facilities raises a number of concerns, including safety, and not least, the extra treatment cost associated with transportation. Bioreactors, however, can be used in some cases for treatment of contaminated groundwater.

Landfarming

The basis of landfarming is the controlled application of waste on a soil surface to allow the indigenous microorganisms to biodegrade the contaminants aerobically (195). Because it is a shallow treatment with a large exposed surface area, the relative contributions of biodegradation and volatilization have always been debated. Volatilization of crude oil in temperate climates is minimal, whereas in hot climates it might approach 40% (219).

Landfarms differ from biopiles and windrow composting technologies in that landfarms are fixed facilities to which the materials for remediation are transported. The oil industry has a long history of using landfarming; landfarms around the world are used to treat various upstream (drilling wastes) and downstream (refinery wastes) materials, including oily sludges.

BOX 5.2
Bioremediation Using Windrows at Newcastle-upon-Tyne

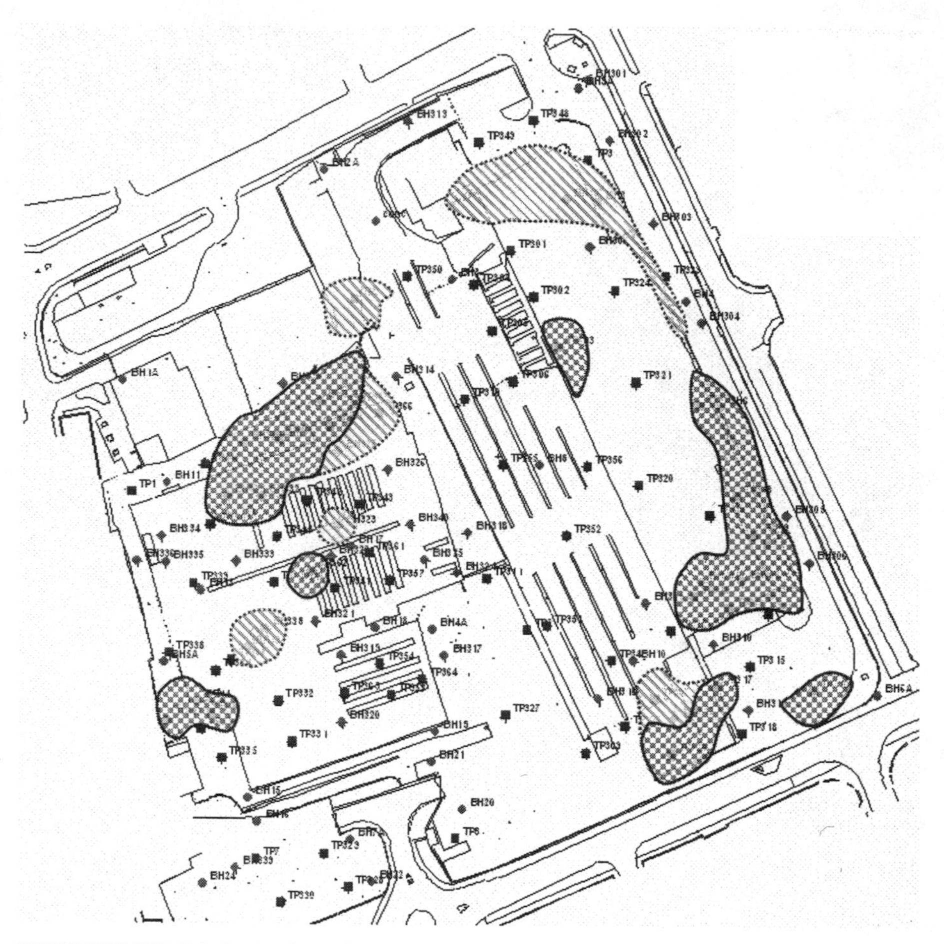

BOX FIGURE 5.2.1 Hydrocarbon contamination at the site. Dark squares are areas of highest contamination.

A former quarry site was to be redeveloped as a shopping complex in Newcastle-upon-Tyne, United Kingdom. The site was an in-filled quarry, containing more than a 10-m depth of incinerator ash. Before development could be done, it was necessary to determine that the land was not contaminated to a level that would present a significant risk to human health.

An extensive site investigation was performed which revealed the following:

- Heavy metals were recorded within the ash;
- Low concentrations of dioxins were recorded within the ash;
- Numerous above- and below-ground storage tanks were identified;
- Up to 1.2 m of phase-separated product was found on the groundwater;
- Asbestos was identified within buildings;
- There were widespread hotspots of hydrocarbon contamination (Box Figure 5.2.1.).

A quantitative risk assessment was carried out; the remediation criteria are shown in Box Table 5.2.1. The quantitative risk assessment revealed that benzene and benzo(a)pyrene were the main drivers ne-

BOX 5.2 *(continued)*

BOX TABLE 5.2.1 Remedial criteria for soils

Contaminant	Maximum identified site concn (mg kg^{-1})	Remedial target concn (mg kg^{-1})	Derivation[a]
TPH	11,749	5,000	Lower threshold of TPH, saturation in soil
C_5–C_7 aromatic hydrocarbons	0.011 (as benzene)	1.9	RBCA—volatilization from soil to indoor air
Benzene	0.011	2.1	RBCA—volatilization from soil to indoor air
PAH	24,000	1,400	Extrapolation of benzo(a)pyrene target by TPHCWG critical fraction method
Benzo(a)pyrene	1,118.9	140	RBCA—to ensure risks to future construction/maintenance workers are acceptable

[a] RBCA, risk-based corrective action; TPHCWG, total petroleum hydrocarbons criteria working group.

cessitating cleanup before construction of the shopping complex could begin. The risk assessment indicated the potential receptors at the site, and risk-based remedial targets were set.

Based upon the quantitative risk assessment, a multifaceted remediation approach was developed. As is often the case, a mixture of contaminants was found, and the remedial remedy had to deal with not only contaminated soil and groundwater, but also infrastructure in the form of unwanted buildings and storage tanks.

The strategy for cleanup is described below.

- All concentrations of heavy metals and dioxins were below risk-based target criteria, so no soil remediation was necessary.
- For all identified asbestos within buildings, controlled removal was necessary prior to demolition. The ubiquitous bête noire asbestos was found in buildings, and this called for specialist removal of the asbestos before structures could be demolished, but no soil remediation was necessary.
- Phase-separated product was identified, so groundwater remediation with product recovery was necessary due to the Environment Agency Statutory Requirement.
- UST removal was required due to the Environment Agency Statutory Requirement.
- Concentrations of hydrocarbons above risk-based target levels were identified, so soil remediation was required.

The site investigation and quantitative risk assessment identified 16,000 m^3 of hydrocarbon-contaminated soil requiring remediation. Approximately 50% of the contaminated soil was granular ash. Most of the remaining contaminated soil was cohesive material or clay. The cohesive material was not suitable for bioremediation. It was excavated and removed for off-site disposal. Laboratory testing revealed that the ash fraction was amenable to treatment by on-site ex situ bioremediation. Thus, about half of the contaminated soil was amenable to ex situ bioremediation.

Windrowing was selected as the method of choice (Box Fig. 5.2.2). Trial pits dug on a 20-m grid across the entire site delineated the regions of hydrocarbon contamination. Excavation and screening of contaminated soils were conducted so that bioremediation would be performed only on material below 50 mm in size. The oversize material was crushed and reused on-site as engineering fill. The material for bioremediation was laid out in windrows in the treatment area, treated with nutrients, and aerated by windrow turning on a weekly basis.

Regular laboratory analyses were performed to monitor removal of contaminants until risk-based target levels were achieved and the material was released for use as general fill on the site. Box Fig. 5.2.3 shows the results of monitoring of windrow 11 during the bioremediation period. This windrow was 80 m long and had a volume of 600 m^3. As indicated by these results, bioremediation achieved target reductions in hydrocarbon levels within a relatively short time.

Box 5.2 continues

BOX 5.2
Bioremediation Using Windrows at Newcastle-upon-Tyne *(continued)*

BOX FIGURE 5.2.2 Field measurements on windrows. Courtesy of WSP Remediation Ltd., Cardiff, Wales, United Kingdom.

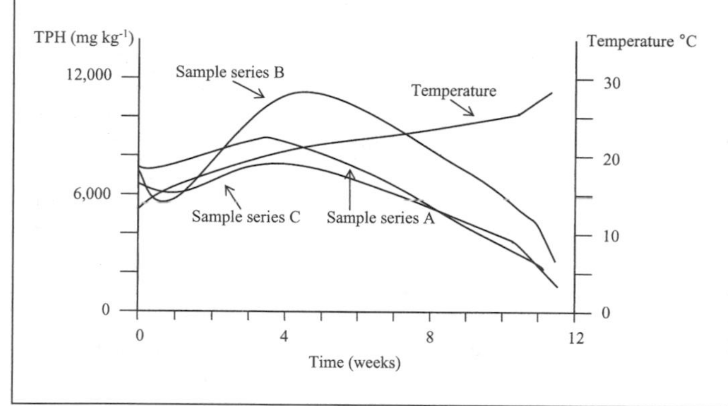

BOX FIGURE 5.2.3 Three-sample series from windrow 11, showing progress of TPH removal during the bioremediation phase of the contract.

Landfarms also have been proven effective in reducing concentrations of all components of fuels found in USTs, including petrol, diesel, and kerosene, as well as primarily nonvolatile oils such as heating and lubricating oils (324). For the oil industry, it has been seen as a relatively cost-effective and simple technique for dealing with refinery wastes, but there are environmental concerns over landfarming operations (11). In terms of total refinery wastes, a small proportion is landfarmed. Between 1986 and 1993, a mere 4.9% of all refinery sludge was disposed of by landfarming from 89 reporting Western European oil refineries (340); despite being the cheapest reported option, very few refineries in Europe use it. In a number of countries, the technique is not permitted at all (77).

LANDFARM DESIGN

The preparation of the site for a landfarm bears similarity to the methods for preparation of the base for biopiling, but of course the area to be prepared has to be much larger. The amount of land required depends on the volume of materials to be treated and the depth of the landfarm soil. The depth of landfarm soils is usually between 30 and 45 cm, although very powerful soil tillers can till down to about 60 cm. An advisable upper sludge loading is 150 g of sludge per kg of soil (13). Waste is typically applied in layers of no more than 20 cm. Time should be invested to maximize the accuracy of the land requirements, as a key to economical landfarm operation is to size the installation properly and then run it close to capacity (W. Younkin, C. Suaznabar, and S. Parsons, paper presented at the SPE International Conference on Health, Safety, and the Environment in Oil and Gas Exploration and Production, Stavenger, Norway, 26 to 28 June 2000). The most difficult and variable part of the calculation is the residence time (the time taken to reach the treatment target). Younkin et al. reported a variation in residence time to reach a 1% total petroleum hydrocarbons target of 139 to 218 days. Given variabilities in soil, climate, and sludge quality, treatability studies are highly recommended (Fig. 5.13).

Landfarm Base. Landfarm surface areas are highly variable but may be of the order of 4 ha (40,000 m^2). The total land area is divided up into treatment cells, generally square and of variable size. A U.S. Army design specifies a treatment cell size of 1 acre (4,047 m^2) (320). A BP Amoco landfarm to treat drilling wastes is a 4-ha site in a rectangle, laid out in 45 treatment cells of 30 by 30 m each. By so dividing the land, wastes can be sequenced in time so that continuous operation can be maintained.

Given the large area and the shallowness of a landfarm, the base has to be properly designed and built to accommodate local climate. Leachate control is essential for wet and temperate climates. To this end, a high-integrity liner is required. A compacted clay layer of 0.6 m with a hydraulic conductivity of 10^{-7} cm s^{-1} is suitable. A 1-mm-thick HDPE liner is a suitable alternative, and at permanent sites, both may be used.

The leachate collection system to sit on top of the liner consists of slotted-pipe laterals embedded in a granular drainage layer (Fig. 5.14). The grade of a treatment cell should be between 0.5 and 2%, depending on local rainfall. The granular drainage layer is formed from compacted, well-sorted gravels and coarse sand, with a minimum compacted hydraulic conductivity of 10^{-2} cm s^{-1}. Gravels greater than 13 mm in diameter may damage the HDPE liner, and it would be advisable in such a case to have a protective layer of sand or geotextile between the granular drainage material and the liner.

Perimeter Dike. The perimeter dike is designed to prevent runoff and storm water run-on based on a 25-year flood. A minimum freeboard of 0.3 m between the top of the dike and the surface of the treatment cells, and also from the top of the dike to the exterior surface of the cell, is recommended.

Stockpile Area. Stockpiling is necessary for the storage of contaminated material for treatment, already treated material awaiting haulage out of the farm, and oversize material. Stockpiling areas should be lined with a chemically resistant impermeable geomembrane to a minimal thickness of 1 mm. Rain is prevented from entering the stockpile with a 0.25-mm-thick impermeable geomembrane.

LANDFARM OPERATION

The main activities of landfarm operation are placing of contaminated material, aeration, watering, fertilization, and removal of treated material. The most important of these is aeration. The soil amendments are as for other ex situ bioremediation technologies. The critical issues, as always, are water and oxygen.

FIGURE 5.13 Photograph of experimental landfarm plots at an oil sludge storage facility, Perm, Russia. The left half of the plot has been sown with common Russian grasses. One plot has been treated with biofertilizer (top), and the other is an untreated control plot (bottom). After Kuyukina et al. (165).

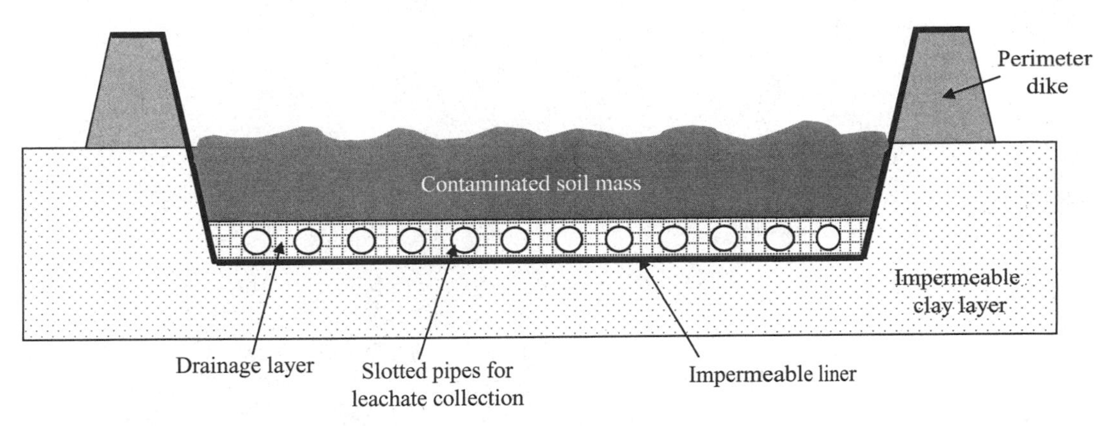

FIGURE 5.14 Diagram showing a landfarm schematic (not to scale).

Irrigation. As in all bioremediation processes, a fine balance has to be sought, between having enough water for essential microbiological needs and not so much that it would fill the soil pores and inhibit oxygen diffusion. A sprinkler-type irrigation system is the norm, with overlap between adjacent sprinkler patterns. The water delivery rate is variable, but the recommended lower rate is 0.7 liter s^{-1} (1,000 m^{-2}). The key issue is that the system should be able to deliver water evenly over the whole surface to prevent short-circuiting. As with other technologies, the moisture should be kept to between 40 and 80% of field capacity (water holding capacity). Unlike for the other technologies, greater effort has to be expended in monitoring the moisture because of the large surface area involved.

Nutrient Addition. Slow-release fertilizers are preferred for landfarming. After a new batch of contaminated soil or sludge is applied on a landfarm, standard practice is to measure ammonia, nitrate, and *ortho*-phosphate at 2-week intervals, especially during the first few weeks of treatment, and thereafter every 6 weeks for the duration of the treatment. A simple rule of thumb is to reapply fertilizer when the level of available nitrogen falls below 50 mg kg of $soil^{-1}$, and the application rate should be adjusted so that the level does not exceed 100 mg kg^{-1}. The trigger level for *ortho*-phosphate is 5 mg kg^{-1}.

Aeration. The technique for aeration is simple and well established. The soil is tilled with agricultural equipment (Fig. 5.15). The timing of tilling is the most important variable. If done too soon after rainfall with a soil containing a significant amount of clay, this can form clay clods, in which the subsequent aeration will be very poor. The U.S. Department of the Army (320) recommends tilling every 2 weeks. Huesemann (138) states that too-frequent tilling may not be advantageous due to deleterious effects on soil structure and increased evaporative water losses.

In a hot climate, Hejazi and Husain (127) noted that of the three main operations (tilling, watering, and nutrient addition), tilling was the activity most responsible for hydrocarbon removal. The majority of the effect could be attributed to weathering (volatilization) and not biodegradation. In fact, they also concluded that the addition of nutrients and water slowed down hydrocarbon removal, probably by compacting the soil and minimizing weathering. However, this was within the extremes of climate experienced in Saudi Arabia. In cooler climates, it is likely that biodegradation has a greater influence. Landfarming is also being considered, along with biopiles, for the treatment of oil-contaminated arid Kuwait soils (Box 5.3).

FIGURE 5.15 Landfarming oil refinery sludges. A disc harrow (top) and a tined tiller (bottom) are being used. Despite proven success in varied climates over several decades, there are concerns over this practice.

BOX 5.3
Bioremediation of Oil-Contaminated Kuwait Soils Using Landfarming and Biopiles

During the past decade, there have been several large-scale oil releases that have contaminated large areas of land. Probably the greatest contamination occurred in Kuwait as a result of the deliberate release of oil by Iraq at the time of the first Gulf War. The discharged oil formed over 300 oil lakes, covering large land areas (Box Fig. 5.3.1). Despite physical recovery of some of the oil, much remains a decade after the contamination occurred. Discussions are continuing to select remediation strategies that are appropriate and cost-effective.

Bioremediation, involving enhanced landfarming and biopiles, has been evaluated at the pilot scale and is one of the remediation techniques under consideration. Al-Awadhi et al. (9) reported on experiments that were initiated in November 1992 at the Burgan oil field, in which 16 landfarming plots of 12 m^2 each were constructed. Experimental plots for landfarming were established using heavily contaminated soils from oil lakes so that the concentrations of oil were 6 to 8%. The experimental plots were amended with nitrogen, phosphorus, and potassium (NPK, 15:15:15) fertilizer to produce carbon-to-nitrogen (C-to-N) ratios of 50:1 and 75:1. All of the plots were also tilled to a depth of 30 cm once every 2 weeks and irrigated with freshwater (2,000 liters per plot). The study was conducted for 18 months, during which time petroleum hydrocarbon concentration, PAH, and heavy metals were monitored regularly. The result obtained showed that landfarming treatment resulted in more than 80% reduction of oil contamination within 15 months. The treatment also resulted in a substantial reduction of the PAH concentrations. Plant toxicity experiments were carried out with bioremediated soil. The yield was almost within the normal range, so the treated soil had been sufficiently

BOX FIGURE 5.3.1 Photograph showing disastrous oil contamination in Kuwait as a result of the Gulf War that resulted in the formation of hundreds of square kilometers of oil lakes.

BOX 5.3 *(continued)*

restored for vegetation. From the plant analysis after vegetation, it was clear that the uptake of PAH, sulfur, and heavy metals was negligibly small. These studies indicate that optimum landfarming conditions are obtained if the soil is well aerated and contains sufficient moisture and nutrients. The indigenous microorganisms will be efficient enough to biodegrade the oil, provided that optimum conditions are maintained during treatment. Overall bioremediation seems to be a practical approach to the restoration of oil-contaminated soils, even over large areas of heavily contaminated desert soils.

Soil Slurry Reactors

Slurry-phase ex situ bioremediation has several apparent advantages. The addition of water makes the materials easier to move, by pumping, and the ability to set up liquid-phase bioreactors means that a greater degree of process control is possible. A great deal of knowledge of the functioning of fermentors and chemostats has been accrued, and the interrogation of kinetics in slurry-phase systems should accordingly be simplified. Another advantage of slurry-phase bioremediation that has been championed is the great degree of flexibility that can be achieved.

A typical setup for a soil slurry reactor is depicted in Fig. 5.16. Having several reactors in series means that the operating conditions can be easily and rapidly modified. Indeed, even the metabolism can be altered drastically. For example, attempting to treat, say, a waste containing a highly chlorinated phenol aerobically may be folly. Since the molecule is already highly oxidized due to the presence of electronegative chlorine atoms, further electrophilic microbial attack may prove futile. Under such circumstances, the first reactor might be better run anaerobically to bring about at least partial reductive dechlorination (163). Removal of chlorines and substitution with hydrogen would make the molecule more amenable to oxidative microbial hydroxylation and subsequent ring cleavage, which could be carried

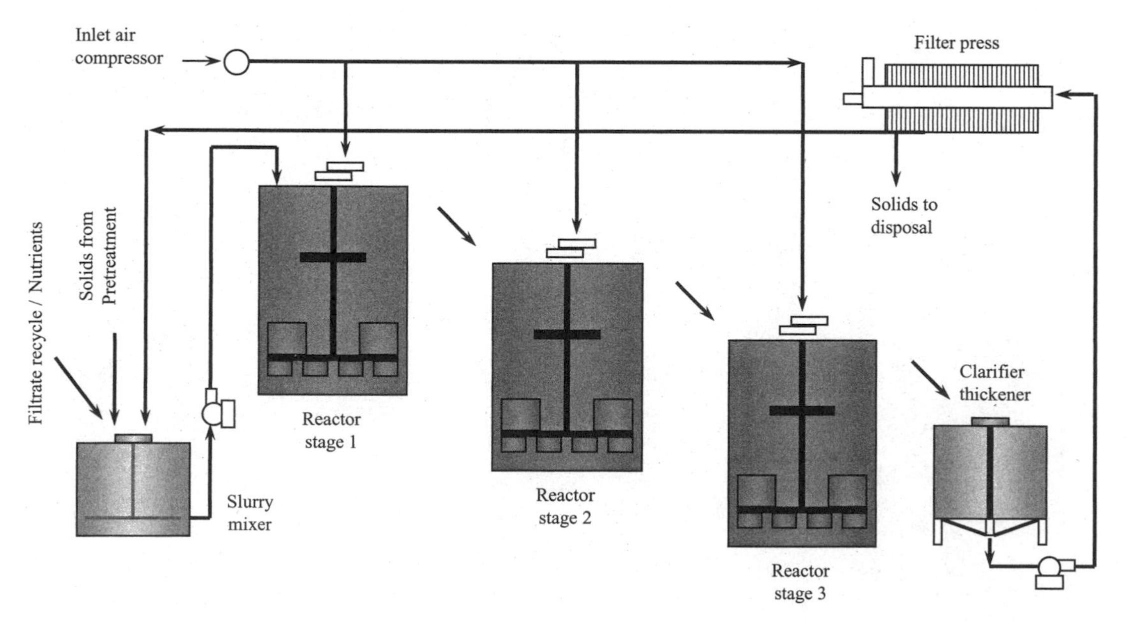

FIGURE 5.16 Soil slurry reactor.

out in a second bioreactor stage. Subsequent stages would involve slurry thickening and then filtration to form a treated cake. Such an arrangement might speed reaction, not only by optimizing growth conditions, but also by increasing the apparent concentrations of contaminants in solution by encouraging desorption from soil to the aqueous phase. Indeed, some studies do convince that this approach is efficacious. For example, Geerdink et al. (107) demonstrated experimental oil biodegradation rates 70 times higher in a slurry process than in a comparable landfarm.

The reasons that soil slurry reactors are not popular for contaminated land bioremediation are due to practical consideration. Large amounts of water would be required, and the engineering is complicated by the need for water recovery and recycling. Water itself is an expensive commodity. Another reason is that the large volumes of soil at many sites mean that the bioreactors would be impractically large. This would virtually dictate that such facilities were fixed and so would require the transportation of large volumes of contaminated soil, most probably by road. Soil slurry bioreactors would have to have a good record of full-scale success before the large investments necessary would become available.

IN SITU BIOREMEDIATION TECHNOLOGIES

The very term bioremediation should tell us that this technology is highly dependent upon external conditions, which is key to determining whether bioremediation can be performed in situ. Are the in situ environmental conditions suitable for the microbial activities necessary for successful bioremediation to occur? Can the environment be modified to create conditions that favor microbial activities and bioremediative removal of the pollutants? The conditions of greatest importance in this consideration are the physicochemical and chemical conditions that exist in the contaminated soil and water. Conditions that have to be optimal or near-optimal to allow bioremediation to proceed at a reasonable rate include the following: dissolved oxygen for aerobic processes; electron acceptors for anaerobic processes; pH; temperature; nutrient availability, especially with regard to nitrogen and phosphorus; water content (for soil); soil composition; alkalinity; salinity; metal concentrations; and concentration bioremediable contaminants (biodegradability versus toxicity). This list is not exhaustive, but already it is obvious that microbial activity in a contaminated site is affected by a wide variety of site conditions.

Often, having to deal with microorganisms and having to consider this complex list of environmental parameters are off-putting to the remediation engineer. However, previous experience can simplify matters. In most cases, it is possible to simplify this list to the most important conditions, which are (in no particular order of importance):

- water content or moisture,
- pH,
- temperature,
- nutrient status, and
- electron acceptor status (usually oxygen).

All of these factors are more or less influenced by a fundamental property of soils that requires some discussion before proceeding further. That property is soil porosity, which is the ratio of the volume of voids to the total volume of a soil and is expressed as a percentage. Above all, soil porosity has a defining influence upon the transport process that will make or break the applicability of bioremediation. Water availability, contaminant availability, gradients of oxygen and biodegradation waste products, pH, and even heat transfer are all strongly influenced by porosity. Porosity also influences the movement of contaminants into underlying groundwater and away from the site of contamination.

While the porosity gives an indication of the applicability of bioremediation to a contaminated site, alone it is not a sufficiently detailed measure. Soil is a three-phase system composed of solids, liquids, and gases. The solid components determine the porosity, and the pores are more or less filled by the liquids and gases. The ease of flow of water through a soil is its hydraulic conductivity, or coefficient of permeability, which in soil science is defined by Darcy's law:

$$Q = AK \cdot d\Psi / dx$$

where Q is flow (amount of water passing per unit of time through cross-sectional area A), K is hydraulic conductivity, and $d\Psi/dx$ is the water potential gradient (change of potential per unit of distance along the direction of flow).

Hydraulic conductivity is a velocity term and is often expressed as centimeters per second. The lower the porosity of a soil, the lower the hydraulic conductivity, and at low values there is a serious inhibition for the application of bioremediation. In particular, the very low hydraulic conductivity associated with clay (Table 5.1) makes clay soils difficult for bioremediation, where soil modification is difficult or impossible.

In situ bioremediation is best performed on sandy soils, with a lower permeability limit of about 10^{-5} cm s^{-1} (42). The hydraulic conductivity of dry, unconfined bentonite is 10^{-6} cm s^{-1}; when the bentonite is saturated, the conductivity drops to less than 10^{-9} cm s^{-1}. While the porosity problem can often be overcome with ex situ bioremediation technologies by adding bulking agents or shredding the soil, even then low clay or silt content is required, with a minimum void volume of 25% recommended (35). Compared to ex situ bioremediation systems, there is very little to see at an in situ site; the bulk of the activity happens underground. Most of the quoted advantages of in situ compared to ex situ bioremediation relate to this fact.

TABLE 5.1 Typical values for hydraulic conductivities of soils[a]

Soil type	Hydraulic conductivity (cm s^{-1})
Clay	10^{-9}–10^{-6}
Silts, sandy silts, clay silts, tills	10^{-6}–10^{-4}
Silty sand and fine sands	10^{-5}–10^{-3}
Well-sorted sands	10^{-3}–10^{-1}
Well-sorted gravels	10^{-2}–10^{0}

[a] Adapted from the work of Cookson (68).

At in situ bioremediation sites, there is very little excavation done, so site disturbance is much less. This should make in situ bioremediation cost-competitive since the bulk of the transportation and excavation cost has been nullified. As there are no biopiles or windrows to be constructed, the construction engineering is less involved, and in situ treatment is highly suited to small, inner-city sites, such as old gas stations. Naturally, it allows treatment of the deep subsurface, which is not feasible with ex situ methods unless the cost and space implications of deep excavation can be borne. A related issue is that in situ bioremediation allows treatment around and under buildings, which means that working sites can be treated as well as abandoned ones. Because there is limited excavation, there is much less exposure of the workforce and others in the vicinity of the site to volatiles.

This lack of excavation also accounts for most of the quoted disadvantages of in situ systems. Because the contractor is effectively working blind in the subsurface, detailed site assessment is required, and the hydrogeology and contaminant distribution must be known as accurately as possible. Gathering the necessary borehole information and laboratory analyses can make the site assessment phase very expensive. Working at depth in very inhomogeneous matrices makes process control difficult: whereas altering the temperature or pH of a biopile is straightforward, this is not the case with in situ technologies. It is also difficult to accurately predict end points, and careful monitoring of the process, while necessary, is difficult, as the subsurface cannot be made homogeneous.

All things considered, as more experience is being gained with in situ treatments, it is becoming more popular, whereas in the early development of bioremediation ex situ treatments were much more common. An interesting illustration of this can be seen in the Superfund program. The 11th edition of the *Treatment Technologies for Site Cleanup: Annual Status Report* reveals the changes in bioremediation approaches that have occurred over the last few decades (331) (Fig. 5.17 and 5.18). During the period from 1982 to 2002, ex situ bioremediation accounted for nearly 11% of source control projects, whereas from 2000 to 2002, there were only 3 of a total of 56 projects using ex situ bioremediation. The shift from ex situ to in situ bioremediation began in 1991. Since then, in situ bioremediation projects have risen from 13.2 to 17.7% of total source control projects (330).

In situ bioremediation technologies are overwhelmingly dominated by soil bioventing.

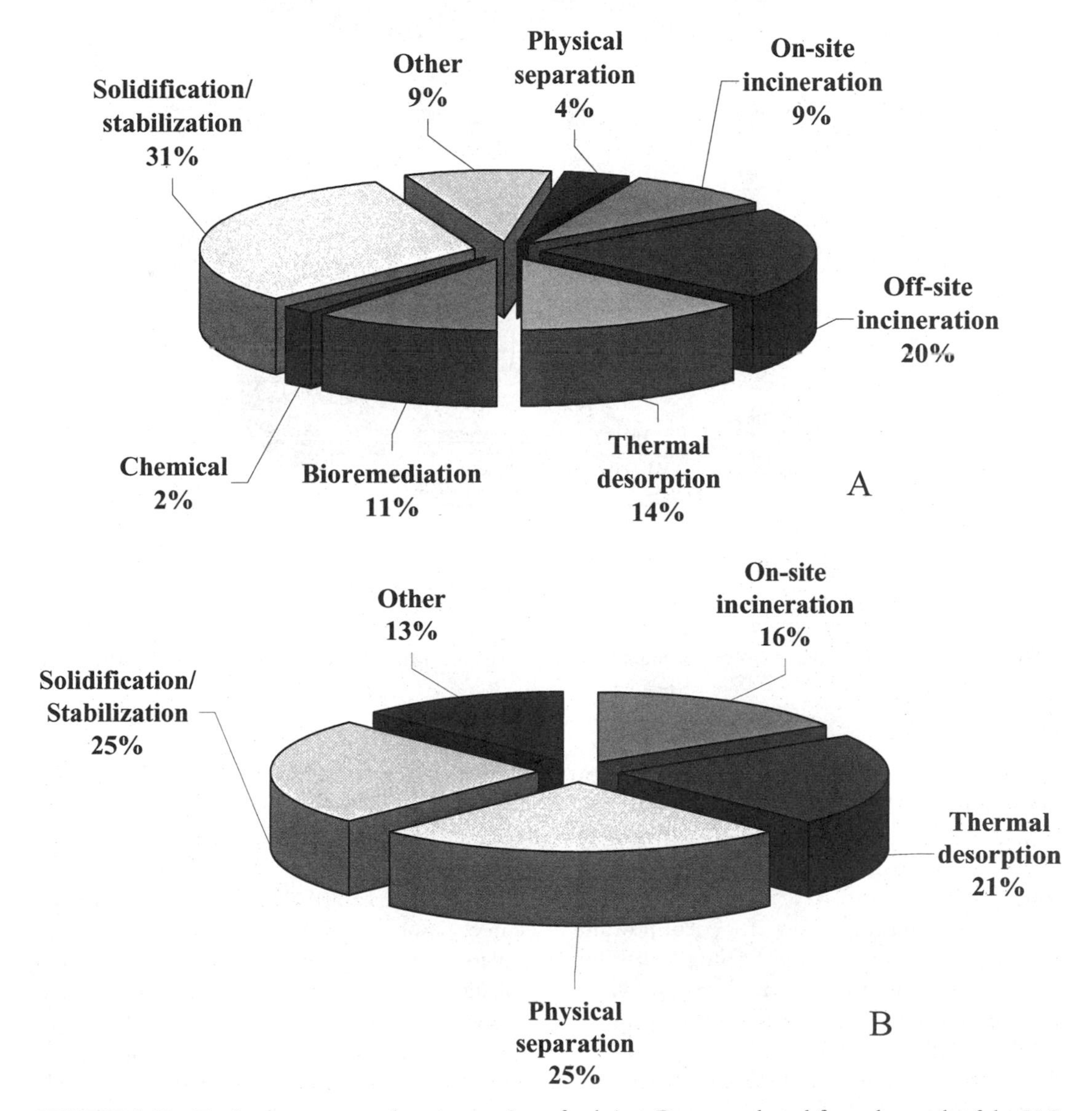

FIGURE 5.17 Ex situ source control projects at Superfund sites. Data are adapted from the work of the U.S. EPA (331). (A) 1982 to 2002. (B) 2000 to 2002.

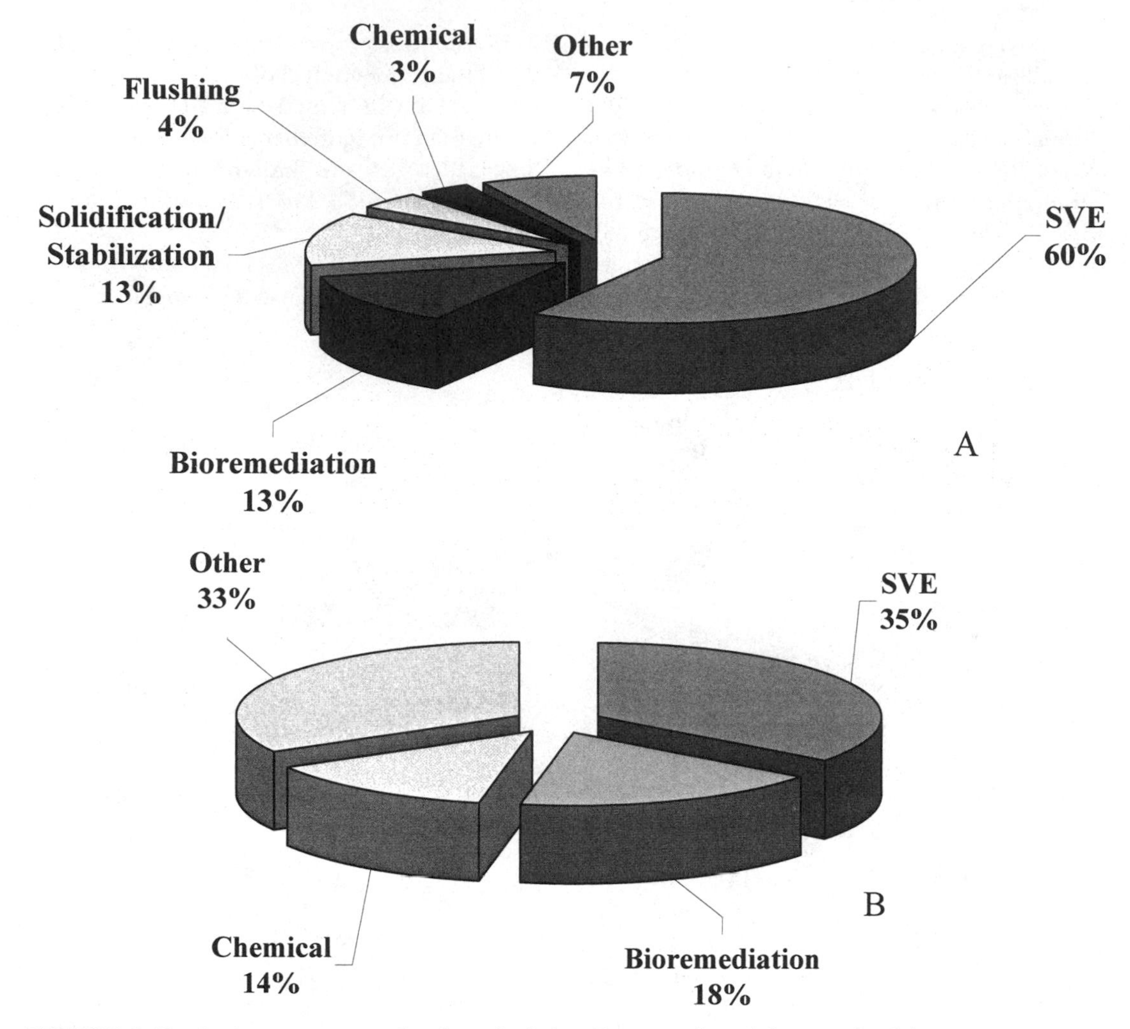

FIGURE 5.18 In situ source control at Superfund sites. Data are adapted from work of the U.S. EPA (331). (A) 1982 to 2002. (B) 2000 to 2002.

Indeed, this is the only in situ bioremediation technology used widely for soil cleanup. The U.S. Air Force listed 45 bioventing projects carried out at Air Force bases alone, ranging in volume from 600 to 311,500 yd^3 (319). The other biotechnologies are for groundwater cleanup. Pump-and-treat technologies are very widely used for contaminated groundwater, but biological enhancements are not common (325), and from 1982 to 2000, only 4% of above-ground pump-and-treat projects involved biological treatment (331). Permeable reactive barriers (PRBs) have traditionally been designed as chemical and physical intervention techniques, with incidental biodegradation taking place, and it is only recently that deliberately turning PRBs into bioremediation technology has arisen. MNA, while gaining acceptance, cannot reasonably be called a technology unless enhancement is practiced by some means. Cumulatively, air sparging continues to represent the in situ technology of choice for the treatment of groundwater at Superfund sites. However, the choice of in situ bioremediation of groundwater projects has risen from 8% in 1997 to 36% in 2002 (331).

Bioventing

Bioventing (Figure 5.19) bears great similarity to the physical extraction technique of soil vapor extraction (SVE). The primary engineering objective of both SVE and bioventing is stimulation of airflow in the vadose zone. However, whereas SVE is designed to maximize contaminant volatilization, bioventing is operated at much lower airflow rates to optimize oxygen transfer and utilization by microorganisms (92). SVE aims to extract volatile compounds from groundwater; it does not involve transformation of the compounds. Bioventing aims to achieve transformation of the contaminant through microbial attack. The two technologies use the same equipment, especially for aeration.

Bioventing differs markedly from SVE in that nutrients and moisture are often added to the subsurface to stimulate biodegradation during bioventing (173). The two functions, however, are not mutually exclusive: SVE always entails a variable component of biodegradation, and likewise bioventing can involve an element of volatilization. Campagnolo and Akgerman (58) concluded that the conservative estimate of 15% aerobic biodegradation during SVE is realistic. Thornton and Wootan (314) first suggested bioventing as a technology because they estimated that 38% of hydrocarbon degradation during soil venting experiments was caused by biodegradation. It is hardly surprising, then, that some studies have suggested a deliberate combined approach when the contaminants favor such a method (e.g., see the study by Malina et al. [189]).

CONTAMINANTS BIODEGRADED BY BIOVENTING

Bioventing has found its niche in the in situ treatment of fuel spills (322). Whereas fresh gasoline may be too volatile to be treated by true bioventing, bioventing has been used many times for the full-scale bioremediation of diesel (88) and kerosene (35) spills. It has also been used successfully on PAHs (187) and a mixture of acetone, toluene, and naphthalene (174). Bioventing is not considered appropriate for the treatment of PCBs and other chlorinated hydrocarbons. However, deliberate en-

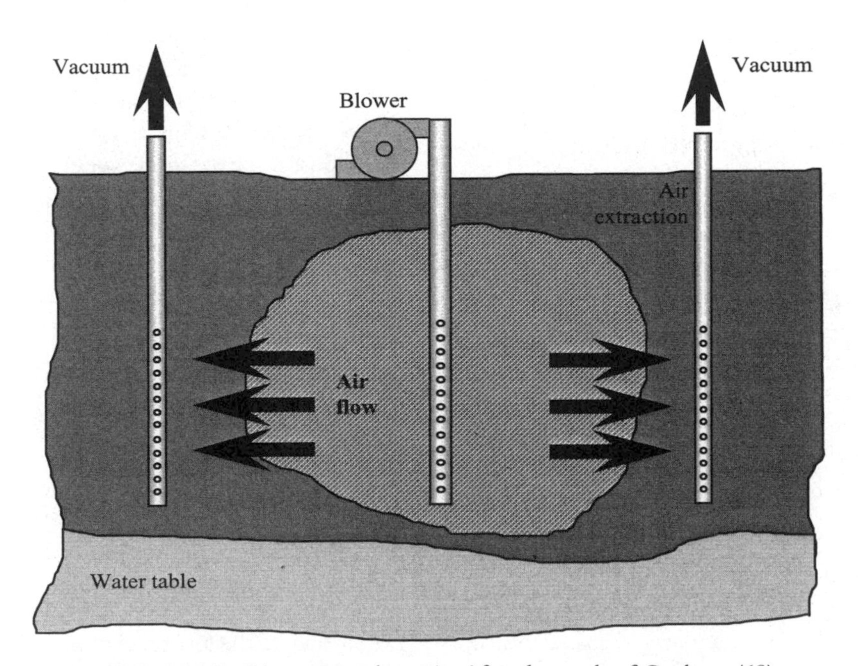

FIGURE 5.19 Bioventing schematic. After the work of Cookson (68).

couragement of cometabolism has proven successful in the biodegradation of trichloroethylene (TCE) by injecting oxygen and phenol (137), and the advent of anaerobic bioventing offers the potential for treatment of chlorinated compounds, such as DDT and energetic compounds (286). Trials are ongoing on the applicability of bioventing to gasoline-ethanol mixtures, which are common in Brazil (239). Modifications such as the use of electrical heating elements have made bioventing of petroleum spills applicable in cold regions (98).

BIOVENTING DESIGN

The design of bioventing systems is described in full by the U.S. Environmental Protection Agency (EPA) (322, 323). As with ex situ bioremediation, bioventing relies crucially on effective aeration. Unlike ex situ treatment, at a potential bioventing site, it is geology that dictates whether it is applicable. Soil gas permeability is once again the governing issue (84), which determines the relationship between applied pressure, or vacuum, and gas flow rate. To achieve the optimum aeration regimen whereby biodegradation is maximized and volatilization is minimized is the ideal, and inducing uniform airflow is the best way to achieve this (87), although in practical terms this may not be possible.

The basic steps involved in designing a bioventing system are (323) as follows.

1. Determine the required airflow system.
2. Determine the required airflow rates.
3. Determine the working radius of influence.
4. Determine well spacing.
5. Determine vent well requirements.
6. Provide detailed design of blower, vent wells, and piping.
7. Determine monitoring point requirements.

There are three primary physical characteristics that are used in bioventing design: soil gas permeability, contaminant distribution, and the radius of influence of oxygen.

Permeability: Gas and Vapor Transfer. Flux of gas through a soil matrix is described by Darcy's law, which shows that flux of gas is proportional to permeability. Generally, a significant flux decrease occurs when the permeability is less than 10^{-4} cm s^{-1}. At this point, the flux drops exponentially with decreasing permeability (112). While bioventing is certainly possible in relatively low-permeability soils (e.g., see the work of Phelps et al. [248]), when the soil gas permeability falls below 10^{-4} cm s^{-1}, gas flow will be through secondary porosity, such as fractures, or through more permeable strata present. In such cases, the feasibility of bioventing is site specific. In higher-permeability soils, the most important bioventing limitation becomes soil moisture. A high level of soil moisture replaces soil gas with water, which drastically reduces air diffusion. Kirtland and Aelion (161) refer to low-permeability sediments as those with a hydraulic conductivity of less than 10^{-3}cm s^{-1}.

Contaminant Distribution. With consideration to partitioning phenomena, a petroleum hydrocarbon spill onto soil will allow contaminants to be present in any or all of the phases: sorbed to soils in the vadose zone; in the vapor phase of the vadose zone; free-floating on the water table as residual saturation of the vadose zone; or in the aqueous phase, either dissolved in the pore water of the vadose zone or dissolved in the groundwater. Due to their higher density, dense non-aqueous-phase liquids (DNAPLs) partition to the vadose and saturated zones, whereas light non-aqueous-phase liquids (LNAPLs) distribute primarily to the vadose zone. As bioventing is really a vadose zone treatment, it is the LNAPLs that are the focus of attention. LNAPLs are more likely to migrate to the capillary fringe relatively uniformly in a sizeable spill. Then they will spread laterally along the surface of the saturated zone. The fluctuating water table allows migration of LNAPLs below the water table, but they cannot permeate the saturated zone unless a critical capillary pressure is exceeded (322).

The fate of individual LNAPLs in the vadose zone is determined by the partition phenomena discussed in chapter 1. They can remain as free product, partition to the vapor phase or the aqueous phase (pore water), or sorb to solids. The equilibrium concentration of most hydrocarbons in the aqueous or vapor phase is determined by the immiscible phase, if present, or the sorbed phase, if an immiscible phase is absent. Only limited oxygenation of the capillary fringe by bioventing is possible due to the limitation of oxygen diffusion into water, and the pore space is water saturated in the capillary fringe. However, when bioventing is operated in air injection mode, the positive pressure depresses the water table. This dewaters the capillary fringe, allowing for more effective treatment.

Airflow Systems. Injection of air is preferred by the U.S. EPA to air extraction. Air injection is easier to operate and maintain and also less expensive. In the presence of surface buildings or basements within the radius of influence of a bioventing system, air extraction may be preferred to prevent the accumulation of gases within these buildings. Air extraction, however, does not move air outwards to create an extended bioreactor zone, and therefore there is a relatively greater contribution from volatilization than biodegradation. Another consequence of the negative pressure created is that with extraction systems it is possible to cause the water table to rise. The effect on biodegradation at the capillary fringe would be the opposite of the situation with air injection: water saturation decreases oxygen diffusion, leading to a drop in biodegradation rate. For air extraction, an explosion-proof blower is required when working with petroleum hydrocarbons, a knockout drum and storage tank are needed upstream of the blower to remove condensates, and the off-gas may require treatment, which will significantly increase the cost of the treatment.

Air injection bioventing may cause soil desiccation to the point that microbial activity would be limited. Experience has shown, however, that moisture loss is minimal, even over a 3-year period. Air injection rates are typically low in bioventing, and drying may be a problem only in the immediate vicinity of the well.

Airflow Requirements. An in situ respiration test is a requirement of bioventing systems to prove that there is an active microbial population consuming oxygen, and to estimate the maximal oxygen demand of that population. This maximum microbial oxygen demand is used to calculate the airflow requirement. A simple calculation for required airflow rate can then be performed:

$$Q = \frac{k_o \times V \times \theta_a}{(20.9\% - 5\%) \times 60\,\mathrm{min/h}} \tag{1}$$

where Q is the flow rate (in cubic meters per minute), k_o is the oxygen consumption rate (percent per hour), V is the volume of contaminated soil (in cubic meters), and θ_a is gas-filled porosity (a fraction; at most bioventing sites, it ranges from 0.1 to 0.4).

Gas-filled porosity is given by

$$\theta_a = \theta - \theta_w$$

where θ is total porosity and θ_w is water-filled porosity (both given in cubic centimeters per cubic centimeter).

Total void volume is estimated by

$$\theta = 1 - \rho_k/\rho_T$$

where ρ_k is soil bulk density (grams of dry soil per cubic centimeter) (323) and ρ_T is soil mineral density (in grams per cubic centimeter), estimated at 2.65.

Water-filled void volume can be calculated as

$$\theta = M(\rho_k/\rho_T)$$

where M is soil moisture (grams of moisture per gram of soil).

The water-filled porosity is often assumed to be 0.2 or 0.3, as it is difficult to measure. Flow rates of 400 to 600 liters min^{-1} might be considered normal. Some literature suggests that a flow rate of 1 pore volume per day to 1 pore volume per week is typical of bioventing. For this, an estimate of the total pore space of the system is required. This can be done by multiplying the total volume of soil at the site by the soil porosity. Then the time to exchange 1 pore volume of soil vapor can be calculated from:

$$E = \frac{\theta V}{Q} \quad (2)$$

where E is pore volume exchange time (in hours), θ is total porosity (in cubic centimeters of vapor per cubic centimeter of soil), and Q is total vapor flow rate (cubic meters of vapor per hour).

Oxygen Radius of Influence. The radius of influence is defined as the maximum radius to which oxygen has to be supplied to sustain maximal biodegradation. Radius of influence is the most important factor that directs engineering design of full-scale bioventing systems once the soil gas permeability and contaminant distribution are known. It is used to space venting wells and to size blowers to guarantee that the entire site gets aeration, albeit that it might not be possible to achieve uniform aeration.

The radius of influence for bioventing is different from that for SVE. In SVE, all that matters is the distance from the air extraction well where vacuum or pressure can be measured; i.e., there is no microbial consumption factor. For bioventing, the radius of influence is a function of both airflow and oxygen consumption rate (323).

The most accurate determination of oxygen radius of influence for a particular site is direct measurement. If the injection mode is used, the blower can be started and changes in oxygen concentration can be measured at distances from the vent well. However, a change in oxygen concentration may take days or more to reach equilibrium. Pressure change occurs more quickly, in a matter of a few hours, but is less satisfactory than oxygen concentration change, and the relationship between them is not well understood. The relationship is strongly influenced by soil permeability; e.g., relatively impermeable soils register a pressure change quickly after a blower is started, even though oxygen permeability is very low. Alternatively, the radius of influence can be estimated from:

$$R_I = \sqrt{\frac{Q(20.9\% - 5\%)}{\pi \times h \times k_o \times \theta_a}} \quad \textbf{(3)}$$

where R_I is the radius of influence (in meters), Q is airflow rate (in cubic meters per day), h is aerated thickness (in meters), and θ_a is air-filled porosity (in cubic centimeters of air per cubic centimeter of soil).

In practice in the field, well spacings are in the region of 1 to 1.5 times the radius of influence (Fig. 5.20). Generally, the radius of influence can range from 1.5 m for fine-grained, relatively low-porosity soils to 30 m for coarse-grained, high-permeability soils (324).

Blower Sizing. Based on the air requirements for a project, the appropriate blower has to be acquired. The two most important considerations are the target airflow rate and the expected pressure drop within the bioventing system. Design should be directed at the upper performance limits required, i.e., maximum oxygen demand and lowest soil gas permeability. That way, flow can be vented at times of lower demand. Using a blower that is too small for the highest demand can shorten the life of

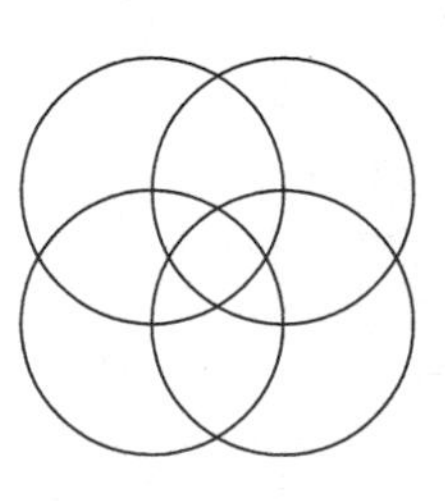

Wells spaced at the radius of influence

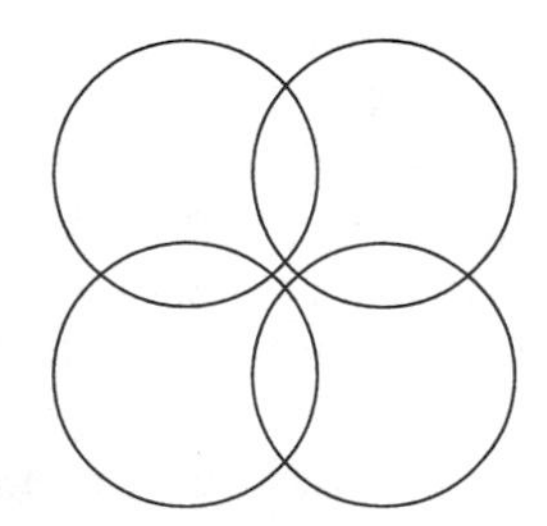

Wells spaced at 1.5 x radius of influence

FIGURE 5.20 Vent well spacing, based on radius of influence.

a blower. With the airflow and expected pressure known, a blower is selected such that those variables are in the middle of its performance range. At the relatively low airflow rates and pressures required in bioventing, regenerative centrifugal blowers operate efficiently. They can be manufactured to be non-sparking explosion proof. This type of blower is a common choice in bioventing projects. In practice, it would be appropriate to test three different-size blowers (based on power rating), one whose median performance is at the desired flow rate and pressure, one above, and one below, before final selection. The blower is at the heart of a bioventing system but is a relatively inexpensive item of equipment. Typical blower sizes in bioventing are 1 to 5 hp (approximately 0.75 to 3.7 kW).

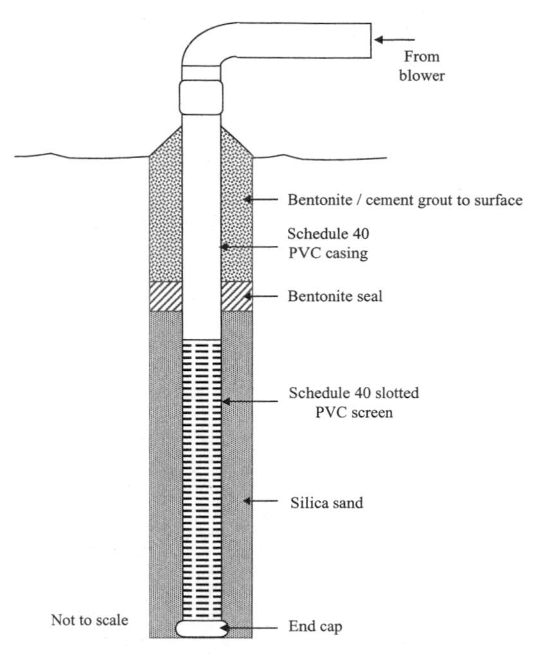

FIGURE 5.21 Bioventing well design. After work of the U.S. EPA (323).

Vent Well Design and Construction. The two orientations of vent wells are vertical and horizontal. Horizontal wells are, for the obvious reason of expense, used only when the contamination is shallow (less than, say, 3 m). Construction of horizontal wells at depths greater than 7.5 m becomes difficult. However, for shallow contamination at less than 3 m, horizontal wells are reported to be more effective than vertical wells. Vertical wells are used more commonly since the construction is simplified.

Construction is a standard process, using commonly available materials. The piping used is the ubiquitous 4-in. (10.2-cm)-diameter schedule 40, slotted PVC pipe that is used in landfill applications (Fig. 5.4), although various diameters are used, ranging from 2 in. in shallow or sandy soils all the way to 12 in., depending on depth and flow. Slot size should be consistent with the soil grain size. Standard 4-in.-diameter PVC pipe is the most common.

A typical injection well construct is shown in Fig. 5.21. Slight differences are allowable for extraction wells. For example, vertical wells can be designed with a longer screen section to improve the prospects of uniform airflow. Many drilling techniques are acceptable, with hollow-stem augering probably the most common. The diameter of the borehole should be at least twice the outside diameter of the pipe used. The drilling is done to the required depth, and the pipe work is then assembled. The vertical pipe consists of the slotted screen in the lower portion, joined to schedule 40 PVC casing of the same diameter. The total length of slotted screen depends on the depth of contamination in the vadose zone. Continuing the well into the saturated zone can cause problems. About 6 ft (1.8 m) of casing should extend into the subsurface from the surface, with a variable length above ground connecting to the header. In fact, it is common practice to conceal the piping manifold below ground in a shallow trench if possible. The slotted screen is terminated with an end cap.

The vertical piping is placed in the hole, screen end first, and held in the center of the borehole. The filter pack material is placed in the annular space created between the borehole wall and the screen. The filter pack material extends about 0.3 m above the screen. The filter pack would normally be silica sand. This is followed by a minimum of 0.6 m of bentonite seal. Above this, 0.5 m of bentonite-cement slurry is poured. This forms the seal that prevents short-circuiting of air back to the surface. The seal is more critical in injection wells than in extraction wells.

Number of Vent Wells. The simplest way to calculate the number of vent wells required at a site is to divide the total surface area of the treatment zone by the area corresponding to the radius of influence of a single well. The wells should then be spaced evenly within the treatment area.

Air Injection and Extraction. If a site is covered with an impermeable cap, such as concrete or buildings, airflow to the surface is prevented during air injection bioventing. This can be overcome by using a combination of injection and extraction wells.

A case study for bioventing design was presented by Pearce and Pretorius (241), based upon a single vent well and three monitoring wells. A soil gas permeability of 3.8 Darcy units and a radius of influence of 9.5 m were measured. A high oxygen utilization rate indicated an active microbial population, and a theoretical biodegradation rate, based on hexane, of 752 mg kg^{-1} $month^{-1}$ was calculated. It was concluded that bioventing was possible at the test site.

Combined Technologies

Bioventing is a technology that lends itself to combination with other soil remediation technologies. The complexity of the subsurface sometimes dictates that no one technology is suitable on its own. The most obvious choice is a combination with SVE, as the technologies are very similar in the hardware used. The main difference is simply in airflow. Malina et al. (189) described the use of sequential vapor extraction and bioventing for the remediation of soil contaminated with toluene and decane. Of the two, toluene is more toxic to microorganisms and is more volatile and therefore more amenable to treatment by SVE to remove the toxicity of toluene to allow subsequent bioventing. Such an approach could be successful at complex industrial sites, such as petrochemical plants, where a range of biodegradable and volatile or toxic compounds are present.

Another obvious combination is the use of bioventing in the extraction mode combined with treatment of the off-gases. Indeed, this is often a legislative requirement. To remediate contaminated land under a paint factory, Origgi et al. (238) opted for venting and biofiltration of the off-gas. Approximately 100 kg of naphtha, toluene, and xylenes was remediated. The bulk of the soil removal was done through venting (approximately 70 kg), and it was judged that biodegradation accounted for a further 18 kg. The biofiltration system produced an outlet gas stream which was acceptable under Italian law.

An interesting case was presented by Dasch et al. (78) at a geologically complex site contaminated with 18,000 liters of mixed gasoline and oil. The remediation strategy was a combination of pump-and-treat method, product removal, vapor extraction, and bioventing. Approximately 16,600 liters was removed in 21 months, 59% by vapor extraction, 28% by bioventing, and 13% by the pump-and-treat method. What is more, synergies between the technologies were obvious, e.g., pumping and treatment of groundwater lowered the water table, improving the efficiency of vapor extraction, which in turn added oxygen for the stimulation of biological activity.

Because only low flow rates are possible in low-permeability soil, some techniques are being investigated and applied to improve bioventing. Fracturing is an enhancement technology designed to increase the efficiency of other in situ technologies in low-permeability conditions. Existing fissures are enlarged and extended, and new fractures are created, primarily horizontally, during fracturing. Pneumatic fracturing is commonly used in soil fracturing. Fracture wells are drilled in the contaminated vadose zone and left uncased for most of their depth. Short bursts of compressed air can be injected to fracture the formation.

This was the approach taken by Venkatraman et al. (342) for the remediation of a gasoline-contaminated low-permeability soil. Fracturing was used not only to enhance airflow, but also for the delivery of the soil amendments phosphate, nitrate, and ammonium salts. The integration of fracturing with bioventing removed 79% of the benzene, toluene, and xylenes (BTX) at the site. It was estimated that 85% of the total BTX removed was attributable to biodegradation.

The injection of pure oxygen may be useful for providing higher oxygen concentrations in low-permeability soils for a given volume than is possible with air injection. Ozonation may be useful to partially oxidize recalcitrant compounds to speed up biodegradation. Air sparging of saturated soil is one approach being investigated to enhance biodegradation and volatilization.

Anaerobic bioventing seems a contradiction in terms, but the use of gaseous hydrogen as a reducing agent in the vadose zone, followed by oxygen in a subsequent aerobic step, has achieved reductive dechlorination of TCE, with rapid oxidation of the reductive by-products to accomplish complete remediation (207).

Biosparging

One way to remove contaminants from groundwater is to pump and capture the contaminants or otherwise separate the contaminants from the water, which is returned to the aquifer. In theory, prolonged pumping could eventually flush out all the contaminants, but the solubility properties of many contaminants of aquifers make reliance on physical flushing alone prohibitively slow and expensive. The widely used pump-and-treat method for the decontamination of groundwater may require long periods of recirculation of groundwater from the aquifer to the surface—to clean an aquifer by simple water flushing may take 15 to 20 years and several thousand times the volume of the contaminated portion of the aquifer.

In situ air sparging (IAS) has emerged as a popular alternative. This method consists of injecting a gas, usually air, into the saturated subsurface, below the lowest point of contamination. This promotes the partition of volatile and semivolatile contaminants from the dissolved and free phases into the vapor phase. Inevitably, as the dissolved oxygen concentration of the groundwater is increased, microbiological activity is stimulated. To turn this into a deliberate bioremediation technology rather than a volatilization one requires process modification (4), and the technique that is emerging has been called biosparging. A schematic of biosparging is shown in Fig. 5.22

In a similar fashion to bioventing, the equipment required is rather simple and mostly inexpensive, and time and money have to be spent to understand the local subsurface in sufficient detail to guarantee end points. Unlike bioventing, of course, biosparging is a technique of the saturated zone, which creates the need for engineered differences from bioventing. The effectiveness of biosparging is governed by two overriding factors: soil permeability, which determines the rate of transfer of oxygen from the gas phase to the aqueous phase and eventually to the microorganisms; and contaminant biodegradability.

Some experts in the field of air sparging regard biosparging as a particular mode of operation of IAS (148, 149). In practice, biosparging is frequently used to refer to an IAS system when the intent is to operate without SVE (148, 149). Moreover, they feel that the lower air injection rates (<0.14 m^3 min^{-1} compared to 0.28 to 0.57 m^3 min^{-1} per vertical well in a typical IAS application) are inefficient and likely also to be ineffective at many contaminated sites. It is conceded, however, that safe operation without SVE is possible at some sites and that with time, after the bulk of the volatile compounds has been removed, biodegradation will become the dominant mechanism of removal, but this can happen only once the dissolved hydrocarbon concentration in water reaches less than 1 mg liter^{-1} (147).

Part of the problem lies in the difficulty in

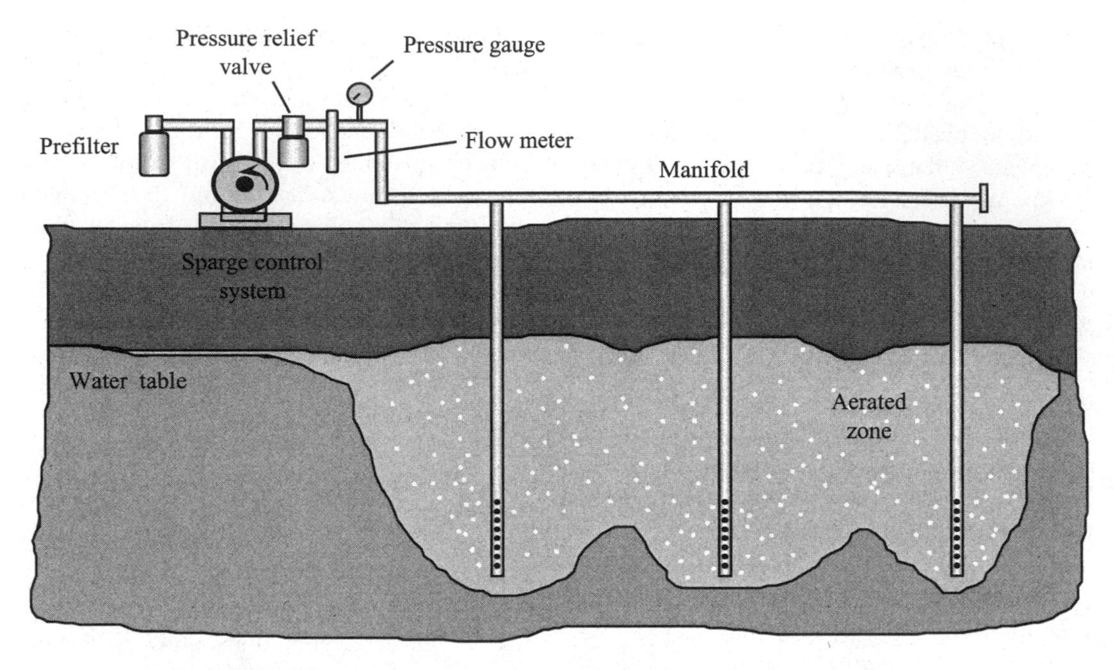

FIGURE 5.22 Biosparging schematic. After the work of Cookson (68).

predicting airflow in the subsurface at the depths at which air sparging and biosparging take place. Bruce et al. (53) conducted a study of a single air injection well in a sandy aquifer contaminated with gasoline. Increasing the airflow from 10 to 20 ft^3 min^{-1} (about 0.28 to 0.57 m^3 min^{-1}) increased the cumulative mass removal by a factor of 2 to 3. That is why it is expected that biosparging is inefficient, and it would certainly require much longer treatment periods than air sparging. Rutherford and Johnson (271) showed that trapped air remaining in an aquifer during pulsed air injection, during which air was supplied at a high flow rate for short periods and then shut off, continued to deliver air to the aquifer during the shutdown periods. This might continue for up to a day. Such a strategy might deliver the high-flow advantages of air sparging and yet also satisfy the lower airflow requirements for biosparging. Johnson et al. (149) suspect that pulsed airflow is thus the more efficient way to run biosparging, and under such conditions it resembles a hybrid of air sparging and biosparging.

CONTAMINANTS TREATED BY BIOSPARGING

In a manner analogous to the comparison between SVE and bioventing, air sparging and biosparging are suited to the treatment of different contaminants, although there is inevitable overlap. Air sparging utilizes higher airflow velocities and the treatment is therefore dominated by volatilization, with some incidental biodegradation. The lower flow velocities of biosparging favor biodegradation over volatilization. Biosparging is most often used at sites with groundwater contamination by middle distillate fuels, such as diesel and kerosene. The more volatile components of gasoline are also more toxic to microorganisms, and fresh gasoline spills in groundwater would be better treated by air sparging. Longer-chain hydrocarbons, such as those present in lubricating and heating oils, are intrinsically biodegradable but at lower rates than, say, medium-chain-length *n*-alkanes. However, these components are still amenable to biosparging, but the process necessarily takes longer. These larger hydrocarbons are also less volatile and therefore less amenable to air sparging. A pilot test to evaluate biosparging for the treatment of 1,2-dichloroethane is being conducted (231). As of 1999, only three Superfund groundwater remediation projects had used biosparging (330). These projects were treating *bis*(2-ethylhexyl)phthalate (DEHP)(a plasticizer used in many plastics), mixed chloroaliphatics and BTX, and petroleum hydrocarbons and TCE.

BIOSPARGING DESIGN

Biosparging is much less established as a treatment technology than bioventing. Detailed engineering design for biosparging is commensurately more difficult to find. Useful information is given by the U.S. EPA (324). Specifically, biosparging should not be used if free product is present in significant quantities; if basements or other confined utilities are located underground at the site, unless another technology such as SVE is being used for vapor control; or if the contaminants are in a confined aquifer, since the sparged air will have no escape path. Setting such exclusion conditions limits the number of suitable sites. Many of the criteria for suitability of bioventing also apply to biosparging.

Permeability. The intrinsic permeability of soil is the single most important factor that determines the suitability of biosparging at a particular site. Measurement of permeability as Darcy units or hydraulic conductivity is common in the literature. A hydraulic conductivity of greater than 10^{-4} cm s^{-1} at 20°C is suitable for biosparging, but at less than 10^{-5} cm s^{-1} it is unlikely to be.

Airflow Systems. An oil-free compressor of suitable size for the flow rate (middle of the operational range) is required. As demand should change through the duration of a biosparging project, the compressor should be sized according to maximum demand and should be rated for continuous duty at this demand. The compressor should be equipped with a particulate filter to prevent downstream contamination. Flow rate and pressure should be measurable.

Typical airflow rates for biosparging are low compared to those for air sparging and would be in the range of 85 to 700 liters min^{-1} per injection well. A pulsed air supply has been suggested as a means of improving mixing and distribution of air in the saturated zone. This has been suggested as a possible reason for observed improvements in remediation rate (363). The increase in hydrostatic pressure with depth of injection must be taken into account, but typically the air pressure will be in the range of approximately 69,000 to 103,000 Pa (roughly 10 to 15 lb/in^2).

Well Design and Construction. The choice of horizontal or vertical wells is governed by the same reasoning as for bioventing. Horizontal wells are best used at shallow sites, if 10 or more sparge points are required, or if the area affected is under a building or some other surface structure. Vertical wells are required for deep contamination (greater than about 8 m) and where only a few wells are required.

Construction is quite standard (Fig. 5.23) and similar to that of bioventing wells but with some significant differences. The wells are usually fabricated from 1- to 5-in.-diameter PVC or steel pipe. The slotted, or perforated, screened interval is usually about 0.3 to 1 m long and is set about 1.5 to 4.5 m below the lowest point of contamination. Proper capping is essential, especially because of the elevated pressures, to prevent air from short-circuiting back to the surface. To enable even distribution of air (or indeed, to divert more air to where it is needed), each well should be fitted with a pressure gauge and flow regulator.

The piping manifold can be buried in a shallow trench, depending on the site. Metal pipe should be connected directly to the compressor because of the elevated temperature of the exit air. At the pressures in the system, PVC pipe can be used otherwise.

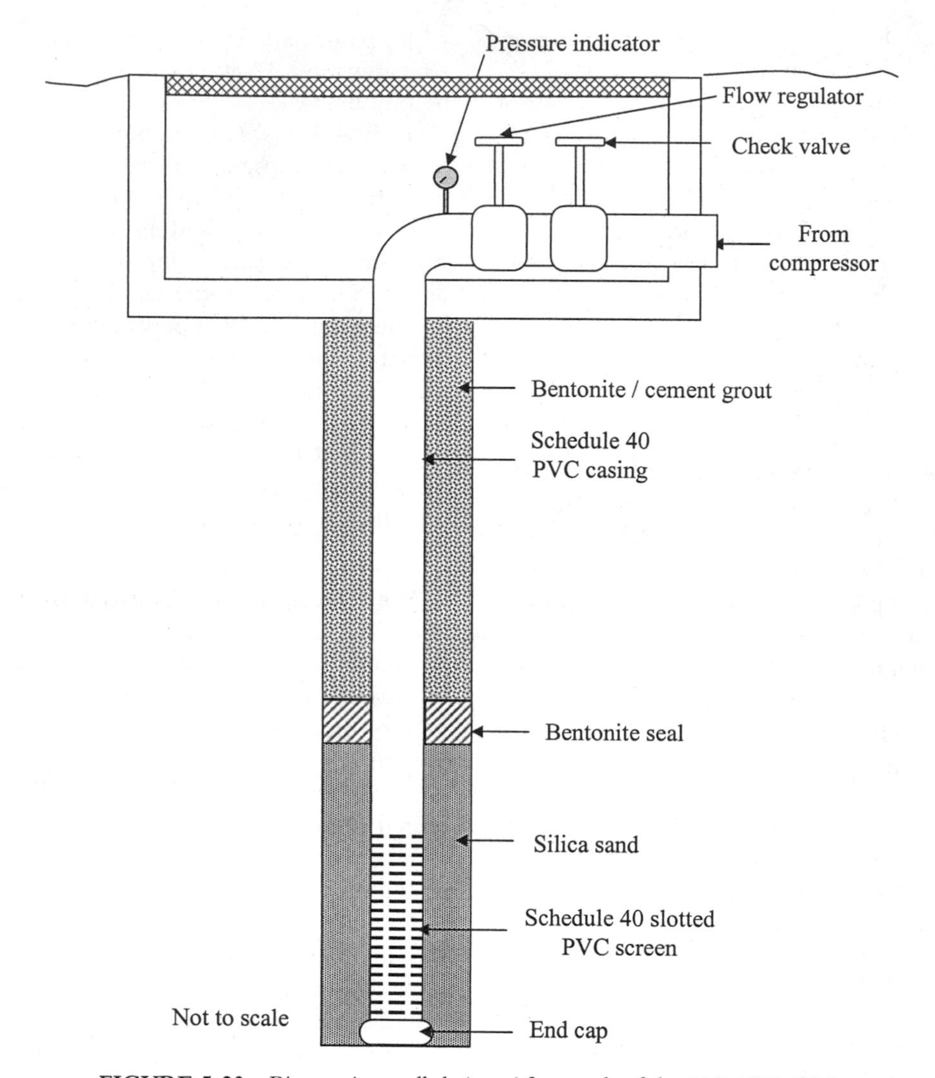

FIGURE 5.23 Biosparging well design. After work of the U.S. EPA (324).

Number and Spacing of Wells. The required number and spacing of wells are defined by the bubble radius. The bubble radius is the greatest distance from a sparging well at which a sufficient sparge pressure and airflow can be induced to enhance the biodegradation of contaminants (324). It is determined mainly by the hydraulic conductivity of the aquifer that is being sparged and is generally in the region of 1.5 m for fine-grained soils to 30 m for coarse-grained soils. Closer well spacing is appropriate in zones of high-level contamination to improve oxygen delivery.

Nutrient Delivery. Laboratory trials are used to determine if nutrients need to be added. In particular, the addition of ammonium ions might be tightly regulated or even specifically banned, depending on local authority regulations.

PRBs

The permeable reactive barrier (PRB) is an interception technology for the remediation of contaminated groundwater (Fig. 5.24). PRBs are also known as passive treatment walls. They are installed across the flow path of the ground-

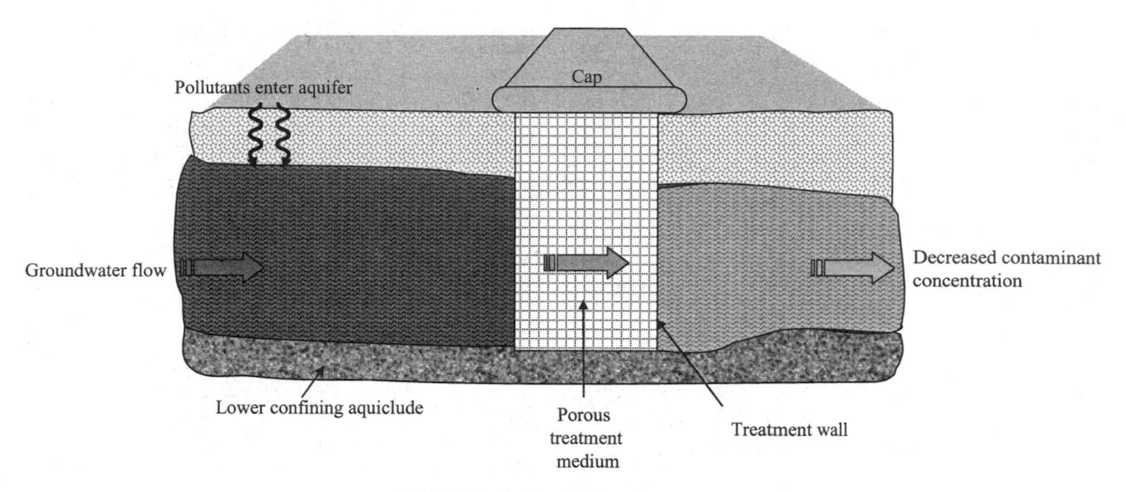

FIGURE 5.24 PRB schematic.

water and are constructed from porous materials, so that the water can pass through the wall. Yet the wall contains materials that prevent the passage of the pollutants. The pollutants are either degraded within the wall or retained in a concentrated form. The wall materials are various and often consist of zero-valent metals [mostly zero-valent iron, Fe(0)], chelators, sorbents, or compost. It is clear that this is not a bioremediation technology, then. However, an inevitable consequence of flowing water containing a low concentration of pollutants entering a porous material of high surface area is that the materials will become colonized by microorganisms. Thus, with time, it is likely that a contribution to the treatment is biological. A short leap of the imagination, then, has led to attempts to deliberately create biological PRBs, which Kalin (153) has termed passive bioreactive barriers. It should be stressed at this point that this is a long way from being proven full-scale technology, but it is worth examining what is being done.

In fact, as most PRBs are not bioreactors, the growth of microorganisms within the barrier material has been perceived as a possible detrimental effect due to the potential for biofouling to decrease the permeability of the barrier. It is necessary to keep the reactive zone permeability greater than or equal to the permeability of the aquifer to avoid diversion of the flowing water around the barrier.

Most PRBs are based on Fe(0) as the reactive material (about 80 laboratory, pilot, and full-scale installations worldwide), mostly due to its ability to reductively dechlorinate troublesome chloroaliphatics such as TCE. They are also able to remove some metals (59). The working lifetime of these PRBs may be limited by precipitation of secondary minerals due to reaction with groundwater (177). This could be exacerbated by microbial growth. It is generally believed that the high pH within the treatment zone of Fe(0) (around 10) would discourage microbial growth, and some studies show the impact of microorganisms on the performance of Fe(0) PRBs to be minimal (326). Data that show that diverse microbial communities can establish in this high-pH, highly reducing environment are emerging (97, 116), but how this will affect the long-term performance of Fe(0) barriers is not yet known.

The role of microorganisms in PRBs is likely to be enhancement, rather than a standalone biotechnology on its own. In this context, bio-enhanced PRBs are being researched for removal of metals from acid mine drainage (14, 186), explosives (236), chlorinated solvents (155), and inorganic pollutants (216). The metal removal is usually mediated by metal sulfide precipitation due to the action of the dissimilatory sulfate-reducing bacteria. Metal sulfides have extremely low-solubility products. Flowthrough columns have shown the feasibility of microbial colonization of Fe(0) to enhance the removal of TCE, sulfate, nitrate, and Cr(VI) (106). Moon et al. (216) created a reactive barrier that used sulfur granules as the electron donor and autotrophic sulfur-oxidizing bacteria as the biological component. More than 90% of the nitrate in the synthetic influent was removed. Other biological materials, although nonliving, are being used, primarily for their sorptive properties, e.g., bone meal apatite and bone charcoal for the removal of uranium (103); peat for the removal of benzene, toluene, ethylbenzene, and xylene (BTEX) (203); and chitosan-coated sand for the removal of copper (352).

An attractive feature of PRBs combined with biological treatment is the possibility for sequenced treatment, with, for example, anaerobic, reductive chemical and biological treatment followed by aerobic biodegradation of dechlorinated hydrocarbons and other, nonhalogenated pollutants that might be present in a pollutant cocktail (86). A full-scale reactive barrier utilizing bacterial sulfate reduction was installed in 1995 at the Nickel Rim mine site in Ontario, Canada. The barrier contains municipal compost to promote sulfate reduction, and the compost also increases the alkalinity of the mine tailing water (pH 5 to 6) that is being treated (38). It is present in a 1:1 ratio with pea gravel. After a 3-year period, the barrier removed > 1 g of sulfate liter^{-1} and >250 mg of iron liter^{-1}, demonstrating the long-term viability of the approach (37). Several full-scale PRBs have been built in Germany since 1998, and there are plans to add a biological zone to one of these for the treatment of PAHs (40).

A related technology is the use of biobarriers to reduce permeability. Cunningham and Hiebert (71) outlined the principle of using a biobarrier to funnel contaminated groundwater to, for example, an Fe(0) PRB. Large numbers of mucoid bacteria are pumped into injection wells. A growth substrate is added to stimulate growth. Bacterial growth and extracellular polymer production substantially reduce the hydraulic conductivity of the formation. Another, related area of research is the incorporation of reactive materials into synthetic membranes to decrease contaminant diffusion through membrane barriers; i.e., the objective is increased containment in this case, not increased permeability (290).

PRB DESIGN

As the PRB is not primarily a biotechnology, there are few design details for bioenhanced PRBs. Design details are available for PRBs (60, 326), to which the reader is directed. One of the great advantages of PRBs is cost savings, because pumping, large-scale excavation, and off-site disposal are eliminated. The conventional construction technique is trench and fill, which is a relatively hazardous and costly way to build a PRB. Day et al. (80) discussed the potential for use of geotechnical methods such as slurry trenching, deep soil mixing, and grouting to simplify the construction, lower costs, and increase safety to site personnel. In the present context, the use of biopolymer slurries shows promise. Reactive Fe(0) barriers can be installed by using biodegradable guar gum slurry without significantly decreasing the reactivity or long-term treatment characteristics of the iron. Biopolymers produce high viscosity for suspending the iron for injection or as a liquid shoring for trenching.

A challenge to bioreactive barriers is the delivery of nutrients, which has been shown to be hydrogeologically difficult and can add considerable expense to a project (153). Kalin gives details of time frames for PRB implementation, from technology selection to operation (see Box 5.4).

Bioslurping

Bioslurping evolved as a combination of several technologies to deal with one of the big challenges to in situ remediation: the presence of free product, specifically LNAPLs. Such is the evolution of the technology that it has at least one book dedicated to it (251). Bioslurping combines vacuum-assisted LNAPL recovery with bioventing and SVE (252). Thus, bioslurper systems simultaneously recover free product and remediate the vadose, capillary, and saturated zones. They use an aboveground vacuum pump to create enough vacuum airlift to draw LNAPLs from the subsurface, along with soil gases and small amounts of groundwater. Bioslurping can greatly enhance LNAPL recovery compared to conventional skimming and pumping technologies.

Most of the equipment involved is similar to that already described in previous sections. The main components of a bioslurper system are (209):

- recovery well,
- vacuum pump capable of extracting liquids and vapors (usually a liquid ring pump),
- liquid-vapor and oil-water separator units, and
- water and vapor treatment systems, if required.

The heart of a bioslurping system is the in-well dual drop tube assembly (Fig. 5.25). The use of the dual drop tube assembly greatly reduces the risk of emulsification of the free product and groundwater, along with solid entrainment, by allowing the separation of the free product and groundwater in the well before vacuum extraction. An above-ground vacuum pump enhances subsurface migration of LNAPL to the extraction well, where it is withdrawn from the upper drop tube (the secondary drop tube), or "slurp tube," in the dual assembly. The primary drop tube, usually a 1-in. (2.54-cm)-diameter PVC pipe, is shielded in the lower end with a larger-diameter, open-ended pipe that extends both above and below the end of the secondary tube. This shield is usually made of 2-in. PVC pipe and is termed the fuel isolation sleeve (252). It is conventionally 4 ft (about 1.2 m) long, extending about 2 ft above and below the end of the slurp tube.

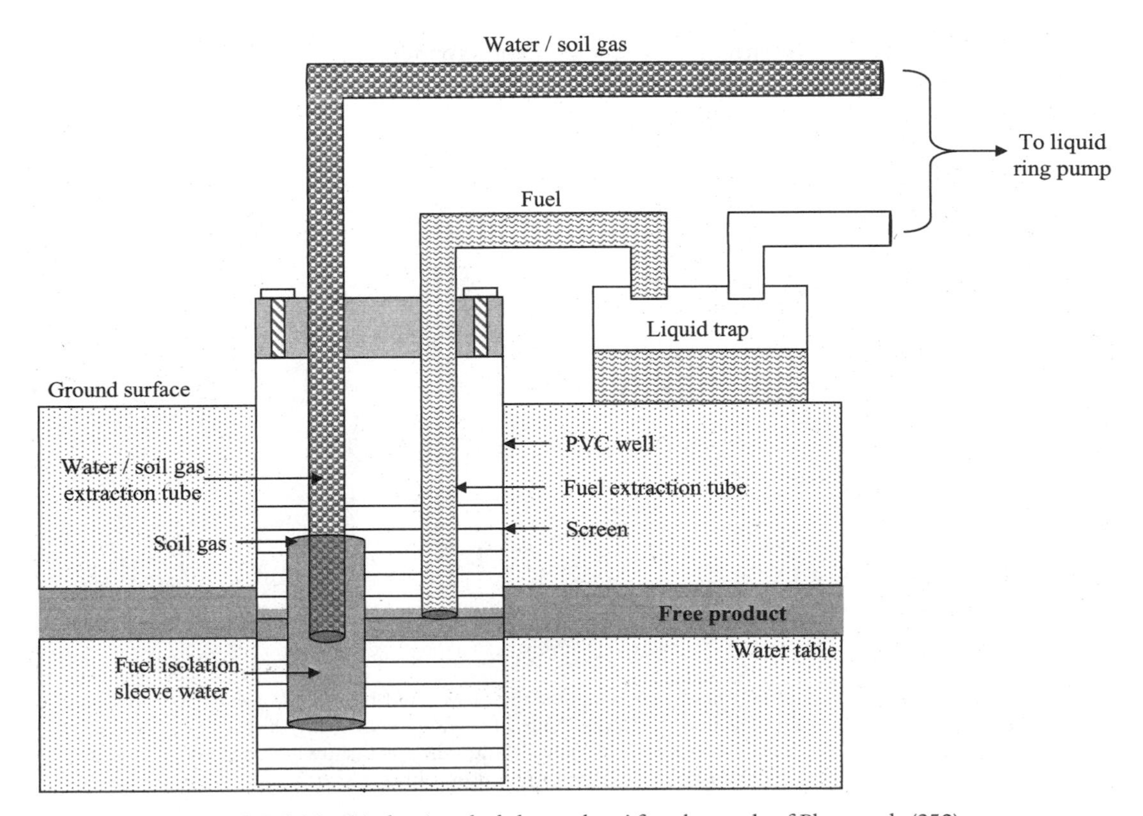

FIGURE 5.25 Bioslurping dual drop tube. After the work of Place et al. (252).

BOX 5.4
PRB

The diversity of potential pollutants at former coal gasification plants and the historical numbers that are now brownfield sites worldwide mean that a variety of remedial solutions are needed to manage the risk associated with historic land use. Contamination of soil and groundwater at these sites often occurs, but the nature and concentrations of hydrocarbons are variable, likely as a result of site-specific conditions. One of the greatest challenges is to remediate soil and groundwater at sites that still have operational activities, since intrusive remediation is usually ruled out because of major disruption and the presence of surface buildings and subsurface infrastructure such as pipelines and storage tanks.

One approach to the latter problem has gone from research to full-scale implementation at an active site in the United Kingdom (153). An intensive site survey identified the risk drivers at the site, and a sequenced PRB (SEREBAR) was designed and installed to intercept and biologically degrade the organic contaminants present in the groundwater (Box Fig. 5.4.1). The risk assessment identified a number of contaminant compounds of concern, which then focused the design (engineering and biodegradation) and implementation of the PRB. The SEREBAR contains stages of anaerobic and aerobic biotransformation of pollutants, and abiotic sorption. As this is a research project within a full-scale working system, the design contains this final contingency to safeguard regulatory compliance.

Early evidence shows that there is complete removal of contaminants of concern from the groundwater across the SEREBAR. As the project progresses, the role of bioremediation within the overall process will be investigated by conventional microbiological and novel molecular biological monitoring techniques, and mathematical models for the remediation process will be developed. The installed SEREBAR is shown in Box Fig. 5.4.2.

(Box provided courtesy of the SEREBAR group, via Bob Kalin. The SEREBAR group consists of SecondSite Property Holdings Ltd., Parsons Brinkerhoff, Queen's University Belfast, CEH Oxford, and University of Surrey.)

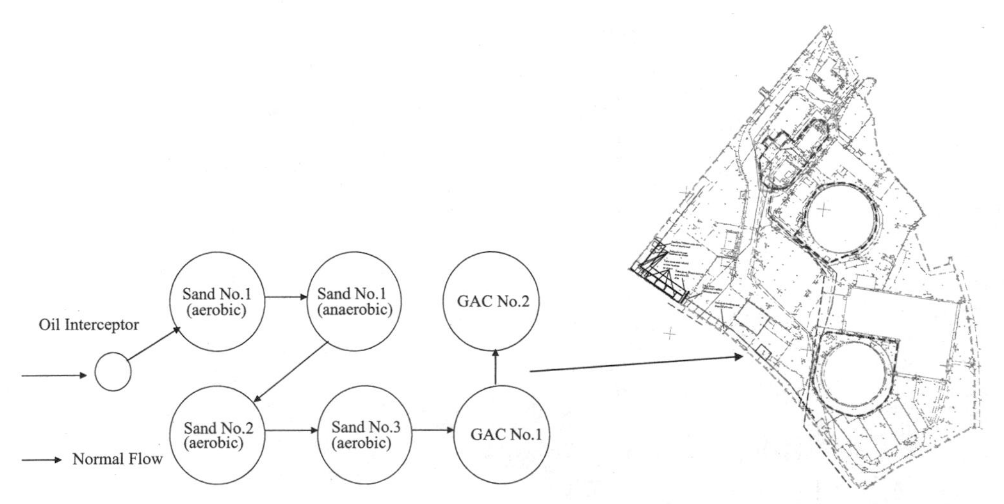

BOX FIGURE 5.4.1 Site and design drawings of the engineered bioreactive barrier for project SEREBAR at a former coal gasification site in the United Kingdom, combining abiotic and anaerobic biotransformation, aerobic biotransformation and abiotic sorption stages showing conceptual reactor design. Reprinted from the work of Kalin (153) with permission from Elsevier.

BOX 5.4 *(continued)*

BOX FIGURE 5.4.2 Photograph of SEREBAR at active site of remediation. Courtesy of SecondSite Property Holdings.

When the vacuum pump is switched on, the negative pressure induced in the well promotes LNAPL flow towards the well and draws LNAPL trapped in pore space above the water table. In response to pumping, the LNAPL level decreases to the point where the slurp tube draws in and extracts vapor, in a manner akin to SVE. This vapor extraction promotes airflow in the unsaturated zone, which stimulates bioremediation in a manner akin to bioventing. When the vacuum causes the water table to rise slightly, the system reverts back to extraction of LNAPL and groundwater. By avoiding high levels of groundwater rebound, large water table fluctuations are prevented, and this greatly reduces the chance of LNAPL being trapped below that water table, the "smearing" problem often encountered with more conventional free-product recovery systems. The drop tube position within the well is altered by using a section of flexible tubing connecting the extraction manifold to the drop tube.

Like other airflow-based technologies, bioslurping is ineffective in low-permeability soils (157) but is reported to also be cost-effective, over and above the advantages cited above. The primary cost saving is due to the reduction in the amount of extracted groundwater, which minimizes storage, treatment, and disposal costs.

MNA

The key to the use of the term MNA is the first word, monitored. Natural attenuation on its own would be a sophisticated way to say "do nothing" or "walk away." A range of alternative names for MNA is in use, including intrinsic remediation, intrinsic bioremediation, bioattenuation, and passive remediation. In addition to being accepted by the U.S. EPA as a potential remedy, many state UST programs now accept MNA at appropriate sites (Box 5.5).

The view of the U.S. EPA is that MNA is not an excuse for not taking action, but rather that it is an alternative approach to achieving remediation goals that may be appropriate to a limited set of site circumstances, where its use meets the applicable statutory and regulatory requirements (327). Thus, if we turn to the EPA (319) for a definition of MNA that will distinguish it from natural attenuation, we find the reliance on natural attenuation processes (within the context of a carefully controlled and monitored approach to site cleanup) to achieve site specific remediation objectives within a time frame that is reasonable compared to other alternatives.

More specifically, natural attenuation can be defined as naturally occurring processes in soil and groundwater environments that act without human intervention to reduce the mass, toxicity, mobility, volume, or concentration of contaminants in those media (235). In nearly all situations, the reduction in contaminant mass in the subsurface is brought about mostly by aerobic and anaerobic biodegradation processes (26). The biggest challenge in MNA is

BOX 5.5
MNA of BTEX-Contaminated Aquifers

The use of MNA (intrinsic bioremediation) as a remediation strategy at contaminated groundwater sites has dramatically increased in recent years (225). Bekins et al. (36) described two case studies of natural attenuation that are based on two U.S. Geological Survey Toxic Substances Hydrology Toxics Program research sites—one involved crude oil contamination in the shallow subsurface at Bemidji, Minn., and the other involved hard rock mining contamination in arid Southwest alluvial basins at Pinal Creek, Ariz.

At the crude oil study site near Bemidji, both aerobic and anaerobic biodegradation contributed to the natural attenuation processes affecting the hydrocarbon plume. The conclusion was that it is possible to rely upon natural biodegradative activities to clean up contaminated groundwater. The spill at Bemidji occurred in 1979 when a pipeline carrying crude oil burst and contaminated the underlying aquifer. U.S. Coast Guard scientists studying the site found that toxic chemicals leaching from the crude oil were rapidly degraded by natural microbial populations (333). It was shown that the plume of contaminated groundwater stopped enlarging after a few years as rates of microbial degradation came into balance with rates of contaminant leaching. This was the first documented example of intrinsic bioremediation in which naturally occurring microbial processes can be used to remediate contaminated groundwater.

Additional studies at Bemidji showed that benzene mineralization occurred in the groundwater under Fe(III)-reducing conditions (16). However, the capacity for benzene oxidation coupled to Fe(III) reduction was not detected at the other petroleum-contaminated aquifers that were evaluated (15). The anaerobic benzene oxidation under in situ conditions was found within only a portion of the Fe(III)-reducing zone of the Bemidji aquifer (16). The anaerobic oxidation of benzene was associated with populations of members of the family *Geobacteraceae*, which are known to couple the oxidation of several monoaromatics, including toluene, with Fe(III) reduction. The results suggested that the microbial community composition in the Bemidji aquifer may have played a key role in anaerobic benzene degradation and that *Geobacteraceae*, in particular, were associated with benzene-degrading activity (16). Molecular community analyses support the conclusion that *Geobacter* spp. play an important role in the anaerobic oxidation of benzene in the Bemidji aquifer (267).

As in the Bemidji studies, a number of investigators have tried to determine the biodegradability of BTEX compounds in aquifers. Vidovich et al. (346) reported that analyses of BTEX concentrations in groundwater beneath two leaking USTs indicated that the dissolved hydrocarbons had been attenuating naturally at both sites, limiting the migration of the dissolved BTEX compounds; i.e., they claimed that the data showed that intrinsic bioremediation of BTEX in these two different aquifers was occurring. Gieg et al. (109) like-

BOX 5.5 *(continued)*

wise found that intrinsic bioremediation contributed to the attenuation of hydrocarbons in an aquifer contaminated by gas condensate hydrocarbons. They were able to show that benzene, toluene, ethylbenzene, and each of the xylene isomers were biodegraded under sulfate-reducing conditions and that toluene also was biodegraded under methanogenic conditions.

Several investigators have focused on the anaerobic biodegradation of BTEX compounds in aquifers and the value of adding nitrate to stimulate denitrifying conditions or molecular oxygen to overcome limitations of anerobic BTEX metabolism by supporting aerobic BTEX biodegradation. Berwanger and Barker (39) found that BTEX biodegradation was not occurring at significant rates in an anaerobic, methane-saturated aquifer beneath a Canadian landfill; however, they were able to demonstrate that in situ bioremediation of aromatic hydrocarbons (BTEX) could be achieved by injecting hydrogen peroxide as an oxygen source. Morgan et al. (218) also found that oxygen was the factor limiting BTEX degradation at a groundwater site contaminated with high levels of gasoline. All compounds were rapidly degraded under natural aerobic conditions. No biodegradation occurred under anaerobic conditions except when nitrate was provided as a terminal electron acceptor for microbial respiration. Under denitrifying conditions, there was apparent biodegradation of benzene, toluene, ethylbenzene, *m*-xylene, and *p*-xylene, but *o*-xylene was not degraded. Degradation under denitrifying conditions occurred at a much lower rate than under oxygenated conditions. Phelps and Young (247) also found that the fate of the different BTEX components in anoxic sediments is dependent on the prevailing redox conditions as well as on the characteristics and pollution history of the sediment.

Studies at the Borden aquifer in Ontario, Canada, showed varying degrees of BTEX biodegradation under denitrifying conditions (32). After an acclimatization period, toluene was biodegraded rapidly in the presence of NO_3^-; however, the xylene isomers and ethylbenzene were biodegraded very slowly, and benzene was recalcitrant. Given the recalcitrance of benzene and high thresholds of the compounds that did biotransform, the addition of NO_3^- as an alternate electron acceptor would not be successful in this aquifer as a remedial measure. Wilson et al. (365) examined rates of biodegradation under mixed oxygen and denitrifying conditions for aromatic hydrocarbons typically present in a manufactured gas processing site groundwater and subsurface sediments. They reported that an in situ bioremediation scheme which combines moderate aerobic (7 mg of O_2 liter^{-1}) and denitrifying conditions would likely prove more successful than solely aerobic remediation for the long-term remediation of aromatic hydrocarbons.

Carbon and sulfur isotopic analyses of groundwater recovered from monitoring wells beneath an unleaded-fuel-containing BTEX-contaminated aquifer in the United Kingdom indicated that sulfate reduction was playing a key role in BTEX degradation at this site (299). Profiles of pore water chemistry beneath the plume source showed distinct zones of biodegradation that could be correlated with the chalk aquifer hydrogeology. Conditions became increasingly reducing as first oxygen, then nitrate, and finally sulfate were consumed via microbial oxidation of BTEX fuel residues. In other aquifers, anaerobic biodegradation of at least benzene can seemingly occur in the absence of alternate electron acceptors (359). Benzene can be converted to methane and carbon dioxide under methanogenic conditions, leading to significant natural attenuation of benzene in some anaerobic petroleum-contaminated aquifers (357).

Kao and Wang (154) described the results of a full-scale intrinsic bioremediation investigation conducted at a gasoline spill site in Dublin, N.C. Due to the occurrence of non-aqueous-phase liquid (NAPL) hydrocarbons beneath the former spill location, dissolved BTEX compounds were being continuously released from NAPL into the groundwater, with a total BTEX concentration of 60 mg liter^{-1}. At this spill site, a cropland extended from the midplume area to the downgradient edge of the plume. Groundwater and microbial analyses indicated that iron reduction was the dominant biodegradation process between the source and midplume area. However, nitrate spill in the cropland area switched the degradation pattern to denitrification and also changed the preferential removal of certain BTEX components. Under iron-reducing conditions, toluene and *o*-xylene declined most rapidly, followed by *m*-xylene and *p*-xylene, benzene, and ethylbenzene. Within the denitrifying zone, toluene and *m*-xylene and *p*-xylene had very rapid degradation rates, followed by ethylbenzene, *o*-xylene, and benzene. The mass flux calculations showed that up to 93.1% of the BTEX compounds was removed within the iron-reducing zone, and 5.6% of the BTEX compounds was degraded within the nitrate spill zone. Results

Box 5.5 continues

BOX 5.5
MNA of BTEX-Contaminated Aquifers *(continued)*

revealed that the mixed intrinsic bioremediation processes (iron reduction, denitrification, methanogenesis, and aerobic biodegradation) had effectively contained the plume, and iron reduction played an important role in the BTEX removal.

Hunkeler et al. (140) examined the value of intrinsic bioremediation following years of active biostimulation. A diesel fuel-contaminated aquifer in Menziken, Switzerland, had been treated for 4.5 years by injection of aerated groundwater, supplemented with potassium nitrate and ammonium phosphate, to stimulate indigenous populations of petroleum hydrocarbon-degrading microorganisms. After dissolved petroleum hydrocarbon concentrations had stabilized at a low level, engineered in situ bioremediation was terminated and natural attenuation was monitored. In the first 7 months of intrinsic in situ bioremediation, redox conditions in the source area became more reducing, as indicated by lower concentrations of sulfate and higher concentrations of Fe(II) and methane. The rate of dissolved inorganic carbon and methane production in the source area was more than 300 times higher than the rate of petroleum hydrocarbon elution, indicating that natural attenuation in which biodegradation was coupled to consumption of naturally occurring oxidants was an important process for removal of petroleum hydrocarbons that remained in the aquifer after termination of engineered measures.

Multiple attempts have been made to engineer bioremediation solutions for BTEX-contaminated soil and groundwaters. Lovley (184) suggested that anaerobic BTEX degradation has the potential to remove significant quantities of BTEX from petroleum-contaminated aquifers; however, he pointed out that engineering anaerobic systems in situ for bioremediation of BTEX compounds is difficult. Enhancement of in situ anaerobic biodegradation of BTEX compounds has been successfully demonstrated at a petroleum-contaminated aquifer in Seal Beach, Calif. (74). Injection of nitrate and sulfate into the contaminated aquifer was used to accelerate BTEX removal compared to remediation by natural attenuation. Nitrate was utilized preferentially over sulfate and was completely consumed within a horizontal distance of 4 to 6 m from the injection well; sulfate reduction occurred in the region outside the denitrifying zone. Benzene degradation also appears to have been stimulated by the nitrate and sulfate injection close to the injection well but only toward the end of the 15-month demonstration. The results are consistent with the hypothesis that benzene can be biodegraded anaerobically after other preferentially degraded hydrocarbons have been removed.

DaSilva and Alvarez (79) investigated the potential of bioaugmentation to enhance anaerobic BTEX biodegradation in groundwater contaminated with ethanol-blended gasoline. They tested the value of injecting methanogenic consortia that had been enriched with benzene or toluene and *o*-xylene. Up to 88% benzene biodegradation was observed in some cases, but benzene removal was hindered by the presence of toluene. Thus, they concluded that the value of bioaugmentation for treating BTEX-contaminated groundwater remains uncertain.

Valo (334) reported on full-scale bioremediation of BTEX compounds in soil and groundwater at a chemical manufacturing site at Keyport, N.J., that had been contaminated by volatile aromatics after a storage tank had been damaged. Groundwater was contaminated with 20 to 130 mg of total BTEX compounds liter^{-1}, and soil was contaminated with 100 to 8,000 mg kg^{-1}. Bioremediation employing biofilters was used for the cleanup. The permit levels were as follows: benzene in soil and in water, 1 and 0.01 ppm, respectively; total xylenes in soil, 10 ppm; ethylbenzene in soil, 100 ppm; and toluene in soil, 500 ppm. Soil BTEX compounds were degraded in compost and partly in air biofilters. A benzene level of 1 ppm was achieved within 10 to 30 days; 60 to 80% of the volatile aromatics in the gas biofilter feed was degraded in the reactors.

The U.S. Geological Survey has studied biostimulation treatments at several sites in the United States that were contaminated by fuel spillages (332). Biostimulation with nutrient addition was used to treat sandy soils and underlying groundwater at Hanahan, S.C., that were contaminated by a leakage of 80,000 gal of kerosene-based jet fuel. The groundwater had been leaching BTEX compounds from the fuel-saturated soils and carrying them toward a nearby residential area. The nutrients were delivered through infiltration galleries, and groundwater was removed by a series of extraction wells. Within a year, contamination in the residential area had been reduced by 75%. Monitoring showed that BTEX concentrations in highly contaminated groundwater nearest the infiltration galleries where the nutrients were injected were totally removed—5,000 ppb of toluene was lowered to no detectable contamination. The conclusion of the U.S. Geological Survey was that bioremediation had worked!

providing unequivocal evidence of the extent of microbial reactions, that these reactions are ongoing and sustainable (297).

There are three criteria listed by the U.S. National Research Council (NRC) that must be met in MNA (226):

1. demonstration of loss of contaminants from the site,

2. demonstration by laboratory analyses that microorganisms obtained from the site have the potential for the transformation of the contaminants at the site, and

3. evidence that biodegradation potential is realized in the field.

The loss of contaminants from the site could clearly be due to several abiotic processes, although evidence for it is easy to gather. Acquiring evidence that biological removal is happening is harder. Similarly, the second NRC criterion can be rather easily demonstrated, usually through column-based microcosm studies. The really difficult NRC criterion to meet is the third. The realization that the majority of microorganisms in the environment are uncultivable with standard laboratory media and conditions (12) means that traditional bacteriology is of little help. What are required are robust techniques that link microbial community structure and function (190, 191) with metabolic activity, and a great deal of effort in recent years has been directed to this area, but to date the techniques are not routinely applicable in most commercial laboratories.

The "lines of evidence" approach of the U.S. EPA is very similar (327). This involves providing three tiers of site-specific information, or lines of evidence, to demonstrate that natural attenuation is actually occurring. These three tiers are (366):

1. documented loss of contaminants at the field site,

2. presence and distribution of geochemical and biochemical indicators of natural attenuation, and

3. direct microbiological evidence.

The U.S. EPA (361) has a six-step process that allows the decision maker to decide between an engineered remediation or to perform a feasibility study into MNA (Fig. 5.26).

The general approach to assessing the feasibility of MNA consists of collecting pollutant and biogeochemical data from groundwater monitoring wells across the site, and then a site-specific conceptual model is developed. The data are correlated in time and space with the chemicals of concern to establish predominant biodegradation mechanisms (235). Modeling is then used to establish the rate of removal of the contaminants and the time taken to nullify the risk, which is then matched to an exposure pathway analysis to see if natural attenuation is suitable for the site. For suitability, the contaminants must be removed before exposure to a receptor can happen. This process has been named risk-based MNA, and Khan and Husain (156) have set out a seven-step method for its evaluation.

1. Review of available data and risk characterization.

2. Development of a conceptual model.

3. Refinement of the conceptual model: filling in the gaps.

4. Modeling of natural attenuation by use of analytical and/or numerical fate-and-transport models.

5. Exposure pathway analysis.

6. Preparation of a long-term monitoring plan.

7. Obtaining of approval for MNA from the regulator.

In the majority of cases in which MNA is proposed as a remedy, its use may be appropriate as one component of the total remedy, that is, either in conjunction with other treatments or as a follow-up measure at the end of active treatment(s). A good example would be a site with mixed contaminants and DNAPL. If the other contaminants can be removed by use of a technology but removal of the DNAPL has been determined to be technically impracticable, then MNA can be considered as a follow-up measure (337).

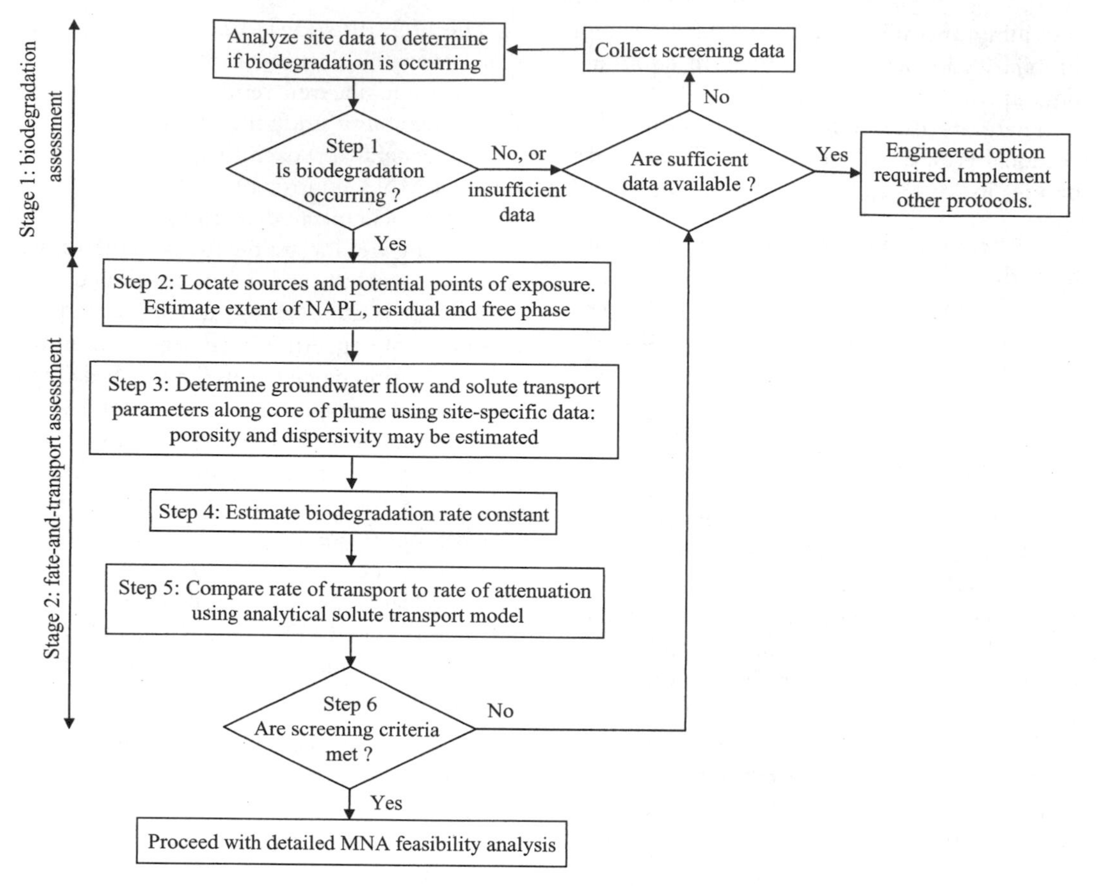

FIGURE 5.26 MNA decision support flowchart. After work of the U.S. EPA (361).

There are some important caveats to the use of MNA for the cleanup of contaminated groundwater which must be emphasized.

1. MNA should be viewed with great caution as the sole remedy at contaminated sites. The percentage of Records of Decision selecting only MNA as a groundwater remedy rose from 6% in 1986 to a peak of 32% in 1998 but dropped to only 4% in 2002 (331).

2. It is by no means a purely biological process. On the contrary, there are chemical and physical phenomena that might be of more or less importance, including dilution, volatilization, adsorption, and chemical reactions with subsurface materials.

3. It cannot be deemed to be a technology per se. In a U.S. EPA document, *Engineered Approaches to* In Situ *Bioremediation of Chlorinated Solvents* (329), MNA is not discussed, as it does not involve enhancement technologies.

4. Even to consider its use at a site requires site-specific modeling to demonstrate that natural processes are occurring that will reduce contaminant concentrations below regulatory standards before any potential exposure pathways are completed. Its use should therefore be considered only in low-risk situations.

5. Above all, there must be documentary evidence that natural attenuation is actually occurring at the site.

All this being said, MNA can work out to be a very inexpensive option where other technologies would struggle to be cost-effective, especially where there are existing buildings, paved areas, and utilities (27). During selection, however, it should be remembered that MNA may take substantially longer to meet targets than other technologies. For example, Khan et al. (157) cited an example in which natural attenuation was chosen at a site contaminated with chloroaromatics at Cape Canaveral, Fla. Two models were run, one with source removal and one without. The expected times to completion were 20 and 40 to 53 years, respectively. However, such long periods need not be a concern as long as the period is consistent with the results of the risk assessment. If the time to remediation is less than the time identified for the pollution to reach an exposure point, then the risk is mitigated (66).

It is becoming clear that other contaminants may be suited to the MNA remedy on a site-specific basis, e.g., explosives (245), chloroaliphatics (366), and petroleum hydrocarbons (156). The gasoline additive methyl *tertiary*-butyl ether (MTBE) is problematic, as it is a very high-production chemical with dubious biodegradability. A review of existing data (261) showed that MTBE does biodegrade in the laboratory and at actual release sites, and the reviewers concluded that MNA might be an appropriate remediation remedy, although on a case-by-case basis.

A variant on MNA in which an integrated technology called barrier-controlled MNA can be used to manage plumes of contaminants has been proposed by Filz et al. (99). The basic concept is that a low-permeability, nonreactive barrier is used to release contaminants into an aquifer at a rate that optimizes natural attenuation. This may save costs since barriers can be constructed at relatively low cost and may speed up source reduction.

While biostimulation and MNA are useful approaches, they also have limitations. The complexity of considering these approaches is well exemplified by the saga of General Electric in seeking to use bioremediation to treat PCB-contaminated sediments of the Hudson River (Box 5.6).

BOX 5.6
Limitations of Bioremediation for Treating PCB-Contaminated Hudson River Sediments

The Hudson River, a major waterway in New York State, has been heavily contaminated with various pollutants, including, notably, PCBs. From the late 1940s until 1977, the General Electric Company (GE) discharged as much as 1.3 million pounds of PCBs into the upper Hudson River. The releases of PCBs into the Hudson River came primarily from two industrial plants involved in production of electrical capacitors that contained PCBs. PCBs are a family of over 200 man-made compounds that were used as complex mixtures known as Aroclors. A variety of Aroclor formulations with different congener composition, chlorination levels, and chemical properties were produced. The widespread use of PCB fluids in industrial applications has resulted in their presence at many locations throughout the world. PCBs have been linked to increased risk of cancer in humans and have multiple impacts upon fish and wildlife, necessitating environmental cleanup.

PCBs are subject to microbial biodegradation (2). Biodegradability depends upon the number of chlorine substituents and their positions in the specific PCB congeners (2, 3). Anaerobic bacteria that dechlorinate PCBs typically attack the more-chlorinated PCB congeners through reductive dechlorination, a process that removes chlorines but leaves the biphenyl rings intact. Thus, PCB congeners with many chlorine substituents are subject to anaerobic dechlorination to form lower-molecular-weight PCB congeners with fewer chlorine substituents. The lower-molecular-weight PCB congeners are subject to aerobic biodegradation. This process has occurred extensively in a number of sediments, which results in a reduction in more-chlorinated PCB congeners and a commensurate increase in less-chlorinated congeners (226).

A variety of cleanup methods, including bioremediation, have been considered to remove the PCBs from the Hudson River (214, 215). In 1991, GE conducted field experiments in the Hudson

Box 5.6 continues

BOX 5.6 *(continued)*

River of the aerobic biodegradation of PCBs by naturally occurring microorganisms (124). A 73-day field study of in situ aerobic biodegradation of PCBs in the Hudson River showed that indigenous aerobic microorganisms can degrade the lightly chlorinated PCBs present in these sediments. Addition of inorganic nutrients, biphenyl, and oxygen enhanced PCB biodegradation, as indicated both by a 37 to 55% loss of PCBs and by the production of chlorobenzoates, intermediates in the PCB biodegradation pathway. Repeated inoculation with a purified PCB-degrading bacterium failed to improve biodegradative activity. Biodegradation was also observed under mixed but unamended conditions, which suggests that this process may occur commonly in river sediments, with implications for PCB fate models and risk assessments.

Despite evidence suggesting that nearly all of the PCB congeners contaminating the Hudson River sediments should be biodegradable, only 60% of the PCBs in the Hudson River sediments were degraded in experimental tests (124). The problem seems to be due primarily to bioavailability; the aerobic bacteria cannot degrade the more lightly chlorinated PCBs because they are sorbed by organic particles in the sediment. The resistant fraction probably consists of PCBs that are dissolved in the polymeric, natural organic matrix of the sediment and must diffuse through this matrix before desorption can occur. It is likely that this resistant PCB fraction is not available to microorganisms in its sorbed state and therefore represents the primary limitation to the biodegradation process. This diffusion process may also limit uptake by other organisms and has important implications for the determination of risk estimates.

These results indicate that bioremediation may be limited and unable to adequately remove PCBs from contaminated river sediments. F. J. Mondello (GE Global Research Technical Report, http://www.crd.ge.com) evaluated the potential effectiveness of bioremediation for the Aroclor 1260 contamination at GE Canada's former transformer-manufacturing facility located in Guelph, Ontario. He concluded that there are several major factors limiting the effectiveness of bioremediation for commercial PCB mixtures, including the inability of organisms to degrade highly chlorinated congeners or to obtain carbon or energy from degrading lightly chlorinated PCBs, the limited bioavailability of PCBs to microbes resulting from PCB sorption to organic compounds in soils and sediments, unfavorable temperature and moisture conditions, and the inability of anaerobic bacteria to completely dechlorinate PCB molecules. Many of these factors are applicable to the GE Canada Guelph site, making it highly unlikely that the level of Aroclor 1260 can be reduced to acceptable levels by bioremediation; these considerations are also applicable to the contaminated Hudson River sediments.

On the basis of the results of numerous laboratory studies and several field tests and after 10 years of research and field testing by GE-funded efforts, the U.S. EPA concluded that biological dechlorination will not naturally remediate contaminated Hudson River sediments to the levels considered to be safe. Although GE and many scientists who have studied PCB biodegradation still contend that natural processes, including reductive dechlorination and aerobic biodegradation, have substantially reduced the risk to humans and the environment and that these natural processes should be allowed to continue, the EPA decided to pursue other methods (175). As early as 1984, the EPA proposed removing the PCBs by dredging the sediment on the bottom of the river. In February 2002, the EPA authorized a cleanup plan for the Hudson that specifies dredging 2.65 million cubic yards of the river bottom over a 3-year period. Intrinsic bioremediation will continue naturally even as sediments are dredged from the river.

BIOAUGMENTATION

Bioaugmentation in the context of bioremediation, as described by Alexander (10), can be considered the inoculation of contaminated soil or water with specific strains or consortia of microorganisms to improve the biodegradation capacity of the system for a specific pollutant organic compound(s).

Several hundred U.S. companies sell microorganisms for environmental cleanup. Many of the products, however, are of dubious value and are little more than "snake oil." Often, naturally occurring microorganisms at a contaminated site are already biodegrading the pollutant, and addition of the microorganisms in these products does not harm, but provides no help, either. Most cultures to biodegrade hydrocarbons in contaminated soil and water do not enhance the rates of biodegradation above those attributable to indigenous microorganisms. Many engineers and industrial plant supervisors nevertheless feel compelled to use such cultures to show they are making an effort to be environmentally responsible (23).

Bioaugmentation often is considered for the bioremediation of compounds that appear to be recalcitrant, i.e., contaminants that persist in the environment and appear to be resistant to microbial biodegradation (Box 5.7). In this context, bioaugmentation may have value; i.e., if indigenous microorganisms lack the capability to degrade the contaminant and if organisms that will be active can be introduced, bioremediation can become useful in that situation. Many still look to modern biotechnology, i.e., the application of recombinant DNA technology, as the panacea to problems of pollution with highly resistant contaminants.

There are three fundamental approaches to bioaugmentation of a contaminated site, although little attention is paid to these during operations. The first is to increase the genetic diversity by inoculation with allochthonous microorganisms. By increasing the genetic diversity of the soil or water, it is assumed that this increases the catabolic potential (81) and thereby the rate of removal of the contaminant(s) by biodegradation will increase.

The second is to take samples from the site

BOX 5.7
Bioremediation of MTBE-Contaminated Aquifers Employing Bioaugmentation

MTBE is a common fuel oxygenate added to reformulated gasoline since the 1970s to reduce air emissions. It has entered the environment at various gasoline- and jet fuel-contaminated sites and threatens groundwater supplies. MTBE poses unique remediation challenges due to its high mobility and low natural degradation potential; it migrates farther and faster than gasoline BTEX compounds, the other hazardous compounds typically found at such sites. From late 1984 to early 1985, approximately 10,800 gal of gasoline leaked from two storage tanks and piping under the Naval Exchange gas station at the Naval Base Ventura County Port Hueneme Site. The MTBE entered a sandy aquifer. Laboratory experiments had indicated that MTBE can be aerobically degraded by a mixed bacterial culture, suggesting bioremediation as a possible treatment strategy at this site. Salanitro et al. (272) were the first to demonstrate complete biodegradation of MTBE; they used a mixed bacterial culture designated BC-1 that had been isolated from industrial sewage sludge enrichment culture on high concentrations of MTBE. However, natural concentrations of MTBE degraders are low, raising the potential for using bioaugmentation with inoculation of MTBE degraders, since natural attenuation rates were very low.

Salanitro et al. (273) performed a field demonstration, using oxygen injection and inoculation with a similar microbial inoculum, designated MC-100, at a national test plot at Port Hueneme, Calif. MTBE concentrations decreased after 30 days and throughout the 261-day experiment eventually decreased to less than 0.001 to 0.01 mg liter^{-1}.Tertiary-butyl alcohol (TBA) concentrations also declined in the bioaugmented plot to <0.01 mg liter^{-1}. The MTBE plume at this site was more than 4,000 ft long. MTBE concentrations ranged from 2 to 9 μg liter^{-1}. According to a report by Naval Facilities Engineering Command (227), intermittent sparging with pure oxygen started 6 weeks before microbial seeding with MC-100 and succeeded in raising dissolved oxygen levels from about 1 μg liter^{-1} to 10 to 20 μg liter^{-1}. Thirty-two days after seeding with MC-100, MTBE levels immediately downgradient dropped 90%. By day 261 in the treated plot, MTBE was not detectable in many sample locales, with 10 to 50 μg liter^{-1} in a few locales. The oxygen-only plot did show some MTBE decreases, apparently due to enhanced natural biodegradation processes after some lag time. The in situ MC-100 biobarrier appears capable of degrading MTBE to <5 μg liter^{-1}. At this site, the combination of bioaugmentation with added oxygen appears to be a feasible in situ MTBE biotreatment option.

In separate tests, Heath and Lory (125) demonstrated that Shell BC-4, also obtained from activated sludge of industrial wastewater treatment plants, could effectively be used to treat MTBE-contaminated groundwater. Shell BC-4 is a mixture of ordinary soil bacteria such as coryneforms, pseudomonads, and *Achromobacter* species that have been accli-

Box 5.7 continues

BOX 5.7
Bioremediation of MTBE-Contaminated Aquifers Employing Bioaugmentation
(continued)

mated to MTBE for more than a year. Bioreactor studies with BC-4 in the presence of oxygen have shown 99% removal of MTBE from groundwater, based on a 25-h retention time. The end products of the degradation are carbon dioxide and water. The field trial focused on the use of Shell BC-4 as an in situ biobarrier to downgradient MTBE migration. Shell Global Solutions has developed proprietary microbial products (BioRemedy and BioGAC) which can be used in bioremediation programs for groundwater contaminated by MTBE and TBA (http://www.shellglobalsolutions.com). BioRemedy uses a proprietary mixture of naturally occurring microorganisms that convert MTBE or TBA to carbon dioxide and water with no adverse environmental effects. BioRemedy can be used in the direct inoculation of contaminated groundwater, intercepting a spreading pollution plume or treatment of groundwater in an above-ground system. Inoculation of soil microcosms with these specialized cultures has shown that they can totally degrade soluble-phase concentrations of at least 5 to 80 mg of MTBE liter^{-1} and 5 to 50 mg of BTEX compounds liter^{-1} at 9 and 25°C. Demonstration projects in Port Hueneme and Tahoe City using BioRemedy have proved it to be a successful, cost-effective method for reducing MTBE and TBA in groundwater and in controlling the migration of MTBE and TBA pollution plumes at a number of retail gasoline sites.

and use them as initial inocula for serial enrichments with the contaminant(s) in question as the sole source of carbon (Figure 5.27). Typically, the selected strains are a subset of the fast-growing microorganisms. During this procedure, each subsequent enrichment should increase the proportion of the biodegradative population compared to the remainder. This inoculum is then returned to the site in large numbers in order to increase the rate of biodegradation. This approach, then, does not rely on increasing genetic diversity. Rather, it solely increases the catabolic potential of the specific strains capable of degrading contaminating chemicals. The strains used as inocula in these cases typically demonstrate exceptional degradative capacities and rapid growth rates; however, upon bioaugmentation, many, if not most, strains provide only a brief burst of pollutant biodegradation, only to decline by several log orders into the background of the indigenous microbial community due to competition. For example, during massive bioreactor bioaugmentation with a denitrifying bacterium, Bouchez et al. (44) observed almost complete disappearance of the strain used as the inoculum from the reactor within 2 days. Another possible reason for this failure is that typically the enrichment technique uses media containing much higher concentrations of the target pollutant and nutrients than the parent ecosystem, resulting in the enrichment of rapidly growing strains, which have every likelihood of being minority strains in the natural population (355).

The third approach is often performed in practice but is not valued in the scientific community. This involves the addition of uncharacterized consortia present in materials such as sewage sludge and compost. These are easy sources of inocula for companies to sell to engineers, who, faced with a contaminated site, often feel compelled to do something and who do not fully evaluate the microbial value of bioaugmentation. Even materials such as garden waste provide extra microbial communities, even though that is not the primary function in the bioremediation, which is normally to provide heat-generating materials during composting.

Bioaugmentation Is Mired in Uncertainty

When bioaugmentation is considered, i.e., whether or not microbes should be added, there is little agreement on strategies, inoculum

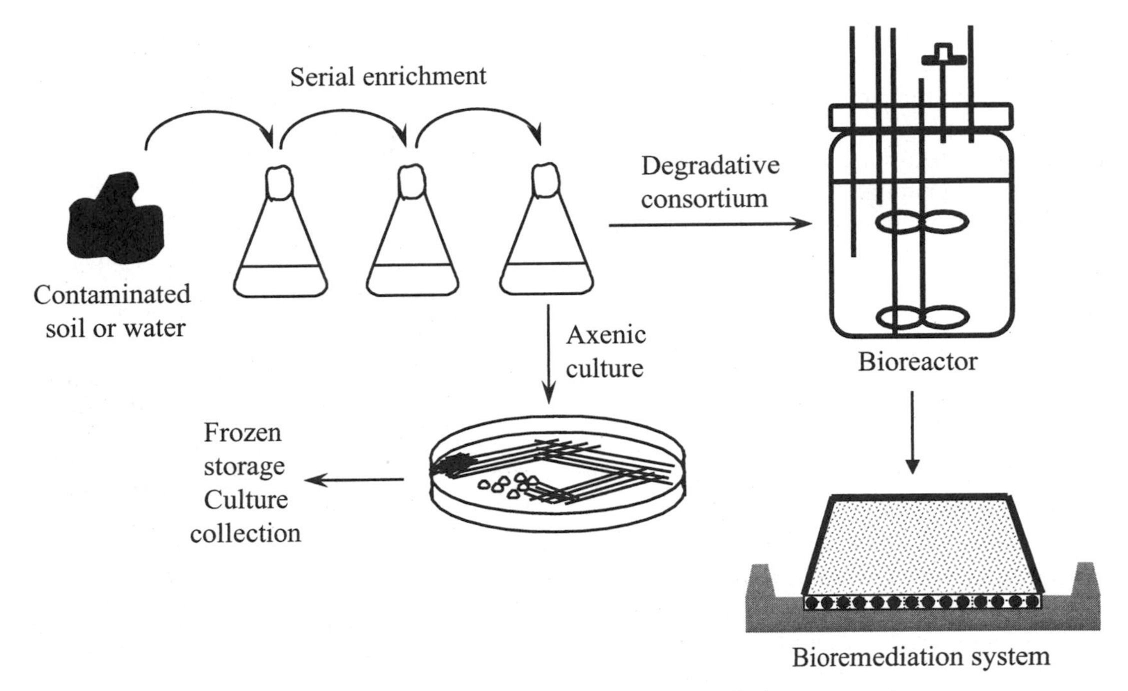

FIGURE 5.27 Typical serial enrichment procedure for bioaugmentation.

size, strain selection, other strain factors to be taken into account, delivery systems, or the efficacy of freeze-drying (264). Indeed, there is little agreement about the whole contentious issue. As a result, the scientific community in bioremediation is intrigued, whereas the engineering community opinion ranges from suspicion to hostility. An opinion that is often expressed, however, is that there may be no need at all for bioaugmentation in projects such as fuel bioremediation, as there should be indigenous microbes at most sites to deal with fuels, and that the role for bioaugmentation should be the remediation of recalcitrant compounds where there might be so few indigenous strains present that augmentation is necessitated (146). A niche area for bioaugmentation for fuel spills has appeared in the literature: cold climates (e.g., see reference 212), since a great deal of crude oil reserves are extracted in cold regions. Arctic (265) and Antarctic (6, 8) soils are considered to be deficient of microorganisms compared to other soils. Low numbers of hydrocarbon-oxidizing bacteria in pristine cold-climate soils and short summer seasons may limit the spontaneous enrichment of oil-contaminated soils with autochthonous hydrocarbon oxidizers when biostimulation alone is practiced.

As exemplified by tests involving the *Exxon Valdez* oil spill, a range of strategies and a large number of microorganisms have been used in bioaugmentation, and yet to date there is no guarantee of success (345), especially with commercial products. Mearns (204) found that bioaugmentation with a microbial product did not significantly enhance oil biodegradation in a contaminated marsh in Galveston Bay. Two different microbial products, which exhibited enhanced biodegradation of Alaska North Slope crude oil in shaker flask tests, did not accelerate biodegradation in a field experiment conducted on an oiled beach in Prince William Sound (343). Thouand et al. (315) found that most commercial inocula that they tested lacked any significant ability to degrade crude oil. Venosa et al. (344) also found that an indigenous microbial inoculum did not increase oil biodegradation in a beach environment. Regardless of such equivocal results, many microbial products have been commercialized. Lin et al. (180) concluded that microbial products which are cultured and selected to enhance oil degradation rates have great uncertainties, especially in systems such as wetlands, where hydrocarbon-degrading bacteria are naturally prevalent. They question the value of bioaugmentation for oil spills in light of the fact that addition of microbes had not been demonstrated to be effective in increasing rates of oil degradation and given the high costs of microbial amendments.

The view that increasing the number of microorganisms within the system should increase the rate of removal of contaminants has proven simplistic. Various factors are known to influence the success or failure, but predictability is beyond our means at present. In the case of soil, the milieu is hostile to the introduction of allochthonous microorganisms, and large numbers will never survive unless they have a selective advantage. Critics of bioaugmentation point to the fact that inocula are developed in the laboratory, far from the hostile environments they are intended to compete in (264). The controlling factors are both biotic and abiotic (for a review, see reference 341). The abiotic factors are numerous: they include inappropriate pH or pH shift, UV irradiation, desiccation, and lack of available inorganic nutrients. The biotic factors are more complex and less well understood.

Biotic factors include competition with the indigenous populations and low levels of available carbon substrates (111). A lower threshold substrate concentration for effective bioremediation is a controversial area. Sims et al. (295) calculated that at least 150 mg of soil per liter of pore water is required. Another view is that simultaneous utilization of a variety of carbon sources present in low concentrations can be performed until substrate exhaustion at micromolar levels by communities of microorganisms. Indeed, this may be possible within individual microorganisms. For example, the apparent lack of catabolite repression within the rhodococci (354) lends credence to this theory.

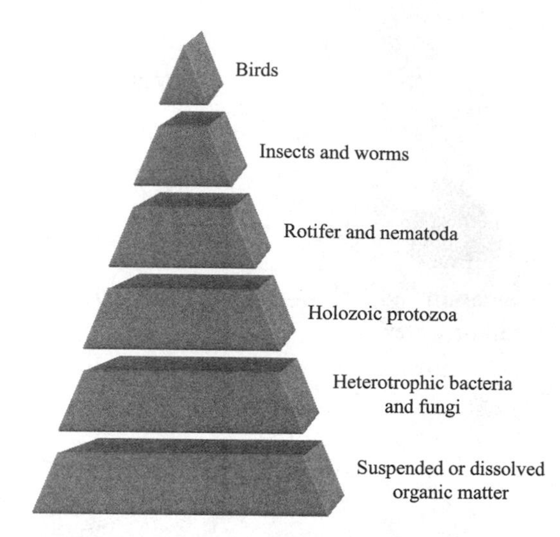

FIGURE 5.28 Food web relationships in a percolating filter. After the work of Wheatley (359). Practitioners should remember that adding microorganisms in a bioremediation project is very different from a chemical addition. It has the potential to shift the balance in the food web, so that simply adding more might make more problems, rather than solve any.

In bioaugmentation studies, the predator-prey interaction is often overlooked (Fig. 5.28). For example, in a wastewater bioaugmentation study, Bouchez et al. (44) observed improved performance only during the first day after inoculation, despite inoculation at massive levels. Fluorescent in situ hybridization revealed that the digestive vacuoles of protozoa gave strong signals with a bacterium-specific probe. This provided direct experimental evidence for a warning often expressed over bioaugmentation: high levels of inoculation can lead to an imbalance in the predator-prey interaction, with the wholesale destruction of the inoculum by protozoa. Moreover, the inoculum may also induce strong grazing pressure on other microbial species present, thus creating the risk that the genetic diversity of the site can actually be decreased.

SUCCESSES AND FAILURES OF BIOAUGMENTATION

The current situation is that there are many recorded successes and failures of bioremediation treatment of soil contaminants (90). Some examples of successful bioaugmentation for degradation of recalcitrant pollutants are use of pure cultures of bacteria for atrazine (287), a consortium for atrazine (230), a pure culture for 2,4,6-trichlorophenol (17), a pure culture for pentachlorophenol (70), and activated soil for pentachlorophenol (33). In some cases where bioaugmentation has been applied to petroleum hydrocarbon contamination, the results have been contradictory and little, if any, benefit has been derived (85, 229). Even with diesel, which is easily biodegraded, the bioaugmentation experience has been contradictory. In none of the five soils studied by Margesin and Schinner (194) did bioaugmentation result in greater diesel decontamination than biostimulation alone. Indeed, some authors have reported negative effects on diesel-contaminated soil bioremediation, with either acclimatized indigenous populations (83) or commercial bioaugmentation products (e.g., see reference 213). Cunningham and Philp (72) conducted successful bioaugmentation, and subsequently attained less success (176), with diesel-contaminated soil at the same source, using the same approach. There are many reasons for the failures, which have been extensively discussed and reviewed (e.g., see references 44, 305, and 348).

Can Strain Selection Be Improved?

From the plethora of studies of bioaugmentation, a trend that should be taken account of is that, despite selection of strains with great catabolic potential, the organisms have often failed to survive in the highly competitive environments of water and soil. Perhaps it is time for a shift of focus to survival rather than catabolic potential in strain selection. Yet the ability to survive is much more difficult to select for than a simple catabolic trait. Molecular biology is starting to furnish more methodical approaches to strain selection in this context. For instance, Goddard et al. (110) monitored 690 isolates of fluorescent pseudomonads from a single field site and determined the genetic composition and dynamics of the population, using restriction fragment length polymorphism rRNA (ribotyping) analysis, over several seasons. The population was found to be highly heterogeneous: the 690 isolates characterized consisted of 385 ribotypes, and most genotypes were transient, being detected only once. However, approximately 26 ribotypes were detected more frequently (spatially and temporally) and 1 (ribotype A) was ubiquitous in the samples analyzed. Ribotype A was subsequently genetically tagged (*lacZY* with kanamycin resistance), used to inoculate laboratory-based field soil, and found to persist in the soil significantly longer than three randomly selected ribotypes, taken from transient populations, which had been tagged in the same manner. These results clearly suggest that different strains have different competencies in terms of survival when introduced into the environment as inocula.

The enrichment bias mentioned above suggests that an improvement in enrichment procedure may be to use media that are much lower in target pollutants and nutrients. In a study to compare the catabolic pathways from 2,4-dichlorophenoxyacetic acid (2,4-D)-degrading bacteria isolated by direct plating and by prior enrichment, Dunbar et al. (91) isolated fewer different strains, representing different populations, by enrichment than by direct plating. Moreover, the strains from enrichment exhibited almost exclusively a single hybridization pattern with 2,4-D gene probes. This is strongly suggestive that enrichment culture seriously underestimates the diversity and distribution of catabolic pathways in nature.

By contrast, Zocca et al. (372) used traditional enrichment techniques for the isolation of PAH-degrading bacteria and showed that the enrichment procedures led to the isolation of taxonomically different bacteria. These authors have suggested that the only way to resolve the argument is through the integration of cultivation and noncultivation techniques. For this to be so, there is a need for the use of more appropriate enrichment media. Bruns et al. (54) were able to cultivate low numbers of bacteria from a nutrient-depleted marine environment using rich media but were able to increase the cultivation efficiency by an order of magnitude when they used a dilute mineral medium. In liquid most probable number procedures, the cultivation efficiency was increased by a further order of magnitude.

Different criteria apply to strain selection for an enclosed bioreactor. For example, van der Gast et al. (336) isolated strains for bioaugmentation of spent metal-working fluids (MWF) in a bioreactor system. The MWF-bioreactor system was characterized by a constant and reliable supply of a defined pollutant mixture within an environmentally controlled system. In this instance, the authors used a triple-selection criterion:

1. dominance of source populations in the target habitat (waste MWF),
2. tolerance to cocontaminants (MWF are chemically mixed), and
3. ability to degrade components of MWF.

Extensive analysis of the community composition and structure of a single MWF formulation, both temporally and spatially at a worldwide scale, was undertaken to identify the ubiquitous microbial populations in waste MWF. Subsequent screening cycles were then based on the ability of isolates to tolerate the toxicity of cocontaminants and to degrade individual chemical constituents of the MWF. The procedure resulted in an assemblage of five isolates that proved to be 85% more effective at processing waste MWF in bioreactors than undefined inocula from sewage and that also proved to possess exceptional resilience in the bioreactor (335, 336).

Similar strain selection approaches applied to a more complex, dynamic system such as soil, however, often fail to achieve reliable success. Blumenroth and Wagner-Dobler (41) tested whether strains originally isolated from a contaminated soil expressed enhanced survival upon reinoculation compared to that of competent nonindigenous strains. They concluded that survival of the reinoculated indigenous microorganisms was not significantly greater than that of a nonindigenous source of inoculum.

SELECTION OF ADDITIONAL TRAITS

A large number of desirable traits could be imagined. Where oil or solvent contamination levels are very high, it is likely that catabolic strains will also possess solvent tolerance mechanisms. The hydrophobicity of many pollutants adds to their toxicity to microorganisms by the nonspecific effects on membranes. The ability to combat this toxicity would give a catabolic strain a competitive edge over others without the ability. Multiple responses to solvent action are known in bacteria (279), among which two of the best known are *cis-trans* isomerization and efflux pumps. The *cis-trans* modification is a rapid response to exposure to solvents that takes place within a minute of exposure, and this is common within the genus *Pseudomonas* (151). Two similar but distinct solvent tolerance efflux pumps in *Pseudomonas putida* were characterized in 1998 (158, 258) (Fig. 5.29). A new gene cluster that confers solvent tolerance to *P. putida* was described by Phoenix et al. (250), and when these genes were transcriptionally fused to *lux* genes, the biosensor was responsive to a wide variety of organic molecules, including BTEX compounds, naphthalene, and complex mixtures of aliphatic and aromatic compounds.

If the theory that many bioaugmentation cultures fail because of limited availability of the pollutant(s) as a carbon source is correct, then the ability to survive starvation would be an important trait to try to select. Starvation survival has been studied in a wide range of bacteria. It is estimated that the expression of more than 70 genes is affected by the stationary-phase sigma factor σ^S, encoding functions leading to altered morphology and metabolism, increased starvation survival, and survival of fluctuating environmental conditions (128, 129), the last two of which are common occurrences in bioremediation. In the enterobacteria, cyclic AMP (cAMP) is involved in the regulation of the majority of the genes expressed during starvation (277). The addition of very low concentrations of cAMP to marine media increased the cultivability of marine bacteria in low-nutrient conditions by nearly

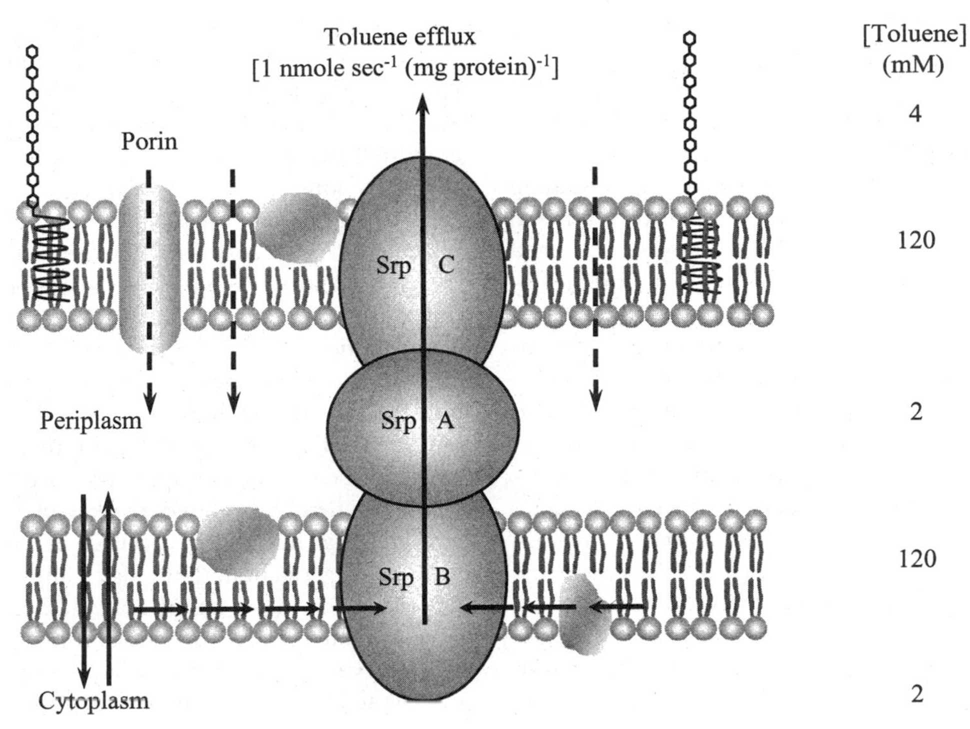

FIGURE 5.29 Solvent efflux pump schematic. The diagram shows how toluene might be transported out of the gram-negative cell by membrane- and periplasm-spanning proteins.

100% (54). Perhaps this is a way to improve the effectiveness of enrichment for catabolic strains. The same authors also observed a pronounced increase in cultivability by the inclusion of 10 μM acyl homoserine lactones to media.

Stach and Burns (302) used biofilm enrichment to generate greater diversity of PAH-degrading bacteria from soil than obtained by using batch culture liquid enrichment, suggesting that the overwhelming predominance of planktonic enrichment procedures may be a major source of bias in recovery of bacterial diversity.

In a study investigating phenol-stimulated TCE bioremediation, Futamata et al. (104) found that an enrichment created by batch phenol feeding had poor ability to degrade TCE. When fed phenol continuously, however, the enrichment had high TCE-degrading activity, and the two different enrichments had shifted in the predominant phylogenetic groups of phenol hydroxylases. This study is one that suggests a role for preferential gene selection during enrichment.

PLANT-ASSOCIATED STRAIN SELECTION AND DELIVERY

The pairing of catabolically relevant microorganisms that can colonize plant roots with a plant circumvents some of the problems faced by traditional bioaugmentation. Kuiper et al. (164) were the first to methodically identify a strain indigenous to a host plant's rhizosphere that exhibited preferential growth on plant root exudates and enhanced pollutant-degrading ability and rhizosphere competence. The approach is aided not only by the abundance of pollutant-degrading microorganisms typically associated with the rhizosphere, but also by the fact that the highly competent pollutant-degrading inoculum is sustained on a carbon source that is constantly being replenished by the plant (296). This would aid survival of strains when the pollutant is present in low available concentrations.

ADAPTATION OF STRAINS: VALUE OF PRIMING OR ACTIVATION

Priming is generally described as predisposing an isolate or population of microorganisms to future conditions in which they are designed to perform a function. Priming for bioremediation might consist of enriching clean soil for particular pollutant-degrading microorganisms by repeated exposure to the relevant pollutant(s). In other words, this bears similarity to the traditional serial enrichment procedure, but it is done in soil, not liquid culture. The resulting soil, with its highly competent degrader microbial community, is then used as the inoculum for the target soil (108). This approach has several advantages: the consortium of indigenous microorganisms within the primed soil are all utilized in the bioaugmentation, not just a laboratory-grown isolate; the primed soil consortium is maintained within its native soil, minimizing the competitive elements (i.e., microorganisms, predation, or chemicals) of the target soil; and otherwise less culturable, yet potentially highly competent, pollutant degraders are included obligatorily. This approach, while logical and easier to set up than the traditional shake flask enrichment, is certainly among the most understudied bioaugmentation strategies. It may not appeal to the purists who wish to work with defined cultures, but it has the potential to make bioaugmentation more reliable as a field practice.

How Many Microorganisms Are Needed?

There is no agreement in the scientific or engineering communities on what would be an appropriate number of microorganisms to apply to a contaminated site. The best that can be said is that the amount of inoculum to add is site specific (255), which is virtually no help at all. Guidelines exist, for example, for delivery of microbes in various plant systems. No such guidelines are available for bioremediation.

Inoculation of contaminated soils with very large numbers of microorganisms with degradative capabilities for the contaminants present ignores a fundamental observation. Dejonghe at al. (81) elaborated on the concept of Corman et al. (69) that populations of microorganisms added to soils stabilize at around 10^3 CFU (g of soil)$^{-1}$, a number which seems to be independent of the ecosystem. Greater numbers merely lead to the unwelcome overproliferation of the protozoa. Furthermore, unless the concentration of the target substrate is very high, there will not be enough carbon to support the growth the division of large numbers of bacteria. This concept, if correct, then militates against the second approach to bioaugmentation described above: the key would lie in increasing the genetic diversity as a means to increasing catabolic potential. Curtis et al. (75) estimated that a ton of soil could contain 4×10^6 different prokaryotic taxa.

However, several authors have postulated that the ability of an inoculum to degrade recalcitrant compounds is related to the size of the inoculum. A *Flavobacterium* sp. was able to mineralize pentachlorophenol (PCP) in freshwater at a density of $5.5. \times 10^7$ cells ml^{-1} (196). Edgehill and Finn (94) also related the removal of PCP by an *Arthrobacter* sp. to inoculum size. Upon observing the same phenomenon with *p*-nitrophenol in lake water, Ramadan et al. (257) argued that the failure of a small inoculum may be due to the failure to survive. They concluded that the low inoculum densities failed to mineralize *p*-nitrophenol when protozoa were active, but upon the suppression of protozoa by use of eukaryotic inhibitors, the mineralization proceeded.

For contaminated soils, the numbers of applied microorganisms have been highly variable. Intuitively, there would be a greater need for higher numbers than for the bioaugmentation of contaminated water, for several reasons: greater competition from indigenous bacteria, greater predation from protozoa and perhaps bacteriophage, and lack of availability of pollutants due to sorption phenomena in soils. Topp (317) showed that 50% of the applied atrazine was degraded when 10^5 CFU of strain C147

g^{-1} was used to inoculate a soil with little capacity for the mineralization of atrazine. By contrast, Rousseaux et al. (269) used only one application of a lower inoculum (10^4 CFU g^{-1}), and this was sufficient to obtain a threefold increase in atrazine mineralization compared to that in uninoculated soil. This inoculum is equivalent to 10^{13} CFU ha^{-1} on the basis of 1,000 tons of soil ha^{-1}. Given that a culture could be grown to a density of, say, 10^9 CFU ml^{-1} in the laboratory, that would imply that a total inoculum volume for 1 ha of 10 liters would be sufficient. If, however, the required inoculum was 10^7 CFU g^{-1}, then the total volume of inoculum required would be 10,000 liters, which is a much more daunting prospect. Given such inconsistencies, it is little wonder that the engineering community can be hostile to bioaugmentation. Such numbers as inocula to soil are not unrealistic. Briglia et al. (48) used an inoculum of 10^7 CFU of *Rhodococcus chlorophenolicus* PCP-1 g^{-1} for the bioremediation of PCP, in a soil where peat apparently limited the PCP availability. Pearce et al. (242) augmented petroleum-contaminated soil to an astounding 10^{12} bacteria g^{-1}.

Bioaugmentation with GMOs

Genetically engineered microorganisms (GEMs), which are also referred to as genetically modified microorganisms or genetically modified organisms (GMOs), have shown great potential for bioremediation applications in soil, groundwater, and activated sludge environments, exhibiting enhanced degradative capabilities encompassing a wide range of chemical contaminants that are found at numerous soil and groundwater sites (275). Various investigators have used modern biotechnology to create recombinant bacteria with potential for environmental applications. The future of bioremediation in part may depend upon the ability to develop real-world applications for these GMOs that can meet regulatory requirements and gain public acceptance (82, 355).

The first living organism ever patented was a hydrocarbon-degrading pseudomonad that was genetically engineered by Ananda Chakrabarty of General Electric (62–64); the bacterium had been designed to degrade components of crude oil for potential applications to oil spills. Chakrabarty was able to fuse plasmids to genetically engineer a strain of *Pseudomonas* having the single-cell capability for multiple separate degradative pathways, including camphor, salicylate, and naphthalene degradative pathways, which gave the organism specific beneficial oil-degrading capabilities. The initial attempt to obtain a patent for the GEM was denied by the U.S. Patent Office on the basis of the argument that living bacteria are not patentable because they are products of nature (63). General Electric appealed the decision, which was eventually decided upon by the U.S. Supreme Court in 1980 in a 5 to 4 decision in the case of Diamond v. Chakrabarty (Sydney Diamond was the Commissioner of Patents who sought to prevent the granting of a patent) (63). This landmark decision said that life could be patented, opening the door for major investments in biotechnology and the possibility of using genetic engineering for creating organisms that could be used for bioremediation (64).

Since the granting of the patent to Chakrabarty, many researchers and start-up bioremediation companies have considered genetically engineering microorganisms for bioremediation applications; none so far have succeeded in achieving real-world success, but efforts continue despite numerous technical and regulatory hurdles (373). Two critical issues are (i) the ability to construct GMOs for release in bioremediation with an acceptable degree of ecological predictability and (ii) the ability to monitor the performance of GMOs within a complicated environment, in particular in terms of their survival, gene transfer potential, and impact on the native microbial population (82). The vast majority of studies pertaining to genetically engineered microbial bioremediation are supported only by laboratory-based experimental data; there are relatively few examples of GEM applications in environmental ecosystems (275). A variety of GMOs that have been designed for bioremediation are still at the laboratory or early field test stage, but there is optimism that in the future GMOs will be used for bioremediation, targeting most recalcitrant pollutants in inhospitable environments at relatively low cost (368).

The GEMs being constructed are being designed to degrade resistant compounds, such as chlorinated aromatics and nitroaromatics. Dutta et al. (93) successfully engineered the nitrogen-fixing, symbiotic soil bacterium *Sinorhizobium meliloti* to enable it to bioremediate in situ 2,4-dinitrotoluene-contaminated soil in the presence of alfalfa plants. Using a combined strategy of random mutagenesis of haloalkane dehalogenase and genetic engineering of a chloropropanol-utilizing bacterium, Bosma et al. (43) constructed an organism that is capable of growth on 1,2,3-trichloropropane, which is a highly toxic and recalcitrant compound generated as a waste product during the manufacture of the industrial chemical epichlorohydrin. Their results demonstrate that directed evolution of a key catabolic enzyme and its subsequent recruitment by a suitable host organism can be used for the construction of bacteria for the degradation of a toxic and environmentally recalcitrant chemical.

Besides strains engineered to have specific metabolic capabilities, GMOs that are able to function in hostile environments, such as those containing high concentrations of heavy metals, have been created. Gray (115) described the potential of fungi for bioremediation applications in soils contaminated with heavy metals and radioactive elements. Brim et al. (50) developed a radiation-resistant bacterium for the treatment of mixed radioactive wastes containing ionic mercury. Brim et al. (49) also engineered *Deinococcus geothermalis* for in situ bioremediation of radioactive wastes. Yoon (370) has constructed multiple-heavy-metal-resistant phenol-degrading pseudomonads. These laboratory studies encourage the development of "designer biocatalysts" that can degrade xenobiotics at sites contaminated with multiple heavy metals. Sriprang et al. (301) developed a novel system for the bioremediation of heavy-metal-contaminated soils; they engineered a strain of *Mesorhizobium huakuii* to express tetrameric human metallothionein. When this bacterial symbiont forms nodules, the host plant accumulates high concentrations of the heavy metal, e.g., cadmium. The plant can then be harvested, thereby removing the heavy metal from the soil.

Although various GEMs that may be useful for bioremediation have been developed, the only field trials conducted have involved the use of reporter genes. Reporter genes, such as the *lux* genes that confer bioluminescence and the green fluorescent protein gene, allow tracking of the environmental fate of GEMs (160, 304). Layton et al. (171), for example, constructed a bioluminescent reporter strain of *Ralstonia eutropha* for the detection of PCBs in the environment. Sayler's research team also conducted field trials with a genetically engineered *Pseudomonas fluorescens* strain that exhibited bioluminescence when degrading naphthalene; they used this strain to examine hydrocarbon-biodegrading activity in the field and to monitor the fate of introduced GMOs (101). In this field release, escape from the application site was detected only sporadically; the highest incidence of bacterial escape occurred when the relative humidity and wind speed were low. Genetically engineered luminescent bacteria can be used to detect environmental pollutants (263). Robert Burlage and colleagues from the U.S. Department of Energy Oak Ridge National Laboratory have developed genetically engineered bacteria with green fluorescent protein in the presence of trinitrotoluene; these bacteria can detect land mines and have been field tested at Edwards Air Force Base in California (18, 19).

The lack of field trials to test the efficacy of biodegradation by GEMs in soils and aquifers is a serious impediment to the development of GEM-based bioremediation applications, since the only manner in which to fully address the competence of GEMs in bioremediation efforts is through long-term field release studies. Therefore, it is essential that field studies be performed to acquire the requisite information for determining the overall effectiveness and risks associated with GEM introduction into natural ecosystems (275). One way to eliminate the regulatory hurdles of releasing GMOs into the environment is to use enzymes, i.e., to eliminate the living organism (256). Shinkyo et al. (292) used site-directed mutagenesis to develop an enzyme that is capable of attacking tetrachlorinated dibenzo-*p*-dioxins that could be applicable to the bioremediation of soils contaminated with dioxin (292). Mixed cultures capable of oxidizing and hydrolyzing endosulfan may be a good source of enzymes for use in enzymatic bioremediation of endosulfan residues (309). Enzymes derived from organophosphate-resistant sheep blowfly have also been found to detoxify orthophosphate insecticides (S. Davidson, unpublished data). Microorganisms have been engineered to degrade organophosphates (351) and to produce enzymes that can detoxify organophosphate-contaminated water (288, 289). These enzymes have been shown to work under field conditions (300) and have the potential to degrade parathion, malathion, and monocrotophos (256); carbaryl; many synthetic pyrethroids; and endosulfan, which are all insecticides (170). Orica conducted a field trial on a cotton farm near Narrabri, Australia, in 2002, treating 80,000 liters of contaminated drainage water with a recombinant organophosphate-degrading enzyme and reducing the organophosphate levels by 90% in only 10 min (170). Thus, recombinant enzymes have great potential for use in selected bioremediation applications.

Despite these novel applications of enzymes and rapid advances in molecular biology, there has been little progress in actually bringing bioremediation employing GMOs to the field. This despite the fact that, as reflected by the Eurobarometer survey on the social perception of biotechnology, biological research for environmental remediation is precisely the application of genetics that Europeans sympathize with the most and are least concerned with regarding risks (82). However, no releases of GMOs into the environment have taken place in the United Kingdom for bioremediation applications (368), nor is the Department of Energy considering such releases in the United States (373). Safety concerns, regulatory hurdles, and anticipated negative public reactions are preventing such field trials and real-world applications.

Novel Delivery Systems

A fundamental observation about delivery of bioaugmentation cultures has to be explained. Given all the trials that an introduced culture has to face, the delivery of the culture to a soil bioremediation project must be done correctly if it is to have a chance of success. In ex situ systems, this is best done by thorough mixing with the soil during the construction. It has been often observed that bacteria do not move appreciably through soil (10). The inoculum therefore must be distributed to nearly all sites immediately adjacent to the pollutants. This lack of movement is due to physical filtration by the solids in the soil and adsorption to soil particles. Therefore, the practice of addition of cultures to, say, a biopile is likely to be futile.

In groundwaters, similarly, there is evidence that bacteria do not travel great distances before becoming attached to a solid surface. Since appreciable biodegradation by an inoculum to groundwater would require the inoculum to travel substantial distances, it is not yet clear whether such an approach is valuable.

Delivery of bioaugmentation cultures in an immobilized form may offer more complete and/or more rapid degradation. In bioremediation applications, the immobilized matrix may also act as a bulking agent in contaminated soil, facilitating the transfer of oxygen, which is crucial for rapid hydrocarbon mineralization. Immobilization is known to reduce competition with indigenous microorganisms (179) and to offer protection from predation (303) and extremes of pH and toxic compounds (255, 312) in the contaminated soil. There is also evidence of increased biological stability, including plasmid stability, in immobilized cells (61).

Straube et al. (307) reported preliminary promise for improved survival of bioaugmentation cultures when immobilized by sorption to vermiculite, both during storage and after deployment to a contaminated soil. However, entrapment in a polymer matrix has been the most common method of whole-cell immobilization (103), and this offers the additional advantage in bioremediation that the environment of the immobilization matrix can be varied greatly to improve the physiological stability of the microorganisms. Cunningham et al. (73) presented evidence that polyvinyl alcohol (PVA) entrapment improved the bioremediation of diesel-contaminated soil at laboratory scale. They also showed an improvement by incorporating a synthetic oil absorbent into the matrix. Naturally occurring oil absorbents are plentiful as waste products, and materials such as sunflower seed husks and sawdust are able to take up several hundred percent of their own weight of oil. When cultures immobilized in PVA hydrogels are incubated at 25°C, the gels have been seen to bulge, presumably as a result of CO_2 production, and eventually burst (Figure 5.30). If this same phenomenon occurs in contaminated soils, then this would act as a means of slow release of cultures, which may prevent overgrazing by protozoa.

Daane and Häggblom (76) described a system that also has the potential for immobilization and slow release by a more natural mechanism. They added *Ralstonia eutropha* (pJP4) to earthworm egg capsules (cocoons), and this initiated biodegradation of 2,4-D in soil. The cocoons were also able to tolerate and biodegrade high levels of 2,4-dichlorophenol, showing that this is also a mechanism of toxicity reduction. Upon hatching, the cocoons released strains to the soil and initiated biodegradation. Thus, the system acted to transport and slowly release the bacteria.

BIOSTIMULATION

An alternative to increasing numbers of biodegrading microorganisms on the site is the practice of biostimulation. Biostimulation aims at

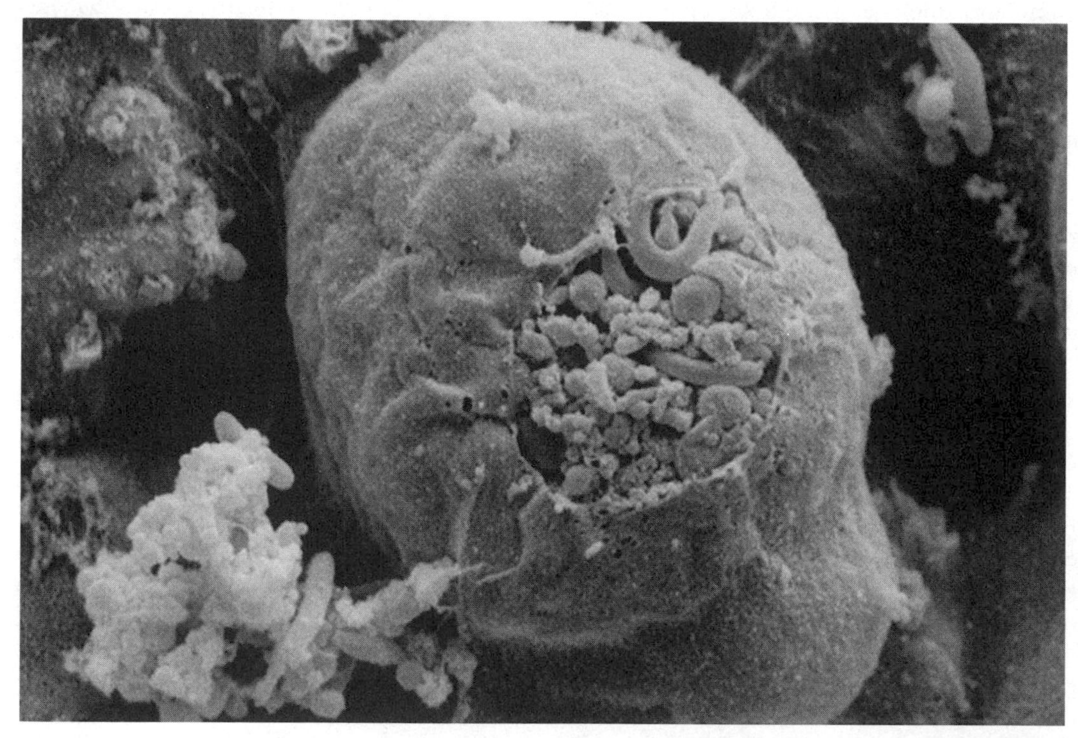

FIGURE 5.30 Scanning electron micrograph of bacteria emerging from bulges in a PVA immobilization gel. Such evidence suggests that it should be possible to manufacture engineered slow-release bioaugmentation.

enhancing the activities of indigenous microorganisms that are capable of degrading the offending contaminant. It is applicable to the bioremediation of oil-contaminated sites, often being viewed as helping nature, i.e., an extension of the natural remediation of soil and groundwater (Box 5.8). In many cases, the addition of inorganic nutrients acts as a fertilizer to stimulate biodegradation by autochthonous microorganisms (24). Biostimulation in some cases involves the addition of inorganic nutrients to stimulate biodegradation by autochthonous microorganisms (24); in other cases, it is the intentional stimulation of resident xenobiotic-degrading bacteria by use of electron acceptors, water, nutrient addition, or electron donors (360).

Although not an approach to bioaugmentation as such, biostimulation leads to an increase in catabolic potential without increasing genetic diversity. In practice, it is a much simpler and less costly approach to bioremediation. The bioremediation community has diverging opinions on the subject. Many practitioners of bioremediation state that biostimulation is all that is required for the mineralization of naturally occurring hydrocarbons such as petroleum mixtures: 3.5 billion years of evolution has provided the genetic diversity required, and all that is needed to stimulate mineralization is the correct balance of carbon with nitrogen and phosphorus. However, it must be noted that bioaugmentation may be required for the biodegradation of more recalcitrant xenobiotic pollutants, where the evolutionary timescale may be limited to the years since the industrial revolution. Indeed, Pritchard (255) opined that future bioremediation projects to remove recalcitrant compounds will undoubtedly involve inoculation to introduce unique and specialized metabolic capabilities into a contaminated matrix.

Oxygen

Due to the limited solubility of oxygen in water, various attempts have been made to increase it. The most obvious route is the injection of pure oxygen rather than air, as this can increase the dissolved oxygen concentration severalfold. More common has been the use of hydrogen peroxide.

BOX 5.8
RENA of Oil-Contaminated Soils in Nigeria

Bioremediation has been used to clean up hundreds of oil spill sites in Nigeria that were contaminated as a result of oil exploration and oil production operations (20). The cleanup, which began in 1999, was developed and performed by Shell Petroleum Development Company of Nigeria. The company's preferred approach for soil and groundwater remediation programs is known as remedial enhanced natural attenuation (RENA). Liquid sludge is removed for ex situ treatment, and the residual hydrocarbon is bioremediated in situ. Application of the RENA process for remediation of hydrocarbon-impacted soil accelerates the natural process by using human intervention to increase access to water, sunlight, oxygen, and nutrients, i.e., in situ bioremediation based upon biostimulation of the indigenous microorganisms. The intervention involves simple tilling of the impacted soils, which aerates the soil and ensures adequate oxygen to support aerobic hydrocarbon biodegradation. It also involves occasional addition of nutrients, water, and sometimes topsoil. These actions increase the microbial population that biodegrades petroleum hydrocarbons and provides optimized conditions for hydrocarbon metabolism by the indigenous microbial communities. Shell reports that the use of RENA is cost-effective and restores the soils and groundwaters to preoiling conditions. RENA is viewed as an environmentally friendly and natural process that has gained acceptance in the communities where it is carried out. It is simple and can be performed by local people, which aids in its acceptance. It is well suited for the tropical climate of the Niger Delta, where soils have diverse microbial populations, including naturally occurring hydrocarbon-degrading microorganisms, warm temperatures, and abundant rainfall.

OXYGEN AVAILABILITY AND TRANSPORT

As most of the bioremediation technologies are aerobic processes, due to the greater efficiency of aerobic biodegradation of organic compounds, the delivery of oxygen to soil and groundwater in ex and in situ bioremediation technologies is crucial to success. Indeed, oxygen should in most cases be regarded as the key component of a bioremediation that will ultimately determine the success or failure of the application. It has been shown on many occasions that hydrocarbon biodegradation will proceed without special soil amendments as long as oxygen is present (139).

The concentration of oxygen dissolved in water is low. Since moisture is critical to microbial activity, the essential interrelationship between oxygen and water is obvious. Oxygen transport from the open air to the pores of the soil bioremediation matrix to the pore water and thence to the microorganisms is a two-component process. Convection and diffusion are the two governing mechanisms, and oxygen transport is inextricably linked to the water content of the matrix since both convection and diffusion are dramatically reduced in water. Water saturation of soil pores slows oxygen transport to very low levels, and quickly the rate of oxygen consumption by microorganisms exceeds the ability of convection and diffusion to replace it, resulting in deoxygenation and, if left unchecked, anaerobiosis.

The low rate of oxygen delivery is a major drawback in bioremediation. Soil contaminated with 10 m^3 of hydrocarbons would require about 2 million m^3 (tons) of water saturated at 10 mg $liter^{-1}$ for its biodegradation.

CONVECTIVE AIR MOVEMENT

In ex situ bioremediation technologies, convection can be passive, making use of the buoyancy of hot air, or forced by using blowers or vacuum pumps. For example, in composting or biopile systems treating high levels of organic contamination, hot air can sometimes be seen rising from the top of the pile (Fig. 5.6 and 5.11), and convection pulls cooler air in from the atmosphere to replace it. The phenomenon is not so noticeable in biopiles with blowers, but in any case convection plays a major role in the delivery of oxygen to the pore space of soil.

Convective air movement relies on soil porosity to maintain a low resistance to flow. Water saturation of the soil pores is a major impediment to convective air transport. Even with forced aeration, air will short-circuit water-saturated zones, causing unsaturated zones to dry further, and force the saturated zones into anaerobiosis. One of the principal reasons for using adapted windrow turner equipment in contaminated land compost systems is to redistribute moisture, as well as to fulfill the primary function of soil aeration. With static biopile systems, this is not possible, so there is greater need to ensure that moisture distribution and uniform mixing are accomplished during the soil preparation phase, prior to biopile construction.

OXYGEN DIFFUSION

Molecular diffusion is the movement of a molecule through a mixture, most commonly caused by a concentration gradient of the diffusing molecule. The molecule moves by random vibrational motion in the direction to equalize the concentrations and thus destroy the gradient. At this equilibrium point, all the molecular species in the mixture are uniformly dispersed and the concentration of any one species is the same throughout the system.

At the microorganism and micropore levels, diffusion dominates over convection as the major oxygen transport mechanism. Gas diffusion is particularly sensitive to water saturation: diffusivities in liquids are generally 4 to 5 orders of magnitude lower than in gases at atmospheric pressure (197), as a result of the large difference in free path length between collisions. With gases, the free path is orders of magnitude greater than the size of the molecule, whereas in liquids the free path is less than the molecular diameter. Calculations of oxygen diffusion in water and in air illustrate the need for air saturation, and not water saturation, of soil pores.

The diffusion of oxygen in air is a function of both temperature and pressure. The diffusion coefficient for a binary pair of gases is calculated from the Chapman-Enskog equation (246):

$$D_{AB} = \frac{0.001858T^{3/2}[(M_A + M_B)/M_A M_B]^{1/2}}{P\sigma_{AB}^2 \, \Omega_D} \quad (4)$$

where D_{AB} is diffusivity (in square centimeters per second); T is temperature (kelvin); M_A and M_B are the molecular weights of components A and B; P is pressure (in atmospheres); $\sigma_{AB} = (\sigma_A + \sigma_B)/2$, which is the effective collision diameter (in angstroms); and Ω_D is the collision integral, which equals $f(kT/\varepsilon_{AB})$, where k is Boltzmann's constant, ε is the Lennard-Jones force constant for common gases, and $\varepsilon_{AB} = \sqrt{\varepsilon_A \varepsilon_B}$.

For the diffusion of oxygen in air at 20°C:

	ε/k	σ	M
Oxygen	106.7	3.467	32
Air	78.6	3.711	29

$$\sigma_{AB} = (3.467 + 3.711)/2 = 3.589$$
$$\varepsilon_{AB/k} = (106.7 \times 78.6)^{0.5} = 91.578$$
$$kT/\varepsilon = 293/91.578 = 3.199$$

From tables (197), $\Omega_D = 0.9328$.

Substituting these values into the Chapman-Enskog equation gives a diffusivity of oxygen in air of 0.198 $cm^2\ s^{-1}$ at 20°C.

The diffusion of oxygen in water is best described by the Wilke-Chang equation (362):

$$D_v = 7.4 \times 10^{-8} \frac{(\psi_B M_B)^{1/2} T}{\mu V_A^{0.6}} \quad (5)$$

where D_v is diffusivity (in square centimeters per second), T is absolute temperature (Kelvin), μ is viscosity of solution (in centipoise), V_A is molar volume of solute as liquid (25.6 $cm^3\ mol^{-1}$ for oxygen [358]), ψ_B is the association parameter for solvent (2.26 for water) (358), and M_B is the molecular weight of solvent.

For the diffusion of oxygen in water at 20°C:

	M	V_A	μ at 20°C
Oxygen	32	25.6	
Water			1.002

Substituting these values into the Wilke-Chang equation above gives a diffusivity of oxygen in water at 20°C of 1.972×10^{-5} $cm^2\ s^{-1}$, which is approximately 10,000 times lower than in air at the same temperature.

OXYGEN DIFFUSION IN SOILS

Ideally, the soil gas diffusion coefficient is measured by laboratory or field methods, but in practice this is likely to be too time-consuming and expensive, so it is simpler to use empirical models that relate the gas tortuosity factor τ_g to specific bulk soil properties that can be readily measured (139).

$$\tau_g = \frac{\alpha^{10/3}}{\theta^2} \quad (6)$$

where θ is porosity, and α is volumetric air content (the volume of gas space per total volume of soil). Porosity and volumetric air content may be computed for a specific soil if the bulk density (ρ_b), particle density (ρ_p), and moisture (μ) are known:

$$\theta = 1 - \rho_b/\rho_p$$

$$\alpha = \theta - \mu\ \rho_b/\rho_{water}$$

Bulk density is defined as the mass of dry soil per total soil volume and may be estimated from the dry weight of a soil core sample. Particle density is the ratio of dry soil mass to unit volume of soil particles, the latter of which may be determined from the volume of water displaced by the soil particles. Moisture content is the mass of water per mass of dry soil and may be determined by gravimetry after oven drying. The density of water is assumed to be 1 g ml^{-1}.

The soil gas diffusion constant (D_s) can now be calculated as a function of soil moisture content and porosity:

$$D_s = \frac{\left[\theta + \frac{\mu\rho_p}{\rho_{water}}(\theta - 1)\right]^{10/3}}{\theta^2} \cdot D_a \quad (7)$$

Predictably, the gas diffusion coefficient in soil decreases with increasing moisture content and decreasing porosity.

Many soils have a porosity of approximately 50%, and since microbial activity is considered optimal at a moisture content of at least 50% of its field capacity (i.e., a μ value of 0.05 to 0.10 g of water g of dry soil^{-1} for most soils), it is expected that D_s will range from 0.005 to 0.04 cm^2 s^{-1} in most moderately moist and microbially active soils. D_s values within this range are used for the computation of oxygen penetration distances and cleanup times (139).

OXYGEN DELIVERY TO GROUNDWATER

There are several ways to deliver oxygen to groundwater during in situ bioremediation. The most appropriate method for a particular project will be dictated by the contaminants and their concentrations which set the oxygen demand, and the hydrogeology of the site. Hydraulic conductivity has an enormous effect on the amount of oxygen that has to be supplied. For a given hydraulic gradient, oxygen supply rate varies by several orders of magnitude from low to high hydraulic conductivity (51).

The simplest, least expensive method is to sparge air into the well bore by using porous bubbler devices such as are in common use in the wastewater treatment industry. Saturating water with air in this way will achieve a dissolved oxygen concentration of around 8 to 10 mg liter^{-1}, depending on salinity and temperature. Saturated water diffuses from the well bore into the subsurface at a rate determined by the hydraulic conductivity. In low-permeability aquifers, pure oxygen will increase the saturation to about 50 mg liter^{-1}, but this is a much more expensive operation. A more common method is to pump oxygen-saturated water into the aquifer. The above comments on concentrations and hydraulic conductivity apply.

The very low diffusivity of oxygen in water compared to that in air threatens failure for many aquifer bioremediation projects. Alternative, more drastic methods of aeration have been investigated. Of these, the injection of hydrogen peroxide (H_2O_2) has been the most successful. It is highly soluble in water and decomposes to water and oxygen. Therein lies the biggest problem with using hydrogen peroxide in situ: the rate of decomposition has to be controlled. Its decomposition is catalyzed by common soil components such as ferric iron and naturally occurring organic materials. For this reason, hydrogen peroxide is not a practical oxygen delivery method for the vadose zone. Too-large amounts of oxygen generated are lost to the gas phase. In the saturated zone, more is dissolved in water. However, even then, if the decomposition cannot be controlled, the zone of influence around the injection well may be too limited. Hopkins and McCarty (135) pointed out that hydrogen peroxide, while possessing advantages over pure oxygen, was even more expensive, and if too much is added, the saturation level in water of oxygen can be exceeded, leading to gas accumulation in the aquifer, with an attendant clogging problem. Hydrogen peroxide can be carried greater distances when stabilized with phosphate, and the phosphate also acts as a nutrient.

Another counterproductive limitation of hydrogen peroxide delivery is that it is in fact biocidal. Lethal concentrations of hydrogen peroxide for soil microorganisms are in the order of 30 to 50 mg liter^{-1} (68). Brown et al. (52) suggested that the practical concentration in water be kept at 100 mg liter^{-1} to avoid gas pocketing and to limit microbial killing.

However, there is no doubt that if delivery conditions can be made optimal, hydrogen peroxide delivers high concentrations of oxygen. It provides 0.47 g of oxygen per g of hydrogen peroxide delivered. The biodegradation of 1 kg of hydrocarbon would require the delivery of 400,000 kg of water containing 8 mg of oxygen liter^{-1}. The same mass of hydrocarbon would be biodegraded by the delivery of 13,000 kg of water containing 500 mg of hydrogen peroxide liter^{-1} (132). Berwanger and Barker (39) investigated the in situ bioremediation of aromatic hydrocarbons in a methane-saturated groundwater, using hydrogen peroxide as the oxygen source. Added at a nontoxic level, it promoted the rapid biodegradation of BTEX compounds. Failures have also been recorded, on two occasions because of aquifer plugging in field trials using hydrogen peroxide to promote in situ biodegradation of jet fuel.

OXYGEN REQUIREMENTS IN BIOREMEDIATION PROJECTS

The calculations of oxygen requirements cannot be exact; rather, they are presented as guidelines that may help in the selection of blower size, for example. They also may be useful in selecting whether to use passive or forced aeration in an ex situ project.

The theoretical oxygen demand (ThOD) of an organic compound can be calculated by either solving a balanced chemical equation or using an arithmetic equation. The aerobic mineralization of a 2,4-dichlorophenol can be represented as

$$C_6H_4Cl_2O + 6O_2 \rightarrow 6CO_2 + H_2O + 2HCl$$

1 mol of dichlorophenol requires 6 mol of oxygen for its mineralization

⇨ 163 g of dichlorophenol requires 192 g of oxygen

⇨ 1 mg of dichlorophenol requires 192/163 mg of oxygen

⇨ ThOD of dichlorophenol = 1.18 mg mg^{-1}

$$\text{The ThOD} = 16\,\frac{\left[2c + 0.5(h - cl - 3n) + 3s + 2.5p + 0.5na - o\right]\text{mg mg}^{-1}}{\text{MW}} \qquad (8)$$

$$\text{For dichlorophenol} = \frac{16\left[12 + 0.5(4 - 2 - 0) + 0 + 0 + 0 - 1\right]}{163}$$

$$= 192/163$$

$$= 1.18 \text{ mg mg}^{-1}$$

There is exact agreement between the two methods.

In planning the bioremediation of a soil contaminated with this compound, either as the sole contaminant or as the contaminant identified by risk assessment to pose the greatest risk to receptors at the site, this figure can then be used to calculate the total oxygen requirement. Given the average level of contamination of the soil, the total volume of soil, and the time period of the project, a figure for the total oxygen requirement can easily be calculated. At most contaminated sites, this situation is unlikely. Most bioremediation projects are performed on petroleum hydrocarbon-contaminated sites, and there may be hundreds of compounds or more present. In this case, a design approach has been elaborated by Cookson (68), in which an "average" contaminant is calculated, and the requirements for oxygen, nitrogen, and phosphorus can subsequently be calculated.

Alternate Electron Acceptors

Under anaerobic conditions, an alternative electron acceptor to oxygen is required. Nitrate is usually the electron acceptor of choice in these conditions, although care has to be taken with the regulatory situation regarding nitrate in groundwater. A very different approach is the selection of alternative electron donors. For example, added toluene, propane, or methane acts as a cometabolic substrate, which is oxidized and also stimulates the oxidation of the chlorinated compound(s). Hydrogen, or a hydrogen-generating substrate, can be added to act as a direct reductant (120).

FORMULATIONS OF INORGANIC NUTRIENTS

Atlas and Bartha (24) observed dramatic mineralization of crude oil upon stimulation with nitrogen and phosphorus, but limited stimulation with either one singly. Subsequent studies have shown the need for nitrogen and phosphorus as the primary nutrient additions in bioremediation. Nearly always, the supply of potassium, sulfur, magnesium, calcium, iron, and trace elements is greater than the demand (10). Common commercial NPK fertilizers are used, but the influence of potassium is often ignored.

Nutrient addition still constitutes a considerable expense in bioremediation. A successful formulation at one contaminated site does not guarantee success at others. Formulations of nutrients for site application vary in the amounts of macronutrients (carbon, potassium, nitrogen, and phosphorus), micronutrients (sulfur, magnesium, and calcium), and trace elements (the most common are iron, manganese, cobalt, copper, and zinc). A list of nutrient solutions used for field and laboratory bioremediation and their formulations is given elsewhere (5).

Soil contaminated with organic pollutants such as oils and fuels is likely to contain an excess of carbon compared to nitrogen and phosphorus, and the indigenous bacteria may be unable to biodegrade the material, even if it is highly biodegradable and is a high-quality source of calories, due to a nutrient limitation. Common assumptions are that only one nutrient is limiting at any one time and that only when one deficiency is overcome does another become limiting. This may prove not to be so (10). Combinations of inorganic nutrients often are more effective than single nutrients (310). Laboratory-based respiration experiments by Liebeg and Cutright (178) showed that a low level of macronutrients and a high level of micronutrients were required to stimulate the activity of indigenous microbes. There seemed to be no difference in effect of changing the dominant macronutrient, but a difference was seen if the macronutrients were omitted completely. The greatest stimulation was achieved with a solution consisting of 75% sulfur, 3% nitrogen, and 11% phosphorus.

Nitrogen is the nutrient most commonly used in bioremediation projects (178). It is used primarily to support biosynthesis (NH_4^+ and NO_3^-) or as an alternative electron acceptor to oxygen (NO_3^-). It is commonly applied in the form of urea, any ammonium salt, or ammonium nitrate. Bacteria readily assimilate all of these, but care would have to be taken in a bioremediation regarding ammonium salts, as they also exert an oxygen demand. Activated sludge has been suggested to be a useful source of nitrogen for PAH biodegradation in soils (152). Slow-release forms of nitrogen allow persistent biostimulation with less chance of water pollution with nitrates. Dried blood acts as a slow-release agent for nitrogen (306), as can a range of natural materials such as peat, compost, and manure (217). Slow-release formulations would be particularly appropriate on beaches (237) to prevent the nutrients from being diluted to extinction by wave action. A slow-release fertilizer can also be made with inorganic nutrients, and it is much easier to make a predictable ratio of nutrients this way. An example is the encasement of ammonium nitrate, calcium phosphate, and ammonium phosphates to give an N-to-P ratio of 10.6:1, in a polymerized vegetable oil (181).

Phosphorus is routinely seen as the next most important nutrient. Phosphorus is used for cell growth and is added as a phosphate salt (sodium, potassium, orthophosphate, or polyphosphates). It can be added as diammonium phosphate, and then its contribution to the nitrogen budget should also be considered (349). The application of phosphate carries implications. Phosphate has a high binding affinity for most soils. If added, say, to a windrow or biopile after construction, its transport will be limited and will result in a high concentration near the surface and a deficit in the interior. This is especially true in calcareous soils, where the high calcium content will result in the majority of the phosphate being adsorbed or precipitated and thus unavailable to microorganisms.

Many nutrient formulations nevertheless contain an excess of phosphate. In particular, formulations for groundwater bioremediation where H_2O_2 is to be added to improve oxygenation often contain excess orthophosphate to decrease the rate of peroxide decomposition. Doing so changes the geochemistry of the aquifer and may result in sufficient precipitation of hydroxides and phosphates to cause aquifer plugging (5). The maximum orthophosphate concentration that may avoid significant precipitation is about 10 mg liter^{-1}, but to achieve this concentration uniformly through the groundwater requires the nutrient formulation to contain a much higher concentration due to the soil sorption. The concentration required to optimize microbial cell growth at the lowered oxygen concentrations typical in bioremediation operations is unknown. However, consider the mineralization of 1 kg of organic carbon. If 300 g is used to make new biomass, and those cells have a C-to-P ratio of 50:1, then 6 g of P is required to be incorporated into the biomass and as such must be present in an available form. At high contaminant concentrations, it may be difficult to overcome phosphorus limitation in soil and in groundwater. When it is understood that very high levels of contamination can occur, e.g., 100 g of TPH per kg of dry soil (211), then this makes effective phosphorus stimulation difficult to deploy.

An interesting case was presented by Silva et al. (293), for the bioremediation of atrazine. Atrazine is still one of the most common pesticides detected in groundwaters in Europe and the United States. This was a combined biostimulation study with bioaugmentation with a *Pseudomonas* strain that degrades atrazine. This strain, however, uses atrazine as a source of nitrogen, and the authors hypothesized that the limitation would be one of carbon, not nitrogen. Thus, they biostimulated with organic acids and concluded that a $C_{substrate}$-to-$N_{atrazine}$ ratio of more than 40 was required for the mineralization of atrazine.

OLEOPHILIC FERTILIZER

When the contaminant is oil or another hydrophobic pollutant, there is the danger that applied water-soluble sources of nitrogen and phosphorus will have poor contact with the material. The objective would be to encourage intimate contact between microorganisms, oil, and nutrients. The obvious way to facilitate this is the incorporation of surfactants into the formulation. To maximize the contact between oil and water, the surfactant of choice should minimize the interfacial tension between the two. When the oil droplets in water, or water droplets in oil, are sufficiently small that the system remains transparent, then the phase containing most of the surfactant and the dispersed droplets is called a microemulsion (67). Microemulsions are characterized by having ultralow interfacial tensions (less than 0.01 mN m^{-1} can readily be achieved in the laboratory with some fairly ordinary and inexpensive surfactants). One such oleophilic fertilizer was used in the beach cleanup operations for *Exxon Valdez*. It was designed to adhere to oil and is a microemulsion of a saturated solution of urea in oleic acid, containing tri(laureth-4)-phosphate and butoxy-ethanol (181).

FAILURES OF NUTRIENT BIOSTIMULATION

Biostimulation has been used successfully in bioremediation much more often than it has failed. The reasons for failure, however, are usually unclear, and understanding these reasons may be crucial to the acceptance of bioremediation as a full-scale technology. Possible reasons advanced by Alexander (10) are high levels of nitrogen and phosphorus already present in the soil which may make further fertilization futile, the presence of nitrogen and phosphorus in the pollutants (also see reference 293), and a low concentration of pollutant(s). Additionally, there may be other limiting factors such as oxygen availability (Box 5.9).

Venosa et al. (344) used a randomized block design to investigate the influence of biostimulation and bioaugmentation on the removal of crude oil contaminating a sandy beach. High levels of oil biodegradation were seen in the untreated plots, and even though nutrient addition enhanced the rate of biodegradation, it was concluded that it might not be appropriate. Significantly, there was no difference between plots treated with nutrients only and those treated with nutrients and bioaugmentation

BOX 5.9
Bioremediation of Salt Marshes

Salt marshes, which are among the ecologically most sensitive ecosystems, occasionally are contaminated by oil spillages. These are very difficult habitats to clean. Several field studies have been carried out in coastal wetlands to evaluate the potential of oil bioremediation (371). The results of these studies have been varied, sometimes showing successful bioremediation and sometimes revealing a lack of stimulated hydrocarbon biodegradation.

Field trials on oil bioremediation in a *Spartina alterniflora* salt marsh in Nova Scotia, Canada, showed that addition of ammonium nitrate and triple super phosphate could stimulate oil biodegradation at low contamination levels, but no enhancement occurred at high concentrations where the oil had penetrated into the deeper reduced soil layers (172). Similarly, nutrient addition to a *Haloscarcia* sp. salt marsh in Gladstone, Australia, showed variable results; biostimulation produced about 20% more oil loss than in the untreated plots for an area contaminated with a medium crude oil, but the nutrient amendment did not significantly increase the rate of loss of Bunker C oil which had also contaminated areas of the salt marsh (55).

A study at San Jacinto Wetland Research Facility in Texas also showed that addition of diammonium phosphate significantly enhanced the biodegradation rates of aliphatic hydrocarbons and PAHs (210). However, in subsequent trials, biostimulation with nutrient addition failed to enhance rates of degradation; bioaugmentation with microbial cultures also did not increase rates of oil biodegradation (294). Addition of nutrients to an oiled *S. alterniflora* Louisiana salt marsh also failed to stimulate hydrocarbon biodegradation; however, the nutrient concentrations in this salt marsh were already high and natural attenuation was occurring at significant rates when the tidal cycle exposed the surface of the marsh to air, indicating that oxygen availability was the major factor controlling rates of oil biodegradation (291).

As summarized by Zhu et al. (371), these studies have shown that oil biodegradation on coastal wetlands is often limited by oxygen, not nutrient availability. If oil penetration into subsurface layers is minor, then biostimulation may be a viable strategy for cleanup, but if penetration of more than a few millimeters has occurred, then biostimulation will have diminished effectiveness due to the increased likelihood of limiting oxygen concentration in the oil impact zone. They conclude that when biostimulation is used, nitrogen concentrations of 2 to 10 mg of N liter^{-1} should be maintained in the pore water to achieve optimal oil biodegradation. Overall determination of whether to rely upon natural attenuation or to employ biostimulation should consider cost, practicality, and potential ecological impacts as well as the likelihood of achieving significantly enhanced rates of oil biodegradation.

with an indigenous inoculum. The high level of biodegradation in the untreated plots was attributed to the relatively high background level of nitrogen at the site. It should also be noted that this work was done on small plots. The same conclusions may not apply to a full-scale spill across a beach. In a similar manner, Takeuchi et al. (311) observed high methanotroph activity and rapid degradation of TCE without the addition of nutrients to contaminated groundwater: the water already contained sufficient nitrogen and phosphorus.

Nyman (232) observed no effect of nitrogen and phosphorus addition to marsh soils contaminated with crude oil other than temporarily accelerated microbial activity. Ruberto et al. (270) performed biostimulation with nitrogen and phosphorus on Antarctic soil contaminated with gas oil and jet fuel hydrocarbons, as this soil had previously been shown to be low in nitrogen and phosphorus. They observed growth inhibition at nitrogen and phosphorus concentrations frequently observed in rich soils. Likewise, Seklemova et al. (280) failed to improve bioremediation of diesel oil in soil by using inorganic fertilizers. Occasionally, laboratory tests on the mineralization of aromatic and aliphatic hydrocarbons indicate inhibition of biodegradation by nitrogen (e.g., reference 192); the reasons for this remain unknown, but it may simply reflect a lack of CO_2 production as carbon is diverted to biomass production (10), rather than true inhibition.

APPLICATION OF NUTRIENTS

Nutrient additions can be calculated by various means. Knowledge of the composition of the average bacterium is the start point. Elemental assay of the dry mass of *Escherichia coli* gives an approximate composition of the protoplasm of 50% carbon; 20% oxygen; 14% nitrogen; 8% hydrogen; 3% phosphorus; 2% potassium; 1% sulfur; 0.05% each calcium, magnesium, and chlorine; 0.2% iron; and a total of 0.3% trace elements, including manganese, cobalt, copper, zinc, and molybdenum (228). The nutrient requirements of the microbial cell are approximately the same as the cell composition except for carbon, which is supplied by the organic contaminant.

A disadvantage of a commercial preparation containing mixed nutrients is that the ratio of the various constituents cannot be modified according to requirements based on the carbon load created by the carbon-containing contaminant. In particular, the ratio of carbon to nitrogen to phosphorus seems to be important, and the ability to alter these components according to site contaminant concentrations would be important.

For engineering calculations, the average composition of a bacterium is often expressed as a chemical formula, which varies in the literature. The most common of these is $C_5H_7NO_2$, as these four elements constitute over 90% of the mass of cells. However, this takes no account of phosphorus, which, with calcium, constitutes about 70% of the remainder, and phosphorus has the potential to be limiting in contaminated soil or water. Another composition that takes account of phosphorus is $C_{42}H_{100}N_{11}O_{13}P$. Any such estimation should be used with caution. The ThOD of 1 g of $C_5H_7NO_2$ is 1.42 g, and that of $C_{42}H_{100}N_{11}O_{13}P$ is 1.71 g (159).

The calculation then depends on the desired ratio of C:N:P. The figures above would imply a ratio of 100:20 to 26:3. In wastewater treatment, the ratios that are published as optimum are usual for BOD:N:P, and 100:5:1 is adequate for aerobic treatment (123), whereas Gray (114) states that 100:6:1 is required to maintain the optimal nutrient balance for heterotrophic activity in conventional activated sludge. Anaerobic processes are much less demanding on nitrogen and phosphorus. For anaerobic digestion, Huss (142) found an optimum of 100:0.5:0.1. For ex situ bioremediation of contaminated soil, the desired C:N:P ratio is 100:15:1 (349).

For an ex situ biopile of 500 m^3 of contaminated soil, containing 10,000 mg of TPH kg^{-1}, the following shows how to calculate both nitrogen and phosphorus needs. The density of the contaminated soil should be measured on a site-specific basis, but for the example calculation, a density of 1.5 tons m^{-3} is assumed, i.e., 1,500 kg m^{-3}. Also, it is assumed that hydrocarbon is 80% carbon.

Assumed carbon concentration = 10,000 mg kg^{-1} × 0.8 = 8,000 mg kg^{-1}

Desired C:N:P ratio = 100:15:1

Required N concentration = (8,000 mg kg^{-1}) × 0.15 = 1,200 mg kg^{-1}

Required P concentration = (8,000 mg kg^{-1}) × 0.01 = 80 mg kg^{-1}

Total kg of soil to treat = 500 m^3 × 1,500 kg m^{-3} = 750,000 kg

Total N required = 750,000 kg × 0.0012 kg kg^{-1} = 900 kg

If urea (CH_4N_2O) is the sole source of nitrogen and is 46.6% nitrogen by weight, then:

Amount of urea needed = 900 kg/0.466 kg kg^{-1} = 1,931.3 kg = 1,932 kg

Total P required = 750,000 kg × 0.00008 kg kg^{-1} = 60 kg

If potassium dihydrogen orthophosphate (KH_2PO_4) is used as the sole source of phosphorus and is 22.8% phosphorus by weight, then:

Amount of KH_2PO_4 required = 60 kg/0.228 = 263.15 kg = 264 kg

Total nutrients required = 1,932 kg of urea and 264 kg of KH_2PO_4

A similar treatment is given by von Fahnestock et al. (349) in which a part of the nitrogen contribution comes from diammonium phosphate.

A more thorough treatment of nitrogen and phosphorus requirements is given by Cookson (68), which requires:

- the estimation of an "average" contaminant in terms of its elemental composition, based on the types and concentrations of individual hydrocarbons present at the site;
- calculation of the redox half-reactions involved, based on a bacterial cell composition of $C_5H_7O_2N$;
- knowledge of the fraction of the contaminant associated with conversion to microbial cells (yield fraction) (f_s); and
- summation of the overall reaction from the half-reactions.

While this approach is more complex, it should be more accurate, and it also quantifies the amount of oxygen required and the amount of CO_2 produced. Submitting the required information to an electronic spreadsheet for future calculations can mitigate the complexity. Whether this level of accuracy is required is debatable, as the exact quantities of soil will not be known, the concentrations of each individual contaminant will vary across a site, and some assumptions about yield values have to be made.

Application Rate. If the application is done as a single dosage during construction of an ex situ biopile, windrow, or even a new cell of a landfarm, then it is quite straightforward. Much more difficulty is experienced in nutrient addition to in situ systems since the mixing is not possible, and surface addition or injection has a very great possibility of uneven distribution of nutrients, especially phosphorus, due to its high tendency for sorption.

Rate of application of urea = 1,932 kg/500 m^3 = 3.86 kg m^{-3}, or 2.57 kg t^{-1}

Rate of application of KH_2PO_4 = 264 kg/500 m^3 = 0.53 kg m^{-3}, or 0.35 kg t^{-1}

However, Hicks (131) recommended that a single dose should be limited to 3 lb per cubic yd (about 1.8 kg m^{-3}) to avoid osmotic effects. Another problem that may occur in ex situ systems is that addition of all the nutrients as a single dose might later lead to nutrient limitation if they are lost in leachate and not recycled back. This is especially true of nitrogen, which is much more mobile than phosphorus. Nutrients should be added as required in a bioremediation project (dictated by respiration measurements). Slow-release nutrients can be added as a single dose, and this simplifies the logistics of future site operations.

Boufadel et al. (45) have suggested an optimum concentration of 2.5 mg of nitrate (N) for the biodegradation of 2 g kg of sand^{-1}, which equates to 180 μM, for bioremediation on beaches. Since the hydrocarbon content of the system being bioremediated changes with time, the determination of nutrient requirements based on initial contaminant concentration may be flawed. In marine spills, it is almost impossible to make accurate determinations of how much oil is on a beach, and it has been suggested that aiming for 100 to 200 μM available nitrogen in the interstitial water is a better approach (254).

Cometabolism and Alternative Electron Donors

Cometabolism is defined as the degradation of a compound only in the presence of other organic material that serves at the primary energy source (198). Essentially, in cometabolism an enzyme for a natural substrate also transforms a pollutant compound. As an example, TCE can be transformed by methane monooxygenase and toluene monooxygenase. Either methane or toluene supplied as a cosubstrate can be used for the bioremediation of TCE-contaminated groundwater (182).

The most favorable outcome of a cometabolic event is that the transformation product can be mineralized, where the initial substrate was not. In other words, the product of the cometabolism becomes an energy source for another microorganism. This, fortunately, often is the case. If not, then the next favored outcome is a reduction in toxicity of the transformation product.

Cometabolism is more important in bioremediation than is generally appreciated. It is known that many microorganisms participate in cometabolic processes, and cometabolism is important for many transformations, including those of some PAHs, halogenated aliphatic and aromatic hydrocarbons, and pesticides. Aerobic or anaerobic cometabolism is possible, depending on the contaminants and the field situation. A number of field demonstrations have shown the value of the cometabolism approach to in situ bioremediation (Box 5.10).

Another approach to a similar problem is the use of an alternative electron donor with the objective of changing redox conditions. Chemical reductants are often toxic and can be quenched quickly in large-scale applications. The injection of molasses into an aquifer would stimulate microbial aerobic activity and thus deplete the oxygen. This approach has been done at the Avco Lycoming Superfund site, Williamsport, Pa. The site is contaminated with not only TCE, DCE, and VC but also hexavalent chromium and cadmium. The full-scale injection of molasses at this site using 20 4-in. (10.2-cm)-diameter injection wells was done to bring about reductive dechlorination of the CAH, both directly by creating anaerobic conditions and by cometabolism, and to reduce the concentration of metals (Cr^{6+} cannot be removed but can be reduced to the much less toxic Cr^{3+}). In an area that was converted to an anaerobic reactive zone, the TCE concentration was reduced by 90%, to within the cleanup goal. The concentration of DCE initially increased, showing that TCE dechlorination was occurring, but subsequently fell to 19 μg liter^{-1}, which was also below the cleanup target. Likewise, hexavalent chromium removal was successful to below target (328, 329). The contractor has quoted that this technology is more cost-effective than the pump-and-treat method at this site.

Other electron donors being used at Superfund groundwater bioremediation sites include acetate, lactate, benzoate, and methanol, in all cases for the bioremediation of chlorinated compounds (330).

ALTERNATIVE ELECTRON ACCEPTORS

Denitrifying bacteria can be rapidly stimulated when supplied with a source of carbon and nitrate. Typically, acetate is supplied as a growth substrate, and nitrate and/or sulfate is supplied as an alternative electron acceptor to oxygen for the bioremediation of CAH (e.g., see references, 284 and 285). Providing nitrate to the subsurface is less expensive than maintaining aerobic conditions, and as nitrate is highly soluble, it is easier to maintain a residual concentration in groundwater (364). Semprini et al. (284, 285) observed a gradual decrease in the concentration of carbon tetrachloride after injection of acetate, nitrate, and sulfate. However, after removal of nitrate from the injection fluid, the carbon tetrachloride transformation rate accelerated, something that the same research group had observed previously. It was concluded that the dehalogenation was being carried out by a different group of organisms, possibly stimulated by the denitrifiers. A field demonstration of nitrate amendment in a recirculating well is being carried out at the Hanford West Area Site, Richmond, Wash. (330).

By use of a mixture of oxygen, at microaerophilic concentrations, and nitrate for the biodegradation of a mixture of BTEX, naphthalene, and phenanthrene, it was observed (364) that:

- toluene and naphthalene biodegradation was favored at microaerophilic oxygen levels (1.5 to 2.0 mg liter^{-1});
- the data suggested that oxygen and nitrate were used sequentially to biodegrade naphthalene and toluene, respectively (i.e., denitrification was inhibited until oxygen was depleted); and
- benzene degradation did not occur under either condition.

BOX 5.10
Cometabolism for In Situ Bioremediation of TCE-Contaminated Aquifers

TCE is a major contaminant of aquifers. TCE is used as a metal degreaser and in various products such as dyes, printing ink, and paint. Its widespread use has resulted in widespread contamination, as evidenced by its presence at over 33% of the Superfund sites in the United States. The most common method for removing TCE from groundwater is to pump the water from the aquifer and treat it above ground, which is costly and time-consuming. Hence, various efforts have been made to develop in situ bioremediation treatments to remove TCE.

Most microorganisms cannot grow on TCE as the sole source of carbon and energy. However, various aerobic microorganisms can cometabolize TCE. Under certain anaerobic conditions, TCE and tetrachloroethylene can be partially dehalogenated to form *cis*-1,2-, *trans*-1,2-, and 1,1-dichloroethylene (*c*-DCE, *t*-DCE, and 1,1-DCE, respectively) and vinyl chloride (VC) (347). Methylotrophs produce methane monooxygenase, an enzyme that can attack TCE; methylotrophs are able to degrade TCE to DCE and VC by cometabolism (100). Various investigators have considered using methylotrophs for the bioremediation of TCE-contaminated sites. McCarty et al. (200) found that they could stimulate indigenous methylotrophic populations and bring about the biodegradation of TCE within an aquifer in situ. In situ bioremediation treatment would require 5,200 kg of methane and 19,200 kg of oxygen in order to convert 1,375 kg of TCE from an aquifer of 480,000 m^3 containing a contaminant load of 1,617 kg of TCE.

The U.S. EPA's Superfund Innovative Technology Evaluation Program conducted a demonstration of the Enhanced In Situ Bioremediation Process at the ITT Industries Night Vision Facility in Roanoke, Va. Groundwater at the site was contaminated with chlorinated and nonchlorinated VOCs due to solvent leaks from storage tanks. The biostimulation process, developed by the U.S. Department of Energy and licensed to Earth Tech, Inc., involves injection of a mixture of air, gaseous-phase nutrients, and/or methane into contaminated groundwater to stimulate and accelerate the growth of existing microbial populations, especially metanotrophs.

At the former Hastings Naval Ammunition Depot, bioremediation based upon methane injection was used to treat a TCE groundwater plume (89). Methane was applied at a concentration of 4% by volume to increase the methanotroph population and degrade the residual TCE by cometabolism. Triethylphosphate was also added at a concentration of 0.03% by volume to supplement low levels of phosphorus present in the aquifer. Air-methane injection occurred for periods of 3.5 days followed by 3.5 days when the sparging was turned off. Increases of dissolved carbon dioxide concentrations and methanotroph densities were observed, providing evidence that the TCE reduction was due to cometabolism by an increased population of methanotrophs.

Several other cosubstrates, such as phenol and toluene, can also be used to support TCE biodegradation (136, 137, 199). Pilot studies conducted at Moffett Federal Airfield at Mountain View, Calif., demonstrated that TCE could be effectively biodegraded cometabolically through the introduction into the subsurface of a primary substrate (such as toluene or phenol) and oxygen to support the growth and energy requirements of a native population of microorganisms (284). The Moffett Federal Airfield site is contaminated with a variety of chlorinated aliphatic hydrocarbons (CAHs). During initial field trials at the site, methane was used as a cometabolic substrate for aerobic metabolism of TCE, *c*-DCE, *t*-DCE, and VC. The methanotrophic culture developed was good at transforming *t*-DCE and VC, but removal was poor for *c*-DCE and TCE. As a consequence, phenol was evaluated over two field seasons and was demonstrably superior to methane, removing 90% of TCE and *c*-DCE in a single pass at concentrations of CAH up to 1 mg liter^{-1} (136, 137).

A follow-up study was initiated to investigate the efficacy of toluene as an alternative primary source of energy for the removal of 1,1-DCE and VC and also to examine the effectiveness of hydrogen peroxide as an alternative to pure oxygen. In this study (135), 9 mg of phenol liter^{-1} and 12.5 mg of toluene liter^{-1} were used and removed to less than 1 μg liter^{-1} each, and highly efficient removal of TCE was noted with both additions. The authors cautioned, however, that the cometabolic approach has to be evaluated in the laboratory before full-scale trials are undertaken. There are many microorganisms that can use phenol or toluene as the primary substrate, and many different pathways for TCE transformation.

A full-scale remediation system was installed at a TCE-contaminated site at Edwards Air Force Base (201, 202). The remediation system consisted of two wells designed to pump out water and to allow the introduction of toluene, oxygen gas, and hydrogen peroxide (as a source of oxygen) into the discharge

BOX 5.10 *(continued)*

water. This permitted the development of a biodegradation zone with water circulating between the two wells to clean the aquifer. Successful evaluation was completed in March 1997. With pumping at 25 liters per min at each well and the introduction of 9 mg of toluene liter^{-1}, 30 mg of dissolved oxygen liter^{-1}, and 41 mg of hydrogen peroxide liter^{-1} for fouling control and additional oxygen, 83 to 85% TCE biodegradation was achieved with each pass through a treatment well. An estimated 60-m width of the TCE-contaminated plume was treated with this system, reducing its upgradient TCE by about 98% from 1,200 to 25 μg liter^{-1}. The toluene concentration was reduced to 1.4 + 0.6 μg liter^{-1} at the 22- by 22-m boundary of the steady zone. Potential clogging was successfully controlled.

While aerobic cometabolic transformations of TCE may be a suitable approach in some cases for the bioremediative cleanup of TCE-contaminated aquifers, anaerobic dehalogenation may also be a useful approach (122). Palumbo et al. (240), for example, found that the presence of perchloromethene inhibited the methylotrophs, suggesting that anaerobic perchloromethene removal would be necessary prior to stimulation of methylotrophs to remove TCE. Several anaerobic methods for removal of chlorinated solvents from aquifers have been investigated (329). Semprini et al. (283) found that carbon tetrachloride-, trichloroethane-, and Freon-contaminated sites could be bioremediated by stimulating indigenous denitrifying populations through the addition of acetate. Anaerobic TCE degradation occurs by reductive dechlorination with TCE being sequentially reduced to DCE, VC, and ethene. Because VC is a potent human carcinogen, it is critical to ensure that it does not accumulate in this process. Fumarate injection has been tested as a suitable substrate for anaerobic dehalogenation in TCE-contaminated sites (25). Rapid fumarate reduction to succinate was observed in wells where TCE reductive dechlorination was occurring. It appears that fumarate amendment has the potential to stimulate reductive dechlorination, even in aquifers for which no reductive dechlorination activity has been reported previously.

Injection of lactate has been tested for use in a deep, fractured rock aquifer contaminated with a TCE plume at the Idaho National Engineering and Environmental Laboratory. After 8 months of lactate addition, complete dechlorination was occurring at all monitoring points from 200 to 400 ft below ground within 100 ft of the injection well, and ethene was present in higher concentrations than any of the chlorinated ethene compounds (298). In situ anaerobic biodegradation enhanced with the injection of lactate has also been used to treat a large TCE plume that was due to historical injection of sludge waste into a basalt aquifer at the Pinellas Northeast Site, Largo, Fla. (105). The innovative remedy is known as reductive anaerobic biological in situ treatment technologies. For 8 months, lactate was injected 200 to 300 ft below ground surface. The success of the project in a complex fractured basalt aquifer may be a milestone both for fractured rock remediation and for in situ bioremediation of chlorinated solvent source areas.

Enhancing Bioavailability

SYNTHETIC SURFACTANTS AND BIOREMEDIATION

The lack of available pollutant as the microbial substrate is one of the serious barriers to bioremediation because the timescales become protracted due to mass transfer to the aqueous phase becoming the rate-limiting step. This applies to both contaminated soils and groundwater. Improving this bioavailability at full scale is not an easy task. Some surfactants enhance the solubilization and removal of contaminants (29, 30, 185). At the time of writing of the *Technology Practices Manual for Surfactants and Cosolvents*, there was no full-scale application of surfactant-cosolvent flushing for groundwater remediation (185), and to date it has seen very little development (331). There are major concerns about the large-scale use of surfactants in this manner. In particular, surfactants vary greatly in their toxicity to humans and ecotoxicity, and their resistance to biodegradation may lead to increased pollution (222). This is one of the major barriers to the development of the technique. For soil remediation systems, another technology-inhibiting observation is that the addition of surfactants to soils can form highly viscous emulsions that are difficult to remove (46, 244). Large quantities of surfactants are also required (193), and in a soil system, large quantities of aqueous chemicals can ruin soil permeability.

The application of surfactants to release hydrophobic pollutants, with the objective of increasing their bioavailability and biodegradability, has had mixed results (168). For example, there is evidence that some surfactants can enhance the aqueous solubility of PAHs (113). However, in a survey of eight synthetic surfactants (316), the ability to solubilize PAH was variable. The more hydrophobic surfactants exhibited better solubilizing properties. Contradictory results have been obtained from tests on the biodegradability of PAHs solubilized by synthetic surfactants. The more hydrophobic surfactants were less tolerated by bacteria. Liu et al. (183) showed that Brij 30 was degraded along with naphthalene but that Triton X-100 was not. Yet in both cases the addition of surfactant did not significantly affect either the rate of naphthalene degradation or the eventual amount degraded, despite improved bioavailability. Indeed, many synthetic surfactants are known to inhibit PAH-degrading microorganisms (318). The addition of synthetic nonionic surfactants to soil-water systems has proven to be completely inhibitory to biodegradation of phenanthrene at doses in excess of the critical micelle concentration (166). By contrast, surfactant-solubilized alkanes have been shown to be substantially more biodegradable (56).

The problem of large volumes and soil saturation may be overcome by the use of surfactant foams. Large volumes of air per unit volume of foam are injected into soil (268), and the foam contains 70 to 90% air. In a laboratory study that compared Triton X-100 liquid and foam injection into soil contaminated with PCP, twice as much PCP was removed by the foam than with liquid surfactant solution (221).

Various field-scale tests have been done on surfactant flushing for the removal of a variety of pollutants and have met with variable success. The results are tabulated in the work of Mulligan et al. (222).

A POTENTIAL ROLE FOR BIOSURFACTANTS

Many, but not all, biosurfactants are produced as a response to the low water solubility of *n*-alkanes as growth substrates. Oxygen is more soluble in the oil phase than the aqueous phase, and the low water solubility of alkanes places an additional pressure on microbial growth: that of mass transfer from the oil into the aqueous phase (121). Biosurfactants may offer several advantages over synthetic surfactants. They are generally less toxic and more biodegradable than synthetic surfactants (253). They are very effective surfactants, having about a 10- to 40-fold-lower critical micelle concentration than synthetic surfactants (168). Their production by microbes is widespread: in the last decade at least 13 new glycolipid producers and 13 new lipopeptide producers have been reported (57).

Types of Biosurfactant. Diverse molecules are produced as biosurfactants by prokaryotic and eukaryotic microorganisms, but they can be divided into groups based on overall structure: glycolipids, lipopeptides, and high-molecular-weight biopolymers (308). Those most commonly associated with oil-degrading bacteria are the glycolipids, and the hydrophilic head group is normally rhamnose or trehalose. The trehalose glycolipids have trehalose, a nonreducing disaccharide, linked by an ester bond to long-chain fatty acids. Among the latter, α-branched β-hydroxy fatty acids play a preferential role (Fig. 5.31). The trehalose glycolipids are common in the genus *Rhodococcus*. These compounds vary in overall chain length, with the largest and most complex being found in the mycobacteria (C_{20} to C_{90}). This variability in structure means that the molecules have different hydrophile-lipophile balances and therefore might be useful in solubilizing a wide range of pollutants.

One group of biosurfactants that has been studied extensively is the rhamnolipids from *Pseudomonas aeruginosa* (133). Up to seven homologs have now been identified (1). Two types of rhamnolipids are produced: they have

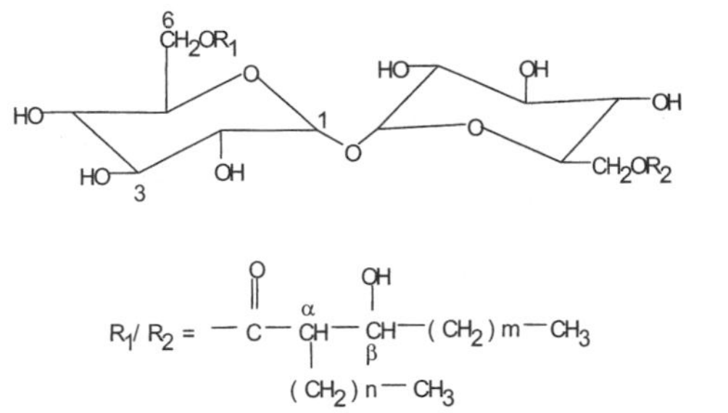

FIGURE 5.31 A trehalose glycolipid from *Rhodococcus ruber*. After the work of Philp et al. (249). The trehalose imparts the hydrophilic group, and the long-chain fatty acids impart the hydrophobic group.

either two rhamnoses attached to β-hydroxydecanoic acid or one rhamnose connected to the identical fatty acid.

Biosurfactants and Alkane Metabolism. Three types of alkane uptake response have been postulated (134), namely, uptake of mono-dispersed dissolved alkanes in the aqueous phase (minimal for alkanes above C_{10}), direct contact of cells with large oil droplets, and contact with fine oil droplets (pseudosolubilization) (Fig. 5.32).

An interesting comparison can be drawn between pseudomonads and rhodococci in their biosurfactant physiology. In the rhodococci, the nonionic glycolipids are not excreted from the cell, and they render the cell surface hydrophobic, which may then facilitate the attachment and subsequent passive transport of alkanes into the cell (134), requiring neither energy nor specialized membrane components. This is consistent with the observation that rhodococci almost always are highly adherent to alkanes and seldom render the aqueous phase of dual-phase culture media turbid. Being cell wall associated, the nonionic trehalose lipids may also impede the wetting of the cell surface. This would increase the direct cell-hydrocarbon contact at the aqueous-hydrocarbon interface (350). In contrast, the anionic rhamnolipids from some *Pseudomonas* strains are released into the culture medium and cause turbidity by emulsification. The third mechanism is likely to be quantitatively more important in pseudomonads, since these cells do not have a highly hydrophobic surface.

The application of biosurfactants to contaminated soil remediation is limited. Oberbremer and Müller-Hurtig (233) observed two degradation phases for a model oil system containing 10% soil. The second phase was quantitatively the more important and relied upon the production of glycolipids characteristic of

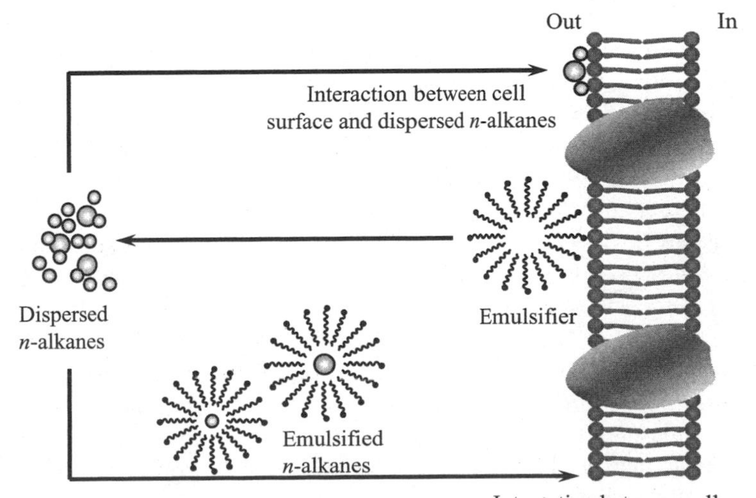

FIGURE 5.32 Influence of biosurfactants on alkane metabolism. After the work of Hommel (134).

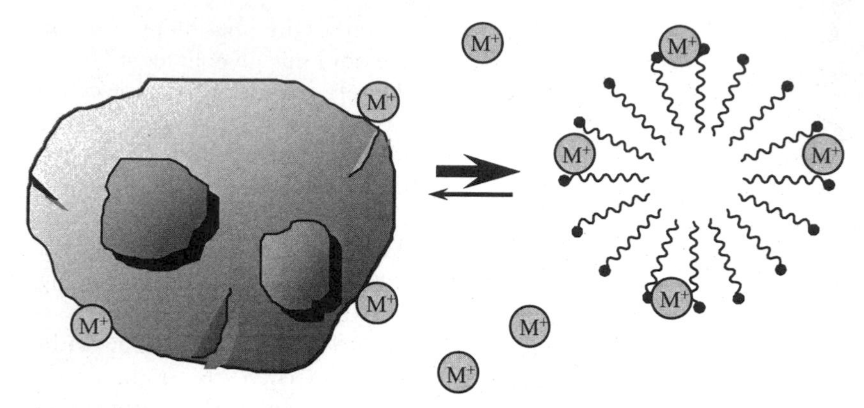

FIGURE 5.33 Positively charged metals may be released from soils by binding to negatively charged head groups in anionic (bio)surfactants.

Rhodococcus erythropolis to lower the interfacial tension between the oil and aqueous phases. All of the bacterial colony types isolated produced trehalose tetraesters similar to those of *R. erythropolis*. In a later study (234) using the same model system, purified glycolipids were added, and all types added decreased the adaptation times for the model oil degradation and increased the hydrocarbon degradation efficiency.

Jain et al. (145) reported that the addition of *P. aeruginosa* UG2 biosurfactants significantly enhanced the biodegradation of the aliphatic tetradecane, pristane, and hexadecene, but not that of the aromatic 2-methyl naphthalene. However, Scheibenbogen et al. (276) demonstrated the removal of both aliphatic and aromatic hydrocarbons with these biosurfactants. To identify the most efficient biosurfactant for desorption of hydrophobic pollutants from soil (338), several classes of microbially produced surfactants were tested, e.g., *Acinetobacter calcoaceticus* RAG-1, *Bacillus subtilis* Suf-1, *P. aeruginosa* UG2, and the trehalose mycolates from *R. erythropolis*. The biosurfactants of *Candida lipolytica*, *Candida tropicalis*, and *R. erythropolis* performed poorly in these tests. However, crude extracts of biosurfactants from various rhodococci have been used to remove crude oil from contaminated sand (144). Greatest efficiency was obtained with crude oils of lower asphaltene and higher paraffin content, for which almost total removal was possible.

The objectives of biosurfactant addition to metal-contaminated soils are similar: to increase the solubility of the metal in the aqueous phase and thereby facilitate the removal of the metal from the soil (Fig. 5.33). The principal removal mechanism with metals is flushing, since they are not biodegradable. They are also charged, whereas the organic pollutants are largely neutral. There are two proposed mechanisms for biosurfactant-enhanced removal of metals from soils (208). The first is complexation of metals already in the aqueous phase, thus promoting more metal desorption to restore chemical equilibrium. The second is increased solubilization as a result of direct contact between metal and biosurfactant due to lowering of the interfacial tension.

The first report of metal complexation by a biosurfactant (312) showed efficient complexation of cadmium (Cd^{2+}) by rhamnolipid from *P. aeruginosa*. This work was not done with soils, and little is known of how biosurfactants will behave in this substrate. Herman et al. (130) demonstrated strong sorption of rhamnolipids to soils, either by cation bridging of the anionic polar group and sorbed cations in soil or by hydrophobic interaction between the nonpolar tails and hydrophobic sites in the soil organic matter.

Another concern with the use of microorganisms to remediate contaminated soil is microbial mobility. Several reports have shown that bacteria are not mobile in soils and sand; this imposes limitations on in situ bioremediation techniques. Recent work (29, 30) has shown that rhamnolipids can prevent the irreversible adsorption of bacteria to soil particles, thereby increasing their mobility. Three possible mechanisms have been proposed: an increase in the negative charge density of the soil preventing bacterial adhesion through charge repulsion, solubilization of the extracellular polysaccharides which bind the microorganisms and allow biofilm formation, and sorption of biosurfactant to the soil, which may physically prevent bacteria from colliding with the soil surface. If these mechanisms do operate, then the highly hydrophobic rhodococcal cell surface may offer a distinct advantage over the use of other organisms.

Field tests of biosurfactant-enhanced bioremediation are rare. A biofertilizer based on rhodococcal biosurfactants was developed (144) and tested at pilot scale in the field on soils highly contaminated with crude oil wastes (165). Following a treatment in a slurry bioreactor for 2 months, the contents were then transferred to landfarming cells for further treatment, resulting in a TPH concentration of 1 to 1.5 g kg^{-1} in a further 5 to 7 weeks.

The principal reason for lack of studies with biosurfactants is production cost, which has in turn meant that until relatively recently there has been no commercial availability of glycolipid biosurfactants. Now, a company (Jeneil Biosurfactants) is producing rhamnolipids. Gu and Chang (117) used this biosurfactant to greatly enhance the extraction of phenanthrene from artificially contaminated soils. However, the biosurfactant is still too expensive for large-scale field use in bioremediation. The use of unconventional substrates, particularly renewables (188), should make their production less expensive. The synthesis of biosurfactants using microbial enzymes makes continuous production and recovery easy (162), and this has advanced to the preparative scale with lipase-catalyzed synthesis of glucose esters (369).

Cyclodextrins. Cyclodextrins have a hydrophilic shell and are highly water soluble, but they also have a hydrophobic cavity within which hydrophobic organic compounds can form inclusion complexes (Fig. 5.34). Hydroxypropyl-β-cyclodextrin (HPCD) has higher water solubility than other cyclodextrins, and HPCD has been shown to increase the solubility of naphthalene and phenanthrene by 20- and 90-fold, respectively, in the presence of 50 g of HPCD $liter^{-1}$(28). Moreover, this is achieved without micellization, which removes the problem of formation of viscous emulsions within soil.

Biodegradation of toluene and *p*-toluic acid has been enhanced in the presence of β-cyclodextrin due to alleviation of the toxicity of the substrates (278). It has also been demonstrated that HPCD can enhance the biodegradation of phenanthrene (353). Bardi et al. (34) accelerated the biodegradation of dodecane, tetracosane, naphthalene, and anthracene by adding β-cyclodextrin to liquid shake flask cultures. The ability of various cyclodextrins to remove two PAHs, naphthalene and phenanthrene, from soils has been shown previously (28). HPCD had greater capacity for removal of these compounds from soil than β-cyclodextrin, with 80% recovery of naphthalene and 64% recovery of phenanthrene. For both compounds, the slowest desorption was for the soil with the highest organic carbon content, with competitive interaction between the organic matter and cyclodextrins being the suspected desorption inhibition mechanism. As cyclodextrins are biodegradable and nontoxic (259), and their prices are approaching those of surfactants, they are beginning to look like feasible materials for enhancing field bioremediation.

CONCLUDING REMARKS

As the real-world applications of bioremediation continue to expand, so will the examples of specific successful approaches. Cost, efficacy (especially the ability to meet regulatory targets),

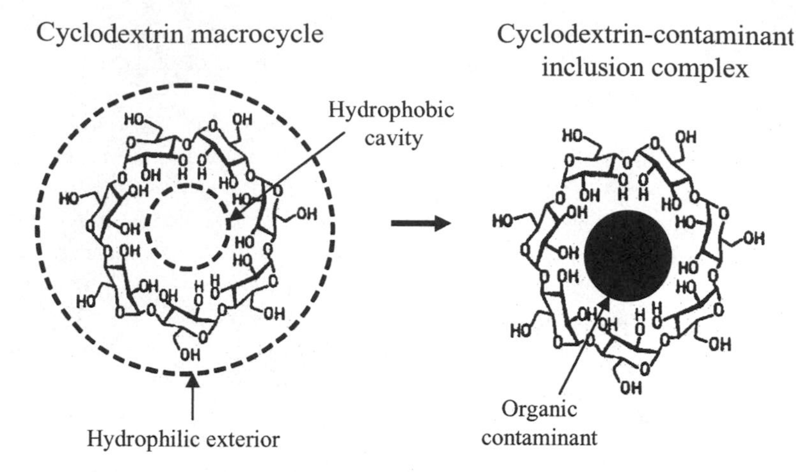

FIGURE 5.34 The cyclodextrin molecule. Courtesy of Brian Reid, University of East Anglia, East Anglia, United Kingdom.

and predictability—the ability to apply engineering principles to biological processes in complex environments—are key factors in the ability of bioremediation to become more significant in the highly competitive remediation industry. For the immediate future, emphasis likely will be placed on ex situ treatments that can be better controlled and hence are more predictable and on MNA (intrinsic bioremediation) with its inherently lower costs. The longer-term success of bioremediation, though, may well depend upon developing in situ treatments that can greatly accelerate the rates of degradation of contaminants, especially in groundwater, in a predictable, cost-effective manner.

The efficacy of bioaugmentation remains critical to the question of the future success of in situ bioremediation. Despite the widespread commercialization of microorganisms for environmental applications and the willingness of environmental engineers and managers to apply microorganisms, bioaugmentation often is of dubious value; scientific proof for the efficacy of in situ bioremediation is much greater for biostimulation than for bioaugmentation. Yet the aspirations for in situ bioremediation, as evidenced by many of the academic research and development efforts, rest with finding or developing microorganisms that can be used for bioaugmentation—in particular whether GMOs can be applied to deal with the more significant residual contamination problems. In this context, biotechnology holds great promise that has yet to be realized for environmental applications.

REFERENCES

1. **Abalos, A., A. Pinaso, M. R. Infante, M. Casals, F. Garcia, and A. Manresa.** 2001. Physicochemical and antimicrobial properties of new rhamnolipids by *Pseudomonas aeruginosa* AT10 from soybean oil refinery wastes. *Langmuir* **17:**1367–1371.
2. **Abramowicz, D. A.** 1990. Aerobic and anaerobic biodegradation of PCBs. *Crit. Rev. Biotechnol.* **10:**241–251.
3. **Abramowicz, D. A., M. J. Brennan, H. M. Van Dort, and E. L. Gallagher.** 1993. Factors influencing the rate of polychlorinated biphenyl dechlorination in Hudson River sediments. *Environ. Sci. Technol.* **27:**1125–1131.
4. **Adams, J. A., and K. R. Reddy.** 2003. Extent of benzene biodegradation in saturated soil column during air sparging. *Groundwater Monitor. Remed.* **23:**85–94.
5. **Aggarwal, P. K., J. L. Means, and R. E. Hinchee.** 1991. Formulation of nutrient solutions for *in situ* bioremediation, p. 51–66. *In* R. E. Hinchee and R. F. Olfenbuttel (ed.), In Situ *Bioreclamation. Applications and Investigations for Hydrocarbon and Contaminated Site Remediation.* Butterworth-Heinemann, Stoneham, Mass.
6. **Aislabie, J. M., M. McLeod, and R. Fraser.**

1998. Potential for biodegradation of hydrocarbons in soil from Ross Dependency, Antarctica. *Appl. Microbiol. Biotechnol.* **49:**210–214.

7. **Aislabie, J. M., M. R. Balks, J. M. Foght, and E. J. Waterhouse.** 2004. Hydrocarbon spills on Antarctic soils: effects and management. *Environ. Sci. Technol.* **38:**1265–1274.
8. **Aislabie, J. M., R. Fraser, S. Duncan, and R. L. Farrell.** 2001. Effects of oil spills on microbial heterotrophs in Antarctic soils. *Polar Biol.* **24:** 308–313.
9. **Al-Awadhi, N., R. Al-Daher, A. ElNawawy, and M. T. Balba.** 1996. Bioremediation of oil-contaminated soil in Kuwait. I. Landfarming to remediate oil-contaminated soil. *J. Soil Contam.* **5:**243–260.
10. **Alexander, M.** 1999. *Biodegradation and Bioremediation*, 2nd ed. Academic Press, San Diego, Calif.
11. **Alfke, G., G. Bunch, G. Crociani, D. Dando, M. Fontaine, P. Goodsell, A. Green, W. Hafker, G. Isaak, J. Marvillet, B. Poot, H. Sutherland, A. van der Rest, J. van Oudenhoven, T. Walden, E. Martin, and H. Schipper.** 1999. *Best Available Techniques to Reduce Emissions from Refineries—Introduction.* CONCAWE 99/01-1. [Online.] CONCAWE, Brussels, Belgium. http://aaa.am.lt/naudinga.
12. **Amann, R., H. Lemmer, and M. Wagner.** 1998. Monitoring the community structure of wastewater treatment plants: a comparison of old and new techniques. *FEMS Microbiol. Ecol.* **25:** 205–215.
13. **American Petroleum Institute.** 1983. *Land Treatment Practice in the Petroleum Industry.* Report prepared by Environmental Research and Technology Inc., Washington, D.C.
14. **Amos, R. T., K. U. Mayer, D. W. Blowes, and C. J. Ptacek.** 2004. Reactive transport modelling of column experiments for the remediation of acid mine drainage. *Environ. Sci. Technol.* **38:**3131–3138.
15. **Anderson, R. T., and D. R. Lovley.** 1997. Ecology and biogeochemistry of *in situ* groundwater bioremediation. *Adv. Microb. Ecol.* **15:** 289–350.
16. **Anderson, R. T., J. N. Rooney-Varga, C. V. Gaw, and D. R. Lovley.** 1998. Anaerobic benzene oxidation in the Fe(III) reduction zone of petroleum-contaminated aquifers. *Environ. Sci. Technol.* **32:**1222–1229.
17. **Andreoni, V., G. Baggi, M. Colombo, L. Cavalca, M. Zangrossi, and S. Bernasconi.** 1998. Degradation of 2,4,6-trichlorophenol by a specialised organism and by indigenous soil microflora: bioaugmentation and self-remediability for soil restoration. *Lett. Appl. Microbiol.* **27:** 86–92.
18. **Anonymous.** 2000. Science offers glowing microbes to detect explosives at Edwards. [Online.] http://www.edwards.af.mil.
19. **Anonymous.** 2001. TNT-munching microbes get makeover at Edwards. [Online.] http://www.edwards.af.mil.
20. **Anonymous.** 2004. Helping nature to clean up the Niger Delta. *Impact.* **1:**15. [Online.] http://www.shellglobalsolutions.com/impact.
21. **Antizar-Ladislao, B., J. M. Lopez-Real, and A. J. Beck.** 2004. Bioremediation of polycyclic aromatic hydrocarbons (PAH)-contaminated waste using composting approaches. *Crit. Rev. Environ. Sci. Technol.* **34:**249–289.
22. **Atagana, H. I., R. J. Haynes, and F. M. Wallis.** 2003. Co-composting of soil heavily contaminated with creosote with cattle manure and vegetable waste for the bioremediation of creosote-contaminated soil. *Soil Sed. Contam.* **12:** 885–899.
23. **Atlas, R. M.** 1995. Bioremediation. *Chem. Eng. News* **73:**32–42.
24. **Atlas, R. M., and R. Bartha.** 1972. Degradation and mineralization of petroleum in seawater: limitation by nitrogen and phosphorus. *Biotech. Bioeng.* **52:**149–156.
25. **Avakian, M. D.** 2004. *Quantifying Enhanced* In Situ *TCE Biodegradation.* Research Brief 109. [Online.] NIEHS/EPA Superfund Basic Research Program. http://www-apps.niehs.nih.gov.
26. **Azadpour-Keeley, A., J. W. Keeley, H. H. Russell, and G. W. Sewell.** 2001. Monitored natural attenuation of contaminants in the subsurface: processes. *Groundwater Monitor. Remed.* **21:**97–107.
27. **Azadpour-Keeley, A., J. W. Keeley, H. H. Russell, and G. W. Sewell.** 2001. Monitored natural attenuation of contaminants in the subsurface: applications. *Groundwater Monitor. Remed.* **21:**136–143.
28. **Badr, T., K. Hanna, and C. de Brauer.** Enhanced solubilization and removal of naphthalene and phenanthrene by cyclodextrins from two contaminated soils. *J. Haz. Mat.*, in press.
29. **Bai, G., M. L. Brusseau, and R. M. Miller.** 1997. Biosurfactant enhanced removal of residual hydrocarbon from soil. *J. Contam. Hydrol.* **25:** 157–170.
30. **Bai, G., M. L. Brusseau, and R. M. Miller.** 1997. Influence of rhamnolipid biosurfactant on the transport of bacteria through a sandy soil. *Appl. Environ. Microbiol.* **63:**1866–1873.
31. **Balba, M. T., R. Al-Daher, and N. Al-Awadhi.** 1998. Bioremediation of oil-contami-

nated desert soil: the Kuwaiti experience. *Environ. Int.* **24:**163–173.

32. **Barbaro, J. R., J. F. Barker, L. A. Lemon, and C. I. Mayfield.** 1992. Biotransformation of BTEX under anaerobic, denitrifying conditions: field and laboratory observations. *J. Contam. Hydrol.* **11:**245–272.

33. **Barbeau, C., L. Deschenes, D. Karamenev, Y. Comeau, and R. Samson.** 1997. Bioremediation of pentachlorophenol-contaminated soil by bioaugmentation using activated soil. *Appl. Microbiol. Biotechnol.* **48:**745–752.

34. **Bardi, L., A. Mattei, S. Steffan, and M. Marzano.** 2000. Hydrocarbon degradation by a soil microbial population with β-cyclodextrin as surfactant to enhance bioavailability. *Enzyme Microb. Technol.* **27:**709–713.

35. **Battelle Memorial Institute.** 1994. *Bioremediation of Hazardous Wastes at CERCLA and RCRA Sites: Hill AFB 280 Site, Low Intensity Bioreclamation.* U.S. Environmental Protection Agency, Washington, D.C.

36. **Bekins, B. A., B. E. Rittmann, and J. A. MacDonald.** 2001. Natural attenuation strategy for groundwater cleanup focuses on demonstrating cause and effect. *Eos Trans. AGU* **82(5):**53:57–53:58. [Online.] http://toxics.usgs.gov.

37. **Benner, S. G., D. W. Blowes, J. C. Ptacek, and K. U. Mayer.** 2002. Rates of sulfate reduction and metal sulfide precipitation in a permeable reactive barrier. *Appl. Geochem.* **17:** 301–320.

38. **Benner, S. G., D. W. Blowes, W. D. Gould, R. B. Herbert, Jr., and J. C. Ptacek.** 1999. Geochemistry of a permeable reactive barrier for metals and acid mine drainage. *Environ. Sci. Technol.* **33:**2793–2799.

39. **Berwanger, D. J., and J. F. Barker.** 1988. Aerobic biodegradation of aromatic and chlorinated hydrocarbons commonly detected in landfill leachate. *Water Pollut. Res. J. Can.* **23:** 460–475.

40. **Birke, V., H. Burmeier, and D. Rosenau.** 2003. Permeable reactive barrier technologies for groundwater remediation in Germany: recent progress and new developments. *Fresenius Environ. Bull.* **12:**623–628.

41. **Blumenroth, P., and I. Wagner-Dobler.** 1998. Survival of inoculants in polluted sediments: effect of strain origin and carbon source competition. *Microb. Ecol.* **35:**279–288.

42. **Bollati, A., and C. Luzi.** 1997. Appendix 3, p. 167–185. *In* P. Lecomte and C. Mariotti (ed.), *Handbook of Diagnostic Procedures for Petroleum-Contaminated Sites.* John Wiley and Sons Ltd., Chichester, United Kingdom.

43. **Bosma, T., J. Damborsky, G. Stucki, D. Janssen, and B. Dick.** 2002. Biodegradation of 1,2,3-trichloropropane through directed evolution and heterologous expression of a haloalkane dehalogenase gene. *Appl. Environ. Microbiol.* **68:** 3582–3587.

44. **Bouchez, T., D. Patureau, P. Dabert, S. Juretschko, J. Doré, P. Delgenès, R. Moletta, and M. Wagner.** 2000. Ecological study of a bioaugmentation failure. *Environ. Microbiol.* **2:**179–190.

45. **Boufadel, M. C., P. Reeser, M. T. Suidan, B. A. Wrenn, J. Cheng, X. Du, T. H. L. Huang, and A. D. Venosa.** 1999. Optimal nitrate concentration for the biodegradation of *n*-heptadecane in a variably-saturated sand column. *Environ. Technol.* **20:**191–199.

46. **Boving, T. B., X. Wang, and M. L. Brusseau.** 2000. Solubilization and removal of residual trichloroethene from porous media: comparison of several solubilization agents. *J. Contam. Hydrol.* **42:**51–67.

47. **Breitung, J., D. Bruns-Nagel, K. Steinbach, L. Kaminski, D. Gemsa, and E. von Löw.** 1996. Bioremediation of 2,4,6-trinitrotoluene-contaminated soils by two different aerated compost systems. *Appl. Microbiol. Biotechnol.* **44:** 795–800.

48. **Briglia, M., P. J. M. Middeldorp, and M. S. Salkinoja-Salonen.** 1994. Mineralization performance of *Rhodococcus chlorophenolicus* strain PCP-1 in contaminated soil simulating conditions. *Soil Biol. Biochem.* **26:**377–385.

49. **Brim, H., A. Venkateswaran, H. M. Kostandarithes, J. K. Fredrickson, and M. J. Daly.** 2003. Engineering *Deinococcus geothermalis* for bioremediation of high-temperature radioactive waste environments. *Appl. Environ. Microbiol.* **69:**4575–4582.

50. **Brim, H., S. C. McFarlan, J. K. Fredrickson, K. W. Minton, M. Zhai, L. P. Wackett, and M. J. Daly.** 2000. Engineering *Deinococcus radiodurans* for metal remediation in radioactive mixed waste environments. *Nature Biotechnol.* **18:** 85–90.

51. **Brown, R. A., and J. Crosbie.** 1989. Oxygen sources in *in situ* bioremediation. *In Proceedings of the 10th National Conference, HMCRI,* Washington, D.C.

52. **Brown, R. A., R. D. Norris, and R. L. Raymond.** 1984. Oxygen transport in contaminated aquifers, p. 441–450. *In Proceedings of the Conference on Petroleum Hydrocarbons and Organic Chemicals in Groundwater—Prevention, Detection and Restoration.* National Water Well Association, Worthington, Ohio.

53. **Bruce, C. L., I. L. Amerson, R. L. Johnson, and P. C. Johnson.** 2001. Use of an SF_6-based diagnostic tool for assessing air distributions and

oxygen transfer rates during IAS operation. *Bioremed. J.* **5**:337–347.

54. **Bruns, A., H. Cypionka, and J. Overmann.** 2002. Cyclic AMP and acyl homoserine lactones increase the cultivation efficiency of heterotrophic bacteria from the central Baltic Sea. *Appl. Environ. Microbiol.* **68**:3978–3987.

55. **Burns, K. A., S. Codi, and N. C. Duke.** 2000. Gladstone, Australia field studies: weathering and degradation of hydrocarbons in oiled mangrove and salt marsh sediments with and without the application of an experimental bioremediation protocol. *Mar. Pollut. Bull.* **41:** 392–402.

56. **Bury, S. J., and C. A. Miller.** 1993. Effect of micellar solubilization on biodegradation rates of hydrocarbons. *Environ. Sci. Technol.* **27:** 104–110.

57. **Cameotra, S. S., and R. S. Makkar.** 2004. Recent applications of biosurfactants as biological and immunological molecules. *Curr. Opin. Microbiol.* **7**:262–266.

58. **Campagnolo, J. F., and A. Akgerman.** 1995. Modeling of soil vapor extraction (SVE) systems. Part II. Biodegradation aspects of soil vapor extraction. *Waste Manage.* **15**:391–397.

59. **Cantrell, K. J., D. I. Kaplan, and T. W. Wietsma.** 1995. Zero-valent iron for the *in situ* remediation of selected metals in groundwater. *J. Haz. Mat.* **42**:201–212.

60. **Carey, M. A., B. A. Fretwell, N. G. Mosley, and J. W. N. Smith.** 2002. *Guidance on the Use of Permeable Reactive Barriers for Remediating Contaminated Groundwater.* National Groundwater and Contaminated Land Centre Report NC/01/51. Environment Agency, Bristol, United Kingdom.

61. **Cassidy, M. B., H. Lee, and J. T. Trevors.** 1996. Environmental applications of immobilised microbial cells: a review. *J. Ind. Microbiol.* **16**:79–101.

62. **Chakrabarty, A. M.** March 1981. Microorganisms having multiple compatible degradative energy-generating plasmids and preparation thereof. U.S. patent 4,259,444.

63. **Chakrabarty, A. M.** 1998. Diamond v. Chakrabarty: a historical perspective. *In* D. S. Chisum, C. A. Nard, H. F. Schwartz, P. Newman, and S. F. Kief (ed.), *Principles of Patent Law.* Foundation Press, New York, N.Y.

64. **Chakrabarty, A. M.** 2000. Moving ahead with pseudomonad genes, p. 235–242. *In* R. M. Atlas (ed.), *Many Faces, Many Microbes.* ASM Press, Washington, D.C.

65. **Civilini, M.** 1994. Fate of creosote compounds during composting. *Microbiol. Eur.* **2**:16–24.

66. **Clement, T. P., M. J. Truex, and P. Lee.** 2002. A case study for demonstrating the application of U.S. EPA's monitored natural attenuation screening protocol at a hazardous waste site. *J. Contam. Hydrol.* **59**:133–162.

67. **Clint, J. H.** 1992. *Surfactant Aggregation.* Blackie and Son Ltd., Glasgow, United Kingdom.

68. **Cookson, J. T., Jr.** 1995. *Bioremediation Engineering: Design and Application.* McGraw-Hill, New York, N. Y.

69. **Corman, A., Y. Crozat, and J. C. Cleyet-Marel.** 1987. Modelling of survival kinetics of some *Bradyrhizobium japonicum* strains in soils. *Biol. Fertil. Soils* **4**:79–84.

70. **Crawford, R. L., and W. W. Mohn.** 1985. Microbiological removal of pentachlorophenol from soil using a *Flavobacterium. Enzyme Microb. Technol.* **7**:617–620.

71. **Cunningham, A., and R. Hiebert.** 2000. Subsurface biofilm barriers for contaminated groundwater containment, p. 3–4. *In Ground Water Currents.* EPA 542-N–00–004. U.S. Environmental Protection Agency, Washington, D.C.

72. **Cunningham, C. J., and J. C. Philp.** 2000. Comparison of bioaugmentation and biostimulation in *ex situ* treatment of diesel contaminated soil. *Land Contam. Reclam.* **8**:261–269.

73. **Cunningham, C. J., I. B. Ivshina, V. I. Lozinsky, M. S. Kuyukina, and J. C. Philp.** 2004. Bioremediation of diesel-contaminated soil by microorganisms immobilised in a polyvinyl alcohol. *Int. Biodeterior. Biodegrad.* **54**:167–174.

74. **Cunningham, J. A., H. Rahme, G. D. Hopkins, C. Lebron, and M. Reinhard.** 2001. Enhanced *in situ* bioremediation of BTEX-contaminated groundwater by combined injection of nitrate and sulfate. *Environ. Sci. Technol.* **35:** 1663–1670.

75. **Curtis, T. P., W. T. Sloan, and J. W. Scannell.** 2002. Estimating prokaryotic diversity and its limits. *Proc. Natl. Acad. Sci. USA* **99:** 10494–10499.

76. **Daane, L. L., and M. M. Häggblom.** 1999. Earthworm egg capsules as vectors for the environmental introduction of biodegradative bacteria. *Appl. Environ. Microbiol.* **65**:2376–2381.

77. **Dando, D. A., and D. E. Martin.** 2003. *A Guide for Reduction and Disposal of Waste from Oil Refineries and Marketing Installations.* CONCAWE 6/03. [Online.] CONCAWE, Brussels, Belgium. http://aaa.am.lt/naudinga.

78. **Dasch, J. M., A. S. Abdul, D. N. Rai, T. L. Gibson, and N. Grosvenor.** 1997. Synergistic application of four remedial techniques at an industrial site. *Groundwater Monitor. Remed.* **17:** 194–209.

79. **DaSilva, M. L., and P. J. Alvarez.** 2004. En-

hanced anaerobic biodegradation of benzene-toluene-ethylbenzene-xylene-ethanol mixtures in bioaugmented aquifer columns. *Appl. Environ. Microbiol.* **70**:4720–4726.

80. **Day, S. R., S. F. O'Hannesin, and L. Marsden.** 1999. Geotechnical techniques for the construction of reactive barriers. *J. Haz. Mat.* **B67:** 285–297.

81. **Dejonghe, W., N. Boon, D. Seghers, E. M. Top, and W. Verstraete.** 2001. Bioaugmentation of soils by increasing microbial richness: missing links. *Environ. Microbiol.* **3**:649–657.

82. **de Lorenzo, V.** 2001. Cleaning up polluted environments: how microbes can help. *In European Commission-Sponsored Research on Safety of Genetically Modified Organisms—A Review of Results: Bioremediation.* [Online.] http://europa.eu.int.

83. **Demque, D. E., K. W. Biggar, and J. A. Heroux.** 1997. Land treatment of diesel contaminated soil. *C. Geotech. J.* **34**:421–431.

84. **DePaoli, D. W.** 1996. Design equations for soil aeration via bioventing. *Separations Technol.* **6:** 165–174.

85. **Dernbach, L. S.** 1999. Failure of non-indigenous micro-organisms to remediate petroleum hydrocarbon contaminated sites, p. 415–420. *In* B. C. Alleman and A. Leeson (ed.), In Situ *Bioremediation of Petroleum Hydrocarbons and Other Organic Compounds.* Battelle Press, Columbus, Ohio.

86. **Devlin, J. F., D. Katic, and J. F. Barker.** 2004. *In situ* sequenced bioremediation of mixed contaminants in groundwater. *J. Contam. Hydrol.* **69**:233–261.

87. **Diele, F., F. Notarnicola, and I. Sgura.** 2002. Uniform air velocity field for a bioventing system design: some numerical results. *Int. J. Eng. Sci.* **40**:1199–1210.

88. **Downey, D. C., P. R. Guest, and J. W. Ratz.** 1995. Results of a two-year *in situ* bioventing demonstration. *Environ. Prog.* **14:** 121–125.

89. **Drinkwine, A. D.** 2001. Remediation of TCE contaminated ground water using methane injection to enhance cometabolism demonstration project at the Hastings former naval ammunition depot. [Online.] http://hq.environmental.usace.army.mil.

90. **Dua, M., A. Singh, N. Sethunathan, and A. K. Johri.** 2002. Biotechnology and bioremediation: successes and failures. *Appl. Microbiol. Biotechnol.* **59**:143–152.

91. **Dunbar, J., S. White, and L. Forney.** 1997. Genetic diversity through the looking glass: effect of enrichment bias. *Appl. Environ. Microbiol.* **63**:1326–1331.

92. **Dupont, R. R.** 1993. Fundamentals of bioventing applied to fuel contaminated sites. *Environ. Prog.* **12**:45–53.

93. **Dutta, S. K., G. P. Hollowell, F. M. Hashem, and L. D. Kuykendall.** 2003. Enhanced bioremediation of soil containing 2,4-dinitrotoluene by a genetically modified *Sinorhizobium meliloti. Soil Biol. Biochem.* **35**:667–675.

94. **Edgehill, R. U., and R. K. Finn.** 1983. Microbial treatment of soil to remove pentachlorophenol. *Appl. Environ. Microbiol.* **45**:1122–1125.

95. **Eggen, T.** 1999. Application of fungal substrate from commercial mushroom production—*Pleurotus ostreatus*—for bioremediation of creosote contaminated soil. *Int. Biodeterior. Biodegrad.* **44:** 117–126.

96. **Fehmidakhatun, A., A. Mesania, and A. A. Jennings.** 2000. Modeling soil biopile bioremediation. *Environ. Modelling Software* **15:** 411–424.

97. **Fernandez-Sanchez, J. M., E. J. Sawvel, and P. J. J. Alvarez.** 2004. Effect of Fe^0 quantity on the efficiency of integrated microbial-Fe^0 treatment processes. *Chemosphere* **54**:823–829.

98. **Filler, D. M., and R. F. Carlson.** 2000. Thermal insulation systems for bioremediation in cold regions. *J. Cold Regions Eng.* **14**:119–129.

99. **Filz, G. M., M. A. Widdowson, and J. C. Little.** 2001. Barrier-controlled monitored natural attenuation. *Environ. Sci. Technol.* **35:** 3225–3230.

100. **Fogel, M. M., A. R. Taddeo, and S. Fogel.** 1986. Biodegradation of chlorinated ethanes by a methane-utilizing mixed culture. *Appl. Environ. Microbiol.* **51**:720–724.

101. **Ford, C. Z., G. S. Sayler, and R. S. Burlage.** 1999. Containment of a genetically engineered microorganism during a field bioremediation application. *Appl. Microbiol. Biotechnol.* **51:** 397–400.

102. **Fukui, S., and A. Tanaka.** 1982. Immobilised microbial cells. *Annu. Rev. Microbiol.* **36:** 145–172.

103. **Fuller, C. C., J. R. Bargar, and J. A. Davis.** 2003. Molecular-scale characterization of uranium sorption by bone apatite materials for a permeable reactive barrier demonstration. *Environ. Sci. Technol.* **37**:4642–4649.

104. **Futamata, H., S. Harayama, and K. Watanabe.** 2001. Group-specific monitoring of phenol hydroxylase genes for a functional assessment of phenol-stimulated trichloroethylene bioremediation. *Appl. Environ. Microbiol.* **67:** 4671–4677.

105. **Gallardo, V.** 2000. Enhanced *in situ* bioremediation demonstrated in fractured bedrock. *EPA Groundwater Curr.* **38**:2–3.

106. **Gandhi, S., B. T. Oh, J. L. Schnoor, and**

P. J. J. Alvarez. 2002. Degradation of TCE, Cr(VI), sulfate, and nitrate mixtures by granular iron in flow-through columns under different microbial conditions. *Water Res.* **36:**1973–1982.

107. **Geerdink, M. J., R. H. Kleijntjens, M. C. M. van Loosdrecht, and K. C. A. M. Luyben.** 1996. Microbial decontamination of polluted soil in a slurry process. *J. Environ. Eng. ASCE* **122:**975–982.

108. **Gentry, T. J., K. L. Josephson, and I. L. Pepper.** 2004. Functional establishment of introduced chlorobenzoate degraders following bioaugmentation with newly activated soil. *Biodegradation* **15:**67–75.

109. **Gieg, L. M., R. V. Kolhatkar, M. J. McInerney, R. S. Tanner, S. H. Harris, Jr., K. L. Sublette, and J. M. Suflita.** 1999. Intrinsic bioremediation of petroleum hydrocarbons in a gas condensate-contaminated aquifer. *Environ. Sci. Technol.* **33:**2550–2560.

110. **Goddard, V. J., M. J. Bailey, P. Darrah, A. K. Lilley, and I. P. Thompson.** 2001. Monitoring temporal and spatial variation in rhizosphere bacterial population diversity: a community approach for the improved selection of rhizosphere competent bacteria. *Plant Soil* **232:**181–193.

111. **Goldstein, R. M., L. M. Mallory, and M. Alexander.** 1985. Reasons for possible failure of inoculation to enhance biodegradation. *Appl. Environ. Microbiol.* **50:**977–983.

112. **Grasso, D.** 1993. *Hazardous Waste Site Remediation. Source Control.* CRC Press, Boca Raton, Fla.

113. **Grasso, D., K. Subramaniam, J. J. Pignatello, Y. Yang, and D. Ratte.** 2001. Micellar desorption of polynuclear aromatic hydrocarbons from contaminated soil. *Colloids Surf. A Physicochem. Eng. Aspects* **194:**65–74.

114. **Gray, N. F.** 2004. *Biology of Wastewater Treatment*, 2nd ed. Imperial College Press, London, United Kingdom.

115. **Gray, S. N.** 1998. Fungi as potential bioremediation agents in soil contaminated with heavy or radioactive metals. *Biochem. Soc. Trans.* **26:**666–670.

116. **Gu, B. H., D. B. Watson, L. Y. Wu, D. H. Phillips, D. C. White, and J. Z. Zhou.** 2002. Microbiological characteristics in a zero-valent iron reactive barrier. *Environ. Monitor. Assess.* **77:**293–309.

117. **Gu, M. B., and S. T. Chang.** 2001. Soil biosensor for the detection of PAH toxicity using an immobilized recombinant bacterium and a biosurfactant. *Biosens. Bioelectron.* **16:**667–674.

118. **Guerin, T. F.** 2001. Co-composting of pharmaceutical wastes in soil. *Lett. Appl. Microbiol.* **33:**256–263.

119. **Guerin, T. F.** 2001. Co-composting of residual fuel contamination in soil. *Soil Sed. Contam.* **10:**659–673.

120. **Haas, J. E., and D. A. Trego.** 2001. A field application of hydrogen-releasing compound (HRC (TM)) for the enhanced bioremediation of methyl tertiary butyl ether (MTBE). *Soil Sed. Contam.* **10:**555–575.

121. **Haferburg, D., R. K. Hommel, R. Claus, and H. P. Kleber.** 1986. Extracellular microbial lipids as biosurfactants. *Adv. Biochem. Eng. Biotechnol.* **33:**53–93.

122. **Hageman, K. J., J. Istok, J. A. Field, T. E. Buscheck, and L. Semprini.** 2001. *In situ* anaerobic transformation of trichlorofluoroethene in trichloroethene-contaminated groundwater. *Environ. Sci. Technol.* **35:**1729–1735.

123. **Hammer, M. J., and M. J. Hammer, Jr.** 1996. *Water and Wastewater Technology*, 3rd ed. Prentice-Hall, Englewood Cliffs, N. J.

124. **Harkness, M. R., J. B. McDermott, D. A. Aramowicz, J. J. Salvo, W. P. Flanagan, M. L. Stephens, F. J. Mondello, R. J. May, J. H. Lobos, K. M. Carroll, M. J. Brennan, A. A. Bracco, K. M. Fish, G. L. Warner, P. R. Wilson, D. K. Dietrich, D. T. Lin, C. B. Morgan, and W. L. Gately.** 1993. *In situ* stimulation of aerobic PCB biodegradation in Hudson River sediments. *Science* **259:**503–507.

125. **Heath, J., and E. Lory.** 1999. *In situ* bioremediation of methyl tertiary butyl ether (MTBE)-advanced fuel hydrocarbon remediation national test location. [Online.] http://www.stormingmedia.us.

126. **Hejazi, R. F., and T. Husain.** 2004. Landfarm performance under arid conditions. 1. Conceptual framework. *Environ. Sci. Technol.* **38:**2449–2456.

127. **Hejazi, R. F., and T. Husain.** 2004. Landfarm performance under arid conditions. 2. Evaluation of parameters. *Environ. Sci. Technol.* **38:**2457–2469.

128. **Hengge-Aronis, R.** 1993. Survival of hunger and stress: the role of *rpoS* in early stationary phase gene regulation in *Escherichia coli*. *Cell* **72:**165–168.

129. **Hengge-Aronis, R.** 2002. Signal transduction and regulatory mechanisms involved in control of the σ^S (RpoS) subunit of RNA polymerase. *Microbiol. Mol. Biol. Rev.* **66:**373–395.

130. **Herman, D. C., J. F. Artiola, and R. M. Miller.** 1995. Removal of cadmium, lead and zinc from soil by a rhamnolipid biosurfactant. *Environ. Sci. Technol.* **29:**2280–2285.

131. **Hicks, R. J.** 1993. Above ground bioremediation: practical approaches and field experiences.

In Proceedings of Applied Bioremediation, 25 to 26 October, Fairfield, N. J.

132. **Hinchee, R. E., R. N. Miller, and R. R. DuPont.** 1991. Enhanced biodegradation of petroleum hydrocarbons: an air-based *in situ* process, p. 177–183. *In* H. M. Freeman and P. R. Sferra (ed.), *Biological Processes: Innovative Hazardous Waste Treatment Technology Series*. Technomic Publishing, Lancaster, Pa.
133. **Hitsatsuka, K., T. Nakahara, N. Sano, and K. Yamada.** 1971. Formation of a rhamnolipid by *Pseudomonas aeruginosa* and its function in hydrocarbon fermentation. *Agric. Biol. Chem.* **35:** 686–692.
134. **Hommel, R. K.** 1990. Formation and physiological role of biosurfactants produced by hydrocarbon-utilising microorganisms. *Biodegradation* **1:**107–119.
135. **Hopkins, G. D., and P. L. McCarty.** 1995. Field evaluation of *in situ* aerobic co-metabolism of trichloroethylene and three dichloroethylene isomers using phenol and toluene as the primary substrates. *Environ. Sci. Technol.* **29:**1628–1637.
136. **Hopkins, G. D., J. Munakata, L. Semprini, and P. L. McCarty.** 1993. Trichloroethylene concentration effects on pilot field scale *in situ* groundwater bioremediation by phenol-oxidizing microorganisms. *Environ. Sci. Technol.* **27:** 2542–2547.
137. **Hopkins, G. D., L. Semprini, and P. L. McCarty.** 1993. Microcosm and in situ field studies of enhanced biotransformation of trichloroethylene by phenolutilizing microorganisms. *Appl. Environ. Microbiol.* **59:**2277–2285.
138. **Huesemann, M. H.** 1994. Guidelines for landtreating petroleum hydrocarbon-contaminated soils. *J. Soil Contam.* **3:**299–318.
139. **Huesemann, M. H., and M. J. Truex.** 1996. The role of oxygen diffusion in passive bioremediation of petroleum contaminated soils. *J. Haz. Mat.* **51:**93–113.
140. **Hunkeler, D., P. Hohener, and J. Zeyer.** 2002. Engineered and subsequent intrinsic *in situ* bioremediation of a diesel fuel contaminated aquifer. *J. Contam. Hydrol.* **59(3–4):**231–245.
141. **Hur, J. M., and J. A. Park.** 2003. Effect of sewage sludge mix ratio on the biodegradation of diesel-oil in a contaminated soil composting. *Korean J. Chem. Eng.* **20:**307–314.
142. **Huss, L.** 1977. The ANAMET process for food and fermentation industry effluent. *Tribune CEBEDEAU* **30:**390–396.
143. **Illmer, P.** 2002. Backyard composting: general considerations and a case study, p. 3–4. *In* H. Insam, N. Riddech, and S. Klammer (ed.), *Microbiology of Composting*. Springer-Verlag, Heidelberg, Germany.
144. **Ivshina, I. B., M. S. Kuyukina, M. I. Ritchkova, J. C. Philp, C. J. Cunningham, and N. Christofi.** 2001. Oleophilic biofertilizer based on a *Rhodococcus* surfactant complex for the bioremediation of crude oil-contaminated soil. *AHES Soil Sed. Water* **August:**20–24.
145. **Jain, D. K., H. Lee, and J. T. Trevors.** 1992. Effect of addition of *Pseudomonas aeruginosa* UG2 inocula or biosurfactants on biodegradation of selected hydrocarbons in soil. *J. Ind. Microbiol.* **10:**87–93.
146. **Jansson, J. K., K. Björklöf, A. M. Elvang, and K. S. Jørgensen.** 2000. Biomarkers for monitoring efficacy of bioremediation by microbial inoculants. *Environ. Pollut.* **107:**217–223.
147. **Johnson, P. C.** 1998. Assessment of the contributions of volatilization and biodegradation to *in situ* air sparging performance. *Environ. Sci. Technol.* **32:**276–281.
148. **Johnson, P. C., A. Leeson, R. L. Johnson, C. M. Vogel, R. E. Hinchee, M. Marley, T. Peargin, C. L. Bruce, I. L. Amerson, C. T. Coonfare, and R. D. Gillespie.** 2001. A practical approach for the selection, pilot testing, design, and monitoring of *in situ* air sparging/biosparging systems. *Bioremed. J.* **5:**267–281.
149. **Johnson, P. C., R. L. Johnson, C. L. Bruce, and A. Leeson.** 2001. Advances in *in situ* air sparging/biosparging. *Bioremed. J.* **5:**251–266.
150. **Jorgensen, K. S., J. Puustinen, and A. M. Suortti.** 2000. Bioremediation of petroleum hydrocarbon-contaminated soil by composting in biopiles. *Environ. Pollut.* **107:**245–254.
151. **Junker, F., and J. L. Ramos.** 1999. Involvement of the *cis/trans* isomerase Cti in solvent resistance of *Pseudomonas putida* DOT-TIE. *J. Bacteriol.* **181:**5693–5700.
152. **Juteau, P., J. G. Bisaillon, F. Lepine, V. Ratheau, R. Beaudet, and R. Villemur.** 2003. Improving the biotreatment of hydrocarbons-contaminated soils by addition of activated sludge taken from the wastewater treatment facilities of an oil refinery. *Biodegradation* **14:**31–40.
153. **Kalin, R. M.** 2004. Engineered passive bioreactive barriers: risk-managing the legacy of industrial soil and groundwater pollution. *Curr. Opin. Microbiol.* **7:**227–238.
154. **Kao, C. M., and C. C. Wang.** 2000. Control of BTEX migration by intrinsic bioremediation at a gasoline spill site. *Water Res.* **34:**3413–3423.
155. **Kao, C. M., and L. Yang.** 2000. Enhanced bioremediation of trichlorethylene contaminated by a barrier system. *Water Sci. Technol.* **42:** 429–434.
156. **Khan, F. I., and T. Husain.** 2001. Risk-based monitored natural attenuation—a case study. *J. Haz. Mat.* **B85:**243–272.

157. **Khan, F. I., T. Husain, and R. Hejazi.** 2004. An overview and analysis of site remediation technologies. *J. Environ. Manage.* **71:**95–122.

158. **Kieboom, J., J. J. Dennis, J. A. M. de Bont, and G. J. Zylstra.** 1998. Identification and molecular characterization of an efflux pump involved in *Pseudomonas putida* S12 solvent tolerance. *J. Biol. Chem.* **273:**85–91.

159. **Kiely, G.** 1996. *Environmental Engineering.* McGraw-Hill Publishing Company, Maidenhead, United Kingdom.

160. **King, J. M. H., P. M. DiGrazia, B. Applegate, R. Burlage, J. Sanseverino, P. Dunbar, F. Larimer, and G. S. Sayler.** 1990. Rapid, sensitive bioluminescent reporter technology for naphthalene exposure and biodegradation. *Science* **249:**778–781.

161. **Kirtland, B. C., and C. M. Aelion.** 2000. Petroleum mass removal from low permeability sediment using air sparging/soil vapor extraction: impact of continuous or pulsed operation. *J. Contam. Hydrol.* **41:**367–383.

162. **Kitamoto, D., H. Isoda, and T. Nakahara.** 2002. Functions and potential applications of glycolipid biosurfactants—from energy-saving materials to gene delivery systems. *J. Biosci. Bioeng.* **94:**187–201.

163. **Kodres, C. A.** 1999. Coupled water and air flows through a bioremediation soil pile. *Environ. Modelling Software* **14:**37–47.

164. **Kuiper, I., G. V. Bloemberg, and B. J. J. Lugtenberg.** 2001. Selection of a plant-bacterium pair as a novel tool for rhizostimulation of polycyclic aromatic hydrocarbon-degrading bacteria. *Mol. Plant-Microbe Interact.* **14:**1197–1205.

165. **Kuyukina, M. S., I. B. Ivshina, M. I. Ritchkova, S. M. Kostarev, J. C. Philp, C. J. Cunningham, and N. Christofi.** 2003. Bioremediation of crude oil contaminated soil using slurry-phase biological treatment and landfarming techniques. *Soil Sed. Contam.* **12:**85–99.

166. **Laha, S., and R. G. Luthy.** 1992. Effects of nonionic surfactants on the solubilization and mineralization of phenanthrene in soil-water systems. *Biotechnol. Bioeng.* **40:**1367–1380.

167. **Laine, M. M., and K. S. Jørgensen.** 1997. Effective and safe composting of chlorophenol-contaminated soil in pilot scale. *Environ. Sci. Technol.* **31:**371–378.

168. **Lang, S., and J. C. Philp.** 1998. Surface-active lipids in rhodococci. *Antonie Leeuwenhoek Int. J. Gen. Mol. Microbiol.* **74:**59–70.

169. **Lau, K. L., Y. Y. Tsang, and S. W. Chiu.** 2003. Use of spent mushroom compost to bioremediate PAH-contaminated samples. *Chemosphere* **52:**1539–1546.

170. **Lawrence, L.** 2002. Enzymes that can clean up pesticide residues. CSIRO Entomology. http://www.ento.csiro.au.

171. **Layton, A. C., M. Muccini, M. M. Ghosh, and G. S. Sayler.** 1998. Construction of a bioluminescent reporter strain to detect polychlorinated biphenyls. *Appl. Environ. Microbiol.* **64:**5023–5026.

172. **Lee, K., and E. M. Levy.** 1991. Bioremediation: waxy crude oils stranded on low-energy shorelines, p. 541–547. *In Proceedings of the 1991 Oil Spill Conference.* American Petroleum Institute, Washington, D. C.

173. **Lee, M. D., and C. M. Swindoll.** 1993. Bioventing for *in situ* remediation. *Hydrol. Sci. J. J. Sciences Hydrologiques* **38:**273–282.

174. **Leeson, A., and R. E. Hinchee.** 1994. *Field Treatability Study at the Greenwood Chemical Superfund Site, Albermarle County, Virginia.* Report prepared for the U. S. Environmental Protection Agency, Cincinnati, Ohio.

175. **Lewis, R.** 2001. PCB dilemma: government, industry, and public debate dredging vs. bioremediation in the Hudson River. *Scientist* **15(6):** 1.

176. **Li, L., C. J. Cunningham, V. Pas, J. C. Philp, D. A. Barry, and P. Anderson.** 2004. Field trial of a new aeration system for enhancing biodegradation in a biopile. *Waste Manage.* **24:** 127–137.

177. **Liang, L. Y., A. B. Sullivan, O. R. West, G. R. Molin, and W. Kamolpornwijit.** 2003. Predicting the precipitation of mineral phases in permeable reactive barriers. *Environ. Eng. Sci.* **20:** 635–653.

178. **Liebeg, E. W., and T. J. Cutright.** 1999. The investigation of enhanced bioremediation through the addition of macro and micro nutrients in a PAH contaminated soil. *Int. Biodeterior. Biodegrad.* **44:**55–64.

179. **Lin, J. E., and H. Y. Wang.** 1991. Use of co-immobilised biological systems to degrade toxic organic compounds. *Biotechnol. Bioeng.* **38:** 273–279.

180. **Lin, Q., I. A. Mendelssohn, C. B. Henry, Jr., E. B. Overton, R. J. Portier, P. O. Roberts, and M. M. Walsh.** 2003. Effects of bioremediation agents on oil degradation on mineral and sandy salt marsh sediments, p. 47–57. *In* I. A. Mendelssohn and Q. Lin (ed.), *Coastal Marine Institute Development of Bioremediation for Oil Spill Cleanup in Coastal Wetlands.* U.S. Department of the Interior Cooperative Agreement Minerals Management Service Coastal Marine Institute, Gulf of Mexico, OCS Region, Louisiana State University.

181. **Lindstrom, J. E., R. C. Prince, J. C. Clark, M. T. Grossman, T. R. Yeager, J. F. Brad-**

dock, and E. J. Brown. 1991. Microbial populations and hydrocarbon biodegradation potentials in fertilized shoreline sediments affected by the T/V *Exxon Valdez* oil spill. *Appl. Environ. Microbiol.* **57:**2514–2522.

182. **Little, C. D., A. V. Palumbo, S. E. Herbes, M. E. Lidstrom, R. L. Tyndall, and P. J. Gilmer.** 1988. Trichloroethylene biodegradation by a methane-oxidizing bacterium. *Appl. Environ. Microbiol.* **54:**951–956.
183. **Liu, Z., A. M. Jacobson, and R. G. Luthy.** 1995. Biodegradation of naphthalene in aqueous sulfate surfactant systems. *Appl. Environ. Microbiol.* **61:**145–151.
184. **Lovley, D. R.** 1997. Potential for anaerobic bioremediation of BTEX in petroleum-contaminated aquifers. *J. Ind. Microbiol. Biotechnol.* **18(2–3):**75–81.
185. **Lowe, D., B. Kueper, M. Pitts, K. Wyatt, T. Simpkin, and T. Sale.** 1997. *Technology Practices Manual for Surfactants and Cosolvents.* CH2Mhill. [Online.] http://clu-in.org.
186. **Ludwig, R. D., R. G. McGregor, D. W. Blowes, S. G. Benner, and K. Mountjoy.** 2002. A permeable reactive barrier for treatment of heavy metals. *Groundwater* **40:**59–66.
187. **Lund, N. C., J. Swinianski, G. Gudehus, and D. Maier.** 1991. Laboratory and field tests for a biological *in situ* remediation of a coke oven plant, p. 396–412. *In* R. E. Hinchee and R. F. Olfenbuttel (ed.), In Situ *Bioreclamation: Applications and Investigations for Hydrocarbon and Contaminated Site Remediation.* Butterworth-Heinemann Publishing, Stoneham, Mass.
188. **Makkar, R. S., and S. S. Cameotra.** 2002. An update on the use of unconventional substrates for biosurfactant production and their new applications. *Appl. Microbiol. Biotechnol.* **58:**428–434.
189. **Malina, G., J. T. C. Grotenhuis, and W. H. Rulkens.** 2002. Vapor extraction/bioventing sequential treatment of soil contaminated with semivolatile hydrocarbon mixes. *Bioremed. J.* **6:**159–176.
190. **Manefield, M., A. S. Whiteley, N. Ostle, P. Ineson, and M. J. Bailey.** 2002. Technical considerations for RNA-based stable isotope probing: an approach in associating microbial diversity with microbial function. *Rapid Commun. Mass Spectrom.* **16:**2179–2183.
191. **Manefield, M., A. S. Whiteley, R. I. Griffiths, and M. J. Bailey.** 2002. RNA stable isotope probing: a novel means of linking microbial community function to phylogeny. *Appl. Environ. Microbiol.* **68:**5367–5373.
192. **Manilal, V. B., and M. Alexander.** 1991. Factors affecting the microbial degradation of phenanthrene in soils. *Appl. Microbiol. Biotechnol.* **35:**401–405.
193. **Mann, M., J. D. Dahlstrom, P. Esposito, G. Everett, G. Peterson, and R. P. Traver.** 1993. *Innovative Site Remediation Technology: Soil Washing/Soil Flushing.* American Academy of Environmental Engineers, Annapolis, Md.
194. **Margesin, R., and F. Schinner.** 1997. Efficiency of indigenous and inoculated cold-adapted soil microorganisms for biodegradation of diesel oil in Alpine soils. *Appl. Environ. Microbiol.* **63:**2660–2664.
195. **Martin, E.** 2001. Environmental protection, p. 197–210. *In* A. G. Lucas (ed.), *Modern Petroleum Technology*, 6th ed., vol. 2. *Downstream.* John Wiley and Sons, Chichester, United Kingdom.
196. **Martinson, M. M., J. G. Steiert, D. L. Saber, W. W. Mohn, and R. L. Crawford.** 1986. Microbiological contamination in pentachlorophenol-contaminated natural waters, p. 529–534. *In* G. C. Llewellyn and C. E. O'Rear (ed.), *Biodeterioration 6.* CAB International Mycological Institute, Slough, United Kingdom.
197. **McCabe, W. L., J. C. Smith, and P. Harriott.** 2001. *Unit Operations of Chemical Engineering*, 6th ed. McGraw-Hill, New York, N. Y.
198. **McCarty, P. L.** 1987. Bioengineering issues related to *in situ* remediation of contaminated soils and groundwater, p. 143–162. *In* G. S. Omenn (ed.), *Environmental Biotechnology.* Plenum Press, New York, N. Y.
199. **McCarty, P. L.** 1995. Field evaluation of *in situ* aerobic cometabolism of trichloroethylene and three dichloroethylene isomers using phenol and toluene as the primary substrates. *Environ. Sci. Technol.* **29:**1628–1637.
200. **McCarty, P. L., L. Semprini, M. E. Dolan, T. C. Harmon, C. Tiedeman, and S. M. Gorelick.** 1991. *In situ* methanotrophic bioremediation for contaminated groundwater at St. Joseph, Michigan, p. 16–40. *In* R. E. Hinchee and R. F. Olfenbuttel (ed.), *On-Site Bioreclamation: Processes for Xenobiotic and Hydrocarbon Treatment.* Butterworth-Heinemann, Boston, Mass.
201. **McCarty, P. L., M. N. Goltz, and G. D. Hopkins.** 1999. Full-scale evaluation of *in situ* bioremediation of chlorinated solvent groundwater contamination. [Online.] http://www.hsrc.org.
202. **McCarty, P. L., M. N. Goltz, G. D. Hopkins, M. E. Dolan, J. P. Allan, B. T. Kawakami, and T. J. Carrothers.** 1998. Full scale evaluation of *in situ* cometabolic degradation of trichloroethylene in groundwater through toluene injection. *Environ. Sci. Technol.* **32:**88–100.
203. **McGovern, T., T. F. Guerin, S. Horner, and B. Davey.** 2002. Design, construction and

operation of a funnel and gate *in situ* permeable reactive barrier for remediation of petroleum hydrocarbons in groundwater. *Water Air Soil Pollut.* **136:**11–31.

204. **Mearns, A.** 1991. Observations of an oil spill bioremediation activity in Galveston Bay, Texas, p. 38. *In* NOAA Technical Memorandum NOS OMA 57. National Oceanic and Atmospheric Administration, Seattle, Wash.

205. **Michel, F. C., J. Quensen, and C. A. Reddy.** 2001. Bioremediation of a PCB-contaminated soil via composting. *Compost Sci. Utiliz.* **9:** 274–284.

206. **Michel, F. C., Jr., C. A. Reddy, and L. J. Forney.** 1995. Microbial degradation and humification of the lawn care pesticide 2,4-dichlorophenoxyacetic acid during composting of yard trimmings. *Appl. Environ. Microbiol.* **61:** 2566–2571.

207. **Mihopoulos, P. G., M. T. Suidan, and G. D. Sayles.** 2001. Complete remediation of PCE contaminated unsaturated soils by sequential anaerobic-aerobic bioventing. *Water Sci. Technol.* **43:**365–372.

208. **Miller, R. M.** 1995. Biosurfactant-facilitated remediation of metal-contaminated soils. *Environ. Health Perspec.* **103:**59–62.

209. **Miller, R. R.** 1996. *Bioslurping.* Technical overview report TO-96-05. Ground-Water Remediation Technologies Analysis Center (GWRTAC), Pittsburgh, Pa.

210. **Mills, M. A., J. S. Bonner, M. A. Simon, T. J. McDonald, and R. L. Autenrieth.** 1997. Bioremediation of a controlled oil release in a wetland, p. 606–616. *In Proceedings of the 20th Arctic and Marine Oilspill Program (AMOP) Technical Seminar.* Environment Canada, Ottawa, Canada.

211. **Mohn, W. W., and G. R. Stewart.** 2000. Limiting factors for hydrocarbon biodegradation at low temperature in Arctic soils. *Soil Biol. Biochem.* **32:**1161–1172.

212. **Mohn, W. W., C. Z. Radziminski, M. C. Fortin, and K. J. Reimer.** 2001. On site bioremediation of hydrocarbon-contaminated Arctic tundra soils in inoculated biopiles. *Appl. Microbiol. Biotechnol.* **57:**242–247.

213. **Moller, J., H. Gaarn, T. Steckel, E. B. Wedebye, and P. Westermann.** 1995. Inhibitory effects on degradation of diesel oil in soil microcosms by a commercial bioaugmentation product. *Bull. Environ. Contam. Toxicol.* **54:**913–918.

214. **Mondello, F. J., D. A. Abramowicz, and J. R. Rhea.** 1998. Natural restoration of PCB contaminated Hudson River sediments, p. 303–326. *In* G. A. Lewandowski and L. J. DeFilippi (ed.), *Biological Treatment of Hazardous Waste.* John Wiley and Sons, New York, N. Y.

215. **Mondello, F. J., R. J. May, J. L. Lobos, K. M. Carroll, M. J. Brennan, A. A. Bracco, K. M. Fish, G. L. Warner, P. R. Wilson, D. K. Dietrich, D. T. Lin, C. B. Morgan, and W. L. Gately.** 1993. *In situ* stimulation of aerobic PCB biodegradation in Hudson River sediments. *Science* **259:**503–507.

216. **Moon, H. S., K. H. Ahn, S. Lee, K. Nam, and J. Y. Kim.** 2004. Use of autotrophic sulphate-oxidizers to remove nitrate from bank filtrate in a permeable reactive barrier system. *Environ. Pollut.* **129:**499–507.

217. **Moorman, T. B., J. K. Cowan, E. L. Arthur, and J. R. Coats.** 2001. Organic amendments to enhance herbicide biodegradation in contaminated soils. *Biol. Fertil. Soils* **33:**541–545.

218. **Morgan, P., S. T. Lewis, and R. J. Watkinson.** 1993. Biodegradation of benzene, toluene, ethylbenzene and xylenes in gas-condensate-contaminated ground-water. *Environ. Pollut.* **82:**181–190.

219. **Morgan, P., and R. J. Watkinson.** 1989. Hydrocarbon degradation in soils and methods for soil biotreatment. *Crit. Rev. Biotechnol.* **8:** 305–333.

220. **Morris, B. L., A. R. L. Lawrence, P. J. C. Chilton, B. Adams, R. C. Calow, and B. A. Klinck.** 2003. *Groundwater and Its Susceptibility to Degradation: a Global Assessment of the Problem and Options for Management.* Early Warning and Assessment Report Series RS. 03-3. United Nations Environment Programme, Nairobi, Kenya.

221. **Mulligan, C. N., and F. Eftekhari.** 2003. Remediation with surfactant foam of PCP-contaminated soil. *Eng. Geol.* **70:**269–273.

222. **Mulligan, C. N., R. N. Yong, and B. F. Gibbs.** 2001. Surfactant remediation of contaminated soil: a review. *Eng. Geol.* **60:**371–380.

223. **Namkoong, W., E. Y. Hwang, J. S. Park, and J. Y. Choi.** 2002. Bioremediation of diesel-contaminated soil with composting. *Environ. Pollut.* **119:**23–31.

224. **Nano, G., A. Borroni, and R. Rota.** 2003. Combined slurry and solid-phase bioremediation of diesel contaminated soils. *J. Haz. Mat.* **B100:** 79–94.

225. **National Research Council.** 2000. *Natural Attenuation for Groundwater Remediation.* National Academy Press, Washington, D. C.

226. **National Research Council.** 1993. In Situ *Bioremediation; When Does it Work?* National Academy Press, Washington, D.C.

227. **Naval Facilities Engineering Command.** 2000. *Methyl Tertiary Butyl Ether (MTBE) Bio-*

remediation—Tech Data Sheet. Naval Facilities Engineering Command, Washington, D. C.

228. **Neidhardt, F. C., J. L. Ingraham, and M. Schaechter.** 1990. *Physiology of the Bacterial Cell: a Molecular Approach*. Sinauer Associates, Sunderland, Mass.
229. **Nestler, C., L. D. Hansen, D. Ringelberg, and J. W. Talley.** 2001. Remediation of soil PAH: comparison of biostimulation and bioaugmentation, p. 43–51. *In* V. S. Magar, M. von Fahnestock, and A. Leeson (ed.), Ex Situ *Biological Treatment Technologies*. Battelle Press, Columbus, Ohio.
230. **Newcombe, D. A., and D. E. Crowley.** 1999. Bioremediation of atrazine-contaminated soil by repeated applications of atrazine-degrading bacteria. *Appl. Microbiol. Biotechnol.* **51:** 877–882.
231. **Nobre, R. C. M., and M. M. M. Nobre.** 2004. Natural attenuation of chlorinated organics in a shallow sand aquifer. *J. Haz. Mat.* **110:** 129–137.
232. **Nyman, J. A.** 1999. Effect of crude oil and chemical additives on metabolic activity of mixed microbial populations in fresh marsh soil. *Microb. Ecol.* **37:**152–162.
233. **Oberbremer, A., and R. Müller-Hurtig.** 1989. Aerobic, step-wise hydrocarbon degradation and formation of biosurfactants by an original soil population in a stirred reactor. *Appl. Microbiol. Biotechnol.* **31:**582–586.
234. **Oberbremer, A., R. Müller-Hurtig, and F. Wagner.** 1990. Effect of the addition of microbial surfactants on hydrocarbon degradation in a soil population in a stirred reactor. *Appl. Microbiol. Biotechnol.* **32:**485–489.
235. **Odencrantz, J. E., R. A. Vogl, and M. D. Varljen.** 2003. Natural attenuation rate clarifications: the true picture is in the details. *Soil Sed. Contam.* **12:**663–672.
236. **Oh, B. T., and P. J. J. Alvarez.** 2002. Hexahydro-1,3,5-trinitro-1,3,5-triazine (RDX) degradation in biologically-active iron columns. *Water Air Soil Pollut.* **141:**325–335.
237. **Olivieri, R., P. Bacchin, A. Robertiello, N. Oddo, L. Degen, and A. Tonolo.** 1976. Microbial degradation of oil spills enhanced by a slow-release fertilizer. *Appl. Environ. Microbiol.* **31:**629–634.
238. **Origgi, G., M. Colombo, F. DePalma, M. Rivolta, P. Rossi, and V. Andreoni.** 1997. Bioventing of hydrocarbon-contaminated soil and biofiltration of the off-gas: results of a field scale investigation. *J. Environ. Sci. Health A* **32:** 2289–2310.
239. **Österreicher-Cunha, P., E. A. Vargas, Jr., J. R. D. Guimarães, T. M. P. de Campos, C. M. F. Nunes, A. Costa, F. S. Antunes, M. I. P. da Silva, and D. M. Mano.** 2004. Evaluation of bioventing on a gasoline-ethanol contaminated undisturbed residual soil. *J. Haz. Mat.* **110:**63–76.
240. **Palumbo, A. V., W. Eng, P. A. Boerman, G. W. Strandberg, T. L. Donaldson, and S. E. Herbes.** 1991. Effects of diverse organic contaminants on trichloroethylene degradation by methanotrophic bacteria and methane-utilizing consortia, p. 77–91. *In* R. E. Hinchee and R. F. Olfenbuttel (ed.), *On-Site Bioreclamation: Processes for Xenobiotic and Hydrocarbon Treatment*. Butterworth-Heinemann, Boston, Mass.
241. **Pearce, K., and W. A. Pretorius.** 1998. A bioventing feasibility test to aid remediation strategy. *Water SA* **24:**5–9.
242. **Pearce, K., H. G. Snyman, R. A. Oellermann, and A. Gerber.** 1995. Bioremediation of petroleum-contaminated soil, p. 71–76. *In* R. E. Hinchee, J. Fredrickson, and B. C. Alleman (ed.), *Bioaugmentation for Site Remediation*. Battelle Press, Columbus, Ohio.
243. **Peltola, R. J., and M. S. Salkinoja-Salonen.** 2003. Improving biodegradation of VOCs in soil by controlling volatilization. *Bioremed. J.* **7:** 129–138.
244. **Pennell, K. D., L. M. Abriola, and W. J. Weber.** 1993. Surfactant-enhanced solubilization of residual dodecane in soil columns. 1. Experimental investigation. *Environ. Sci. Technol.* **27:**2332–2340.
245. **Pennington, J. C., J. M. Brannon, D. Gunnison, D. W. Harrelson, M. Zakikhani, P. Miyares, T. F. Jenkins, J. Clarke, C. Hayes, D. Ringleberg, E. Perkins, and H. Fredrickson.** 2001. Monitored natural attenuation of explosives. *Soil Sed. Contam.* **10:**45–70.
246. **Perry, R. H., and D. W. Green.** 1997. *Perry's Chemical Engineers' Handbook*, 7th ed. McGraw-Hill, New York, N. Y.
247. **Phelps, C. D., and L. Y. Young.** 1999. Anaerobic biodegradation of BTEX and gasoline in various aquatic sediments. *Biodegradation.* **10:** 15–25.
248. **Phelps, M. B., F. T. Stanin, and D. C. Downey.** 1995. Long term bioventing performance in low permeability soils, p. 277–282. *In* R. E. Hinchee, R. N. Miller, and P. C. Johnson (ed.), In situ *Aeration: Air Sparging, Bioventing and Related Remediation Processes*. Battelle Press, Columbus, Ohio.
249. **Philp, J. C., M. S. Kuyukina, I. B. Ivshina, S. A. Dunbar, N. Christofi, S. Lang, and V. Wray.** 2002. Alkanotrophic *Rhodococcus ruber* as a biosurfactant producer. *Appl. Microbiol. Biotechnol.* **59:**318–324.

250. **Phoenix, P., A. Keane, A. Patel, H. Bergeron, S. Ghoshal, and P. C. K. Lau.** 2003. Characterization of a new solvent-responsive gene locus in *Pseudomonas putida* F1 and its functionalization as a versatile biosensor. *Environ. Microbiol.* **5:**1309–1327.

251. **Place, M. C., C. T. Coonfare, A. S. C. Chen, R. E. Hoeppel, and S. H. Rosansky.** 2001. *Principles and Practice of Bioslurping.* Battelle Press, Columbus, Ohio.

252. **Place, M. C., R. E. Hoeppel, R. Chaudhry, S. McCall, and T. Williamson.** 2003. Application guide for bioslurping principles and practices of bioslurping: use of pre-pump separation for improved bioslurper system operation. [Online]. http://www.estcp.org.

253. **Poremba, K., W. Gunkel, S. Lang, and F. Wagner.** 1991. Marine biosurfactants. III. Toxicity testing with marine microorganisms and comparison with synthetic surfactants. *Z. Naturforsch.* **46:**210–216.

254. **Prince, R. C., and R. M. Atlas.** 2005. Bioremediation of marine oil spills, p. 269–292. *In* R. M. Atlas and J. C. Philp (ed.), *Bioremediation: Applied Microbial Solutions for Real-World Environmental Cleanup.* ASM Press, Washington, D.C.

255. **Pritchard, P. H.** 1992. Use of inoculation in bioremediation. *Curr. Opin. Biotechnol.* **3:** 232–243.

256. **Qiao, C.-L., J. Huang, X. Li, B.-C. Shen, and J.-L. Zhang.** 2003. Bioremediation of organophosphate pollutants by a genetically-engineered enzyme. *Bull. Environ. Contam. Toxicol.* **70:**455–461. [Online.] http://www.ioz.ac.cn/department/agripest.

257. **Ramadan, M. A., O. M. El-Tayeb, and M. Alexander.** 1990. Inoculum size as a factor limiting success of inoculation for biodegradation. *Appl. Environ. Microbiol.* **56:**1392–1396.

258. **Ramos, J. L., E. Duque, P. Godoy, and A. Segura.** 1998. Efflux pumps involved in toluene tolerance in *Pseudomonas putida* DOT-TIE. *J. Bacteriol.* **180:**3323–3329.

259. **Reid, B. J., K. T. Semple, C. J. Macleod, H. Weitz, and G. I. Paton.** 1998. Feasibility of using prokaryotic biosensors to assess acute toxicity of polycyclic aromatic hydrocarbons. *FEMS Microbiol. Lett.* **169:**227–233.

260. **Reid, R. C., J. M. Praunsnitz, and T. K. Sherwood.** 1977. *The Properties of Gases and Liquids*, 3rd ed. McGraw-Hill, New York, N. Y.

261. **Reisinger, H. J., and J. B. Reid.** 2001. Methyl-tertiary butyl ether natural attenuation case studies. *Soil Sed. Contam.* **10:**21–43.

262. **Reisinger, H. J., S. A. Mountain, G. Andreotti, G. DiLuise, A. Porta, A. S. Hullman, V. Owens, D. Arlotti, and J. Godfrey.** 1998. Bioremediation of a major inland oil spill using a comprehensive integrated approach. *Civil Eng. ASCE* **68:**52–55.

263. **Ren, S., and P. D. Frymier.** 2003. The use of a genetically engineered *Pseudomonas* species (Shk1) as a bioluminescent reporter for heavy metal toxicity screening in wastewater treatment plant influent. *Water Environ. Res.* **75:**21–29.

264. **Rittmann, B. E., and R. Whiteman.** 1994. Bioaugmentation: a coming of age. *Water Qual. Int.* **1:**12–16.

265. **Robinson, C. H., and P. A. Wookey.** 1997. Microbial ecology, decomposition and nutrient cycling, p. 41–68. *In* S. J. Woodin and M. Marquiss (ed.), *Ecology of Arctic Environments.* Blackwell, Oxford, United Kingdom.

266. **Romich, M. S., D. C. Cameron, and M. R. Etzel.** 1995. Three methods for the large-scale preservation of a microbial inoculum for bioremediation, p. 229–235. *In* R. E. Hinchee, J. Fredrickson, and B. C. Alleman (ed.), *Bioaugmentation for Site Remediation.* Battelle Press, Columbus, Ohio.

267. **Rooney-Varga, J. N., R. T. Anderson, J. L. Fraga, D. Ringelberg, and D. R. Lovley.** 1999. Microbial communities associated with anaerobic benzene degradation in a petroleum-contaminated aquifer. *Appl. Environ. Microbiol.* **65:**3056–3063.

268. **Rothmel, R. K., R. W. Peters, E. St. Martin, and M. F. Deflaun.** 1998. Surfactant foam/bioaugmentation technology for *in situ* treatment of TCE-DNAPLs. *Environ. Sci. Technol.* **32:** 1667–1675.

269. **Rousseaux, S., A. Hartmann, B. Lagacherie, S. Piutti, F. Andreux, and G. Soulas.** 2003. Inoculation of an atrazine-degrading strain, *Chelatobacter heintzii* Cit1, in four different soils: effects of different inoculum densities. *Chemosphere* **51:**569–576.

270. **Ruberto, L., S. C. Vazquez, and W. P. MacCormack.** 2003. Effectiveness of the natural bacterial flora, biostimulation and bioaugmentation on the bioremediation of a hydrocarbon contaminated Antarctic soil. *Int. Biodeterior. Biodegrad.* **52:**115–125.

271. **Rutherford, K., and P. C. Johnson.** 1996. Effects of process control changes on aquifer oxygenation rates during *in situ* air sparging in homogeneous aquifers. *Groundwater Monitor. Remed.* **16:**132–141.

272. **Salanitro, J. P., L. A. Diaz, M. P. Williams, and H. L. Wisniewski.** 1994. Isolation of a bacterial culture that degrades MTBE. *Appl. Environ. Microbiol.* **60:**2593–2596.

273. **Salanitro, J. P., P. C. Johnson, G. E. Spin-**

nler, P. M. Maner, H. L. Wisniewski, and C. Bruce. 2000. Field-scale demonstration of enhanced MTBE bioremediation through aquifer bioaugmentation and oxygenation. *Environ. Sci. Technol.* **34:**4152–4162.

274. **Sasek, V., M. Bhatt, T. Cajthaml, K. Malachova, and D. Lednicka.** 2003. Compost-mediated removal of polycyclic aromatic hydrocarbons from contaminated soil. *Arch. Environ. Contam. Toxicol.* **44:**336–342.

275. **Sayler, G. S., and S. Ripp.** 2000. Field applications of genetically engineered microorganisms for bioremediation processes. *Curr. Opin. Biotechnol.* **11:**286–289.

276. **Scheibenbogen, K., R. G. Zytner, H. Lee, and J. T. Trevors.** 1994. Enhanced removal of selected hydrocarbons from soil by *Pseudomonas aeruginosa* UG2 biosurfactants and some chemical surfactants. *J. Chem. Technol. Biotechnol.* **59:** 53–59.

277. **Schultz, J. E., G. I. Latter, and A. Matin.** 1988. Differential regulation by cyclic AMP of starvation protein synthesis in *Escherichia coli. J. Bacteriol.* **170:**3903–3909.

278. **Schwartz, A., and R. Bar.** 1995. Cyclodextrin-enhanced degradation of toluene and *p*-toluic acid by *Pseudomonas putida. Appl. Environ. Microbiol.* **61:**2727–2731.

279. **Segura, A., E. Duque, G. Mosqueda, J. L. Ramos, and F. Junker.** 1999. Multiple responses of Gram-negative bacteria to organic solvents. *Environ. Microbiol.* **1:**191–198.

280. **Seklemova, E., A. Pavlova, and K. Kovacheva.** 2001. Biostimulation-base bioremediation of diesel fuel: field demonstration. *Biodegradation* **12:**311–316.

281. **Semple, K. T., and T. R. Fermor.** 1997. Enhanced mineralization of UL-^{14}C-pentachlorophenol by mushroom composts. *Res. Microbiol.* **148:**795–798.

282. **Semple, K. T., B. J. Reid, and T. R. Fermor.** 2001. Impact of composting strategies on the treatment of soils contaminated with organic pollutants. *Environ. Pollut.* **112:**269–283.

283. **Semprini, L., G. D. Hopkins, P. V. Roberts, and P. L. McCarty.** 1991. *In situ* biotransformation of carbon tetrachloride, Freon-113: Freon-11 and 1,1,1-TCA under anoxic conditions, p. 41–58. *In* R. E. Hinchee and R. F. Olfenbuttel (ed.), *On-Site Bioreclamation: Processes for Xenobiotic and Hydrocarbon Treatment.* Butterworth-Heinemann, Boston, Mass.

284. **Semprini, L., G. D. Hopkins, P. L. McCarty, and P. V. Roberts.** 1992. *In situ* transformation of carbon-tetrachloride and other halogenated compounds resulting from biostimulation under anoxic conditions. *Environ. Sci. Technol.* **26:**2454–2461.

285. **Semprini, L., G. D. Hopkins, P. V. Roberts, and P. L. McCarty.** 1992. Pilot scale field studies of *in situ* bioremediation of chlorinated solvents. *J. Haz. Mat.* **32:**145–162.

286. **Shah, J. K., G. D. Sayles, M. T. Suidan, P. Mihopoulos, and S. Kaskassian.** 2001. Anaerobic bioventing of unsaturated zone contaminated with DDT and DNT. *Water Sci. Technol.* **43:**35–42.

287. **Shapir, N., and R. T. Mandelbaum.** 1997. Atrazine degradation in subsurface soil by indigenous and introduced microorganisms. *J. Agric. Food Chem.* **45:**4481–4486.

288. **Shimazu, M., A. Mulchandani, and W. Chen.** 2001. Simultaneous degradation of organophosphorus pesticides and *p*-nitrophenol by a genetically engineered *Moraxella* sp. with surface-expressed organophosphorus hydrolase. *Biotechnol. Bioeng.* **76:**318–324.

289. **Shimazu, M., W. Chen, and A. Mulchandani.** 2004. Biodegradation of organophosphate nerve agents, p. 629–642. *In* E. Valsami-Jones (ed.), *Phosphorus in Environmental Technologies: Principles and Applications.* IWA Publishing, London, United Kingdom.

290. **Shimotori, T., E. E. Nuxoll, E. L. Cussler, and W. A. Arnold.** 2004. A polymer membrane containing Fe^0 as a containment barrier. *Environ. Sci. Technol.* **38:**2264–2270.

291. **Shin, W. S., J. H. Pardue, and W. A. Jackson.** 2000. Oxygen demand and sulphate reduction in petroleum hydrocarbon contaminated salt marsh soils. *Water Res.* **34:**1345–1353.

292. **Shinkyo, R., T. Sakaki, T. Takita, M. Ohta, and K. Inouye.** 2003. Generation of 2,3,7,8-TCDD-metabolizing enzyme by modifying rat CYP1A1 through site-directed mutagenesis. *Biochem. Biophys. Res. Commun.* **308:**511–517.

293. **Silva, E., A. M. Fialho, I. Sa-Correia, R. G. Burns, and L. J. Shaw.** 2004. Combined bioaugmentation and biostimulation to clean up soil contaminated with high concentrations of atrazine. *Environ. Sci. Technol.* **38:**632–637.

294. **Simon, M., R. L. Autenrieth, T. J. McDonald, and J. S. Bonner.** 1999. Evaluation of bioaugmentation for remediation of petroleum in a wetland. *In Proceedings of the 1999 International Oil Spill Conference.* American Petroleum Institute, Washington, D.C.

295. **Sims, G. K., M. Radosevich, X. T. He, and S. J. Traina.** 1991. The effects of sorption on the bioavailability of pesticides, p. 120–137. *In* W. B. Betts (ed.), *Biodegradation: Natural and Synthetic Materials.* Springer-Verlag, Heidelberg, Germany.

296. **Singer, A. C., I. P. Thompson, and M. J. Bailey.** 2004. The tritrophic trinity: a source of pollutant-degrading enzymes and its implications for phytoremediation. *Curr. Opin. Microbiol.* **7:** 239–244.
297. **Smets, B. F., and P. H. Pritchard.** 2003. Elucidating the microbial component of natural attenuation. *Curr. Opin. Biotechnol.* **14:**283–288.
298. **Sorenson, K. S.** 2000. Biodegradation of TCE improved with lactate injection in deep, fractured rock. *EPA Groundwater Curr.* **38:**1–2. [Online.] http://www.clu-in.org.
299. **Spence, M. J., S. F. Thornton, K. H. Spence, S. H. Bottrell, and H. H. Richnow.** 2003. Natural attenuation of BTEX/MTBE in a dual porosity chalk aquifer. *In* V. S. Magar and M. E. Kelley (ed.), In Situ *and On-Site Bioremediation.* Battelle Press, Columbus, Ohio. [Online.] http://www.shef.ac.uk.
300. **Sreenivasulu, C., and Y. Aparna.** 2001. Bioremediation of methylparathion by free and immobilized cells of *Bacillus sp.* isolated from soil. *Bull. Environ. Contam. Toxicol.* **67:**98–105.
301. **Sriprang, R., M. Hayashi, M. Yamashita, H. Ono, K. Saeki, and Y. Murooka.** 2002. A novel bioremediation system for heavy metals using the symbiosis between leguminous plant and genetically engineered rhizobia. *J. Biotechnol.* **99:**279–293.
302. **Stach, J. E., and R. G. Burns.** 2002. Enrichment versus biofilm culture: a functional and phylogenetic comparison of polycyclic aromatic hydrocarbon-degrading microbial communities. *Environ. Microbiol.* **4:**169–182.
303. **Steenson, L. R., T. R. Klaenhammer, and H. E. Swaisgood.** 1987. Calcium alginate-immobilized cultures of lactic streptococci are protected from bacteriophages. *J. Dairy Sci.* **70:** 1121–1127.
304. **Steinberg, S. M., E. J. Poziomek, W. H. Englemann, and K. R. Rogers.** 1995. A review of environmental applications of bioluminescence measurements. *Chemosphere* **30:** 2155–2197.
305. **Stephenson, D., and T. Stephenson.** 1992. Bioaugmentation for enhancing biological waste-water treatment. *Biotechnol. Adv.* **10:** 549–559.
306. **Straube, W. L., C. C. Nestler, L. D. Hansen, D. Ringleberg, P. H. Pritchard, and J. Jones-Meehan.** 2003. Remediation of polyaromatic hydrocarbons (PAHs) through landfarming with biostimulation and bioaugmentation. *Acta Biotechnologica* **23:**179–196.
307. **Straube, W. L., J. Jones-Meehan, P. H. Pritchard, and W. R. Jones.** 1999. Bench-scale optimization of bioaugmentation strategies for treatment of soils contaminated with high molecular weight polyaromatic hydrocarbons. *Resources Conservation Recycling* **27:**27–37.
308. **Sullivan, E. S.** 1998. Molecular genetics of biosurfactant production. *Curr. Opin. Biotechnol.* **9:** 263–269.
309. **Sutherland, T. D., I. Horne, M. J. Lacey, R. L. Harcourt, R. J. Russell, and J. G. Oakeshott.** 2000. Enrichment of an endosulfan-degrading mixed bacterial culture. *Appl. Environ. Microbiol.* **66:**2822–2828.
310. **Swindoll, C. M., C. M. Aelion, and F. K. Pfaender.** 1988. Influence of inorganic and organic nutrients on aerobic biodegradation and on the adaptation response of subsurface microbial communities. *Appl. Environ. Microbiol.* **54:** 212–217.
311. **Takeuchi, M., K. Nanbu, K. Furuya, H. Nirei, and M. Yoshida.** 2004. Natural groundwater of a gas field utilizable for a bioremediation of trichloroethylene contamination. *Environ. Geol.* **45:**891–898.
312. **Tan, H., J. T. Champion, J. F. Artiola, M. L. Brusseau, and R. M. Miller.** 1994. Complexation of cadmium by a rhamnolipid biosurfactant. *Environ. Sci. Technol.* **28:**2402–2406.
313. **Tanaka, H., T. Ohta, S. Harada, J. C. Ogbonna, and M. Yajima.** 1994. Development of a fermentation method using immobilized cells under unsterile conditions. 1. Protection of immobilized cells against anti-microbial substances. *Appl. Microbiol. Biotechnol.* **41:**544–550.
314. **Thornton, J. S., and W. L. Wootan.** 1982. Venting for the removal of hydrocarbon vapors from gasoline contaminated soil. *J. Environ. Sci. Health A* **17:**31–44.
315. **Thouand, G., P. Bauda, J. Oudot, G. Kirsch, C. Sutton, and J. F. Vidalie.** 1999. Laboratory evaluation of crude oil biodegradation with commercial or natural microbial inocula. *Can. J. Microbiol./Rev. Can. Microbiol.* **45:** 106–115.
316. **Tiehm, A.** 1994. Degradation of polycyclic aromatic hydrocarbons in the presence of synthetic surfactants. *Appl. Environ. Microbiol.* **60:**258–263.
317. **Topp, E.** 2001. A comparison of three atrazine-degrading bacteria for soil bioremediation. *Biol. Fertil. Soils* **33:**529–534.
318. **Tsomides, H. J., J. B. Hughes, J. M. Thomas, and C. H. Ward.** 1995. Effect of surfactant addition on phenanthrene biodegradation in sediments. *Environ. Toxicol. Chem.* **14:** 953–959.
319. **U.S. Air Force.** 1996. *Bioventing Performance and Cost Results from Multiple Air Force Test Sites, Technology Demonstration, Final Technical Memoran-*

dum. AFCEE Technology Transfer Division. U.S. Air Force, Washington, D.C.

320. **U.S. Department of Agriculture.** 1997. *Bioremediation Using Landfarm Systems. Guide Specification for Military Construction.* U.S. Army Corps of Engineers CEGS-02287. U.S. Department of Agriculture, Washington, D.C.

321. **U.S. Environmental Protection Agency.** 1995. *Bioremediation Field Evaluation.* EPA/540/R-95/533. U.S. Environmental Protection Agency, Eielson Air Force Base, Alaska.

322. **U.S. Environmental Protection Agency.** 1995. *Bioventing Principles and Practice,* vol. I. *Bioventing Principles.* Manual. EPA/540/R-95/534A. U.S. Environmental Protection Agency, Washington, D.C.

323. **U.S. Environmental Protection Agency.** 1995. *Bioventing Principles and Practice,* vol. II. *Bioventing Design.* Manual. EPA/540/R-95/534A. U.S. Environmental Protection Agency, Washington, D.C.

324. **U.S. Environmental Protection Agency.** 1995. *How to Evaluate Alternative Cleanup Technologies for Underground Storage Tank Sites: a Guide for Corrective Action Plan Reviewers.* EPA 510-B-95-007. U.S. Environmental Protection Agency, Washington, D.C.

325. **U.S. Environmental Protection Agency.** 1996. *Pump-and-Treat Ground-Water Remediation. A Guide for Decision Makers and Practitioners.* EPA/625/R-95/005. U.S. Environmental Protection Agency, Washington, D.C.

326. **U.S. Environmental Protection Agency.** 1998. *Permeable Reactive Barrier Technologies for Contaminant Remediation.* EPA/600/R-98/125. U.S. Environmental Protection Agency, Washington, D.C.

327. **U.S. Environmental Protection Agency.** 1999. *Use of Monitored Natural Attenuation at Superfund, RCRA Corrective Action, and Underground Storage Tank Sites.* Office of Solid Waste and Emergency Response Directive OSWER 9200.4–17P. U.S. Environmental Protection Agency, Washington, D.C.

328. **U.S. Environmental Protection Agency.** 2000. *Abstracts of Remediation Case Studies,* vol. 4. EPA 542-R-00-006. U.S. Environmental Protection Agency, Washington, D.C.

329. **U.S. Environmental Protection Agency.** 2000. *Engineered Approaches to* In Situ *Bioremediation of Chlorinated Solvents: Fundamentals and Field Applications.* EPA 542-R-00-008. U.S. Environmental Protection Agency Office of Solid Waste and Emergency Response Technology Innovation Office, Washington, D.C. http://cluin.org.

330. **U.S. Environmental Protection Agency.** 2001. *Use of Bioremediation at Superfund Sites.* EPA 542-R-01–019. U.S. Environmental Protection Agency, Washington, D.C.

331. **U.S. Environmental Protection Agency.** 2004. *Treatment Technologies for Site Cleanup: Annual Status Report,* 11th ed. EPA-542-R-03–009. U.S. Environmental Protection Agency, Washington, D.C.

332. **U.S. Geological Survey.** 1997. Bioremediation: nature's way to a cleaner environment. [Online.] http://water.usgs.gov.

333. **U.S. Geological Survey.** 2004. Crude oil contamination in the shallow subsurface: Bemidji, Minnesota. [Online.] http://toxics.usgs.gov.

334. **Valo, R.** 1998. Application scale bioremediation of contaminated soil and ground water. *In Proceedings of Second Maj and Tor Nessling Foundation Symposium on Bioremediation of Contaminated Soil and Groundwater: Traditional Methods and Possibilities for Gene-Technology.* Helsinki, Finland. [Online.] http://www.datauniversum.fi.

335. **van der Gast, C. J., A. S. Whiteley, and I. P. Thompson.** 2004. Temporal dynamics and degradation activity of a bacterial inoculum for treating waste metal-working fluid. *Environ. Microbiol.* **6:**254–263.

336. **van der Gast, C. J., C. J. Knowles, M. Starkey, and I. P. Thompson.** 2002. Selection of microbial consortia for treating metal-working fluids. *J. Ind. Microbiol. Biotechnol.* **29:**20–27.

337. **Van Deuren, J. V., T. Lloyd, S. Chhetry, R. Liou, and J. Peck.** 2002. *Remediation Technologies Screening Matrix and Reference Guide,* 4th ed. [Online.] http://www.frtr.gov.

338. **van Dyke, M. I., S. L. Gulley, H. Lee, and J. T. Trevors.** 1993. Evaluation of microbial surfactants for recovery of hydrophobic pollutants from soil. *J. Ind. Microbiol.* **11:**163–170.

339. **Van Gestel, K., J. Mergaert, J. Swings, J. Coosemans, and J. Ryckeboer.** 2003. Bioremediation of diesel-contaminated soil by composting with biowaste. *Environ. Pollut.* **125:**361–368.

340. **van Oudenhoven, J. A. C. M., G. R. Cooper, G. Cricchi, J. Gineste, R. Pötzl, J. Vissers, and D. E. Martin.** 1995. Oil refinery waste disposal methods, quantities and costs 1993 survey. CONCAWE 1/95. CONCAWE, Brussels, Belgium.

341. **van Veen, J. A., L. S. van Overbeek, and J. D. van Elsas.** 1997. Fate and activity of microorganisms introduced into soil. *Microbiol. Mol. Biol. Rev.* **61:**121–135.

342. **Venkatraman, S. N., J. R. Schuring, T. M. Boland, I. D. Bossert, and D. S. Kosson.**

1998. Application of pneumatic fracturing to enhance *in situ* bioremediation. *J. Soil Contam.* **7:** 143–162.

343. **Venosa, A. D., J. R. Haines, and D. M. Allen.** 1992. Efficacy of commercial inocula in enhancing biodegradation of weathered crude oil contaminating a Prince William Sound beach. *J. Ind. Microbiol.* **10:**1–11.

344. **Venosa, A. D., M. T. Suidan, B. A. Wrenn, K. L. Strohmeier, J. R. Haines, B. L. Eberhart, D. King, and E. Holder.** 1996. Bioremediation of an experimental oil spill on the shoreline of Delaware Bay. *Environ. Sci. Technol.* **30:**1764–1775.

345. **Verstraete, W., and E. M. Top.** 1999. Soil clean-up: lessons to remember. *Int. Biodeterior. Biodegrad.* **43:**147–153.

346. **Vidovich, M. M., J. A. McConchie, and S. Schiess.** 2001. The effectiveness of natural attenuation to remediate BTEX contamination in unconfined sand/gravel aquifers: an investigation of two sites. *J. Hydrol.* **40:**205–217.

347. **Vogel, T. M., C. S. Criddle, and P. L. McCarty.** 1987. Transformations of halogenated aliphatic compounds. *Environ. Sci. Technol.* **21:**722–736.

348. **Vogel, T. M.** 1996. Bioaugmentation as a soil bioremediation approach. *Curr. Opin. Biotechnol.* **3:**311–316.

349. **von Fahnestock, F. M., L. A. Smith, G. B. Wickramanayake, M. C. Place, R. J. Kratzke, W. R. Major, A. M. Walker, J. P. Wollenberg, M. J. Carlsey, P. V. Dinh, and N. T. Ta.** 1996. *Biopile Design and Construction Manual.* Technical Memorandum TM-2189-ENV. Battelle, Columbus, Ohio.

350. **Wagner, F., U. Behrendt, H. Bock, A. Kretschmer, S. Lang, and C. Syldatk.** 1983. Production and chemical characterization of surfactants from *Rhodococcus erythropolis* and *Pseudomonas* sp. MUB grown on hydrocarbons, p. 55–60. *In* J. E. Zajic, D. G. Cooper, T. R. Jack, and N. Kosaric (ed.), *Microbial Enhanced Oil Recovery.* Penn Well Publishing Company, Tulsa, Okla.

351. **Walker, A. W., and J. D. Keasling.** 2002. Metabolic engineering of *Pseudomonas putida* for the utilization of parathion as a carbon and energy source. *Biotechnol. Bioeng.* **78:**715–721.

352. **Wan, M. W., I. G. Petrisor, H. T. Lai, D. Kim, and T. F. Yen.** 2004. Copper adsorption through chitosan immobilized on sand to demonstrate the feasibility for *in situ* soil decontamination. *Carb. Polymers* **55:**249–254.

353. **Wang, J. M., E. M. Marlowe, R. M. Miller-Maier, and M. L. Brusseau.** 1998. Cyclodextrin-enhanced biodegradation of phenanthrene. *Environ. Sci. Technol.* **32:**1907–1912.

354. **Warhurst, A. W., and C. A. Fewson.** 1994. Biotransformations catalysed by the genus *Rhodococcus. Crit. Rev. Biochem.* **14:**29–73.

355. **Watanabe, K., and N. Hamamura.** 2003. Molecular and physiological approaches to understanding the ecology of pollutant degradation. *Curr. Opin. Biotechnol.* **14:**289–295.

356. **Watanabe, M. E.** 2001. Can bioremediation bounce back? *Nature Biotechnol.* **19:**1111–1115.

357. **Weiner, J. M., and D. R. Lovley.** 1998. Rapid benzene degradation in methanogenic sediments from a petroleum-contaminated aquifer. *Appl. Environ. Microbiol.* **64:**1937–1939.

358. **Welty, J. R., C. E. Wicks, and R. E. Wilson.** 1984. *Fundamentals of Momentum, Heat, and Mass Transfer,* 3rd ed. John Wiley & Sons, New York, N. Y.

359. **Wheatley, A. D.,** 1985. Wastewater treatment and by-product recovery, p. 68–106. *In* J. M. Sidwick (ed.), *Topics in Wastewater Treatment.* Critical Reports on Applied Chemistry, vol. 11. Blackwell, Oxford, United Kingdom.

360. **Widada, J., H. Nojiri, and T. Omori.** 2002. Recent developments in molecular techniques for identification and monitoring of xenobiotic-degrading bacteria and their catabolic genes in bioremediation. *Appl. Microbiol. Biotechnol.* **60:** 45–59.

361. **Wiedmeier, T. H., M. A. Swanson, D. E. Moutoux, E. K. Gordon, J. T. Wilson, B. H. Wilson, D. H. Kampbell, P. E. Hass, R. M. Miller, J. E. Hansen, and F. H. Chapelle.** 1998. *Technical Protocol for Evaluating Natural Attenuation of Chlorinated Solvents in Ground Water.* EPA/600/R-98/128. U.S. Environmental Protection, Agency, Washington, D.C.

362. **Wilke, C. R., and P. Chang.** 1955. Correlation of diffusion coefficients in dilute solutions. *Am. Inst. Chem. Eng. J.* **1:**264–270.

363. **Wilson, D. J., R. D. Norris, and A. N. Clarke.** 1998. Groundwater cleanup by *in situ* sparging. XIV. An air channeling model for biosparging with a horizontal pipe. *Separation Sci. Technol.* **33:**97–118.

364. **Wilson, L. P., P. C. D'Adamo, and E. J. Bouwer.** 1995. Aromatic hydrocarbon biotransformation under mixed oxygen/nitrate electron acceptor conditions, p. 24–25. *In Bioremediation of Hazardous Wastes: Research, Development and Field Evaluations.* EPA/540/R-95/532. U.S. Environmental Protection Agency, Washington, D.C.

365. **Wilson, L. P., P. C. D'Adamo, and E. J. Bouwer.** 1997. Bioremediation of BTEX,

naphthalene, and phenanthrene in aquifer material using mixed oxygen/nitrate electron acceptor conditions. [Online.] U.S. Environmental Protection Agency National Risk Management Research Laboratory. http://www.epa.gov.

366. **Witt, M. E., G. M. Klecka, E. J. Lutz, T. A. Ei, N. R. Grosso, and F. H. Chapelle.** 2002. Natural attenuation of chlorinated solvents at Area 6, Dover Air Force Base: groundwater biogeochemistry. *J. Contam. Hydrol.* **57:**61–80.

367. **Wong, J. W. C., C. K. Wan, and M. Fang.** 2002. Pig manure as a cocomposting material for biodegradation of PAH-contaminated soil. *Environ. Technol.* **23:**15–26.

368. **WS Atkins Environment.** 2002. *Genetically Modified Organisms for the Bioremediation of Organic and Inorganic Pollutants. Final Report.* [Online.] http://www.defra.gov.uk.

369. **Yan, Y., U. T. Bornscheuer, G. Stadler, S. Lutz-Wahl, R. T. Otto, M. Reuss, and R. D. Schmid.** 2001. Regioselective lipase-catalyzed synthesis of glucose ester on a preparative scale. *Eur. J. Lipid Sci. Technol.* **103:**583–587.

370. **Yoon, K. P.** 2003. Construction and characterization of multiple heavy metal-resistant phenol-degrading pseudomonads strains. *J. Microbiol. Biotechnol.* **13:**1001–1007.

371. **Zhu, X., A. D. Venosa, M. T. Suidan, and K. Lee.** 2004. *Guidelines for the Bioremediation of Oil-Contaminated Salt Marshes.* [Online.] National Risk Management Research Laboratory Office of Research and Development, U.S. Environmental Protection Agency, Cincinnati, Ohio. http://www.epa.gov.

372. **Zocca, C., S. D. Gregorio, F. Visentini, and G. Vallini.** Biodiversity amongst cultivable polycyclic hydrocarbon-transforming bacteria isolated from an abandoned industrial site. *FEMS Microbiol. Lett.*, in press.

373. **Zwillich, T.** 2000. Hazardous waste cleanup: a tentative comeback for bioremediation. *Science* **289:**2266–2267.

MONITORING BIOREMEDIATION

Jim C. Philp, Andrew S. Whiteley, Lena Ciric, and Mark J. Bailey

6

MONITORING NEEDS TO SUPPORT BIOREMEDIATION

Performance monitoring is a critical part of remediation effort. The needs for monitoring of bioremediation projects clearly start even before technology selection. First the nature of the contamination must be determined, in terms of the specific contaminants and their concentrations. Then the nature of the environmental matrix containing those contaminants must be considered. That will determine whether bioremediation can be considered as a possible remediation strategy. Pilot projects in which numerous parameters are measured so as to be able to optimize the rates of contaminant removal and to scale up to full field operations may be critical. Once a full-scale bioremediation effort has begun, however, the number of parameters that must be monitored often are minimal.

As in any remediation effort, one must be able to establish the initial parameters and to show that in the end the target goal of reducing the contaminant to a safe level has been achieved. In some cases, only the starting and end points need be measured to demonstrate success, but in other cases it is critical to follow the progress of the remediation effort and to monitor critical parameters, including those that may need to be modified to optimize the remediation process. In the case of bioremediation, monitoring may involve measuring not only concentrations of contaminants, but also the microbial populations involved in the degradation or transformation of those contaminants and the environmental parameters that influence rates of microbial metabolism.

Monitored natural attenuation (MNA) is a special case, since evidence of microbial activity is essential in determining the feasibility of MNA as a full-scale remedy at a particular site. The U.S. Navy Engineering Service Center has summarized the monitoring requirements for petroleum-contaminated soil and groundwater (see the website http://enviro.nfesc.navy.mil/). Preliminary screening to determine if intrinsic bioremediation is applicable is followed by a detailed evaluation to determine if it will be effective at meeting remediation goals. A detailed evaluation of intrinsic bioremediation should answer two questions. (i) Is biodegradation of the contaminants already occurring? (ii) Is it occurring rapidly enough

Jim C. Philp, Department of Biological Sciences, Napier University, Merchiston Campus, 10 Colinton Road, Edinburgh EH10 5DT, Scotland, United Kingdom. *Andrew S. Whiteley, Lena Ciric, and Mark J. Bailey*, NERC Centre for Ecology and Hydrology, Mansfield Road, Oxford OX1 3SR, United Kingdom.

Bioremediation: Applied Microbial Solutions for Real-World Environmental Cleanup
Edited by Ronald M. Atlas and Jim C. Philp © 2005 ASM Press, Washington, D.C.

to be protective of potential receptors? The focus of this discussion is on evaluating intrinsic bioremediation in groundwater. If intrinsic bioremediation is demonstrated to be a viable remedial option for a site, a long-term monitoring plan must be designed and implemented to verify ongoing effectiveness until remediation goals are met and to detect unexpected contaminant migration away from the site that could impact potential receptors in the area. The monitoring strategy should consider appropriate sampling locations, frequency, and parameters to be measured. At a minimum, the monitoring network should include wells at these locations: upgradient of the plume to monitor background water quality, within the plume to monitor changes in contaminant concentrations, immediately downgradient of the plume to detect contaminant migration, and at a compliance point (e.g., property boundary) upgradient of any potential receptors to provide early detection of contamination before the receptors are reached. Semiannual or annual sampling may be sufficient if contaminant concentrations have been relatively stable during initial site monitoring. More frequent (e.g., monthly or quarterly) sampling may be required to resolve trends in the data when initial monitoring results fluctuate significantly. Parameters to measure should include, at a minimum, contaminant concentrations and water levels to track changes in the plume and groundwater flow direction. Electron acceptors, metabolic byproducts, and general water quality parameters (e.g., temperature, pH, alkalinity, hardness, and redox potential) may be measured to monitor changes in ambient water quality and provide further evidence of ongoing remediation.

CHEMICAL ANALYSES

For full-scale soil or groundwater bioremediation projects, the chemical analyses that are needed effectively aim at quantifying the pollutants of interest at critical points in the process to check that they are being removed from the soil at the expected rate. This may be as simple as the start and end points, although in some remediation efforts extensive monitoring of the pollutants is necessary to comply with regulatory requirements. Sufficient analyses must be performed on appropriate representative samples to establish statistical reliability. Selection of sampling sites and frequency of sample analyses are critical and can greatly impact the cost of the remediation project. For groundwater remediation projects, sampling wells are usually needed; they must be positioned to capture the flow of the contaminants. Regardless of which samples are collected for analyses, they must be handled in ways that ensure preservation of the contaminants for the analyses. Often samples are frozen until they reach the laboratory for chemical analysis, especially if they cannot be transported and analyzed very quickly. Other measures may be needed to preserve samples during transport from the field to the analytical laboratory, especially if microbial analyses are to be performed in addition to chemical analyses of the samples.

As commercial bioremediation contracts are now being awarded with a timescale for completion, the contractor will want to know with best certainty the rate of removal of the contaminants from the soil or groundwater. This is exclusively the role of analytical chemistry. Thankfully, there is no shortage of analytical laboratories equipped to do this kind of work. Therefore, this section will not dwell on the detail of analytical chemistry but will give an overview of the types of techniques used to determine the identity and quantity of pollutants present in a contaminated soil or water. Keith (57) lists 178 U.S. Environmental Protection Agency (EPA)-approved methods, with more than 1,300 method-analyte summaries. This reference is a self-contained source that eliminates lengthy searches of large numbers of EPA documents. It is organized by the type of compound or material, with prescribed levels of precision and accuracy, collection and handling methods, sample preparation, and type of instrumental analysis.

Colorimetric Tests and Field Test Kits

Field test kits do not give the sensitivity or resolution of instruments used in analytical chemistry laboratories, but often they can be used to indicate trends on-site. This is a lower-cost way of generating data and can be integrated with laboratory testing to optimize testing with cost-effectiveness. For example, there are colorimetric test kits for the analysis of hydrocarbons in water and soil over a wide range of concentrations. A sample of water or soil is mixed with a solvent to extract the pollutants under investigation. After separation, a colorimetric reagent is added to identify the pollutant by means of color. The intensity of color provides information on the concentration of pollutants present, but the figure is a rough approximation. It may be useful in identifying a downward trend at the site but is far from definitive.

In a colorimetric approach, certain compounds and elements can react with a complexing agent to give a colored solution. The intensity of the color is proportional to the concentration of analyte, and absorption at a specified wavelength can be measured. Typical analytes that can be determined in this way include nitrate, phosphate, ammonia, cyanide, phenols, and aluminum. In commercially available kits, powders or tablets are mixed with the sample, a color is developed, and the absorbance is measured with a small colorimeter, ideal for field use. In laboratory-based systems, the mixing and measuring are fully automated, either in a batch process or in continuous-flow mode. Detection limits are typically in the tens to hundreds of micrograms per liter.

Metal Analysis

EXTRACTION PROCEDURES FOR METALS

For the determination of metal in soils and other solid media, acid digestion is commonly used to break down the sample matrix. The acids commonly used are nitric, hydrochloric, sulfuric, and perchloric acid, and for organic-rich matrices, an oxidizing agent such as hydrogen peroxide can be applied. Samples can be digested either under reflux, on a hot plate, or with the aid of microwave heating. After digestion is complete, samples are filtered through hardened ashless filter paper prior to analysis.

Atomic Absorption Spectroscopy. Atomic absorption spectroscopy is a common and well-established quantitative technique for detecting metals in a wide range of sample types, including natural waters, soils, sediments, and plant materials. The method is based on the fact that when metals are converted to their atomic state, light of the appropriate wavelength is absorbed and the amount of absorption can be measured (a different light source is required for each element). A calibration curve is obtained by analyzing standard solutions of the metal under investigation, thus allowing conversion of the absorbance measured for test samples into concentration units.

In flame atomic absorption spectrometry (FAAS), the sample solution is sprayed as a fine mist into a flame, commonly air-acetylene or nitrous oxide-acetylene, where it is vaporized at temperatures in excess of 2,400°C into constituent atoms prior to light absorption. Typical levels of detection are in the milligrams per liter range. In graphite furnace atomic absorption spectrometry, also known as electrothermal atomic absorption spectrometry, microliter volumes of sample are deposited into a small graphite tube, which is heated in controlled steps up to atomization temperatures of typically 2,500 to 3,000°C. Detection limits, superior to those of FAAS, are in the micrograms per liter range.

ICP Optical Emission Spectrometry. Inductively coupled plasma (ICP) optical emission spectrometry is a fast, multielement technique with a dynamic linear range and moderate to low detection limits (0.2 to 100 μg $liter^{-1}$). The instrument uses an ICP source (usually argon) to dissociate aspirated samples and standards into their constituent atoms or ions, exciting them to a level where they emit light of a characteristic wavelength detected electronically (e.g., with a photomultiplier tube or charge-coupled device). Up to 60 elements can be screened per single sample run of <1 min, and both aqueous and organic medium samples can be analyzed. In ICP mass spectrometry, a mass spectrometer is used as a detector, giving very low detection limits (0.0005 to 1.0 μg

liter^{-1}) and the ability to undertake high-precision isotope ratio studies.

Hydride Generation. A number of elements, such as arsenic, antimony, bismuth, and tellurium, can be reduced and separated as their volatile hydrides. The technique involves reacting a sample with sodium borohydride (reducing agent) to form the hydride and sweeping it by a carrier gas to a quartz tube mounted on the burner of an FAAS. The heat from the flame breaks down the hydride, releasing the element, and the absorption signal is measured. The useful working range of hydride generation is limited to low concentrations (parts per billion level). At higher concentrations, it is usually necessary to revert to other methods of measurement.

Cold Vapor. Mercury is unique among the metallic elements because of its high vapor pressure. Mercury ions in solution can be reduced by tin chloride or sodium borohydride to metallic mercury. Mercury is swept out of solution by an inert gas (e.g., argon) to a long-path glass absorption cell, where the absorption signal is measured (normally at 253.7 nm). Detection limits in the range below micrograms per liter to nanograms per liter are obtained.

Organic Contaminant Analyses

EXTRACTION PROCEDURES FOR ORGANIC POLLUTANTS

When analytical laboratory procedures are used to measure the concentrations of organic contaminants, the first task for the laboratory analyst is the extraction of the pollutants from the samples supplied. The objective is to transfer the pollutants from the soil or water, which to the analyst are dirty matrices, into a mobile gas, liquid, or supercritical phase that can be subjected to various analytical procedures. These are the most common techniques employed, although others are used in specific situations, e.g., microwave or simply shaking with heating. The extraction, while rather a mundane procedure, lies at the heart of good analysis, since the final result can be only as good as the quality of the extraction. Efficient extraction is critical for accurate monitoring.

For organic compounds, solvent extraction is most often employed to recover the contaminants from the soil or water samples. In liquid-phase transfer, samples are agitated (e.g., shaker) or heated (e.g., Soxhlet extraction) in the presence of an appropriate organic solvent (e.g., dichloromethane, hexane, or ethyl acetate) to remove the compounds of interest. The solvent is separated from the solid matrix by centrifugation and/or filtration, reduced to dryness by use of a rotary evaporator, transferred to suitable vials, and refrigerated prior to analysis. Extracts can be cleaned up to remove interfering components or selectively isolate compounds of interest by using carbon-based media (e.g., C_{18}) and carefully selected solvents.

In gas-phase transfer, the objective is to transfer all the pollutants of interest into the gas phase for subsequent instrumental analysis. It is best applied to contaminants that are only weakly linked to a soil matrix. There are generally two ways to do this. In the first, purge and trap, an inert gas is passed over the sample. In the second, the sample and the gas phase above it are allowed to equilibrate (the headspace analysis method).

The supercritical state of matter is characterized by low viscosity, and thus high diffusivity, and consequently very high solvent power. The fluid has characteristics intermediate between a liquid and a gas. Mineral salts are insoluble in the supercritical fluid, while organic compounds are very soluble. The supercritical fluid of most interest in the extraction of organic pollutants is CO_2. It is an apolar solvent in this regard, and the addition of methanol, for example, as a polar solvent can improve the total extraction power.

GC AND GC-MS. Gas chromatography (GC) and mass spectrometry (MS) analyses are widely used for the determination of volatile and semivolatile organic compounds in environmental media. A GC consists of a flowing mobile phase (such as helium), a heated injection port (consisting of a rubber septum through which a syringe needle is inserted to inject the sample), a separation column supporting the stationary phase (contained in a thermostat-controlled oven) and a detector. Components with a wide range of boiling points are separated by starting at a low oven temperature and increasing it over time to elute the high-boiling-point components. Standards of known composition and concentration are injected for both qualitative (comparison of retention times) and quantitative (peak area) analysis.

Different detectors can be used, depending on the nature of the compound to be determined. For example, for petroleum hydrocarbons (e.g., diesel or BTEX compounds [benzene, toluene, ethylbenzene, and xylenes]), a flame ionization detector is used, whereas for halogenated species (e.g., polychlorinated biphenyls or pesticides) an electron capture detector would be more suitable. Detection limits and linear ranges vary enormously, depending on the analyte of interest and detector used. Coupling an MS onto the end of a GC as a detector allows more detailed and sensitive data to be collected. In an MS, gaseous molecules from the GC are ionized, accelerated by an electric or magnetic field, and then separated according to their mass. A mass spectrum showing the relative abundance of each fragment striking the detector of the spectrometer is generated.

Headspace GC is a technique used for the analysis of volatile organics in solid, liquid, and gas samples. A headspace sample is normally prepared in a vial containing the sample, the dilution solvent, and the headspace. Volatile components from complex sample mixtures can be extracted from nonvolatile sample components and isolated in the headspace or gas portion of a sample vial. Once the sample phase is introduced into the vial and the vial is sealed, volatile components diffuse into the gas phase until the headspace has reached a state of equilibrium. After controlled heating of the vial, a subsample of the gas in the headspace is then injected into a GC system for separation and quantification of the volatile components. Detection limits are typically in the low parts per billion range.

HIGH-PERFORMANCE LIQUID CHROMATOGRAPHY. In high-performance liquid chromatography, mixtures of semivolatile and nonvolatile compounds are separated by pumping the sample through a column (containing a stationary phase), using a pressurized flow of a liquid mobile phase. Components in the sample interact with and migrate through the column at different rates due to differences in solubility, adsorption, size, or charge and pass through a detector that measures a response depending on the properties of the compounds of interest. For example, for the determination of common anions, a conductivity detector is used, whereas for the determination of phenols a UV-visible spectrometer is applied. As with GC, liquid chromatography instrumentation can be interfaced with an MS, yielding better sensitivity, and is a more powerful tool for qualitative and diagnostic investigations.

ISOTOPIC SHIFT. MNA requires the development of special techniques since it is only partly a biodegradation technique, albeit that in the vast majority of cases it is expected to be the major route of pollutant destruction. It requires the development of techniques that can differentiate between microbial removal and physical-chemical removal. Carbon-based stable isotope analysis is possible because ^{13}C exists alongside ^{12}C isotopes of compounds, although the former exist in a very small minority compared to the latter. However, for all organic compounds, there exists a particular ^{13}C:^{12}C ratio. Because ^{13}C typically creates stronger molecular bonds than ^{12}C, biological degradation of compounds generally results in the enrichment of the remaining molecule with ^{13}C isotopes and the depletion of ^{13}C in the formed product (110). Deuterium also exists naturally, and similar observations have been made with the D:H ratio. Thus, isotopic fractionation has the potential to inform on the progress of biotransformation of a pollutant, provided the initial isotopic composition is known.

Studies are now being directed at using isotopic fractionation as a tool to monitor natural attenuation Kolhatkar et al. (63) produced strong evidence for natural attenuation of methyl *tertiary*-butyl ether under anaerobic conditions, using stable isotope analysis. Similarly, carbon and hydrogen isotope analysis has the potential to provide direct evidence of anaerobic biodegradation of benzene in the field (67). Carbon isotopic enrichment factors for anaerobic benzene biodegradation were comparable to those previously published for aerobic benzene biodegradation, but hydrogen enrichment factors were significantly larger. Isotopic fractionation requires highly specialized equipment and in the end may not have the resolution required to identify in situ transformation rates, but the technique deserves further investigation.

MICROBIAL METHODS FOR MONITORING BIOREMEDIATION

For bioremediation to be considered as a remediation technology, it is critical to establish that there is an adequate active microbial population that is capable of attacking the specific contaminant(s). If the site has the requisite values for water content and pH, has porosity within the desirable range, and is contaminated with only petroleum hydrocarbons, then there is a very good chance that there will be an active population of hydrocarbon-oxidizing microorganisms in the soil and that bioremediation may be able to succeed.

Use of fundamental chemical analyses for pollutant identification and standard microbiological techniques for quantification of viable populations of microorganisms is the starting point for monitoring. The techniques for determining the presence of hydrocarbon-degrading microorganisms are routine, inexpensive, and relatively rapid. If the contaminants are more recalcitrant, then the soil may not contain large numbers of microbes that can degrade them, and enumeration might not be possible in this circumstance. A decision whether bioaugmentation may be needed may then be made. Even when it is difficult to cultivate specific microbial populations, however, the technique of enrichment culture can still reveal the presence of the necessary degrading microbes and establish that they have the necessary potential to degrade the contaminant at an acceptable rate of performance. There also are emerging techniques of molecular microbial ecology that do not rely on cultivation which are useful for determining whether bioremediation is feasible. At field trial or demonstration, the accumulation of more types of data than this would be wise. Microbial population changes might be investigated, along with more detailed analytical work, e.g., laboratory tests on the fate of ^{14}C-radiolabeled substrates, to identify specifically whether mineralization of the substrate is taking place or a mere transformation to a more or less toxic, more or less mobile metabolite.

Once a bioremediation project has begun, quantification of bacterial populations generally will not give much additional useful information. The real question relates to activities and the rates of disappearance of the problem contaminant(s). An exception may be in cases of bioaugmentation where ensuring that the added microorganisms persist and remain viable may be needed. Yet even when enumeration of microbial populations is not critical for monitoring the progress of a bioremediation effort, measurement of microbial oxygen consumption and/or CO_2 production in aerobic bioremediation can serve as a "health check" to establish that the process is proceeding according to plan; since gas respirometry can be done by relatively simple techniques on-site, these data can be acquired more often than the analytical chemistry data that inform on the pollutant removal. Metabolic biomarkers and isotopic shift techniques are also being developed and deployed to answer the special-case questions that arise from MNA.

Monitoring Hydrocarbon-Oxidizing Bacteria (HOB)

Hydrocarbons are chemically heterogeneous and almost ubiquitous in the environment. Not only are they found at sites of oil pollution, but chemical analysis has revealed their presence, both aliphatic and aromatic, in most soils and sediments. However, they are present in unpolluted sites in low concentrations. The probable origins of these low concentrations of hydrocarbons are seepage of hydrocarbons from natural deposits, especially gaseous hydrocarbons, and ongoing synthesis of some hydrocarbons by plants and microorganisms.

It is therefore not surprising that HOB are widely distributed in nature. On the order of 1 to 10% of bacteria isolated from uncontaminated soils may be capable of growing on hydrocarbons. Several investigations have demonstrated an increase in numbers of HOB in habitats that suffer from oil pollution. Also, it appears that the addition of an artificial oil slick causes a shift to the isolation of a greater percentage of HOB. However, recent work suggests that despite an increase in this percentage, the biodiversity of the bacterial community may be dramatically reduced (95). Thus, the presence of hydrocarbons in an environment frequently leads to selective enrichment of hydrocarbon-utilizing microorganisms, to the relative detriment of biodiversity.

Hydrocarbons as substrates for microbial growth present special difficulties (105). Many hydrocarbons act as solvents (especially aromatic hydrocarbons), and this can have gross effects on the cell exterior. The cell membrane contains a relatively large lipid content, about 40%. Lipids are soluble in many of these hydrocarbons—the classic lipid extraction procedure uses a chloroform-methanol mixture. Therefore, one effect of hydrocarbons is destruction of the semipermeable nature of the cell membrane, making cells leaky. Second is the lack of a site of weakness in a molecule. Many hydrocarbons, particularly the alkanes, alkenes, and alkynes, consist of chains of carbon atoms with pendant hydrogen atoms. The distribution of electrons and energy is even; all have roughly the same bond strength and length. Many hydrocarbons are extremely insoluble in water. There is now convincing evidence that bacterial hydrocarbon metabolism occurs in the aqueous phase (20).

In an attempt to define a petroleum-degrading bacterium, Rosenberg and Gutnick in 1977 (97) postulated three distinguishing traits or specifications:

1. An efficient hydrocarbon uptake system (special receptor sites for binding hydrocarbons and/or production of unique chemicals which assist the emulsification and transport of hydrocarbons into the cell).

2. Group-specific oxygenases, whose function is to activate molecular oxygen and incorporate it directly into the chemical structure of the hydrocarbon substrate.

3. Inducer specificity, i.e., the positive response of the organism to petroleum and its constituents in inducing the above two systems.

A number of general nutritional requirements are needed to achieve hydrocarbon utilization in bacteria. Hydrocarbons, as their name implies, are composed of hydrogen and carbon; there is a need to supply all other elements required for growth in the growth medium. These include molecular oxygen for the oxygenases; nitrogen, phosphorus, and sulfur; and metals, e.g., K^+ and Na^+, and trace metals.

The limitation for oxygen is easily overcome at the laboratory scale by having the oil-water interface in contact with air at all times. The usual strategy is to pump in sterile air or simply to have the flasks of liquid medium on an orbital shaker at 120 rpm or more. In theory, approximately 150 mg of nitrogen and 30 mg of phosphorus are consumed in the conversion of 1 g of hydrocarbon to cell material. These requirements can be satisfied by the provision of ammonium phosphate in the aqueous phase, or a combination of other salts, such as $(NH_4)_2SO_4$, K_3PO_4, NH_4Cl, $Ca_3(PO_4)_2$, and NH_4NO_3.

In all, about 35 genera of bacteria, 40 genera of fungi, and 9 genera of algae biodegrade oil. Degraders have been isolated from soils, oceans, coastal waters, freshwater and marine sediments, lakes, ponds, and estuaries. Hydrocarbon oxidizers have been isolated from temperate, tropical, and Arctic environments and from a range of shoreline types from Spitsbergen to Australia, including sand, cobble, pebble, and mangrove (37). Liquid hydrocarbons are the most easily degraded, but gaseous alkanes can also be oxidized (6).

The technique chosen for isolation of HOB depends on the hydrocarbon(s) in question. Isolation techniques have in common their need for a solid surface upon which individual colonies of bacteria can grow. This is required for their purification to axenic culture. For the microbiologist, investigation of the biochemistry or genetics of a HOB requires this initial step of isolation to purity. Therefore, the start point is the need for an agar-based medium for isolation. To isolate HOB requires that they grow specifically on the hydrocarbon provided in the growth medium as the source of carbon and energy. This necessitates that there is no other source of carbon in the medium, certainly no preferable source. A huge variety of these mineral media have been formulated. A useful medium for HOB is a modification of that of Goodhue (45), in which most of the desired salts can be 100 times concentrated for convenience. This medium is also without precipitating phosphates. Precipitate can be confused with microbial growth. That is the basal medium without any hydrocarbon. The addition of the hydrocarbon is the more complex issue.

ENUMERATION OF HYDROCARBON-OXIDIZING MICROORGANISMS

A common assumption is that because a contaminated soil or groundwater can be contaminated to high concentrations with hydrocarbons that represent high-calorie sustenance for bacteria, then during bioremediation there should be seen a mass proliferation of bacteria. This assumption could be wrong. Before a heterotrophic bacterium can start to grow and then multiply, it must expend energy for maintenance of viability, termed maintenance energy. This energy is derived from the oxidation of organic compounds. When the supply of the organic compound is large, then total available energy is way in excess of maintenance energy, and growth and multiplication occur. This might appear to be the situation in soil or water that is contaminated to the level of, say, tens of grams per kilogram or liter. However, since most hydrocarbon contaminants have low water solubility, then the actual supply of carbon is limited by diffusion from the liquid or solid to the cell surface. At low available carbon concentrations, a relatively larger amount of that carbon is required for maintenance energy, and a relatively smaller amount is available for cell growth and division. (This is also a reason why bioaugmentation with huge numbers of bacterial cells is flawed: if the introduced bacteria are limited in available carbon, they have a high likelihood of starving to death.) At the threshold level (2), all of the carbon that reaches and enters the cell is used for maintenance, and none is used for growth and division. The analytical chemistry laboratory results would show that the contaminants are being consumed, but microbiology laboratory results would show that the population size and biomass are not increasing.

Despite many efforts, enumeration of the HOB presents problems, which may lead to erroneous assumptions about the progress in bioremediation projects. The problems stem from the fact that most hydrocarbons of interest are either non-aqueous-phase liquids or insoluble solids (e.g., polycyclic aromatic hydrocarbons [PAHs]) and as such in normal bacteriological media they are not bioavailable because of their water insolubility. Early attempts to overcome the problem usually involved trying to incorporate hydrocarbons into agar media (7, 53, 114), but this approach has been criticized frequently (32, 75). Use of purified alkanes, purified mixtures, or distilled fuels gives variable results. Short-chain alkanes (pentane to nonane) have appreciable water solubility, so they can be incorporated into agars. Yet they also have high vapor pressures (pentane and nonane, 426 and 3.22 mm Hg at 20°C, respectively) (125) and are rapidly lost from agar during plate incubation. The medium-chain alkanes, however, the ones of greatest interest, have very low water solubility and are expelled on the surface of agars.

Venkateswaran et al. (123) incorporated crude oil into agar medium by preparing an emulsion of oil in water (5 g liter^{-1}) by sonication and then adding it to sterilized agar medium. Crude oil is more suitable in such tests because of the presence of natural emulsifying agents. With this technique, the emulsion often breaks before colonies grow, and the oil appears on the surface of the medium, interfering with colony formation.

Fuels such as gasoline and diesel contain insufficient natural surfactant material to emulsify the fuel in agar. Nontoxic emulsifying agents, or a low-toxicity solvent such as dimethyl sulfoxide, can be used. Negative controls must be included to ensure that there is no growth on the emulsifier. Also, normal emulsions do not maximize the surface area for bacterial attack, as the interfacial tension is still appreciable. A relatively reliable technique is to add 20% (vol/vol) hydrocarbon, 0.05% (vol/vol) Tween 80, and 80% (vol/vol) sterile medium and ultrasonicate the mixture for at least 5 min. This mixture needs to be kept cold by holding the vessel on ice. Sufficient amount of this emulsion is added to molten agar medium to give a final concentration of 0.5 to 1.0% hydrocarbon. The plates are inoculated after solidification and surface drying. Adjustment of the concentration of Tween 80 may be necessary if it is toxic (determined with a suitable control).

A technique for the growth of oil-degrading bacteria on volatile hydrocarbons in which a mineral salts agar is inoculated and a small tube containing a volatile hydrocarbon, or mixture of hydrocarbons, is placed in the lid of the plate was described by Rosenberg and Gutnick (97). The technique was revisited recently by Kleinheinz and Bagley (62): they used a filter pad in the lid of the petri dish instead (Fig. 6.1). Bacterial growth occurs on the hydrocarbons in the vapor phase. An obvious limitation of such a technique is that only a proportion of the total population, that which can utilize the vapor-phase hydrocarbons, is selected. Thus, the technique is likely to underestimate total oil-degrading populations, selecting for those able to utilize volatiles and semivolatiles. However, a further criticism of the technique is that it cannot be assumed that colonies that appear on such agar plates are oil degraders, as agar can contain enough impurities to allow growth; agar can also absorb volatile nutrients from the air in amounts sufficient to support growth of many bacteria. A recent critical review strongly suggests that this technique is not selective for oil-degrading bacteria (89), and false positives are common (92,120). Additionally, the longer medium-chain *n*-alkanes have insufficient vapor pressure to support growth in the vapor phase, even though it is known that long-chain alkanes can be biodegraded (43).

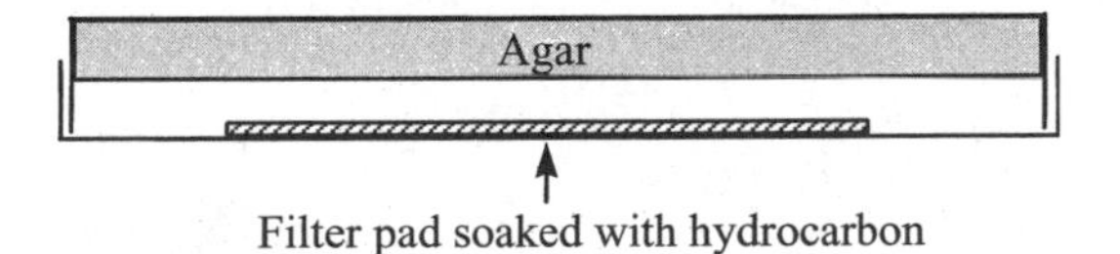

FIGURE 6.1 The filter pad method for growing HOB.

Jorgensen et al. (56) reported different methods for addition of oil to agar: sonication of hot agar solution after autoclaving, application of hexane-acetone-solubilized oil to solidified plates, and the filter pad technique. They were able to discern no pattern in the counts. Unreliable counts, irregular agar surface, and high background counts in the controls with no oil were all noted.

GROWTH OF HYDROCARBON-OXIDIZING MICROORGANISMS ON NONSOLUBLE SOLID HYDROCARBONS

Growth is limited by substrate transfer from the solid phase to the aqueous phase. Maximization of the surface area of the solid substrate is one strategy to maximize contact of the cells with substrate. Kiyohara et al. (59) dissolved phenanthrene in ether to a concentration of 10% (wt/vol) and sprayed the liquid onto the surface of agar plates. The ether immediately vaporizes and leaves a fine layer of the solid substrate on the surface of the plate. If the plate was previously inoculated, the ether is toxic to the cells. Inoculation after formation of the solid substrate layer can be difficult, as the film frequently breaks if an inoculating loop is used. Problems are encountered when the inoculum is pipetted to the solid layer since the layer is hydrophobic. When successful, however, it is immediately apparent which colonies that develop are able to metabolize the hydrocarbon; a cleared zone is easily seen around the colony (Fig. 6.2). A similar approach was adopted by Bogardt and Hemmingsen (15), but soil and water dilutions were incorporated into an agarose overlayer with fine particles of phenanthrene. The overlayer was poured onto an agar mineral salts underlayer, and phenanthrene-degrading colonies were also recognized by a halo of clearing.

To overcome the problems with the spray-plate method, Alley and Brown (3) described a technique in which the solid substrate was sublimed onto the surface of the agar plate after inoculation. This technique is more controlled than the spray-plate technique and allows a reproducible quantity of substrate to be delivered without the use of toxic solvents.

MICROTITER PLATE-BASED MPN TECHNIQUES

As a result of the limitations of traditional solid-agar-based enumeration methods, liquid culture methods have been developed by employing the most-probable-number (MPN)

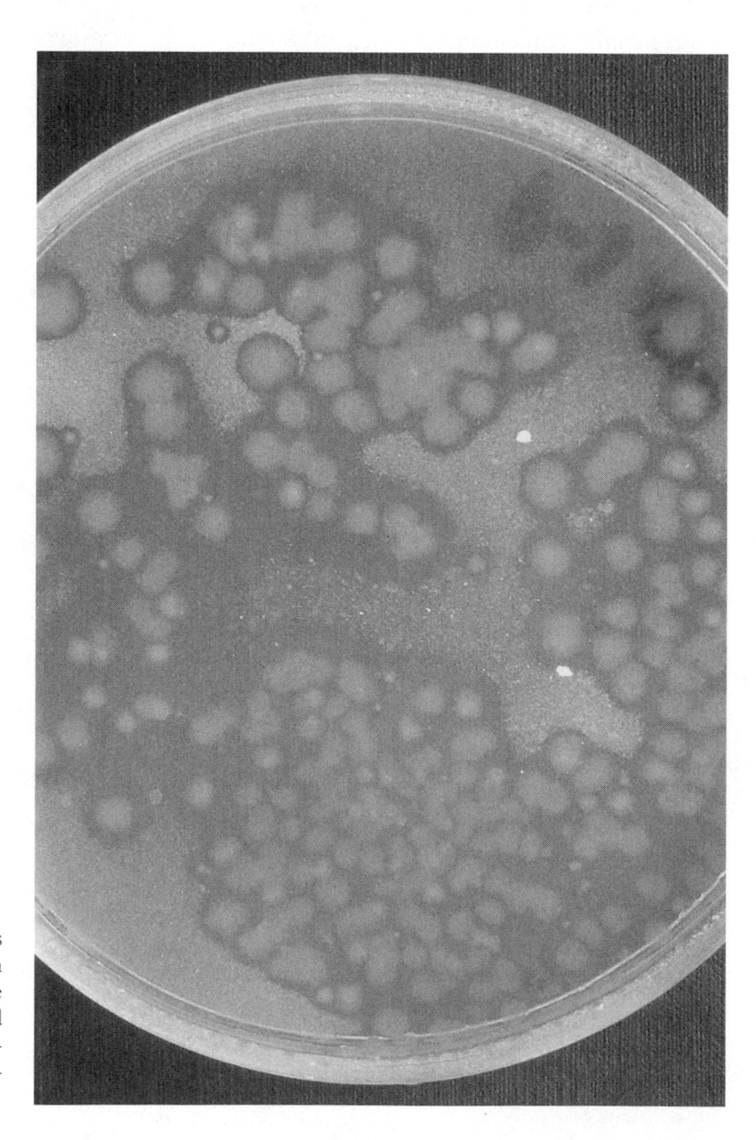

FIGURE 6.2 Bacterial colonies growing on an agar plate with a thin film of phenanthrene applied by the spray-plate technique. The cleared halo surrounding colonies is evidence of phenanthrene biotransformation.

procedure (32). MPN is a statistical method based upon dilution of a sample to the point of extinction; i.e., multiple replicates of various dilutions of a sample are analyzed and the results are compared with statistical tables to determine the MPN of microorganisms in the original sample. The development of 96-well microtiter plates gave the opportunity to miniaturize such procedures. The sheen screen method, introduced by Brown and Braddock (23), represented the start of the miniaturized MPN method for oil degraders. This method was specific for crude oil as a substrate (Fig. 6.3), and a similar method was published for no. 2 fuel oil as the substrate (49). Wrenn and Venosa (131) developed a 96-well microtiter plate MPN procedure to separately enumerate aliphatic and aromatic hydrocarbon degraders in separate plates. The alkane-degrader MPN method uses *n*-hexadecane as the carbon source, growth is scored by turbidity, and the reduction of iodonitrotetrazolium violet to iodonitrotetrazolium-formazan (red precipitate) is used as an indicator of electron transport activity. A recent variant on the technique (21) is similar to that of Wrenn and Venosa but purports to be simpler because of the inclusion of diphenyl tetrazolium violet in the growth medium.

An inherent limitation in all the 96-well microtiter plate MPN techniques is also one of lack of bioavailability. Given the long, narrow

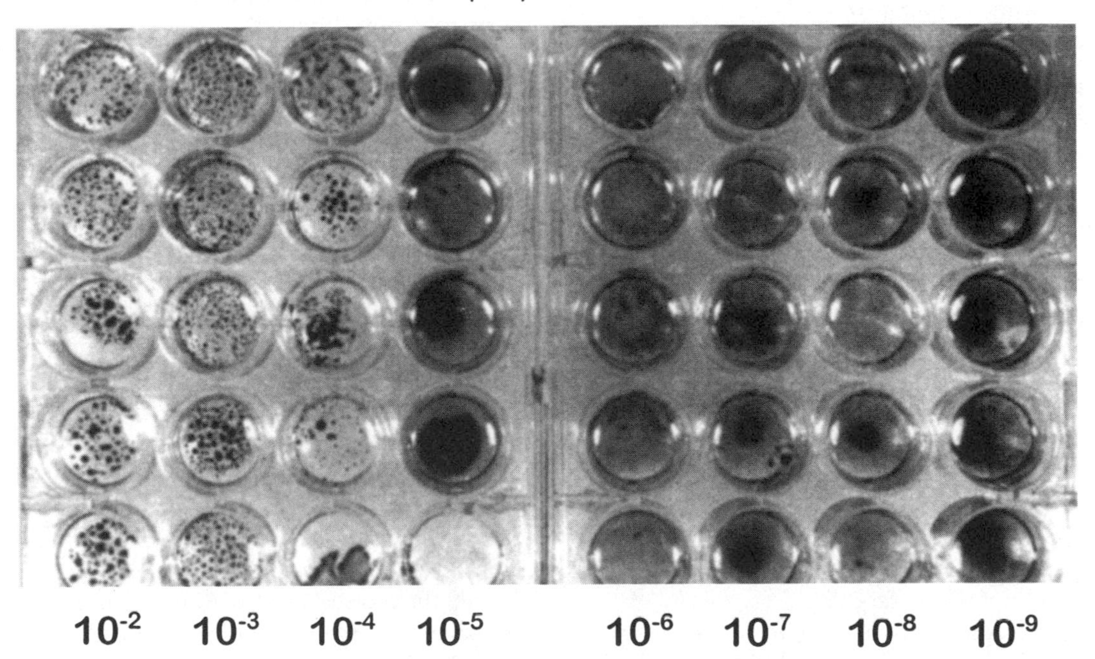

FIGURE 6.3 Illustration of a five-"tube" MPN result for a marine sediment sample obtained by the sheen screen method. The 10^{-2} and 10^{-3} dilutions are all positive for crude oil emulsifications, four of five of the 10^{-4} dilution tubes are scored positive, and one of five of the 10^{-5} dilution tubes scored positive. Courtesy of Ed Brown, University of Northern Iowa.

aspect ratio of the incubation wells, both oxygen mass transfer and mixing of the oil and aqueous phases are limited. This means that the incubation periods are long (typically 2 weeks), and there is a high risk of wells drying out, or at least of the medium becoming very saline through evaporation.

SCREENING STRAINS FOR RELEVANT ENZYME ACTIVITIES

Very often the enzymes of greatest interest are dioxygenases because of their pivotal roles in substrate activation and aromatic ring cleavage. Dioxygenase activity can be screened for by the inclusion of indole in agar plates. Dioxygenases convert indole to indigo (41), and the presence of blue colonies is the selection criterion. A more specific enzyme screen is for catechol 2,3-dioxygenase. Catechol is an extremely important and common intermediate in aromatic catabolism (see chapter 1). Colonies can be sampled by filter lift from plates and sprayed with catechol. The appearance of a yellow-brown pigment within 10 min of incubation at room temperature implies catechol 2,3-dioxygenase activity (128).

RESPIROMETRY

The data generated by analytical chemistry to monitor the removal of pollutants from a site do not provide specific information on biodegradation, as they take no account of nonbiological losses such as volatilization. For mathematical models, there is no substitute since the equations contain biodegradation rate parameters. Clearly, determination of pollutant concentrations and bacterial counts are unsatisfactory as the sole tools for bioremediation monitoring, as too many extraneous factors influence counts and chemical data can be highly variable due to soil heterogeneity. Analytical chemistry is also prohibitively expensive if enough samples are to be analyzed to generate meaningful kinetic data.

Metabolic gas respirometry is a technique that gives a rate of reaction, by measuring either O_2 consumption or CO_2 production. Radiorespirometry using ^{14}C-labeled hydrocarbons, labeled at the most recalcitrant part of the molecule, while more sensitive, is more demanding technically, is more expensive, uses specific hydrocarbons, and generates a radioactive waste. It is also intrusive to the experimental systems, causing unwanted perturbation.

Metabolic gas respirometry is much more flexible in operation because of fewer technical constraints. However, doubts over the sensitivity of respirometry, especially at sites where the initial oxygen concentration is low, have been raised (134). Whereas respirometry is a proven technique for determining biokinetic parameters for degradation of contaminants in groundwater, it remains to be proven in soils (111). Oxygen consumption is less sensitive than CO_2 production (87). Yet O_2 consumption is preferred for in situ experiments (see reference 40), as it is not susceptible to sinks and sources in soils and sediments in the same way that CO_2 is (103). Carbon dioxide is also water soluble, which may limit its in situ use to porous soils and sediments (118). However, CO_2 production actually provides data on mineralization, whereas oxygen consumption may arise from biotransformation, not necessarily mineralization. Carbon dioxide measurement is commonly used for assessing biodegradability in solid media (54).

Valuable data can be obtained when both are measured (e.g., see references 66 and 121). Simultaneous determination of oxygen and CO_2 can be done by GC, using a GC equipped with two concentric columns (100). This permits online monitoring, and data can be used to establish O_2 uptake rate or the CO_2 production rate and then the respiratory quotient (the ratio of CO_2 produced to O_2 consumed) (12). Lors and Mossman (66) concluded that respirometry alone could not be relied upon. For monitoring the biodegradation of PAHs, they recommended the use of O_2 and CO_2 respirometry along with analytical chemistry: having degraded the most available PAHs, the microbial populations may shift to other sources of carbon. Moller et al. (76) found variable agreement between respirometry data and actual diesel degradation rates in a bioventing study. The best agreement was 23 mg kg^{-1} day^{-1}

for actual oil content of the soil and 33 mg kg^{-1} day^{-1} for respirometry.

Some studies have demonstrated the applicability of metabolic gas respirometry when compared with radiorespirometry. Miles and Doucette (73) assessed microcosm-respirometry for measuring the aerobic biodegradability of 14 hydrocarbons. They concluded that the oxygen consumption technique was accurate and reliable compared with radiorespirometry, and reproducibility was good. Gejlsbjerg et al. (44) came to similar conclusions when comparing oxygen consumption with $^{14}CO_2$ evolution in soil and sludge-soil mixtures. Padmanabhan et al. (82) are seeking to develop a field soil biodegradation assay based on respiration of ^{13}C-labeled compounds and monitoring of $^{13}CO_2$ evolution by GC-MS.

Carbon dioxide is most commonly monitored by infrared spectrometry. Given the poor solubility of oxygen in water, it is easier to measure it in the vapor phase or in the headspace of a paramagnetic analyzer, and the latter would be the method of choice for soils. Commercial respirometers with a paramagnetic oxygen analyzer with 0.001% resolution, which frees oxygen measurement from the constraints of manometry, are available.

MOLECULAR BIOLOGY TOOLS FOR BIOREMEDIATION MONITORING

In recent years, molecular methods have been used for the study of microbial community structure and function because of the realization that the majority of microbes in the environment are unculturable by use of standard laboratory agars and conditions (4). It is thought that the percentage of culturable organisms is something in the range of 1 to 10% of the total community; in fact, the figure for the number of bacteria described so far is thought to be approximately 1% (33).

A study in 1994 showed the potential for molecular biological techniques to be used in the field of bioremediation. Investigating oil-degrading bacterial populations of Alaskan sediments, Sotsky et al. (112) found that up to 40% of the hydrocarbon-degrading bacteria hybridized to a gene probe for the *alkB* gene; moreover, hexadecane mineralization was highly correlated with the presence of *alkB*, suggesting the coexistence of degradation genes for low- and medium-molecular-weight *n*-alkanes in a variety of alkanotrophic bacteria.

The methods available now can be used in various ways to provide separate pieces of information. Some are used in order to give a profile of the total community present at a site, while others identify the organisms performing a specific function. Another set of methods focuses on the fluorescent probing of specific whole cells. More recently, new technologies which are substrate based, labeling organisms according to their catabolic potential, have emerged.

Community Profiling Methods

Community profiling methods produce rapid surveys, which provide us with a phylogenetic profile of the microbial population present at a particular site. The speed of application and specificity of these techniques can be utilized to assess the community composition across space, through time, down pollution gradients, and under various treatments.

PCR and 16S Ribosomal DNA (rDNA)

Methodologies that provide molecular fingerprints are most commonly based upon PCR amplification of 16S rRNA genes. The 16S rRNA gene is essential, as it encodes a subunit of the prokaryotic ribosome and is therefore present in all prokaryotic life forms (130).

Several criteria that make the 16S rRNAs and their genes the most widely studied phylogenetic markers have been identified (113):

- the function of ribosomes has not changed for about 3.8 billion years;
- the 16S rRNA genes are universally present among all cellular life forms;
- the size of 1,540 nucleotides makes them easy to analyze;
- the primary structure is an alternating sequence of invariant DNA, more or less conserved to highly variable regions; and
- lateral gene transfer is either totally absent or exceedingly rare.

PCR and rDNA sequencing are used worldwide in assessing the phylogenetic position of novel strains and microbial communities. The main reason for the routine application of this method is the presence of a set of conservative nucleotide stretches, which are scattered over the rRNA genes, serving as target sites of oligonucleotide primers (usually 14 to 20 bases in length). These primers are needed for amplification and subsequent sequence analysis. A set of not more than 10 primers is sufficient to analyze a wide spectrum of phylogenetically diverse organisms. Sequence analysis can be performed on both purified nucleic acid preparations and crude extracts of bacterial cells.

Phylogenetic relationships can be assessed by pairwise similarities. One hundred percent similarity found between a pair of 16S rDNA sequences by different methods indicates very close relatedness, if not identity, of the investigated organisms. The lower the value, the more unrelated the compared organisms. These phylogenetic distances form the basis for phylogenetic trees or dendrograms. The role of PCR in the detection and quantification of gene expression in environmental applications has been reviewed previously (104). The main techniques of community analysis and their relation to bioremediation have also been reviewed (e.g., see references 51 and 55). A comprehensive review of methods for studying soil microbial diversity examined the pros and cons of techniques currently in use (58).

DGGE

Denaturing gradient gel electrophoresis (DGGE) is based on the analytical separation of DNA fragments of identical or near-identical length based upon their sequence composition (78). Separation is based on the changing electrophoretic mobilities of DNA fragments migrating in a gel containing a linearly increasing gradient of DNA denaturants (Fig. 6.4). Changes in fragment mobility are associated with partial melting of the double-stranded DNA in discrete regions, the so-called melting domains. Each band shown on the gel represents a taxonomic unit present in the environment, and the band intensity can be associated with the species' abundance within it. Once the gel has been visualized, it is also possible to directly cut out bands for sequencing.

The method derives from one used in the medical sciences, which was subsequently modified for microbial community analyses

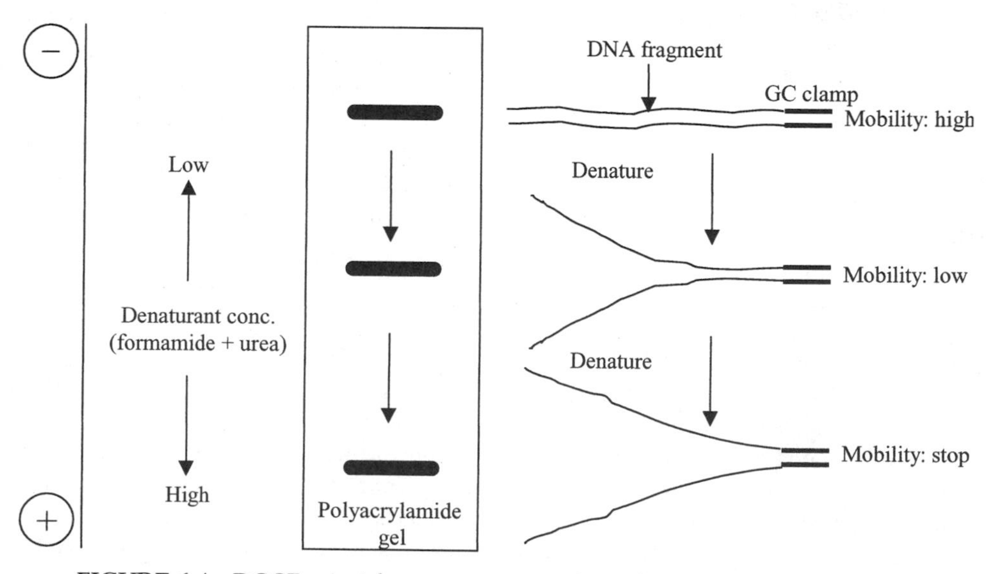

FIGURE 6.4 DGGE principle. conc., concentration. After Iwamoto and Nasu (55).

(77), in which the procedure is performed on the total community nucleic acid. An example of a DGGE gel is shown in Fig. 6.5 (30). The communities shown were extracted from various points of a diesel-contaminated groundwater remediation system.

DGGE has been used successfully in many investigations of community structure. It is now one of the most widespread and well-established methods used to obtain culture-independent microbial profiles and is starting to be used in bioremediation studies. For example, the DGGE community profiles of shoreline plots containing buried oil showed that controls and plots amended with liquid fertilizer had similar patterns (96). However, DGGE revealed that the bacterial community in plots treated with oil and slow-release fertilizer changed rapidly, probably as a result of the higher concentrations of nutrients available in interstitial water. Whiteley and Bailey (127) published evidence for highly structured bacterial communities within different compartments of a phenolic-remediating activated sludge plant. Fluorescent in situ hybridization (FISH) whole-cell targeting mirrored gross changes in community structure shown by DGGE. An issue in bioavailability during bioremediation has been addressed by using DGGE as a tool (42): different phenanthrene-utilizing bacteria inhabiting the same soils may be adapted for different phenanthrene bioavailabilities. The genus *Burkholderia* is itself diverse and is implicated regularly in bioremediation processes. DGGE analysis of PCR products has shown that there were sufficient differences in migration behavior to distinguish the majority of 14 *Burkholderia* species tested (99).

These sorts of studies reveal insights that either are impossible with traditional microbiological studies or else would involve much greater levels of effort, and the place of DGGE

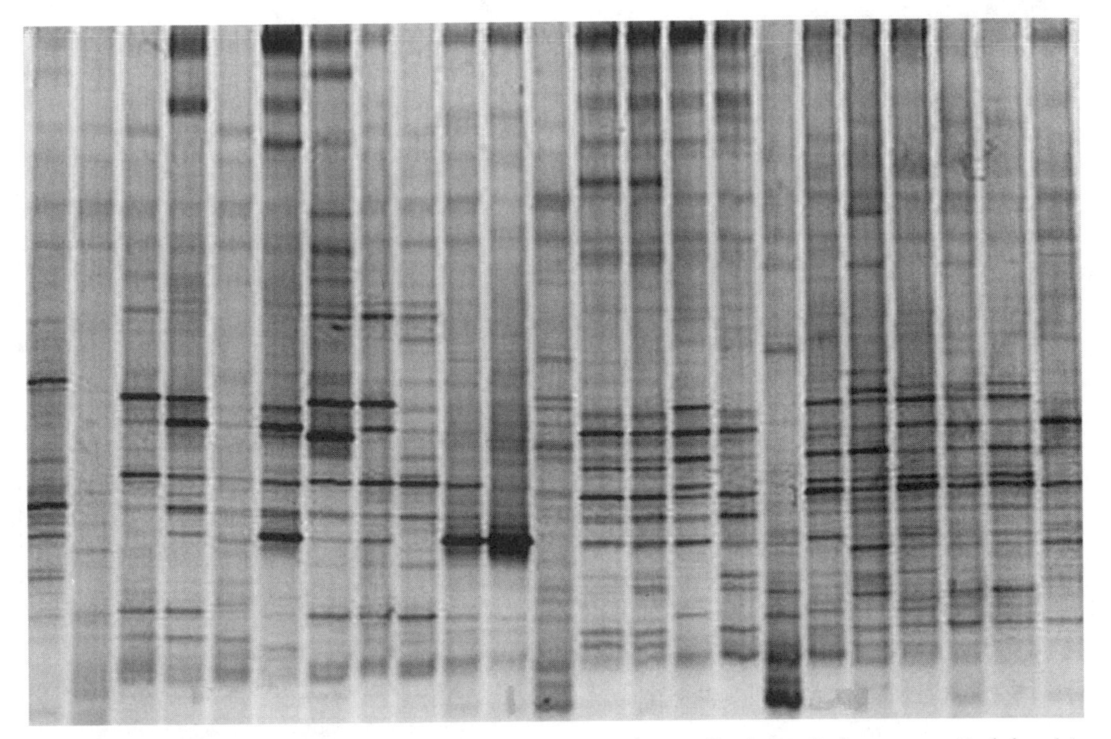

FIGURE 6.5 DGGE profile for samples taken from a groundwater diesel remediation system. Each band in theory represents a single taxonomic unit. Courtesy of Lena Ciric.

as a tool in bioremediation monitoring is assured. DGGE is an efficient and inexpensive method of analyzing community structure and diversity. It also allows greater phylogenetic resolution than many other community analysis methods. An added advantage is the ability to excise bands for sequencing purposes.

LH-PCR

More recently, other methods have become available for community-level analyses, such as length heterogeneity PCR (LH-PCR) (117). LH-PCR works by exploiting the natural variation in length of the 16S rRNA gene due to group-specific variable regions of the microbial community members (126). The PCR is carried out using one fluorescently labeled primer and the amplicon lengths are then separated on a sequencer, where the resultant peaks, comparable with DGGE gel bands, represent various phylogenetic groups. An example of an LH-PCR output can be seen in Fig. 6.6, in which the peaks represent various phylogenetic groups present in a community found in a diesel-contaminated groundwater sample (30).

One drawback of LH-PCR is that organisms can be effectively identified only down to class or subclass level, and single peaks may represent multiple bacterial genera within the same taxonomic subclass (e.g., the γ-*Proteobacteria*). LH-PCR is especially suitable for the tracking of known community members through time or during different treatments. In time, this method will become more reliable, when the databases available for various environmental samples are improved.

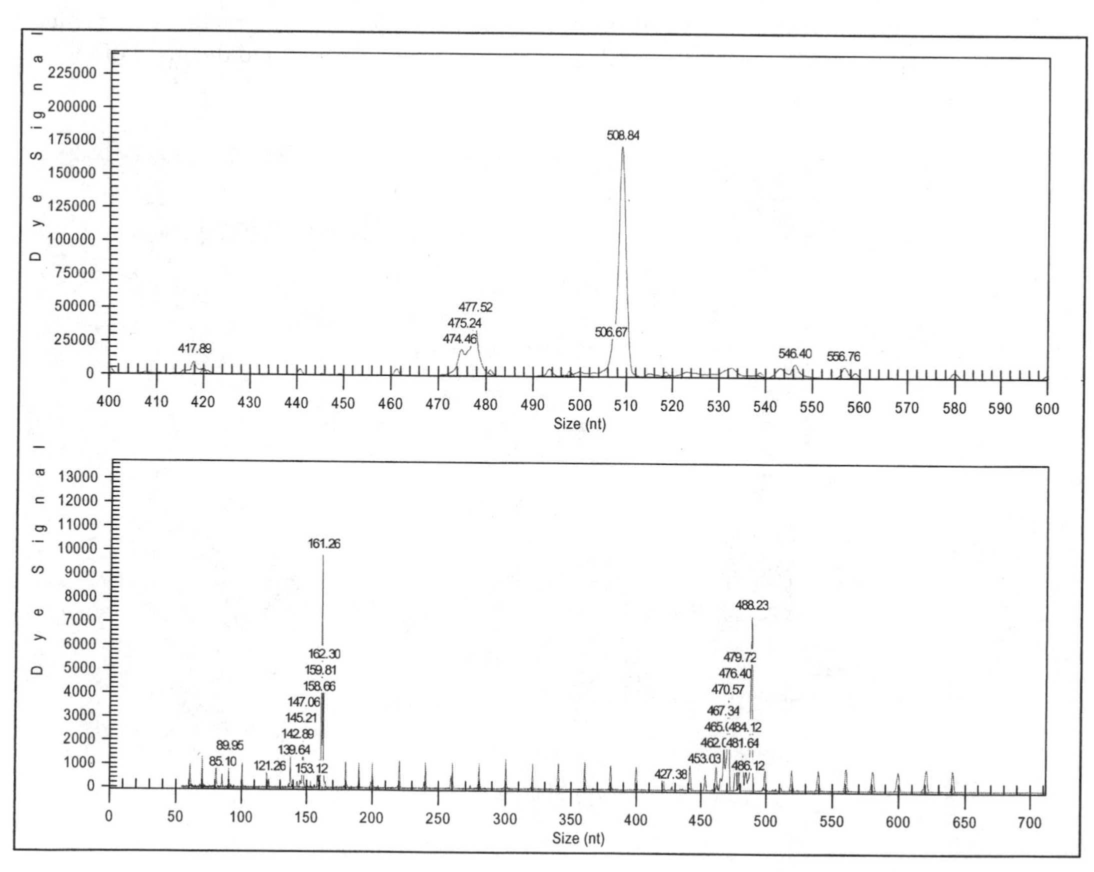

FIGURE 6.6 Output from LH-PCR (top) and tRFLP (bottom). Both outputs correspond to lane 20 on the DGGE gel in Fig. 6.5. Courtesy of Lena Ciric.

tRFLP

Terminal restriction fragment length polymorphism (tRFLP) has been used as a reliable community profiling technique by sizing variable-length restriction fragment digest patterns of amplified 16S rDNA (Fig. 6.7) (65). The method is very similar to LH-PCR but involves an endonuclease restriction step and gives a higher degree of phylogenetic resolution based upon the specific sequence variance within the 16S rRNA gene. tRFLP is reliable for determining the identity of organisms down to group level, and in parallel with LH-PCR it allows the processing of large sample sets in which samples are directly comparable. An example is shown in Fig. 6.6. The problem of identifying organisms below group level is somewhat mitigated by tRFLP. It must be emphasized that the choice of enzyme for digests in order to resolve bacterial taxonomic groups during tRFLP analyses is critical and requires some a priori knowledge of the bacterial groups which are present in the samples for effective phylogenetic targeting.

A notable example of the correlation of community profile change with function in bioremediation was given by Ayala-del-Río et al. (8). Community analysis by tRFLP of 16S rRNA genes showed that there were marked changes in a phenol-plus-trichloroethene (TCE)-fed reactor during the first 100 days but relative stability afterwards, which corresponded to the period of stable function. In a corresponding phenol-fed reactor, however, the community structure changed periodically, and the changes corresponded with the periodicity of TCE transformation rates.

Bacteria of the genus *Syntrophus* are difficult to culture in the laboratory as axenic cultures because they exist syntrophically with methanogenic bacteria. In a study of microbial diversity of a hydrocarbon-and chlorinated-solvent-contaminated aquifer using tRFLP, Dojka et al. (38) found sequence types characteristic of

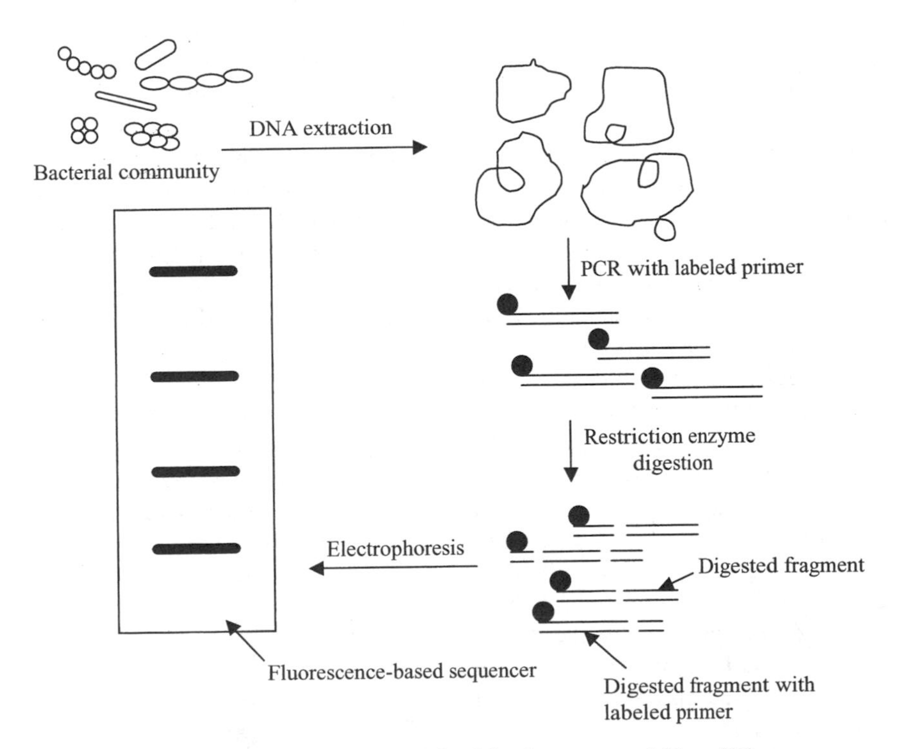

FIGURE 6.7 tRFLP principle. After Iwamoto and Nasu (55).

Syntrophus spp. and *Methanosaeta* spp. They hypothesized from this that the terminal step of hydrocarbon degradation in the methanogenic zone of the aquifer was aceticlastic methanogenesis, with these organisms existing in a syntrophic relationship.

MOLECULAR PROBES IN COMMUNITY ANALYSIS

Rather than providing a community fingerprint, molecular probe methods are used to probe for certain known community members. Some a priori knowledge of the population is required, or can be used to determine whether an organism is present. A molecular probe which hybridizes with the DNA or RNA of the specific organism is applied. The probe can be applied in situ or ex situ. These methods are powerful tools which avoid the possible biases of cloning and PCR amplification techniques and can yield a more direct measure of the target groups of interest.

DNA Microarrays

A DNA microarray (a DNA microchip or DNA chip) is an orderly, high-density matrix of hundreds (or thousands) of individual long cDNA probes or short oligonucleotides bound directly or indirectly to a solid surface (102). Unlike the membrane hybridization format, the chip is a high-density format that allows for simultaneous hybridization of a labeled DNA or RNA target to a large set of probes, thus providing high throughputs. Most applications have been in cell biology, such as for drug discovery or monitoring gene expression patterns in pure culture, but more recently the technology has been used successfully in environmental studies (48). Microarrays are a useful tool for the high-turnover screening of large numbers of samples. They are, however, rather expensive to construct.

Small et al. (109) reported the development of a simple microarray method for the detection of intact 16S rRNA from unpurified soil extracts, and this opened up the possibility of microbial detection without using PCR. This has been extended to the use of DNA microarrays for the detection of specific genes involved in bioremediation (36). Rhee et al. (91) have taken a combined approach, using DNA microarrays to monitor biodegrading populations and to detect biodegradation genes. This area is in its infancy, and more work on detection sensitivity is required, but the implication is that this may be a specific, sensitive, and, importantly, quantitative tool of the future for monitoring bioremediation.

Reverse Sample Genome Probing

Reverse sample genome probing makes use of the entire genome of a microorganism as a specific probe that allows its detection in the environment. Whole-genome genome probes have been used for the detection of *Mycobacterium*, *Mycoplasma*, *Chlamydia*, *Bacteroides*, *Pseudomonas*, *Sphingomonas*, and *Campylobacter* species. A good example of the use of this method was a search for sulfate-reducing organisms in contaminated oil fields (124).

FISH of Whole Cells

Whole-cell in situ hybridization with fluorescently labeled oligonucleotides, in community analysis, was first developed in the late 1980s (35). The procedure involves fixing the environmental sample to permeabilize the cells while maintaining their morphological integrity. The cells are then immersed in hybridization solution containing fluorescently labeled oligonucleotide. After being washed to remove unbound probe, the sample is viewed by epifluorescence microscopy and cells are counted. Limitations associated with the technique (e.g., poor sensitivity for target cells, background fluorescence) may be overcome by a variation that uses peptide nucleic acid (132), in which kinetics of hybridization, signal-to-noise ratio, and specificity were reported to be much improved.

Tani et al. (119) used PCR to detect the phenol hydroxylase gene and used FISH targeting 16S rRNA after injection of *Ralstonia eutropha* KT1 into a TCE-contaminated aquifer after activation with toluene. The numbers of bacteria detected by in situ PCR and FISH were similar up to 4 days, but thereafter FISH detected fewer bacteria than in situ PCR. In petroleum-contaminated aquifers, it has been assumed that dissimilatory sulfate reduction plays an important role. Kleikemper et al. (60) used FISH and DGGE to characterize the population of sulfate-reducing bacteria (SRB) in one such aquifer. Both techniques confirmed that sulfate reduction was enhanced by the addition of carbon source additions, and specific SRB genera and a high level of diversity of SRB were observed.

Labeled Substrates To Probe for Organisms

In addition to the community profiling methods and the use of specific oligonucleotide probes, substrate-based procedures that are culture independent and can be employed in situ have been developed. These methods in particular employ stable isotopes, such as ^{13}C, to determine exactly which organisms are involved in the breakdown of specific contaminants. These methods allow the identification of the organisms involved in the processing of the contaminating compound and are some of the most sophisticated methods used in the field.

PLFA-SIP

^{13}C-labeled substrate is pulsed into the microbial community occupying the environment, resulting in the labeling of polar lipid-derived fatty acids (PLFAs) from assimilating organisms. PLFAs are then extracted, separated, and analyzed for ^{13}C enrichment by isotope ratio MS. Because specific phylogenetic groups produce signature PLFA profiles, the stable isotope enrichment of certain PLFAs reveals which organisms were dominating the metabolism of the labeled substrate. The first PLFA-based stable isotope probing (PLFA-SIP) investigation was carried out by Boschker et al. (19), who identified microorganisms responsible for the oxidation of the greenhouse gas methane in a freshwater sediment environment.

DNA-SIP

DNA stable isotope probing (DNA-SIP) methods are more user-friendly because of sequence information being so widely available, as well as the ease of nucleic acid extraction from environmental samples. Stable isotope-labeled DNA can be isolated from mixed microbial communities on the basis of the increase in buoyant density associated with isotopic enrichment. Density centrifugation in cesium chloride gradients was used to separate heavy DNA from natural DNA, and 16S rDNA clone libraries constructed from heavy DNA were sequenced to obtain the identity of organisms assimilating the ^{13}C-labeled substrate used in the study. This technique has been used to attribute particular members of the *Proteobacteria* with methanol utilization in an oak forest soil (88).

RNA-SIP

Due to the fact that DNA synthesis is associated only with cell replication, the amounts of DNA synthesized in the duration of a ^{13}C pulse may be small in natural environments, where replication is not optimal. The use of RNA in SIP, however, offers the same sequence-based resolution but avoids the limitations of labeling due to its high turnover rate (69). The procedure is performed as for DNA-SIP, with the addition of reverse transcription of the heavy RNA in order to identify organisms. RNA-SIP has been applied to an operative industrial phenol-degrading wastewater treatment system to identify organisms responsible for the metabolism of phenol (70). Figure 6.8 shows the fractionation of natural and heavy ^{13}C-labeled RNA. The fractionation, extraction, and subsequent sequencing of the heavy RNA resulted in the identification of the dominant phenol degrader in a wastewater system as a *Thauera* species.

FISH AND SECONDARY ION MS

Recently, investigations in which individual cells or mixed aggregates of cells in methane-consuming communities were identified by FISH and subsequently analyzed for ^{13}C content by secondary ion MS were conducted (80). The natural abundance of ^{13}C in methane is low, thus enabling the association of

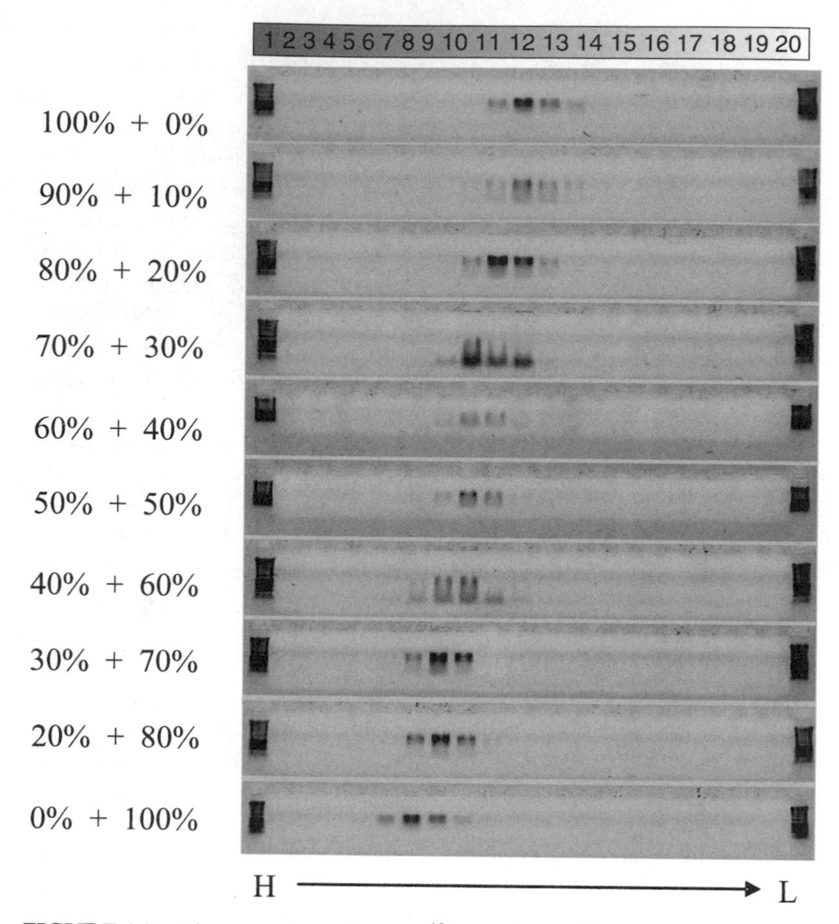

FIGURE 6.8 The appearance of natural (^{12}C) and heavy (^{13}C-labeled) RNA shown in RNA-SIP. The top gradient fractions contain the natural-weight RNA, and the heavy RNA becomes apparent as one moves down the fractions (68). H, heavy; L, light. Courtesy of Andrew S. Whiteley.

cells harboring depleted ^{13}C signatures with methane consumption. All the SIP techniques outlined above are very sophisticated methods of associating an organism with a specific function, namely, its catabolic potential. The one problem associated with SIP is that the labeled substrates are hard to come by and may require custom synthesis, which is a rather costly process.

Metabolic Biomarkers

Metabolic biomarkers provide some of the best evidence available so far for anaerobic in situ biodegradation. A good biomarker should (85):

- be formed during biodegradation of the target compound,
- be an intermediate in the biodegradation pathway,
- be highly specific to the process monitored,
- be water soluble, and
- be biodegradable (thereby informing on recent enzyme activity, not past).

The search for metabolic biomarkers has focused on environments where anaerobic conditions prevail, which is typical of most MNA processes (110). Arylsuccinate derivatives have emerged as potentially universal metabolic biomarkers for the anaerobic biodegradation of toluene, xylene, and ethylbenzene (e.g., see reference 90). The biggest challenge to this technique is that aerobic biodegradation pathways are much better known than anaerobic ones, and a great deal of data on anaerobic biodegradation remain to be acquired.

Detection and Quantification of Specific Catabolic Genes

Real-time PCR-based enumeration of specific catabolic genes may correlate with pollutant transformation rate (11). Competitive PCR can quantify specific genes by using a competitor sequence that is amplified with the same primers. This has been done to accurately quantify catechol 2,3-dioxygenase genes in petroleum-contaminated soil (72), and primer sets for various other mono- and dioxygenases have been used (9). The most obvious limitations are the incomplete record of anaerobic biodegradation pathways and the requirement of potentially very large numbers of primers due to the widely divergent gene sequences for enzymes with similar catabolic functions.

BIOSENSORS FOR TOXICITY AND DETECTION OF BIOAVAILABLE COMPOUNDS

Bioavailability is a central theme in bioremediation. The process of aging of pollutants in contaminated soils applies to organic pollutants in particular, but the metals also bind tightly to soil in the clay fraction. Thus, we have a contradiction for risk assessment: if these pollutants are tightly bound to soil, then they apparently do not constitute a risk. If the partitioning of organic pollutants to humus is an irreversible process, then the simple extraction of pollutants from soil by vigorous methods and subsequent pollutant concentration determination by analytical chemistry is missing a point: the analysis does not inform on the bioavailability of the pollutants. This has been the driving force for the development of a number of techniques that attempt to measure bioavailability. Some of the techniques are mentioned in chapter 1. This section concentrates on the development of whole-cell bioluminescent biosensors (Fig. 6.9) for toxicity testing and for the determination of bioavailable pollutants, because they offer the promises of being less expensive than other techniques and of being deployable in the field (Box 6.1).

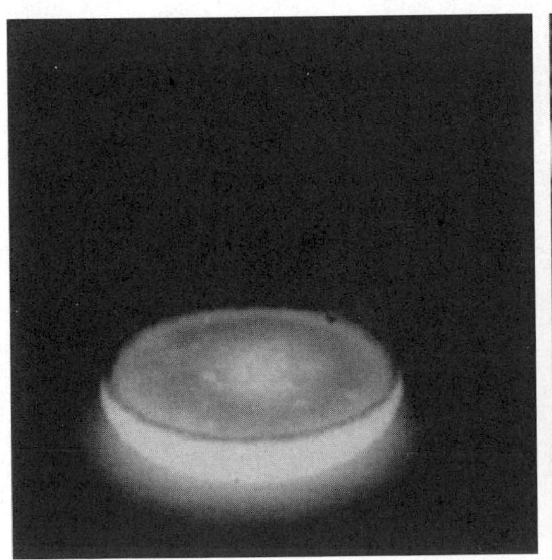

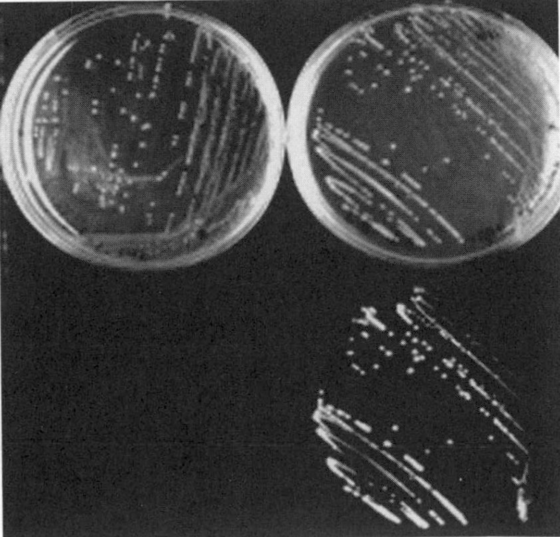

FIGURE 6.9 Bioluminescent bacteria. (a) A shake flask liquid culture. (b) Bioluminescent transposon mutants in daylight (top right) and in the dark room (bottom right). Courtesy of Andrew S. Whiteley.

BOX 6.1
Use of Toxicity Biosensors in Contaminated-Site Mapping

Toxicity biosensors have been proposed for a role in the assessment of contaminated sites. If the site can be mapped for toxicity before samples are sent for analytical chemistry, perhaps some expense can be saved by first identifying the areas of a site that need attention. The technique would involve sampling the soil or sediment in a grid and making extracts for exposure to biosensors. The data would then be plotted on a map of the grid to identify the pattern of toxicity on the site.

This approach was taken on a diesel-impacted site on the River Dee estuary, in the northeast of Scotland (83). Forty coordinates were selected and mapped in a 42-m^2 area within the contaminated zone by use of a differential global positioning system. Kriging is a gridding method that expresses trends suggested in the data across the site. Based on a kriging algorithm, contour plots of the distribution of toxicity data in the study area were produced with the Surfer 7.0 program. When toxicity results determined by the responses of a bioluminescent *E. coli* strain were mapped, defined zones of elevated toxicity could be seen within the site, but these did not correlate with other measured variables. In particular, there was no correlation with areas of elevated total petroleum hydrocarbon levels, indicating the low level of acute toxicity associated with such material.

In a similar manner, a site in the Midlands of the United Kingdom that is contaminated with hydrocarbons and metals was mapped by using toxicity biosensors. This approach was used in conjunction with a risk-based framework for site assessment. The site is close to a river, and the toxicity mapping (Box Fig. 6.1.1) showed that a gradient of toxicity was apparent on the site (shown in grayscale), radiating from one area of highest toxicity (black on the map). The lighter the shade of gray, the lower the toxicity.

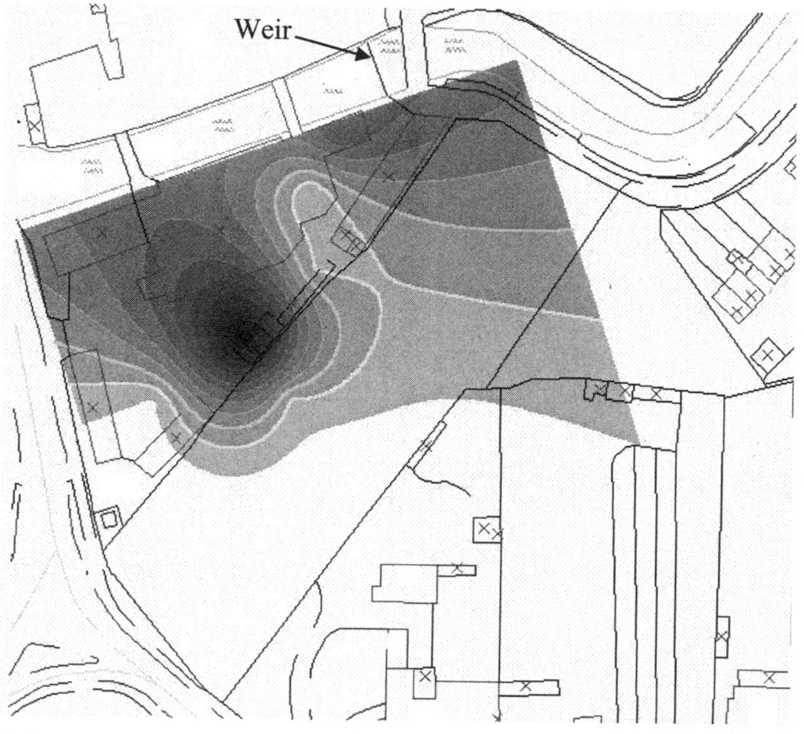

BOX FIGURE 6.1.1 Toxmap, a toxicity map for a contaminated site generated by using toxicity biosensors. Courtesy of Remedios Ltd., Aberdeen, United Kingdom.

Bioluminescence-Based Biosensors

Central to the development of these environmental biosensors was the use of bacterial bioluminescence as a surrogate measure of toxicity. The basis of the test was first described in 1979 (26). This test uses the naturally bioluminescent marine bacterium *Vibrio fischeri*. Of crucial importance to the commercial development of the bacterial bioluminescence toxicity tests was the fact that the change in light output is quantitatively proportional to the concentration of the toxicant. Thus, it was possible to develop quantitative tests and compare the results to those of toxicity tests with higher organisms.

This has obvious advantages over the use of higher organisms in acute toxicity tests, such as convenience, ethical considerations, and cost savings (122). The Microtox acute test has been the subject of many reproducibility evaluations. These studies have confirmed coefficient of variation values averaging 20%, which is better than for other whole-organism bioassays. Meanwhile, the Microtox system has developed to include the Microtox chronic test (25) and the Mutatox genotoxicity test (52). With the revolution in molecular biology and the ability to manipulate the nucleic acids of bacteria routinely, a major focus area is the development of genetically modified bacteria and other organisms for toxicity testing due to some specific limitations of the *V. fischeri* Lux system (129).

Genetically modified microorganisms can also be used to detect and quantify specific pollutants. This type of biosensor is based on the highly specific genetic control mechanisms used by microorganisms to ensure that specific proteins are expressed only when they are needed, for example, for the detoxification of a particular toxic substance. This control is exerted by inducible promoters, consisting of a specific DNA sequence upstream of the genes to be controlled, and a DNA-binding protein that either activates or prevents transcription in response to the presence or absence of the target compound. Biosensors of this type are easily generated by fusing such a controllable promoter to a reporter gene, which generates a detectable signal when the promoter is activated. (For reviews, see references 10, 50, and 34.)

Testing of contaminated water lends itself well to these biosensors, but it should be stressed that a major barrier to their development with contaminated soils has been the meaningful extraction of the pollutants from the soils. This ranges from a simple water shake to the use of nontoxic, mild solvents. However, this methodology has yet to be accepted as a standard technique.

Reporter Genes

A reporter gene may encode an enzyme that generates a color change in the presence of a chromogenic substrate. The best example is β-galactosidase (lactase, LacZ, encoded by the reporter gene *lacZ*, and usually derived from *Escherichia coli*), which hydrolyzes X-Gal (5-bromo-4-chloro-3-indolyl-β-D-galactopyranoside), releasing an indoxyl derivative, which dimerizes in the presence of oxygen to form a blue pigment related to indigo. This can be developed as an agar plate assay as a "detection stick" (61) or can be quantified colorimetrically (74). A major disadvantage of *lac* genes, however, is that there is significant background β-galactosidase in many bacteria.

Another example is the *xylE* gene, encoding catechol 2,3-dioxygenase (metapyrocatechase, XylE, often derived from *Pseudomonas putida*), which acts on catechol to produce a fluorescent yellow compound (2-hydroxy-*cis,cis*-muconic semialdehyde) (133). Unlike β-galactosidase, there is little problem from background activity, as the *xylE* gene is rarely found in microorganisms that have not been exposed to aromatic hydrocarbons (71). Unfortunately, other factors such as oxygen and cell physiology can interfere, thus increasing the technical difficulty in working with it as a reporter.

Bioluminescence represents the best candidate available for environmental reporting, for several reasons. In practical terms, light production is easy to detect and quantify. Most importantly, there is an exceedingly low likelihood of background interference because bioluminescence in bacteria is very rare and restricted to a small number of known habitats. As a result of the discovery of bacterial extracellular signaling (quorum sensing) (39) in bioluminescent bacteria, the genetics of bioluminescence, particularly in *V. fischeri*, is reasonably well characterized and the transfer of the bioluminescence phenotype to other bacteria has become routine.

Although more is known about *V. fischeri* bioluminescence, in practice this luciferase has several disadvantages compared to others. When expressed in *E. coli*, it is unstable at temperatures above 30°C, whereas the *Vibrio harveyi* luciferase is stable up to 37°C (29). The most thermostable luciferase, however, comes from *Photorhabdus luminescens* (formerly *Xenorhabdus luminescens*) (14), an entomopathogenic symbiont with soil nematodes, with a half-life of 3 h at 45°C (31). This can be a matter of great import in transfer of the phenotype to other bacteria, while other disadvantages are specific to toxicity testing. Sample data are shown in Fig. 6.10, in which a genetically modified bacterium expressing the *Photorhabdus luminescens luxCDABE* genes constitutively was exposed to increasing concentrations of 3,5-dichlorophenol. Increasing chlorination of the phenol ring increases the toxicity of the halogenated phenols by a variety of mechanisms (86).

Contaminant-Specific Biosensors

MerR AND RELATED SYSTEMS

Mercuric salts are likewise highly toxic due to their affinity for sulfhydryl groups on proteins; ingestion by humans, after concentration through the food chain, results in neurotoxicity. In many bacteria, mercuric salts are detoxified by mercuric reductase, MerA, which reduces mercuric ions to volatile elemental mercury, which can diffuse out of the cell (Fig. 6.11). The promoter for the gene *merA* is controlled by the activator protein, MerR, which is able to bind mercuric ions. In contrast to ArsR-like repressors, MerR binds to its operator site in both the absence and the presence of mercuric ions; however, when bound to mercuric ions, MerR undergoes a conformational change which distorts the DNA and greatly increases the affinity of the promoter for RNA polymerase, thus activating transcription. Numerous reports have described reporter organisms using this mercury-activated promoter system from various sources. Bontidean et al. (17) measured the concentration of mercury in contaminated soil, using a protein-based biosensor (16), a bacterial cell-based biosensor, and a plant sensor. Both protein and

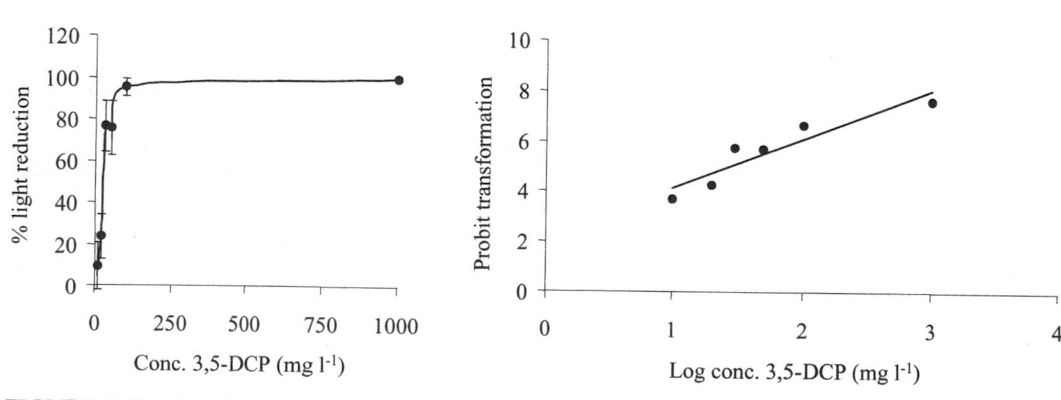

FIGURE 6.10 Sample output data for a genetically modified toxicity biosensor. On contact with 3,5-dichlorophenol (3,5-DCP), this biosensor shows increasing light reduction with concentration (conc.), which can be transformed to a linear plot by probit transformation, allowing the user to input data to software to compute an effective concentration that reduces light output by 50%. After Philp et al. (86).

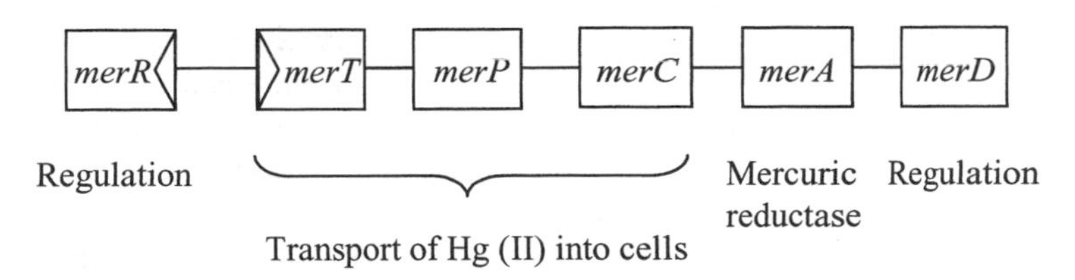

FIGURE 6.11 Mercury detoxification system.

bacterial cell sensors gave accurate mercury concentration responses proportional to the total concentration measured by atomic absorption, while the plant sensor did not.

Many other detoxification systems are controlled by similar mechanisms (24). For example, in *E. coli*, the major chromosomal copper and zinc efflux pumps, CopA and ZntA, are controlled by the activator proteins CopR (CueR) and ZntR, both of which are related to MerR (13, 22, 81, 84, 116). Both of these promoters have been used in luminescent biosensor organisms (for example, see reference 93). Other examples include lead resistance in *R. eutropha* CH34 (18) and cadmium resistance in *P. putida* (64). Many other metal-regulated promoters are known (for reviews, see references 106 and 107). Many are related to either ArsR or MerR; others are of unknown mechanism.

NAPHTHALENE DETECTION

The same principle can be applied to certain organic pollutants. In this case, the promoters used are generally not related to detoxification, but rather related to biodegradation of organic compounds as sources of carbon. The most fruitful sources of regulated degradative operons have been soil bacteria such as *Pseudomonas* spp. One well-studied example is the promoter of the *nah* operon of *Pseudomonas* catabolic plasmids (Fig. 6.12). Fusions of *lux* to this promoter induce luminescence in response to naphthalene, salicylate, and some other aromatic compounds (27). One such example, *Pseudomonas fluorescens* HK44, has been released into the field for in situ monitoring of a bioremediation process (94); the authors described this as the first genetically modified microorganism approved for field

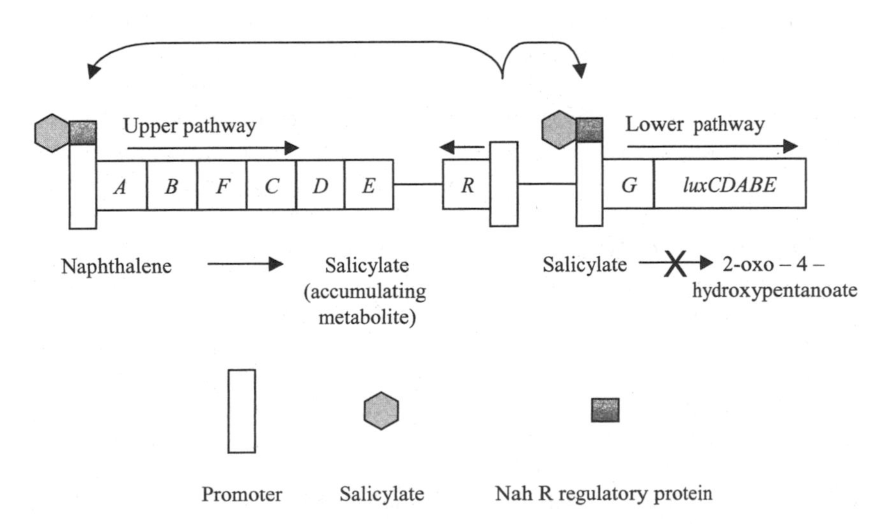

FIGURE 6.12 The naphthalene regulatory system of *P. fluorescens* HK44. After Simpson et al. (108).

testing in bioremediation experiments in the United States.

OTHER ORGANICS

Other examples include detection of hydrophobic compounds by using the promoter of the *P. putida* isopropylbenzene catabolism operon (101) and detection of bioavailable BTEX compounds by using the promoter of a toluene degradative operon (5, 115). Detection of chlorobenzoic acid by using the *fcbA* promoter from an *Arthrobacter* sp. (98), detection of chlorocatechol by using the *clc* promoter from *P. putida* (47), and detection of phenol by using an *Acinetobacter* sp. (1) have been reported.

Toxicity Biosensors for Contaminated Soils

To overcome the lack of availability, a biosensor to detect toxicity in soils associated with PAH was developed to include a biosurfactant (46). The biosurfactant was shown to enhance the bioavailability of phenanthrene via an increase in the rate of mass transfer from soil to the aqueous phase, and the concentration of phenanthrene in the aqueous phase was found to correlate well with the corresponding toxicity data. The authors opined that this biosensor system may be applied as an in situ toxicity detector in soils. In a later study (28), they included transparent glass beads in the immobilization matrix, which is reported to enhance the stability and sensitivity of the biosensor. Their minimum detectable concentration of phenanthrene in soils was approximately equal to 30 mg kg^{-1}.

Remote Sensing in Contaminated Aquifers

The construction of a bioluminescent bioreporter integrated circuit (BBIC) (108) opened up the possibility that bioluminescent biosensors could be used in locations remote from the place of sensor deployment. A BBIC is a device consisting of a bioreporter(s) sustained within a controlled environment, an integrated circuit microluminometer, and a light-tight enclosure (79). Complementary metal oxide semiconductor technology has the potential to miniaturize bioluminescent biosensors, and the development of an off-chip wireless BBIC probe allows the transmission of data over relatively long distances (e.g., around 100 m into a metal-structured building) (79). This technology may allow future deployment of bioluminescent toxicity or detection biosensors into contaminated aquifers.

Biosensor or bioreporter systems based on genetically modified microorganisms provide a versatile and sensitive method for the detection of almost any stimulus which can cause a biological response and are thus in principle highly suited to determination of environmental toxicity. As Belkin (10) has pointed out, chemical analysis can always provide a more accurate quantification of any given chemical species, but biosensors have the advantage of detecting only the bioavailable portion. However, potential problems related to specificity of induction, the limited range of concentrations over which a linear response is likely to be obtained, innate variability and difficulties related to the use of living organisms, and regulatory factors associated with genetically modified microorganisms must be borne in mind. Increasing field experience with biosensors of this type should clarify these issues within the next few years.

CONCLUDING REMARKS

Given that the success of bioremediation depends upon meeting specific performance criteria and demonstrating that the reduction of contaminant concentrations to safe levels is the result of biological activities, monitoring is a critical aspect of all bioremediation projects. This is especially true for MNA, in which the monitoring is the crux of the project. It is also key for bioaugmentation, in which demonstrating the activity of the added organisms in the environment is important. Modern monitoring approaches employ both traditional chemical analyses and a variety of biological methods, ranging from community structure analyses to the use of reporter genes that can visually demonstrate the presence of specific microorganisms and their metabolic activities in field samples. By employing such monitoring tools, it is possible to manage and design more effective bioremediation technologies. Thus, the further development and application of monitoring methodologies will be critical to the future development of bioremediation and its broader application.

REFERENCES

1. **Abd-El-Haleem, D., S. Ripp, C. Scott, and G. S. Sayler.** 2002. A *luxCDABE*-based bioluminescent bioreporter for the detection of phenol. *J. Ind. Microbiol. Biotechnol.* **29:**233–237.
2. **Alexander, M.** 1999. *Biodegradation and Bioremediation*, 2nd ed. Academic Press, San Diego, Calif.
3. **Alley, J. F., and L. R. Brown.** 2000. Use of sublimation to prepare solid microbial media with water-insoluble substrates. *Appl. Environ. Microbiol.* **66:**439–442.
4. **Amann, R. I., W. Ludwig, and K. H. Schleifer.** 1995. Phylogenetic identification and *in situ* detection of individual microbial cells without cultivation. *FEMS Microbiol. Rev.* **59:** 143–169.
5. **Applegate, B. M., S. R. Kehrmeyer, and G. S. Sayler.** 1998. A chromosomally based *tod-luxCDABE* whole-cell reporter for benzene, toluene, ethybenzene, and xylene (BTEX) sensing. *Appl. Environ. Microbiol.* **64:**2730–2735.
6. **Ashraf, W., A. Mihdhir, and J. C. Murrell.** 1994. Bacterial oxidation of propane. *FEMS Microbiol. Lett.* **122:**1–6.
7. **Atlas, R. M., and R. Bartha.** 1973. Abundance, distribution and oil biodegradation potential of microorganisms in Raritan bay. *Environ. Pollut.* **4:**291–300.
8. **Ayala-del-Río, H. L., S. J. Callister, C. S. Criddle, and J. M. Tiedje.** 2004. Correspondence between community structure and function in phenol- and phenol-plus-trichloroethene-fed sequencing batch reactors. *Appl. Environ. Microbiol.* **70:**4950–4960.
9. **Baldwin, B. R., C. H. Nakatsu, and L. Nies.** 2003. Detection and enumeration of aromatic oxygenase genes by multiplex and real-time PCR. *Appl. Environ. Microbiol.* **69:**3350–3358.
10. **Belkin, S.** 2003. Microbial whole-cell sensing systems of environmental pollutants. *Curr. Opin. Microbiol.* **6:**206–212.
11. **Beller, H. R., S. R. Kane, T. C. Legler, and P. J. Alvarez.** 2002. A real-time polymerase chain reaction method for monitoring anaerobic, hydrocarbon-degrading bacteria based on a catabolic gene. *Environ. Sci. Technol.* **36:**3977–3984.
12. **Bellon-Maurel, W., O. Orliac, and P. Christen.** 2003. Sensors and measurements in solid state fermentation: a review. *Proc. Biochem.* **38:** 881–896.
13. **Binet, M. R. B., and R. K. Poole.** 2000. Cd(II), Pb(II) and Zn(II) ions regulate expression of the metal-transporting P-type ATPase ZntA in *Escherichia coli*. *FEBS Lett.* **473:**67–70.
14. **Boemare, N. E., R. J. Akhurst, and R. G. Mourant.** 1993. DNA relatedness between *Xenorhabdus* spp. (Enterobacteriaceae), symbiotic bacteria of entomopathogenic nematodes, and a proposal to transfer *Xenorhabdus luminescens* to a new genus, *Photorhabdus* gen. nov. *Int. J. Syst. Bacteriol.* **43:**249–255.
15. **Bogardt, A. H., and B. B. Hemmingsen.** 1992. Enumeration of phenanthrene-degrading bacteria by an overlayer technique and its use in evaluation of petroleum-contaminated sites. *Appl. Environ. Microbiol.* **58:**2579–2582.
16. **Bontidean, I., C. Berggren, G. Johansson, E. Csöregi, B. Mattiason, J. R. Lloyd, K. J. Jakeman, and N. L. Brown.** 1998. Detection of heavy metal ions at femtomolar levels using protein-based biosensors. *Anal. Chem.* **70:** 4162–4169.
17. **Bontidean, I., A. Mortari, S. Leth, N. L. Brown, U. Karlson, M. M. Larsen, J. Vangronsveld, P. Corbisier, and E. Csöregi.** 2004. Biosensors for detection of mercury in contaminated soils. *Environ. Pollut.* **131:** 255–262.
18. **Borremans, B., J. L. Hobman, A. Provoost, N. L. Brown, and D. van der Lelie.** 2001. Cloning and functional analysis of the *pbr* lead resistance determinant of *Ralstonia metallidurans* CH34. *J. Bacteriol.* **183:**5651–5658.
19. **Boschker, H. T. S., S. C. Nold, P. Wellsbury, D. Bos, W. de Graaf, R. Pel, R. J. Parkes, and T. E. Cappenberg.** 1998. Direct linking of microbial populations to specific biogeochemical processes by ^{13}C-labelling of biomarkers. *Nature* **392:**801–805.
20. **Bouchez, M., D. Blanchet, and J. P. Vandecasteele.** 1995. Substrate availability in phenanthrene biodegradation: transfer mechanism and influence on metabolism. *Appl. Microbiol. Biotechnol.* **43:**952–960.
21. **Braddock, J. F., and P. H. Catterall.** 1999. A simple method for enumerating gasoline- and diesel-degrading microorganisms. *Bioremed. J.* **3:** 81–84.
22. **Brocklehurst, K. R., J. L. Hobman, B. Lawley, L. Blank, S. J. Marshall, N. L. Brown, and A. P. Morby.** 1999. ZntR is a Zn(II)-responsive MerR-like transcriptional regulator of zntA in *Escherichia coli*. *Mol. Microbiol.* **31:** 893–902.

23. **Brown, E. J., and J. F. Braddock.** 1990. Sheen screen, a miniaturized most-probable-number method for enumeration of oil-degrading microorganisms. *Appl. Environ. Microbiol.* **56:** 3895–3896.
24. **Brown, N. L., J. V., Stoyanov, S. P. Kidd, and J. L. Hobman.** 2003. The MerR family of transcriptional regulators. *FEMS Microbiol. Rev.* **27:**145–163.
25. **Bulich, A. A., and G. Bailey.** 1995. Environmental toxicity assessment using luminescent bacteria, p. 29. *In* M. Richardson (ed.), *Environmental Toxicology Assessment.* Taylor & Francis, London, United Kingdom.
26. **Bulich, A. A., and D. L. Isenberg.** 1981. Use of the luminescent bacterial system for the rapid assessment of aquatic toxicity. *ISA Transact.* **20:** 29–33.
27. **Burlage, R. S., G. S. Sayler, and F. Larimer.** 1990. Monitoring of naphthalene catabolism by bioluminescence with *nah-lux* transcriptional fusions. *J. Bacteriol.* **172:**4749–4757.
28. **Chang, S. T., H. J. Lee, and M. B. Gu.** 2004. Enhancement in the sensitivity of an immobilized cell-based soil biosensor for monitoring PAH toxicity. *Sensors Actuators* B **97:**272–276.
29. **Chatterjee, J., and E. A. Meighan.** 1995. Biotechnological applications of bacterial bioluminescence (*lux*) genes. *Photochem. Photobiol.* **62:** 641–650.
30. **Ciric, L., A. S. Whiteley, M. Manefield, R. Griffiths, J. C. Philp, and M. J. Bailey.** 2003. Approaches for analyzing the diversity of diesel degrading microbial communities, p. 449–452. *In Proceedings of the 2nd European Bioremediation Conference*, 30 June to 4 July, Chania, Crete.
31. **Colepicolo, P., K. Cho, G. O. Poinar, and J. W. Hastings.** 1989. Growth and luminescence of the bacterium *Xenorhabdus luminescens* from a human wound. *Appl. Environ. Microbiol.* **55:**2601–2606.
32. **Colwell, R. R.** 1978. Enumeration of specific populations by the most probable number (MPN) method, p. 56–61. *In* J. W. Costerton and R. R. Colwell (ed.), *Native Aquatic Bacteria: Enumeration, Activity and Ecology.* ASTM, Philadelphia, Pa.
33. **Cox, C. B., and P. D. Moore.** 2000. *Biogeography: An Ecological and Evolutionary Approach*, 6th ed. Blackwell Science, Oxford, United Kingdom.
34. **Daunert, S., G. Barrett, J. S. Feliciano, R. S. Shetty, S. Shrestha, and W. Smith-Spence.** 2000. Genetically engineered whole-cell sensing systems: coupling biological recognition with reporter genes. *Chem. Rev.* **100:**2705–2738.
35. **DeLong, E. F., G. S. Wickham, and N. R. Pace.** 1989. Phylogenetic strains: ribosomal RNA-based probes for the identification of single microbial cells. *Science* **243:**1360–1363.
36. **Dennis, P., E. A. Edwards, S. N. Liss, and R. Fulthorpe.** 2003. Monitoring gene expression in mixed microbial communities by using DNA microarrays. *Appl. Environ. Microbiol.* **69:** 769–778.
37. **Diaz, M. P., S. J. W. Grigson, C. J. Peppiatt, and J. G. Burgess.** 2000. Isolation and characterization of novel hydrocarbon-degrading euryhaline consortia from crude oil and mangrove sediments. *Mar. Biotechnol.* **2:**522–532.
38. **Dojka, M. A., P. Hugenholtz, S. K. Haack, and N. R. Pace.** 1998. Microbial diversity in a hydrocarbon- and chlorinated-solvent-contaminated aquifer undergoing intrinsic bioremediation. *Appl. Environ. Microbiol.* **64:**3869–3877.
39. **Dunny, G. M., and S. C. Winans.** 1999. Bacterial life: neither lonely nor boring, p. 1. *In* G. M. Dunnyand and S. C. Winans (ed.), *Cell-Cell Signaling in Bacteria.* American Society for Microbiology Press, Washington, D. C.
40. **Dupont, R. R., W. J. Doucette, and R. E. Hinchee.** 1991. *In situ* bioremediation potential and the application of bioventing at a fuel-contaminated site, p. 262–282. *In* R. E. Hinchee and R. G. Olfenbuttle (ed.), In Situ *Bioreclamation. Applications and Investigations for Hydrocarbon and Contaminated Site Remediation.* Butterworth-Heinemann, Boston, Mass.
41. **Ensley, B. D., B. J. Ratzkin, T. D. Osslund, and M. J. Simon.** 1983. Expression of naphthalene oxidation genes in *Escherichia coli* results in the biosynthesis of indigo. *Science* **222:**167–169.
42. **Friedrich, M., R. J. Grosser, E. A. Kern, W. P. Inskeep, and D. M. Ward.** 2000. Effect of model sorptive phases on phenanthrene biodegradation: molecular analysis of enrichments and isolates suggests selection based on bioavailability. *Appl. Environ. Microbiol.* **66:**2703–2710.
43. **Frontera-Suau, R., F. D. Bost, T. J. McDonald, and P. J. Morris.** 2002. Aerobic biodegradation of hopanes and other biomakers by crude oil-degrading enrichment cultures. *Environ. Sci. Technol.* **36:**4585–4592.
44. **Gejlsbjerg, B., T. Madsen, and T. T. Andersen.** 2003. Comparison of biodegradation of surfactants in soils and sludge-soil mixtures by use of C-14-labelled compounds and automated respirometry. *Chemosphere* **50:**321–331.
45. **Goodhue, C. T.** 1982. The methodology of microbial transformation of organic compounds, p. 9–44. *In* J. P. Rosazza (ed.), *Microbial Transformation of Bioactive Agents*, vol. 1. CRC Press, Boca Raton, Fla.

46. **Gu, M. B., and S. T. Chang.** 2001. Soil biosensor for the detection of PAH toxicity using an immobilized recombinant bacterium and a biosurfactant. *Biosens. Bioelectron.* **16:**667–674.
47. **Guan, X., S. Ramanathan, J. P. Garris, R. S. Shetty, M. Ensor, L. G. Bachas, and S. Daunert.** 2000. Chlorocatechol detection based on a *clc* operon/reporter gene system. *Anal. Chem.* **72:**2423–2427.
48. **Guschin, D. Y., B. K. Mobarry, D. Proudnikov, D. A. Stahl, B. E. Rittmann, and A. D. Mirzabekov.** 1997. Oligonucleotide microchips as genosensors for determinative and environmental studies in microbiology. *Appl. Environ. Microbiol.* **63:**2397–2402.
49. **Haines, J. R., B. A. Wrenn, E. L. Holder, K. L. Strohmeier, R. T. Herrington, and A. D. Venosa.** 1996. Measurement of hydrocarbon-degrading microbial populations by a 96-well plate most-probable-number procedure. *J. Ind. Microbiol.* **16:**36–41.
50. **Hansen, L. H., and S. J. Sorensen.** 2001. The use of whole-cell biosensors to detect and quantify compounds or conditions affecting biological systems. *Microb. Ecol.* **42:**483–494.
51. **Harayama, S., Y. Kasai, and A. Hara.** 2004. Microbial communities in oil-contaminated seawater. *Curr. Opin. Biotechnol.* **15:**205–214.
52. **Hauser, B., G. Schrader, and M. Bahadir.** 1997. Comparison of acute toxicity and genotoxic concentrations of single compounds and waste elutriates using the Microtox/Mutatox test system. *Ecotoxicol. Environ. Safety* **38:**227–231.
53. **Horowitz, A., and R. M. Atlas.** 1978. Crude oil degradation in the Arctic: changes in bacterial populations and oil composition during one year exposure in a model system. *Dev. Ind. Microbiol.* **19:**517–522.
54. **Itävaara, M., and M. Vikman.** 1996. An overview of methods for biodegradability testing of biopolymers and packaging materials. *J. Environ. Poly. Degrad.* **4:**29–36.
55. **Iwamoto, T., and M. Nasu.** 2001. Current bioremediation practice and perspective. *J. Biosci. Bioeng.* **92:**1–8.
56. **Jorgensen, K. S., J. Puustinen, and A. M. Suortti.** 2000. Bioremediation of petroleum hydrocarbon-contaminated soil by composting in biopiles. *Environ. Pollut.* **107:**245–254.
57. **Keith, L. H.** 1996. *EPA's Sampling and Analysis Methods.* CRC Press, Boca Raton, Fla.
58. **Kirk, J. L., L. A. Beaudette, M. Hart, P. Moutoglis, J. N. Klironomos, H. Lee, and J. T. Trevors.** 2004. Methods of studying soil microbial diversity. *J. Microbiol. Methods* **58:** 169–188.
59. **Kiyohara, H., K. Nagao, and K. Yano.** 1982. Rapid screen for bacteria degrading water-insoluble, solid hydrocarbons on agar plates. *Appl. Environ. Microbiol.* **43:**454–457.
60. **Kleikemper, J., M. H. Schroth, W. V. Sigler, M. Schmucki, S. M. Bernasconi, and J. Zeyer.** 2002. Activity and diversity of sulfate-reducing bacteria in a petroleum hydrocarbon-contaminated aquifer. *Appl. Environ. Microbiol.* **68:**1516–1523.
61. **Klein, J., J. Altenbuchner, and R. Mattes.** 1997. Genetically modified *Escherichia coli* for colorimetric detection of inorganic and organic Hg compounds, p. 133. *In* F. W. Scheller, F. Schubert, and J. Fedrowitz (ed.), *Frontiers in Biosensorics I.* Birkhäuser Verlag, Basel, Switzerland.
62. **Kleinheinz, G. T., and S. T. Bagley.** 1997. A filter-plate method for the recovery and cultivation of microorganisms utilizing volatile organic compounds. *J. Microbiol. Methods* **29:** 139–144.
63. **Kolhatkar, R., T. Kuder, P. Philp, J. Allen, and J. T. Wilson.** 2002. Use of compound-specific stable carbon isotope analyses to demonstrate anaerobic biodegradation of MTBE in groundwater at a gasoline release site. *Environ. Sci. Technol.* **36:**5139–5146.
64. **Lee, S. W., E. Glickman, and D. A. Cooksey.** 2001. Chromosomal locus for cadmium resistance in *Pseudomonas putida* consisting of a cadmium-transporting ATPase and a MerR family response regulator. *Appl. Environ. Microbiol.* **67:** 1437–1444.
65. **Liu, W. T., T. L. Marsh, H. Cheng, and L. J. Forney.** 1997. Characterization of microbial diversity by determining terminal restriction fragment length polymorphisms of genes encoding 16S rRNA. *Appl. Environ. Microbiol.* **63:** 4516–4522.
66. **Lors, C., and J. R. Mossmann.** 2004. Contribution of microcosm and respirometric experiments to PAHS' intrinsic biodegradation in the soil of a former coke works. *Polycyc. Aromat. Comp.* **24:**91–105.
67. **Mancini, S. A., A. C. Ulrich, G. Lacrampe-Couloume, B. Sleep, E. A. Edwards, and B. S. Lollar.** 2003. Carbon and hydrogen isotopic fractionation during anaerobic biodegradation of benzene. *Appl. Environ. Microbiol.* **69:**191–198.
68. **Manefield, M., A. S. Whiteley, N. Ostle, P. Ineson, and M. J. Bailey.** 2002. Technical considerations for RNA-based stable isotope probing: an approach in associating microbial diversity with microbial function. *Rapid Commun. Mass Spectrom.* **16:**2179–2183.
69. **Manefield, M., A. S. Whiteley, and M. J. Bailey.** 2004. What can stable isotope probing

do for bioremediation? *Int. Biodeterior. Biodegrad.* **54:**163–166.

70. **Manefield, M., A. S. Whiteley, R. I. Griffiths, and M. J. Bailey.** 2002. RNA stable isotope probing: a novel means of linking microbial community function to phylogeny. *Appl. Environ. Microbiol.* **68:**5367–5373.

71. **Marlowe, E. M., K. L. Josephson, and I. L. Pepper.** 2000. Nucleic acid-based methods of analysis, p. 287–318. *In* R. M. Maier, I. L. Pepper, and C. P. Gerba (ed.), *Environmental Microbiology.* Academic Press, San Diego, Calif.

72. **Mesarch, M. B., C. H. Nakatsu, and L. Nies.** 2000. Development of catechol 2,3-dioxygenase-specific primers for monitoring bioremediation by competitive quantitative PCR. *Appl. Environ. Microbiol.* **66:**678–683.

73. **Miles, R. A., and W. J. Doucette.** 2001. Assessing the aerobic biodegradability of 14 hydrocarbons in two soils using a simple microcosm/respiration method. *Chemosphere* **45:**1085–1090.

74. **Miller, J. H.** 1972. *Experiments in Molecular Genetics.* Cold Spring Harbor Laboratory, Cold Spring Harbor, N. Y.

75. **Mills, A. L., C. Breuil, and R. R. Colwell.** 1978. Enumeration of petroleum-degrading marine and estuarine microorganisms by the most probable number method. *Can. J. Microbiol.* **24:** 552–557.

76. **Moller, J., P. Winther, B. Lund, K. Kirkebjerg, and P. Westermann.** 1996. Bioventing of diesel oil-contaminated soil: comparison of degradation rates in soil based on actual oil concentration and on respirometric data. *J. Ind. Microbiol.* **16:**110–116.

77. **Muyzer, G., E. C. De Wall, and A. G. Uitterlinden.** 1993. Profiling of complex microbial populations by denaturing gradient gel electrophoresis analysis of polymerase chain reaction-amplified genes coding for 16S rRNA. *Appl. Environ. Microbiol.* **59:**695–700.

78. **Myers, R. M., T. Maniatis, and L. S. Lerman.** 1987. Detection of localisation of single base changes by denaturing gradient gel electrophoresis. *Methods Enzymol.* **155:**501–527.

79. **Nivens, D. E., T. E. McKnight, S. A. Moser, S. J. Osbourn, M. L. Simpson, and G. S. Sayler.** 2004. Bioluminescent bioreporter integrated circuits: potentially small, rugged and inexpensive whole-cell biosensors for remote environmental monitoring. *J. Appl. Microbiol.* **96:** 33–46.

80. **Orphan, V. J., C. H. House, K. U. Hinrichs, K. D. McKeegan, and E. F. DeLong.** 2001. Methane-consuming Archaea revealed by directly coupled isotopic and phylogenetic analysis. *Science* **293:**484–486.

81. **Outten, F. W., C. E. Outten, J. Hale, and T. V. O'Halloran.** 2000. Transcriptional activation of an *Escherichia coli* copper efflux regulon by the chromosomal MerR homologue, CueR. *J. Biol. Chem.* **275:**31024–31029.

82. **Padmanabhan, P., S. Padmanabhan, C. DeRito, A. Gray, D. Gannon, J. R. Snape, C. S. Tsai, W. Park, C. Jeon, and E. L. Madsen.** 2003. Respiration of ^{13}C-labeled substrates added to soil in the field and subsequent 16S rRNA gene analysis of ^{13}C-labeled soil DNA. *Appl. Environ. Microbiol.* **69:**1614–1622.

83. **Paton, G. I., C. O. Iroegbu, and J. J. C. Dawson.** 2003. Microbiological characterisation of a diesel contaminated beach site. *Mar. Pollut. Bull.* **46:**903–906.

84. **Petersen, C., and L. B. Moller.** 2000. Control of copper homeostasis in *Escherichia coli* by a P-type ATPase, CopA, and a MerR-like transcriptional activator, CopR. *Gene* **261:**289–298.

85. **Phelps, C. D., J. Battistelli, and L. Y. Young.** 2002. Metabolic biomarkers for monitoring anaerobic naphthalene biodegradation *in situ. Environ. Microbiol.* **4:**532–537.

86. **Philp, J. C., C. French, S. Wiles, J. M. L. Bell, A. S. Whiteley, and M. J. Bailey.** 2004. Wastewater toxicity assessment by whole cell biosensor, p. 165–225. *In* D. Barceló (ed.), *Handbook of Environmental Chemistry*, vol. 5. *Water Pollution: Emerging Organic Pollutants in Wastewaters.* Springer-Verlag, Berlin, Germany.

87. **Prince, R. C., R. Bare, R. Garrett, M. Grossman, C. Haith, L. Keim, K. Lee, G. Holtom, P. Lambert, G. Sergy, E. Owens, and C. Guénette.** 1999. Bioremediation of a marine oil spill in the Arctic, p. 227–232. *In* B. C. Alleman and A. Leeson (ed.), In Situ *Bioremediation of Petroleum Hydrocarbons and Other Organic Compounds.* Battelle Press, Columbus, Ohio.

88. **Radajewski, S., P. Ineson, N. R. Parekh, and J. C. Murrell.** 2000. Stable-isotope probing as a tool in microbial ecology. *Nature* **403:** 646–649.

89. **Randall, J. D., and B. B. Hemmingsen.** 1994. Evaluation of mineral agar plates for the enumeration of hydrocarbon-degrading bacteria. *J. Microbiol. Methods* **20:**103–113.

90. **Reusser, D. E., J. D. Istok, H. R. Beller, and J. A. Field.** 2002. *In situ* transformation of deuterated toluene and xylene to benzylsuccinic acid analogues in BTEX-contaminated aquifers. *Environ. Sci. Technol.* **36:**4127–4134.

91. **Rhee, S.-K., X. Liu, L. Wu, S. C. Chong, X. Wan, and J. Zhou.** 2004. Detection of genes involved in biodegradation and biotransformation in microbial communities by using 50-mer

oligonucleotide microarrays. *Appl. Environ. Microbiol.* **70:**4303–4317.

92. **Rice, L. E., and B. B. Hemmingsen.** 1997. Enumeration of hydrocarbon-degrading bacteria, p. 99–109. *In* D. Sheehan (ed.), *Bioremediation Protocols*. Humana Press, Totawa, N. J.

93. **Riether, K. B., M. A. Dollard, and P. Billard.** 2001. Assessment of heavy metal bioavailability using *Escherichia coli zntAp*::*lux* and *copAp*::*lux*-based biosensors. *Appl. Microbiol. Biotechnol.* **57:**712–716.

94. **Ripp, S., D. E. Nivens, Y. Ahn, C. Werner, J. Jarrel, J. P. Easter, C. D. Cox, R. S. Burlage, and G. S. Sayler.** 2000. Controlled field release of a bioluminescent genetically engineered microorganism for bioremediation process monitoring and control. *Environ. Sci. Technol.* **34:**846–853.

95. **Röling, W. F. M., M. G. Milner, D. M. Jones, K. Lee, F. Daniel, R. J. P. Swannell, and I. M. Head.** 2002. Robust hydrocarbon degradation and dynamics of bacterial communities during nutrient-enhanced oil spill bioremediation. *Appl. Environ. Microbiol.* **68:**5537–5548.

96. **Röling, W. F. M., M. G. Milner, D. M. Jones, F. Fratepietro, R. P. J. Swannell, F. Daniel, and I. M. Head.** 2004. Bacterial community dynamics and hydrocarbon degradation during a field-scale evaluation on a mudflat beach contaminated with buried oil. *Appl. Environ. Microbiol.* **70:**2603–2613.

97. **Rosenberg, E., and D. L. Gutnick.** 1986. The hydrocarbon-oxidizing bacteria, p. 903–912. *In* M. P. Starr, H. Stolp, H. G. Truper, A. Balows, and H. G. Schlegel (ed.), *The Prokaryotes. A Handbook on Habitats, Isolation, and Identification of Bacteria*. Springer-Verlag, Heidelberg, Germany.

98. **Rozen, Y., A. Nejidat, K. H. Gartemann, and S. Belkin.** 1999. Specific detection of *p*-chlorobenzoic acid by *Escherichia coli* bearing a plasmid-borne *fcbA′*::*lux* fusion. *Chemosphere* **38:** 633–641.

99. **Salles, J. F., F. A. De Souza, and J. D. van Elsas.** 2001. Molecular method to assess the diversity of *Burkholderia* species in environmental samples. *Appl. Environ. Microbiol.* **68:**1595–1603.

100. **Saucedo-Castaneda, G., M. R. Trejo-Hernandez, B. K. Lonsane, J. M. Navarro, S. Roussos, D. Dufour, and M. Raimbault.** 1994. On-line automated monitoring and control systems for CO_2 and O_2 in aerobic and anaerobic solid-state fermentations. *Process Biochem.* **29:**13–24.

101. **Selifonova, O. V., and R. W. Eaton.** 1996. Use of an *ipb-lux* fusion to study regulation of the isopropylbenzene catabolism operon of *Pseudomonas putida* RE204 and to detect hydrophobic pollutants in the environment. *Appl. Environ. Microbiol.* **62:**778–783.

102. **Service, R. F.** 1998. Microchip arrays put DNA on the spot. *Science* **282:**396–399.

103. **Sharabi, N. E. D., and R. Bartha.** 1993. Testing of some assumptions about biodegradability in soil as measured by carbon dioxide evolution. *Appl. Environ. Microbiol.* **59:**1201–1205.

104. **Sharkey, F. H., I. M. Banat, and R. Marchant.** 2004. Detection and quantification of gene expression in environmental biotechnology. *Appl. Environ. Microbiol.* **70:**3795–3806.

105. **Sikkema, J., J. A. de Bont, and B. Poolman.** 1995. Mechanisms of membrane toxicity of hydrocarbons. *Microbiol. Rev.* **59:**201–222.

106. **Silver, S., and L. T. Phung.** 1996. Bacterial heavy metal resistance: new surprises. *Annu. Rev. Microbiol.* **50:**753–789.

107. **Silver, S., and M. Walderhaug.** 1992. Gene regulation of plasmid-determined and chromosome-determined inorganic-ion transport in bacteria. *Microbiol. Rev.* **56:**195–228.

108. **Simpson, M. L., G. S. Sayler, B. M. Applegate, S. Ripp, D. E. Nivens, M. J. Paulus, and G. E. Jellison.** 1998. Bioluminescent-bioreporter integrated circuits form novel whole-cell biosensors. *Trends Biotechnol.* **16:**332–338.

109. **Small, J., D. R. Call, F. J. Brockman, T. M. Straub, and D. P. Chandler.** 2001. Direct detection of 16S rRNA in soil extracts by using oligonucleotide microarrays. *Appl. Environ. Microbiol.* **67:**4708–4716.

110. **Smets, B. F., and P. H. Pritchard.** 2003. Elucidating the microbial component of natural attenuation. *Curr. Opin. Biotechnol.* **14:**283–288.

111. **Smith, K., T. Cutright, and H. Qammar.** 2000. Biokinetic parameter estimation for ISB of PAH-contaminated soil. *J. Environ. Eng. ASCE* **126:**369–374.

112. **Sotsky, J. B., C. W. Greer, and R. M. Atlas.** 1994. Frequency of genes in aromatic and aliphatic hydrocarbon biodegradation pathways within bacterial populations from Alaskan sediments. *Can. J. Microbiol.* **40:**981–985.

113. **Stackebrandt, E.** 1998. Phylogeny based on 16S rRNA/DNA. *In Electronic Encyclopaedia of the Life Sciences*. Nature Publishing Group, London, United Kingdom. http://www.els.net.

114. **Stewart, J. E., and L. J. Marks.** 1978. Distribution and abudance of hydrocarbon-utilizing bacteria in sediments of Chedabucto Bay, Nova Scotia, in 1976. *J. Fisher. Res. Board Can.* **35:** 581–584.

115. **Stiner, L., and L. J. Halverson.** 2002. Development and characterization of a green fluorescent protein-based bacterial biosensor for bioa-

vailable toluene and related compounds. *Appl. Environ. Microbiol.* **68:**1962–1971.
116. **Stoyanov, J. V., J. L. Hobman, and N. L. Brown.** 2001. CueR (Ybbl) of *Escherichia coli* is a MerR family regulator controlling expression of the copper exporter CopA. *Mol. Microbiol.* **39:** 502–511.
117. **Suzuki, M., M. S. Rappé, and S. J. Giovannoni.** 1998. Kinetic bias in estimates of coastal picoplankton community structure obtained by measurements of small subunit rRNA gene PCR amplicon length heterogeneity. *Appl. Environ. Microbiol.* **64:**4522–4529.
118. **Swannell, R. P. J., K. Lee, A. Bassères, and F. X. Merlin.** 1994. A direct respirometric method for *in-situ* determination of bioremediation efficiency, p. 1273–1286. *In Proceedings of the Seventeenth Arctic Marine Oil Spill Program Technical Seminar.* Environment Canada, Ottawa, Canada.
119. **Tani, K., M. Muneta, K. Nakamura, K. Shibuya, and M. Nasu.** 2002. Monitoring of *Ralstonia eutropha* KT1 in groundwater in an experimental bioaugmentation field by in situ PCR. *Appl. Environ. Microbiol.* **68:**412–416.
120. **Thurmann, U., C. Zanto, C. Schmitz, A. Vomberg, W. Puttmann, and U. Klinner.** 1999. Correlation between ex situ activities of two neighbouring uncontaminated and fuel oil contaminated subsurface sites. *Biotechnol. Techniq.* **13:**271–275.
121. **U. S. Environmental Protection Agency.** 1995. *Bioremediation Field Evaluation. Eielson Air Force Base, Alaska.* EPA/540/R-95/533. U. S. Environmental Protection Agency, Washington, D. C.
122. **van der Schalie, W. H., T. R. Shedd, P. L. Knechtges, and M. W. Widder.** 2001. Using higher organisms in biological early warning systems for real-time toxicity detection. *Biosens. Bioelectron.* **16:**457–465.
123. **Venkateswaran, K., S. Kanai, H. Tanaka, and S. Miyachi.** 1993. Vertical distribution and biodegradation activity of oil-degrading bacteria in the Pacific Ocean. *J. Mar. Biotechnol.* **1:**33–39.
124. **Voordouw, G., K. J. Voordouw, T. R. Jack, J. Foght, P. M. Fedorak, and D. W. S. Westlake.** 1992. Identification of distinct communities of sulfate-reducing bacteria in the oil fields by reverse sample genome probing. *Appl. Environ. Microbiol.* **58:**3542–3552.
125. **Watts, R. J.** 1997. *Hazardous Wastes: Sources, Pathways, Receptors.* John Wiley and Sons Inc., New York, N. Y.
126. **Whiteley, A. S., R. I. Griffiths, and M. J. Bailey.** 2003. Analysis of the microbial functional diversity within water stressed soil communities by flow cytometric analysis and CTC + cell sorting. *J. Microbiol. Methods* **54:**257–267.
127. **Whiteley, A. S., and M. J. Bailey.** 2000. Bacterial community structure and physiological state within an industrial phenol bioremediation system. *Appl. Environ. Microbiol.* **66:**2400–2407.
128. **Whiteley, A. S., S. Wiles, A. K. Lilley, J. C. Philp, and M. J. Bailey.** 2001. Ecological and physiological analyses of pseudomonad species within a phenol remediation system. *J. Microbiol. Methods* **44:**79–88
129. **Wiles, S.** 2001. BioMate: development of custom-designed bioluminescent sensors for toxicity testing. Ph.D. thesis. Napier University, Edinburgh, United Kingdom.
130. **Woese, C. R.** 1987. Bacterial evolution. *Microbiol. Rev.* **51:**221–271.
131. **Wrenn, B. A., and A. D. Venosa.** 1996. Selective enumeration of aromatic and aliphatic hydrocarbon degrading bacteria by a most-probable-number procedure. *Can. J. Microbiol.* **42:** 252–258.
132. **Xi, C., M. Balberg, S. A. Boppart, and L. Raskin.** 2003. Use of DNA and peptide nucleic acid molecular beacons for detection and quantification of rRNA in solution and in whole cells. *Appl. Environ. Microbiol.* **69:**5673–5678.
133. **Zukowski, M. M., D. F. Gaffney, D. Speck, M. Kauffmann, A. Findeli, A. Wisecup, and J. P. Lecocq.** 1983. Chromogenic identification of genetic regulatory signals in *Bacillus subtilis* based on expression of a cloned *Pseudomonas* gene. *Proc. Natl. Acad. Sci. USA* **80:**1101–1105.
134. **Zytner, R. G., A. Salb, T. R. Brook, M. Leunissen, and W. H. Stiver.** 2001. Bioremediation of diesel fuel contaminated soil. *Can. J. Civ. Eng.* **28:**131–140.

BIOREMEDIATION OF MARINE OIL SPILLS

Roger Prince and Ronald M. Atlas

7

INTRODUCTION

Bioremediation offers an environmentally appropriate and cost-effective response to marine oil spills that reach shore and is deservedly part of the "tool kit" available to spill responders (29, 114). Huge amounts of oil are produced at sea, and even larger volumes are shipped; unfortunately, there are daily small spills and occasional major ones that require remediation. Given the staggering scale of daily oil usage, which is estimated at close to 3.5 billion gallons (1.2×10^{10} liters) per day (2), even the spillage of a tiny fraction (estimated at <0.0035% in 1997 [46]) means that more than 150 million liters is spilled into the world's oceans each year. Fortunately, despite the increasing volume of transported oil, the amount spilled from catastrophic spills has been generally decreasing (46). Even in the early 1980s, the amount of oil spilled from shipwrecks and oil platform blowouts was a little less than the amount that entered the sea from natural seeps (104), and recent estimates suggest that spills have declined to only some 10% of the natural input (106). The only exception to the general decline of the last few decades was the appalling environmental crime in the Arabian Gulf, where some have estimated that Iraqi forces deliberately released more than 450 million gallons (1.6×10^9 liters) of oil into the sea near Kuwait in 1991. An additional 350 million gallons (1.2×10^9 liters) was deposited in the Gulf as fallout from the smoke plumes of the >700 oil well fires in the Kuwait oil fields (1), making this by far the largest release of oil into the marine environment to date.

Even with continuing success in improving ship safety and operational practices to prevent environmental releases, there remains a continuing need for environmentally appropriate and cost-effective tools for dealing with oil spills, both catastrophic and deliberate, when such spills inevitably occur. Bioremediation offers one approach that works with natural processes to minimize the environmental impacts of oil spills, and it has already been used with notable success in Alaska following the spill from the *Exxon Valdez* (22, 126). Bioremediation relies on oil being consumed by numerous microbial species that over the millennia have evolved to exploit it. Safe stimulation of this biodegradation is the heart of bioremediation. In this chapter, we will give a brief overview of the

Roger Prince, ExxonMobil Research and Engineering Co., 1545 Route 22 East, Annandale, NJ 08801. *Ronald M. Atlas*, Graduate School, University of Louisville, Louisville, KY 40292.

Bioremediation: Applied Microbial Solutions for Real-World Environmental Cleanup
Edited by Ronald M. Atlas and Jim C. Philp © 2005 ASM Press, Washington, D.C.

composition of crude oil and the refined fractions that may be spilled at sea, discuss the diversity of organisms able to degrade oil components, and then describe strategies for encouraging the growth of such organisms. We will describe how bioremediation can be integrated with physical techniques to deliver an optimal cleanup and also discuss the environmental harm that might be done if bioremediation were applied carelessly and how this potential can be minimized.

CRUDE OIL AND REFINED PRODUCTS

Crude oil is a fossil fuel, the result of the burial, diagenesis, and catagenesis of ancient biomass (63, 159). The average age of commercially important crude oils is about 100 million years (71% was laid down between 180 and 85 million years ago [159], during the Jurassic and Cretaceous periods). It is generally accepted that aquatic algae, albeit usually with some terrestrial material, gave rise to petroleum, while terrestrial plants gave rise to the great coal reserves of the world. The oldest commercially valuable oils are from source rocks from the Ordovician period (486 million years old), while others are as young as the late Tertiary period (a few million years old). Unusual conditions, such as those at the Guaymas hydrothermal vent site, can even result in the formation of petroleum from biomass that is only approximately 1,000 years old (146), but they do not seem to give rise to commercially significant amounts of oil.

Petroleum has been a part of the biosphere for millions of years, escaping from underground reservoirs through seeps on land and sea. Some seeps are quite spectacular; for example, the burning gas seeps of Baku seem to have been well known for at least 2,500 years (63). Human exploitation of petroleum has included its use as a hafting material to glue stone tools to wooden handles (19) and for embalming (15), but substantial use awaited the 19th century and the need for lamp oil. Terrestrial seeps were the first locations to be drilled. For example, the first U.S. well, the 1859 Drake well in Pennsylvania, was drilled on a seep site. By the 1920s, most known terrestrial seep sites had been drilled, but marine seeps continue to be explored today. It has been estimated that the annual input of crude oil into the sea from seeps is about 180 million gallons (about 7×10^8 liters), accounting for approximately 45% of the total input (106). Levy and Lee (85) have argued that the biodegradation of this oil contributes to local food webs and to commercial fisheries.

Modern use of petroleum has substantially augmented the amount of petroleum getting into the sea, probably doubling it (106). Most of this anthropogenic input comes from municipal runoff and routine shipping (106). Although major tanker spills have serious local impacts, their total contribution is now substantially less than that of natural seeps. Well blowouts, such as Ixtoc-1 in Ciudad del Carmen, Mexico, that released 140×10^6 gal (500×10^6 liters) in 1979, can dwarf spills from tankers (141). One of the largest tanker spills was that from the *Amoco Cadiz* off Brittany, France, in 1978, amounting to about 7×10^7 gal (2.5×10^8 liters) (13). Although these large spills grasp the public's attention, the U.S. Coast Guard estimates that 98% of spills from vessels are of less than 1×10^5 gal ($<0.4 \times 10^6$ liters) (161).

Crude oils are principally hydrocarbons, molecules composed of only carbon and hydrogen, and the hydrogen-to-carbon ratio is typically between 1.5 and 2 (63, 159). There is thus a mixture of aliphatic species (predominant carbons are $—CH_2—$, either as linear chains, known as paraffins, or as rings known as naphthenes) and aromatic species (some carbons are in rings as $—HC{=}CH—$, although there may well be pendant alkyl groups). Alkenes and alkynes, linear unsaturated molecules, are rare in crude oils, although they can be abundant in some refined products, such as gasoline. Typically, about 15% of a crude oil is comprised of molecules containing heteroatoms such as oxygen, sulfur, and nitrogen. This fraction includes compounds known by a variety of names, including polars, asphaltenes, resins, and NSO (see references 63 and 159). Tissot and Welte (159) indicate that the average composition of 527 crude oil samples is 58.2% saturates, 28.6% aromatics, and 14.2% polar compounds, although the absolute values vary widely for different oils. On average, there is rough parity between paraffins, naphthenes, and aromatics.

The paraffins (Fig. 7.1) typically span the range from a single carbon (methane) up to waxes with at least 40 carbons and often more. Linear alkanes can make up 15 to 20% of an undegraded crude oil, although their content can be essentially undetectable or as high as 35%, depending on the source and reservoir conditions. There are also branched alkanes; most are in the C_6 to C_8 range, but pristane ($C_{19}H_{40}$) and phytane ($C_{20}H_{42}$) (Fig. 7.1), molecular relics of the phytol chains of chlorophylls and perhaps other biomolecules, are usually the most abundant individual branched alkanes. Pristane is thought to be the result of initial partial degradation of phytol in the presence of oxygen, while phytane is thought to be the result of diagenesis in the absence of oxygen (159).

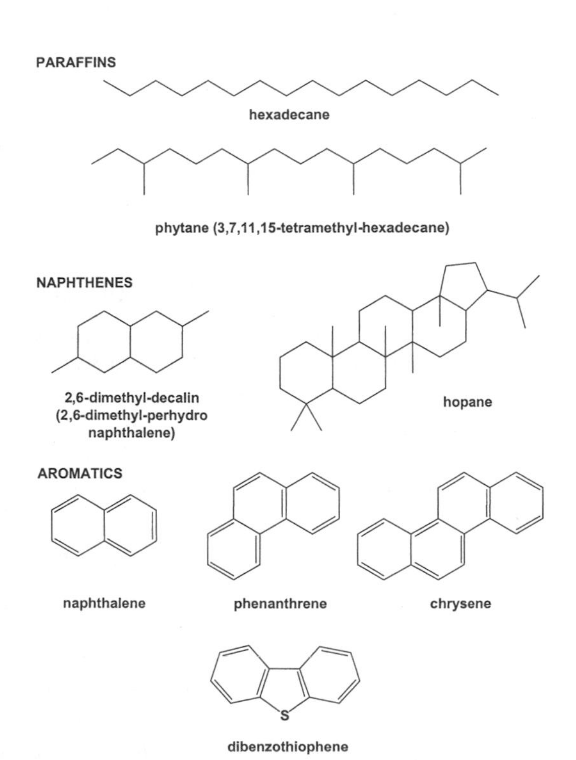

FIGURE 7.1 Representative hydrocarbons in the principal molecular classes in crude oils.

The naphthenes (Fig. 7.1) include parent compounds, such as cyclopentane, cyclohexane, and decalin, together with their alkylated congeners. Tissot and Welte (159) state the average composition of the naphthene fraction of 299 crude oils as 53.9% one- and two-ring naphthenes, 20.4% tricyclic naphthenes, and 24.9% tetra- and pentacyclic naphthenes. These last molecules are among the better understood molecular biomarkers in crude oils, and they are used extensively in correlating reservoirs and source rocks (120), in assigning the depositional environment of source rocks (120), and more recently as conserved internal markers during biodegradation (130).

Because of the separation procedures used in the analysis of crude oils, any molecule containing at least one aromatic ring is included in the "aromatic" fraction, regardless of the presence of saturated rings and/or alkyl substituents. Sulfur aromatic heterocycles, such as thiophenes, benzothiophenes, and dibenzothiophenes (Fig. 7.1), are included in the aromatic category. Indoles and carbazoles, usually the most abundant nitrogen-containing species, and the less-abundant basic nitrogen species such as quinolines are also included in the aromatic category, but they are usually present at much lower concentrations. Alkylated aromatic species are more abundant than their parent compounds, with mono-, di-, and trimethyl derivatives usually being most abundant. Nevertheless, the median aromatic structure probably has one or two methyl substituents, together with a long-chain alkyl substituent (138).

The polar molecules are the most difficult to characterize because they typically cannot be analyzed by gas chromatography, the method of choice for the molecular characterization of hydrocarbons. Petroleum polar compounds contain heteroatoms such as nitrogen, oxygen, and/or sulfur, and the category includes porphyrins, typically with nickel or vanadium in the tetrapyrole, naphthenic acids (Fig. 7.2), and large molecules known as asphaltenes. Some asphaltenes have molecular weights in the thousands and even higher, and many are suspended in the oil rather than dissolved in it (65). The polar fraction of the oil contains the majority of the color centers in crude oil, and in isolation these materials are difficult to distinguish from more recent biological residues, such as the humic and fulvic acids (26, 136), except with sophisticated tools such as isotope analysis.

Crude oils are classified commercially by several criteria, and among the most important is the specific gravity. The oil industry uses a unit known as API (American Petroleum Institute) gravity, defined as [142.5/(specific gravity)] − 131.5, and expressed as degrees. Thus, water has an API gravity of 10°, and denser fluids have lower API gravities. Less dense fluids, for example most hydrocarbons, have API gravities of >10°. For convenience, oils with API gravities of >40° are said to be light oils, while those with API gravities of less than about 16° are said to be heavy. The sulfur content of crude oils is another commercially important parameter; in general, it is inversely proportional to API gravity, typically near 0.1% in light oils to sometimes more than 5% in heavy oils (63). Viscosity is also important; it tends to be proportional to API gravity, but it is also highly dependent on the proportion of waxes in the oil and is also temperature dependent (63).

representative naphthenic acid

putative asphaltene

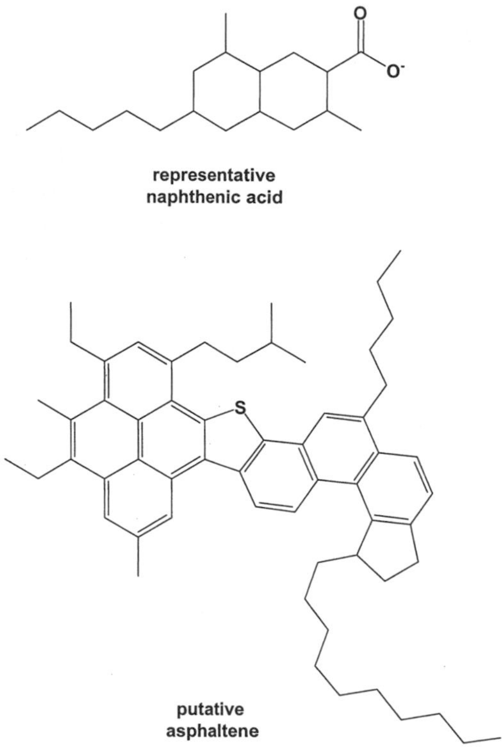

FIGURE 7.2 Representative structures for resin and polar components found in crude oils.

While the largest spills are invariably crude oil, refined products are also transported by sea, and of course all large vessels carry substantial quantities of fuel oil, typically a bunker fuel. Recent spills of refined product include the 1996 spill of 86,000 gal of no. 2 heating oil and 93,000 gal of heavy fuel oil from the *Julie N* in Portland Harbor, Maine (135), and the 2001 spill of 160,000 gal of diesel fuel and 80,000 gal of bunker fuel from the *Jessica* in the Galapagos Islands (33, 44). Bunker fuels are viscous residual fuels, made from what is left after lighter refined products, such as gasoline and diesel fuels, have been removed by distillation.

HYDROCARBON BIODEGRADATION

The biodegradation of hydrocarbons under aerobic conditions, which are characteristic of most marine surface waters and many coastal shorelines, has been studied in some detail, and many reviews that cover the microbiology and biochemistry of the process, on both broad (5, 56, 79, 123, 147) and specific (23, 39, 43, 59, 71, 142, 162, 167) aspects, are available. Here, it suffices to say that the vast majority of hydrocarbons are biodegradable under aerobic conditions, with exceptions being some molecules with a number of tertiary carbons, such as the hopanes (127 [although see references 18 and 50]). Refined products such as gasoline (147) and diesel (45) fuels are essentially completely biodegradable under aerobic conditions. Other oil components, such as some oil resins and polar molecules, seem resistant to biodegradation, and it is obvious that road asphalt is not readily biodegraded, so not all molecules in a crude oil can be expected to disappear by biodegradation. McMillen et al. (96) examined the short-term biodegradability of 17 crude oils in soil microcosms as a function of 78 different parameters that might determine the extent of biodegradation. These included 67 individual chemical species, percent sulfur, etc. They found that the API gravity was the most useful predictor of biodegradability of the most degradable fraction of the oils. At 0.5 wt% oil in a sample with appropriate nutrients, moisture, and aeration, more than 61% of the most degradable oil (API, 46°) was lost in 4 weeks, while only 10% of the least degradable oil (API, 15°) was consumed under the same conditions. Further degradation occurred on a longer timescale, and the literature reports biodegradation potentials as high as 97% for particularly light oils (123).

Many of the concerns about the adverse environmental impacts of a spill are related to the bioavailability of individual hydrocarbons, particularly those on the U.S. Environmental Protection Agency (EPA) list of Priority Pollutants (76), such as the polycyclic aromatic hydrocarbons. Polycyclic aromatic hydrocarbons are a minor constituent of crude oils (159), but they have important toxicological properties (38, 66). There are several pathways for the degradation of such compounds in prokaryotes and eukaryotes (72, 124), but there is an important distinction between biodegradation carried out by many microorganisms and that carried out by animals. Many bacteria degrade polycyclic aromatic hydrocarbons to carbon dioxide, water, and biomass, starting the process with dioxygenase enzymes that insert both atoms of molecular oxygen into the substrate to form first the *cis*-dihydrodiol and then the hydroquinone (Fig. 7.3). The hydroquinone can then be metabolized to carbon dioxide and water, and some is incorporated into biomass. Animals, fungi, plants, and some bacteria initiate the process with rather different enzyme systems, known as cytochrome P450s, from their prominent absorption band when treated with carbon monoxide (112). Cytochrome P450 inserts only one of the two atoms of oxygen in an oxygen molecule into the substrate, forming an epoxide (Fig. 7.3). This may undergo secondary reactions that lead to its excretion as an adduct with sugars, anions, etc., but it may also intercalate and form adducts with DNA (98, 112). It is thus obviously preferable that polycyclic aromatic hydrocarbons be degraded by bacteria rather than eukaryotes, and facilitating such a preference is one of the corollaries of a successful bioremediation protocol.

In the last decade, it has become abundantly clear that the anaerobic biodegradation of hydrocarbons is also a widespread phenomenon. Small water-soluble aromatic compounds, such as benzene and toluene, have been shown to undergo biodegradation under sulfate-reducing, nitrate-reducing, perchlorate-reducing, ferric ion-reducing, humic acid-reducing, and methanogenic conditions (30, 89, 121, 149), and this phenomenon is proving important in remediating terrestrial spills where these compounds have reached groundwater (17, 37, 64, 90). Larger hydrocarbons, such as *n*-alkanes up to nC_{34} (28) and two-, three-, and four-ring aromatic hydrocarbons (36, 97), have also been seen to be biodegraded under anaerobic conditions, and this may be important if oil spills impact anaerobic environments, such as marshes.

Most hydrocarbons are insoluble to any great extent in water (118, 174–176), so their biodegradation must take place at the hydrocarbon-water interface. This leads to a fundamental limitation to hydrocarbon biodegradation in the environment; physical processes such as the formation of tar balls (77, 53) and "pavements" along shorelines (55, 115–117) (where oil and particles become fully saturated with each other and exclude water) sequester oil from the oil-water interface and thereby preserve the oil from biodegradation. This preservation can be furthered by the generation of a photochemically oxidized and polymerized "skin" on tar balls and pavements, identi-

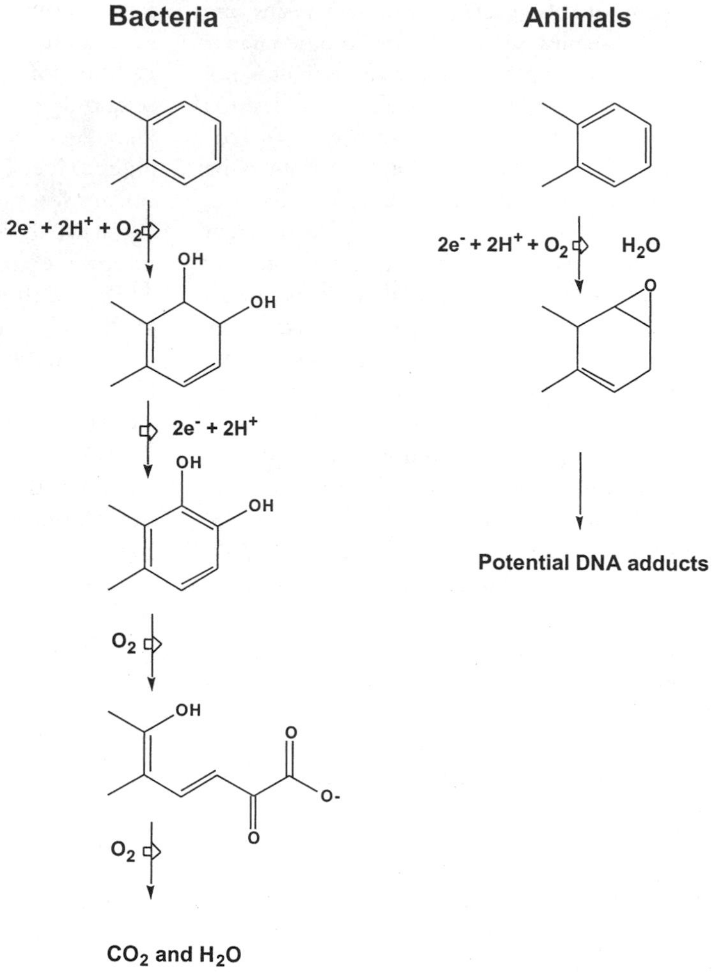

Bacteria

Animals

$2e^- + 2H^+ + O_2$

$2e^- + 2H^+ + O_2$ H_2O

$2e^- + 2H^+$

Potential DNA adducts

O_2

O_2

CO_2 and H_2O

FIGURE 7.3 An overview of the initial reactions of aromatic hydrocarbon activation in hydrocarbon-degrading microorganisms and in animals. In polycyclic aromatic hydrocarbons (76), the aromatic ring shown would be part of a multiring structure. More details can be found elsewhere (124).

fied by loss of polycyclic aromatic hydrocarbons and an increase in polar material in the oil (51). Tar balls and pavements are unlikely to biodegrade at a significant rate until the surface area available for microbial colonization is enhanced, perhaps by physical disruption such as tilling (143).

Even in the absence of the formation of tar balls and pavements, processes that increase the surface area of spilled oil are likely to increase the rate of biodegradation. A natural phenomenon originally termed "clay-oil flocculation" (20) and now termed "oil-mineral fines interactions" (81, 168), in which oil forms stable microaggregates with small mineral particles, is one mechanism whereby the surface area is enhanced. Mineral fines are present in most marine environments, and the formation of oil-mineral microaggregates that can be washed from the shoreline is probably an important contributor to natural cleansing of oiled shorelines, provided that oil does not form pavements (21). Recent work has shown that the process also stimulates biodegradation (168), so not only does oil leave an oiled shoreline, but also it is biodegraded more rapidly. As we shall see below, dispersion of oil with carefully tailored oil dispersants is another way of dramatically increasing the surface area of oil in water. Dispersion has been shown to stimulate biodegradation (152) and hence to lower the environmental impact of an oil spill. Oil from many major spills that has moved away from shorelines and been dispersed at sea due to storms has caused no detectable ecological impacts.

HYDROCARBON-DEGRADING MICROORGANISMS

Aerobic hydrocarbon-degrading microorganisms have been found in almost all ecosystems that have been diligently searched (94, 125). This includes Arctic and Antarctic marine sediments and terrestrial soils and essentially all locations in between. Anaerobic environments have not been studied so thoroughly, but again hydrocarbon-degrading strains seem abundant in most environments. The only exception to date seems to be hyperthermophilic environments, and indeed it is thought that one reason for the preservation of oil in reservoirs is that the oil was sterilized by being heated to 80 to 90°C during burial and diagenesis (170). Nevertheless, *Bacillus thermoleovorans* strains able to grow on alkanes at temperatures from 50 to 80°C have been isolated recently (75), so it may be premature to completely discount the potential existence of hyperthermophilic oil-degrading microbes.

Organisms able to use hydrocarbons as the sole source of carbon and energy have been found in the domains *Bacteria* and *Archaea* and fungi, and some algae are able to degrade at least some hydrocarbons (124). Most members of the *Eukarya* containing cytochrome P450 and peroxidase systems, including fish and mammals, activate and initiate metabolism of polycyclic aromatic hydrocarbons, although this is typically a prelude to excretion rather than mineralization. Unfortunately, as discussed above, this also can lead to the generation of carcinogens (98).

The taxonomy of microorganisms remains only poorly characterized, but to date more than 60 genera of aerobic bacteria, 5 genera of anaerobic bacteria, 2 genera of aerobic archaebacteria, 9 genera of algae, and 95 genera of fungi have been shown to contain species that can degrade at least some hydrocarbons (124). Given the ease of exchange of genes for hydrocarbon degradation between organisms (60, 78), we may confidently expect that many more species will be discovered. The distinguishing feature of oil-degrading microorganisms is their ability to somehow activate the hydrocarbon, either aerobically by inserting one or two oxygen atoms or anaerobically by adding some other moiety, such as fumarate or carbon dioxide (12, 14, 58, 124, 149, 169, 177). Such activation, and perhaps a few subsequent reactions unique to hydrocarbon metabolism, allows the hydrocarbon to enter the standard cellular pathways of metabolism.

Oil-degrading microorganisms may be ubiquitous, but they are typically only a small fraction of the biota of an uncontaminated site, probably because they are substrate limited. An oil spill removes this limitation, and there is generally a bloom of hydrocarbon-degrading organisms so that they become a major fraction of the microbial population. Modern molecular tools allow us to monitor these changes at the species level (32, 54, 67, 68, 70, 73, 74, 150, 156, 163), and it is reasonable to imagine that this approach may be developed to provide indications of when the environmental impact of a spill has diminished to prespill conditions.

Public perception of bioremediation seems to revolve around the use of genetically modified organisms, but this is far from the truth. Perhaps the conviction stems from the fact that the first patent for a generically engineered microorganism was for one that degraded hydrocarbons (31), but to our knowledge no genetically modified organism has been released as part of an oil bioremediation response. Genetically modified organisms are, however, proving useful for monitoring hydrocarbon contamination and its remediation (9, 25, 93, 148, 151, 173). Typically, a light-emitting system, such as the *luc* or *lux* system, is inserted immediately after a gene in the early metabolism of a hydrocarbon. Then when the organism activates its hydrocarbon degradation system, it emits light. In fact, a bioluminescent genetically engineered *Pseudomonas fluorescens* strain has been approved for field testing as a reporter of naphthalene biodegradation (137) and been used in semicontained soil bioreactors. Whether genetically engineered microorganisms will ever play a role in bioremediation in the marine environment, perhaps organisms engineered to degrade particularly recalcitrant chemical species, awaits a resolution of the regulatory uncertainties surrounding this technology (178). Given the ubiquity of oil-degrading microorganisms in the biosphere, however, genetically modified organisms will need to have remarkable properties to enable them to survive and compete with the indigenous biota.

Similar considerations probably apply to addition of unmodified oil-degrading microorganisms to a spill site. As we shall see below, there have been several attempts to add selected strains (139, 140) or indigenous bacteria grown to a high cell density (166) but little evidence indicating success.

BIOREMEDIATION OF OIL-CONTAMINATED MARINE ECOSYSTEMS

Knowing that the majority of molecules in crude oils and most refined products are biodegradable, the question becomes "How can biodegradation be relied on, and even stimulated, to remove hydrocarbons from the environment to minimize the environmental impact of a spill?" It is probably not surprising that there are several answers to the question. Three principal approaches have been suggested, and all have been tried: natural attenuation (reliance on natural biodegradation activities and rates), which is sometimes called intrinsic bioremediation; biostimulation (stimulation of natural activities by environmental modification, such as the addition of a fertilizer to increase rates of biodegradation); and bioaugmentation (addition of exogenous microorganisms to supplant the natural degradative capacity of the impacted ecosystem). Bioaugmentation receives great attention from entrepreneurs but has not been clearly demonstrated to be effective or necessary.

Natural attenuation, which involves a total reliance on naturally occurring biodegradation to remove an environmental pollutant, involves no intervention to alter the natural fate of a pollutant. The problem for oil spill remediation is that the process is often very slow and sometimes incomplete, especially if physical processes lead to the generation of tar balls and pavements. Thus, it does not necessarily remove toxic substances fast enough to preclude adverse ecological consequences. Nevertheless, natural attenuation has become the method of choice in many cases, particularly in Europe, for treating marine crude oil spills. Given the wave action that tends to move oil off many European coasts, this approach has been used in a variety of spills along the British and French coastlines, for example, in the case of most of the shorelines of Brittany impacted by the *Amoco Cadiz* spill (13).

The second approach, biostimulation, aims to accelerate a natural process through environmental modification. The specific treatments for contaminated environments using this approach depend upon the specific rate-limiting factors of that environment. Biostimulation requires that microorganisms capable of degrading the pollutant are already naturally present in the contaminated ecosystem. For example, in the case of an oil spill impacting a shoreline where hydrocarbon-degrading populations are always present, a nitrogen and phosphorus fertilizer can be added to remove rate-limiting factors. This increases the rates of oil biodegradation by allowing greater growth of hydrocarbon-degrading microorganisms, just as addition of an agricultural fertilizer enhances plant growth. Again, this approach relies upon natural biodegradative processes that would occur anyway, but through human intervention the rates of biodegradation are stimulated. This approach to bioremediation can be very beneficial, but it is not instantaneous—the environmental impact of an oil spill can be reduced from decades to years, but the impact cannot be instantly or fully mitigated. It was this approach that was used to treat the *Exxon Valdez* oil spill, in which fertilizer was applied to more than 120 km of oiled shorelines (22, 126).

Finally, bioaugmentation involves the addition of living organisms or enzymes to augment naturally occurring biological populations. A number of companies market cultures of microorganisms for the biodegradation of specific pollutants. In most cases, there are no standardized tests and no licensing requirements, and many of the cultures that are marketed seem to be ineffective and are widely perceived as modern "snake oil." While bioaugmentation may be appropriate to consider in other situations, perhaps with compounds that have not been part of the biosphere for so long, there have been no clear demonstrations that it is useful for treating marine oil spills.

Before the different technologies that might be applied in different situations are addressed, it is appropriate to reiterate some of the findings discussed above.

1. The majority of hydrocarbons in crude oils and refined products are biodegradable, at least under aerobic conditions, and should eventually disappear from the biosphere through biodegradation.

2. Most hydrocarbons are not very soluble in water, and biodegradation is likely to be limited by the surface area available for microbial colonization.

3. Crude oils and refined products are a rich source of carbon and energy for the degrading microorganisms, but they lack other essential nutrients. Oxygen is typically not limiting in most marine, lacustrine, and riverine environments, but bioavailable nitrogen, phosphorus, and perhaps various trace elements may be, especially in the case of large spills. Oxygen is probably limiting in marshes and mangroves.

4. Oil-degrading microorganisms are essentially ubiquitous, so any added as part of a bioaugmentation approach would need to have remarkable properties to survive and compete.

Oil on Water

Oil spilled on water usually floats, and the first imperative is to collect it. This is quite feasible in the case of small spills but unlikely to be effective for larger spills. The usually preferred response is to disperse oil that cannot be collected, using carefully chosen dispersants (35, 47, 83, 105). Modern dispersants have only low to moderate toxicity (52), less than that of the spilled oil, and the environmental benefits from using them instead of allowing oil to reach sensitive shoreline habitats are becoming generally appreciated (119). There are sound theoretical reasons to believe that dispersion should stimulate biodegradation, both because of the larger surface area available for microbial colonization and because some surface-active agents stimulate biodegradation, perhaps by making some components more bioavailable. In this regard, Varadaraj et al. (164) have shown that the sorbitan-based surfactants used in the Corexit range of dispersants and beach cleaners stimulate oil biodegradation. Dispersants do indeed disperse oils into tiny droplets that become broadly scattered in the water column; however, it is difficult to monitor the biodegradation of dispersed oil in a thoroughly realistic way. There is the potential drawback that the dispersant, added at perhaps 5 to 10% of the oil volume, is itself biodegradable and like the spilled oil contains little nitrogen or phosphorus. The biodegradation of the dispersant thus competes with the biodegradation of oil for potentially limiting nitrogen and phosphorus sources, but the dilution that follows successful dispersant application makes it likely that natural background levels of nutrients will be sufficient for rapid biodegradation of both dispersant and oil. Swannell and Daniel (152) demonstrated this in quite large (15-liter) microcosms with low nutrient concentrations typical of United Kingdom coastal and estuarine waters. The obvious next step, of incorporating nitrogen fertilizers into dispersant packages, has been described in several patents (see reference 123), but as far as we know, none of these is commercially available or has been demonstrated to be effective in actually treating an oil spill.

Although not intended to directly stimulate oil biodegradation, dispersants have been widely used in response operations for recent spills, including the 1996 *Sea Empress* spill off the south coast of Wales (91), the 1997 *San Jorge* spill off the east coast of Uruguay (109), and the 1998 IDOHO-QIT offshore pipeline oil spill off the west coast of Nigeria (110). In these cases, the dispersal of the oil minimized coastal impacts, and it seems likely that the increased oil surface area enhanced the rates of biodegradation.

Bioremediation, by the addition of oleophilic fertilizers, has been suggested as a response to floating oil (7, 8), but this remains untested to any great extent. Some have suggested that "seeding" oil slicks with oil-degrading bacteria ought to be a practical way of stimulating the biodegradation of spilled oil (61). It has only been tried once at sea to our knowledge, on the 1990 spill from the *Mega Borg* in the Gulf of Mexico (84), and there was no evidence that biodegradation had been stimulated. This is in line with our expectation, as noted above, but it is only fair to note that logistical problems involved in monitoring the bioremediation were not settled before the spill dispersed. A major issue is the need to have large quantities of bacteria available soon after a spill; otherwise, the spill is likely to disperse or come ashore before the seeding can occur.

Oil on Shorelines

Oil that is not dispersed at sea, or burned (24, 103), usually lands on a shoreline, either as a more or less contiguous slick, as occurred after the *Exxon Valdez* (107), *Sea Empress* (91), and *Amoco Cadiz* (13) spills, or as tar balls as in the Ixtoc-1 well blowout (11). It is important to recognize that amenity beaches, and those that are accessible, will usually be cleaned by the physical removal of as much oil as possible (e.g., see references 91 and 107). The traces of oil left after such operations are prime candidates for bioremediation. The physical cleanup of bulk oil is important, because as discussed above, thick layers of oil are likely to form pavements that will be quite resistant to rapid weathering (115, 116). Oiled sands are typically removed from the shoreline and discarded in secure landfills (e.g., reference 144), although it is unusual for such material to contain more than a few percent oil by weight. Relatively porous cobble beaches are poor candidates for this removal, and washing is often used (99, 107). Nondispersing surfactants significantly improve the effectiveness of shoreline washing and allow the oil to be retained in booms and recovered (99), and this seems a particularly promising approach for areas where human intrusion is difficult or damaging, such as mangroves and swamps (157). Some of these products contain sorbitan-derived surfactants, which themselves stimulate biodegradation (164), so traces that are uncollected may have a secondary beneficial effect on natural recovery.

Bioremediation really comes into play once oil has been reduced to a relatively thin coating on the beach material. Excellent reviews have covered early work in this field (4–6, 57, 80, 82, 96, 123, 154), and we will focus on the most recent work. There is now little doubt that the vast majority of aerobic marine and estuarine environments are nitrogen limited for hydrocarbon degradation if there is a significant spill and that biodegradation will be stimulated if this limitation is overcome by the careful application of fertilizers. This was proved in the case of the bioremediation of the *Exxon Valdez* spill (22) and seems to be generally true (82, 95, 101, 129, 130, 134, 155). Typical enhancements are two- to fivefold (22, 129, 130, 155, 166), and while at first glance this may seem slight, it can mean a difference of years for recovery. Since most marine and estuarine environments are tidal and/or have significant water flow, small oil inputs may not overwhelm the natural delivery of background nutrients. This is especially true if the environment is naturally rich in bioavailable nitrogen, such as estuaries with agriculture in their watersheds. In such situations, it is unsurprising that the biodegradation of more resistant components, such as the polycyclic aromatic hydrocarbons, in a small experimental spill is not markedly stimulated by the addition of nutrients (166), since the natural delivery of nutrients can supply all that is needed for a small plot. It is reasonable to expect that a larger spill could deplete even these environments and that addition of fertilizer could then stimulate biodegradation.

There may be additional limiting nutrients besides nitrogen; phosphate is often a limiting nutrient (41), and it is usually added as part of any fertilizer strategy. Iron is often limiting in marine environments (41), and its addition has been shown to stimulate oil degradation in seawater (39). Addition of low levels of vitamins, perhaps as yeast extract, is a common part of microbial medium preparation, and it has been shown to stimulate oil-degrading microorganisms (133). None of these additional nutrients (phosphate, iron, etc.) has been actually proven to be effective in the field, but the cost of adding them is low, and the potential risks of significant adverse environmental impact are also low, so it seems appropriate to include them in a bioremediation strategy (129, 130).

Several different fertilizer application strategies have been tried, and most have been successful. Oleophilic fertilizers, designed to adhere to the oil and deliver nutrients directly to the oil-degrading microbes (7), were used with success in Alaska (22). These fertilizers were applied with airless sprayers deployed from small boats that could approach the shoreline. Slow-release formulations, again designed to preferentially deliver nutrients to the oil-degrading microbes rather than have them wash out of the shoreline (111), were also used successfully in Alaska (126) and on an experimental spill in Spitsbergen (130, 131). In those cases, they were applied by workers using broadcast spreaders walking on the beaches. Slow-release fertilizers were encased in "socks" to keep them in a steep shingle beach in south Wales following the *Sea Empress* spill (155). Soluble agricultural fertilizers, applied as dry solids, have been used with success by Lee and his colleagues in many years of field work in Canada (see references 80, 82, 124, and 154) and in an experimental spill in Spitsbergen (129, 130). Soluble nutrients were applied in solution every day in a Delaware Bay field trial (166). The most important requirement is that the nitrogen be made available at the oil-water interface for long enough that the oil-degrading microbes have an opportunity to become well established; the precise form of the nitrogen in the fertilizer seems to be of secondary importance. The amount of nitrogen is a matter for careful consideration; too much may have an adverse environmental impact, and too little may have a suboptimum effect. Boufadel et al. (19) have suggested that 2.5 mg of nitrate (N) (180 μM) is optimal for the biodegradation of heptadecane at 2 g kg of sand^{-1}, and this seems a quite reasonable number. It compares with 50 to 200 μM measured in beaches undergoing bioremediation in Alaska (22). An important simplification for monitoring bioremediation was demonstrated in a field trial in Spitsbergen (130, 131), where it was shown that simple colorimetric field test kits could measure the delivered nutrients in beach interstitial water taken from pipe wells installed into the oiled sediment. This allows for optimizing fertilizer application on that beach, and it would also provide guidelines for fine-tuning fertilizer applications on the next section of shoreline to be treated in a full-scale bioremediation operation.

We note that several groups have addressed the determination of the optimum $N_{fertilizer}$:C_{oil} ratio for stimulating biodegradation, but the reality is that it is almost impossible to make accurate estimates of how much oil is on a beach, even if several "representative" samples are analyzed. Most measurements suggest that oil loadings follow a log-normal distribution (see reference 86), and there are dramatic ranges in most studies. It thus seems more appropriate to aim for 100 to 200 μM available nitrogen in the interstitial water rather than an amount proportional to the amount of oil.

Experiments to assess which chemical form of nitrogen might be most effective are complicated by the fact that ammonium fertilizers tend to cause acidification of the medium in closed systems, which in turn inhibits microbial activity (48). Such an effect is unlikely to happen in an open system, and ammonium, nitrate, urea, and polymerized forms of urea have all been used with success. Clearly, the physical release properties will have implications for how frequently fertilizer must be applied; slow-release and oleophilic fertilizers are likely to last longer on a shoreline and thus need to be reapplied less frequently. For example, Swannell et al. (155) saw equivalent results from a weekly application of a soluble fertilizer and a single application of a slow-release fertilizer over 2 months, and clearly the latter is logistically much simpler and less expensive.

As we have discussed above, public perception is that bioremediation involves the addition of oil-degrading microorganisms and that bioaugmentation with genetically engineered microorganisms is being used, but this is untrue. Many studies have found no benefit to inoculating oiled sediment with extra bacteria (5, 6, 57, 80, 82, 123, 154, 158). Perhaps the most startling demonstration was made by the U.S. EPA when testing potential inoculants for stimulation of oil biodegradation in Alaska following the *Exxon Valdez* oil spill (165). Eight products were tested in small laboratory reactors that allowed substantial degradation of oil by the indigenous organisms of Prince William Sound; all eight microbial inocula had a greater stimulatory effect on alkane degradation if they were sterilized by autoclaving prior to addition! This suggests that the indigenous organisms readily outcompeted the added products, but that autoclaving the products released some trace nutrient that was able to stimulate the growth of the endogenous organisms. Nevertheless, commercial inocula are still being used, for example on the *Nakhodka* spill on the northern coast of Japan (62, 160), but convincing chemical analyses to show a stimulation of biodegradation are not yet forthcoming. Rather than adding commercial inocula, an alternative approach is to enrich indigenous organisms from the oiled site and reinoculate them as part of the bioremediation strategy. This was tried by the U.S. EPA on an experimental spill in the Delaware Bay (166) but again without significant benefits.

An approach that combines addition of exogenous bacteria and a polymerized urea fertilizer known as F-1, to which they are uniquely adapted, has been championed by Rosenberg et al. (139, 140). This is an intellectually intriguing approach that attempts to deliver useful nitrogen to only the added oil-degrading bacteria by using a fertilizer that they seem almost uniquely able to use. The combination of fertilizer and inoculum was used on a relatively small spill (100 tons of a heavy crude oil) on a sandy beach in Israel, and significant success was claimed. Unfortunately, as we have discussed before (125), the analytical tools used to assess biodegradation were not the most appropriate, and the success may have been overestimated. No further work on the use of this approach to remediating marine oil spills seems to have been published.

Oil in Marshes and Mangroves

Much less work on the bioremediation of oil in marshes and mangroves has been reported. These environments typically have lush vegetation and anaerobic conditions within a few millimeters of the water-sediment interface. Fortunately, oil does not usually penetrate deeply into such sediments, but cleaning them is a difficult challenge, not least because they are so susceptible to physical damage just by human intrusion. Bioremediation by adding fertilizer seems to be effective in the aerobic parts of these environments (27, 69, 87, 145, 172), but this is not so clear in flooded, anaerobic situations. Nevertheless, even in the absence of a clear demonstration of increased rates of oil removal, there can be clearly beneficial effects of bioremediation on the indigenous plants (42).

While anaerobic bioremediation is becoming well established for terrestrial spills, especially of gasoline, it has yet to be used with success in the marine environment, to our knowledge.

DEMONSTRATING EFFICACY OF BIOREMEDIATION FOR MARINE OIL SPILLS

It has been a major challenge to provide statistically significant evidence that bioremediation is truly effective in removing oil from a shoreline. It sounds relatively simple; surely one measures the amount of oil in two representative sections of shoreline, applies fertilizer to one, and returns after some reasonable time to measure the amount of oil remaining. However, there are several reasons why this is complex. The major one is that the heterogeneity of the natural world leads to huge uncertainties in the amount of oil in a section of beach: the amount of oil in a subsample can be measured with a high degree of precision, but replicate samples can be quite different, and estimates for the amount of oil in a section of shoreline tend to be log-normally distributed (see reference 86). This can be attributed both to the fact that oil rarely arrives on a beach in a uniform manner and to the broad range of sizes of particles, whether sand, gravel, or cobble, even with careful sampling. Furthermore, most beaches are very active environments, and physical movement of oil can occur with each tide. Therefore, more sophisticated approaches must be used, and a tiered approach seems appropriate. We have found that answering the following series of questions can provide the necessary confidence to assess whether bioremediation is indeed working (22, 126, 129–131).

1. Have fertilizer nutrients been delivered to the oiled sediment? Fertilizer is typically applied to the surface of oiled sediments. Does it penetrate to the oiled zone, which is typically beneath the very surface of the shore? We have installed simple wells in some cases to allow sampling of interstitial fertilizer from within the oiled zone and found that simple colorimetric tests are adequate to measure the levels of nitrate and ammonium in the water. A representative data set is shown in Fig. 7.4A. The data come from an experimental spill near Sveagruva, Spitsbergen (approximately 78°N, 17°E), in the summer of 1997 (129, 130). Logistical constraints necessitated the application of fertilizer on two days only 9 days apart. We used a mixture of soluble and slow-release fertilizers because we were concerned that low temperatures (interstitial water, 4 to 9°C) and relatively recently applied oil (7 days prior to the application of fertilizer) might inhibit nutrient release from encapsulated fertilizers. In fact, the data, measured with field test kits (K-6902, K-1510, and K-8510 kits; Chemetrics, Calverton, Va.), showed that nutrients were indeed delivered at approximately the desired levels (100 to 200 μM total nitrogen, approximately equal contributions from nitrate and ammonium).

2. Is microbial activity stimulated by the addition of fertilizer? A decrease in the level of dissolved oxygen in the interstitial water (129–131), an increase in CO_2 evolution from the beach (130, 131, 153), an increase in the size of the population of hydrocarbon-degrading microbes (128, 131), and an increase in the mineralization of radiolabeled hydrocarbon substrates in the laboratory (88) have all been used to address this question. Measurements of dissolved oxygen and CO_2 can be made on the beach and so are by far simplest. Figure 7.4B shows a representative set of measurements of dissolved oxygen in the interstitial water on the fertilized shoreline for which results are shown in Fig. 7.4A (130, 131); it is clear that the dissolved oxygen levels are lower on the portion of the beach that received fertilizer, strongly suggesting increased microbial respiration. Figure 7.4C shows concomitant measurements of the rate of CO_2 evolution from the beach, made with a portable infrared spectrophotometer (153). Logistical constraints allowed measurements on only four days, but it is clear that there was substantially more CO_2 evolution from the fertilized part of the beach than from the unfertilized part. Of course, this does not prove that the CO_2 came from the oil, but it at least demonstrates that microbial activity was enhanced by the fertilizer application.

A more definitive approach is radiorespirometry, although it requires a licensed laboratory and sophisticated equipment. In this technique, a ^{14}C-labeled substrate is added to sediment samples, and the evolution of $^{14}CO_2$ is monitored. Lindstrom et al. (88) used this approach with sediments fertilized as part of the bioremediation operations in Prince William Sound following the spill from the *Exxon Val-*

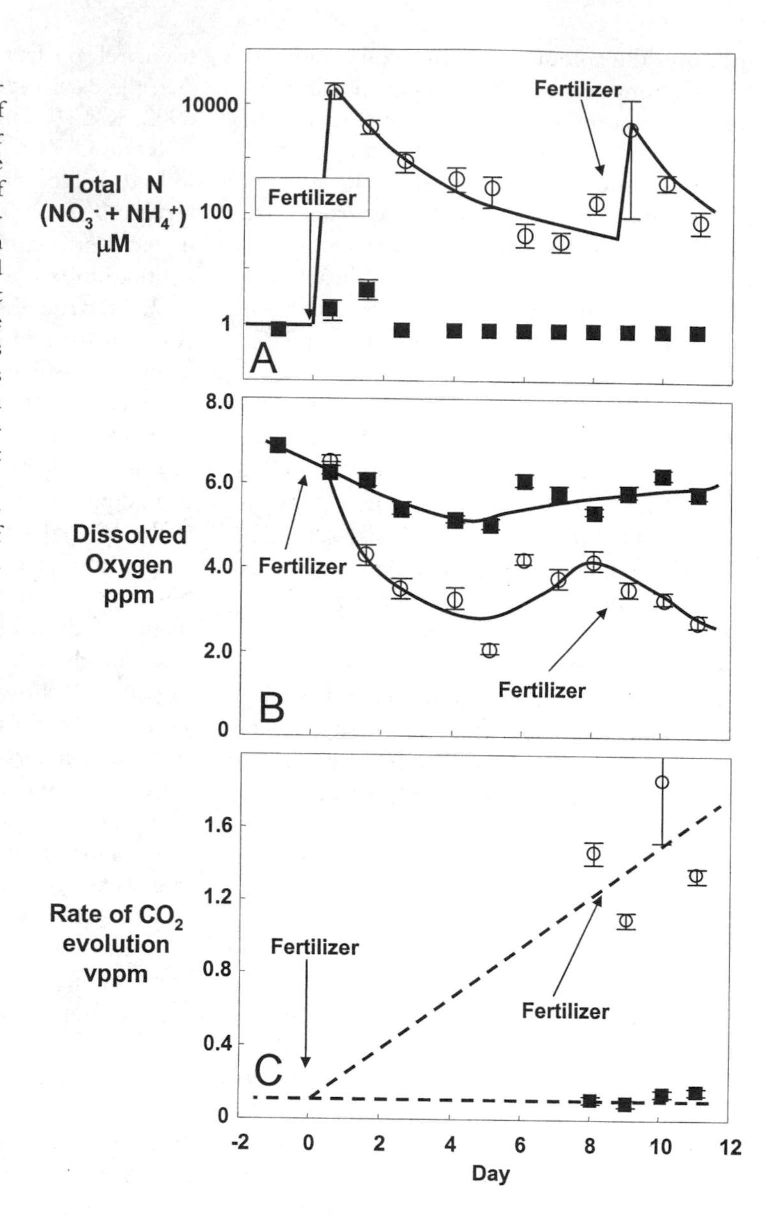

FIGURE 7.4 Some representative data obtained in a field trial of bioremediation on a beach near Sveagruva, on Spitsbergen in the summer of 1997. (A) Amounts of "fixed nitrogen" (nitrate plus ammonium) in interstitial water taken from fertilized and unfertilized portions of the beach. Note that the scale is logarithmic and that the line drawn through the data is merely to guide the eye. Error bars represent standard errors for each point, and those at 1 μM are estimates of the lower detection limit of the colorimetric test kits used. Further details can be found in reference 131. (B) Measurements of dissolved oxygen in that interstitial water. Again, the line is merely to guide the eye, and the error bars are estimates of standard error. (C) Measurements of the rate of CO_2 evolution from undisturbed beach. Data were collected on only four days, but the timescale is the same as for the other panels to allow ready comparison. The dotted lines are shown merely to guide the eye; we have no evidence that the data follow a straight line, but numerous measurements of unoiled, unoiled but fertilized, and oiled unfertilized plots (131) led us to extrapolate to equivalent rates for all sections before fertilizer application. We note that these data demonstrate the variability typically seen in the field and that the simplistic statistical treatment here was verified with more rigorous approaches in the cited references.

dez. Samples were collected from fertilized and unfertilized portions of beaches and incubated with [^{14}C]hexadecane or [^{14}C]phenanthrene, and the evolution of $^{14}CO_2$ was determined after 2 or 3 days, depending on substrate. Figure 7.5A presents the amount of CO_2 evolved by samples from a fertilized part of a beach divided by that from samples from the unfertilized portion. As one might expect, this ratio is close to 1 before the fertilizer was applied, but the ratio steadily increased, for both substrates, after fertilizer application. The ratios remained at about 4 for many days, suggesting a fourfold increase in the rate of biodegradation due to the fertilizer. On day 70, fertilizer was applied to both parts of the beach; this stimulated biodegradation on the part of the beach that had not previously received nutrients, and the ratio fell for both substrates (88). This is very clear evidence that the application of fertilizer stimu-

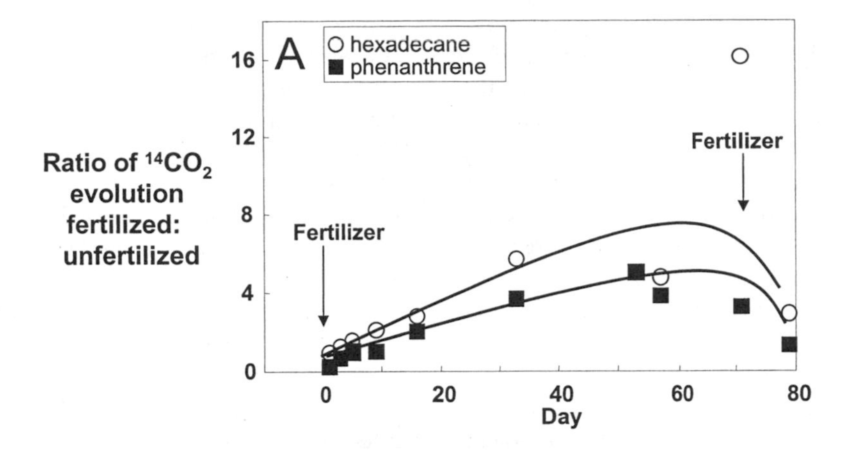

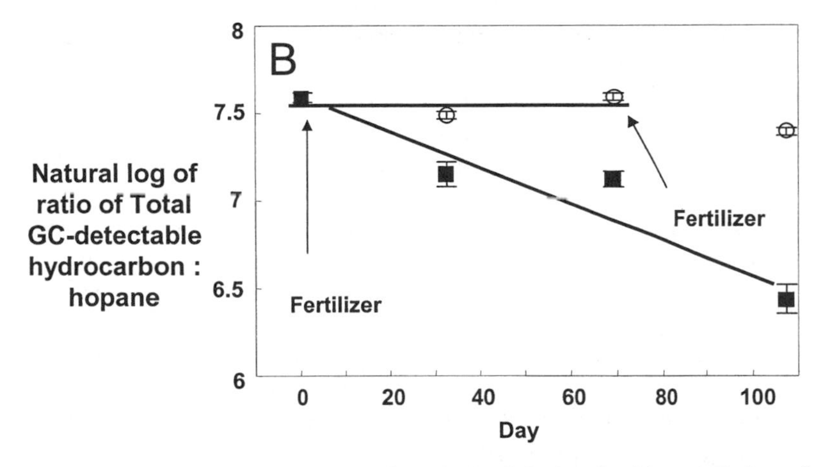

FIGURE 7.5 Some representative data obtained during the bioremediation of the *Exxon Valdez* spill in Alaska in 1990. (A) Representative radiorespirometry data (88). (B) Representative data using hopane as a conserved internal analyte (22) from the same beach. Error bars represent estimates of standard errors, and again we note that these data demonstrate the variability typically seen in the field and that the simplistic statistical treatment here was verified with more rigorous approaches in the cited references. GC, gas chromatography.

lated the biodegradation of hydrocarbons on this beach.

3. Does the chemical composition of residual oil on the beach indicate that it has changed because of biodegradation? It has long been recognized that microorganisms show a preference for biodegrading some compounds in crude oil before others. For example, *n*-alkanes are degraded before the *iso*-alkanes (100), so a decrease in the ratio of *n*-heptadecane to pristane is an early indicator that biodegradation is under way. Pristane itself is readily degraded, however, so it soon fails as a conserved analyte within the oil. Hopanes (120, 127) (Fig. 7.1) are much more resistant to biodegradation, so they serve as excellent conserved reference analytes within the oil. Changes in the relative concentrations of other analytes to an individual hopane can then be attributed to biodegradation (or in some cases to photooxidation [51]), since physical movement of oil should not lead to any change in the ratios of analytes to hopane. A representative data set from the same shoreline as for Fig. 7.5A is shown in Fig. 7.5B. The ratio of the total gas chromatography-detectable hydrocarbons to hopane on fertilized and unfertilized portions of the beach is plotted. The ratio did not change significantly for samples from the unfertilized area, but there was a statistically significant decrease in the ratio on the fertilized part (see reference 22 for a detailed statistical analysis). The data are also suggestive that the ratio began to drop for the unfertilized portion of the beach after fertilizer was added on day 70 (see above). Similar results (22) were seen in the ratio of the *n*-alkanes to hopane and total polycyclic aromatic hydrocarbons on the U.S. EPA priority pollutant list (76), indicating that the bioremediation strategy was indeed leading to the biodegradation of all components of the spilled oil.

In cases where the hopanes are unavailable, such as for diesel, we have used alkylated phenanthrenes as conserved internal analytes, since they are among the most biodegradation-resistant molecules in diesel fuels (40). These also proved useful in the Spitsbergen experiment (130, 131). Since these molecules are known to be biodegradable, it is likely that using them as conserved analytes will actually underestimate the extent of biodegradation, but it seems better to err on the conservative side in assessing field data.

Use of this tiered approach, and comparison of data from fertilized and unfertilized control plots, can give confidence that a bioremediation protocol is indeed effective.

WHEN SHOULD BIOREMEDIATION BE USED FOR TREATING OIL-CONTAMINATED MARINE WATERS AND SEDIMENTS?

Although bioremediation by addition of fertilizers will speed the biodegradation of an oil spill and thereby diminish its environmental impact, it is important to bear in mind that careless application of fertilizers may have unwanted negative impacts on the environment, and they should be used with care. Coastal regions throughout the world are in danger of eutrophication from a wide range of anthropogenic sources (49, 108), and the total amount of nitrogen used in a bioremediation operation should be minimized. Excess inorganic nitrogen is acutely toxic to amphipods and fishes, and it can stimulate unwanted algal and planktonic growth. Ammonia releases into coastal zones from agricultural runoff are sometimes blamed for fish kills. Careful applications, aiming at achieving 100 to 200 μM nitrogen in the interstitial water of an oiled shoreline, should minimize the possibility of these effects, and where they have been monitored there have been no adverse findings (92, 126, 129, 155, 171). Other potential concerns include the possibility that enhanced microbial activity following bioremediation treatment might increase the bioavailability of oil components to other organisms, perhaps by solvation through surfactant release or by the release of partially oxidized intermediates that might be toxic or mutagenic. In fact, no evidence for such effects has been reported (34, 155), and, to the contrary, experiments have shown that bioremediation by nutrient addition leads to a reduction in toxicity (102). Another potential hazard is that increased biodegradation of spilled oil might cause anaerobiosis in previously aerobic sediments. Of course this is a possibility, especially in sediments that are close to anoxic before bioremediation treatment. Nevertheless, it has not been seen in several situations where it might conceivably have occurred (129–131).

As discussed above, there is little evidence to suggest that the asphaltene (polar) fraction of crude oils and heavy fuels is biodegradable, so it is likely to remain even after a successful bioremediation treatment has removed the majority of hydrocarbons from a shoreline. Since the asphaltene fraction is responsible for the color centers in oils, one might expect the black coloration associated with an oil spill to persist. Yet this is not generally the case; in fact, asphaltenes, in the absence of hydrocarbons, lack the oiliness and stickiness of crude oil and no longer adhere to rocks and gravel. Being almost neutrally buoyant, they wash away and become tiny particles, subject to abrasion and dispersion, that are hard to distinguish from humic and fulvic acid residues of more recent biomass.

From the foregoing discussion it is clear that bioremediation through biostimulation of indigenous microbial populations by adding fertilizers can dramatically stimulate the rate of oil biodegradation when the target oil is a relatively thin film of biodegradable material. It is thus quite appropriate that bioremediation should be an important part of the oil spill response "tool kit" for such situations (e.g., see references 29 and 114). While bioremediation clearly works most effectively on highly biodegradable oils and in situations where there are only thin layers of oil on sediment particles, a case can be made for using it more widely during a spill response, even if physical collection of the oil will eventually be achieved.

The principal reason for employing bioremediation is to reduce the ecological impact of a spill and in particular the time during which the ecosystem may be impacted by hydrocarbons. Stimulation of the removal of such compounds by microorganisms before they come in contact with multicellular organisms is bound to decrease the environmental impact of the spill.

Thus, bioremediation should also be considered as part of a cleanup operation even for heavy fuel oils, such as those from the *Erika* (113) and *Prestige* (16), or for Orimulsion (a dispersant-stabilized dispersion of a Cerro Negro bitumen in water) (122, 132), in which only a small fraction of the oil is biodegradable. Removal of the bioavailable fraction more rapidly with a bioremediation treatment should help to minimize the environmental impact of a spill.

CONCLUDING REMARKS

In conclusion, bioremediation is not the panacea for mitigating oil spills—but it is an important tool in reducing the ecological impact of some oil spills. In some cases, it is virtually the only approach for treating a coastal marine oil spill effectively. As evidenced by its use in the *Exxon Valdez* oil spill, bioremediation can be a very effective approach for treating oil-contaminated shorelines, in terms of both cost and efficacy.

REFERENCES

1. **AlGhadban, A. N., F. Abdali, and M. S. Massoud.** 1998. Sedimentation rate and bioturbation in the Arabian Gulf. *Environ. Int.* **24:** 23–31.
2. **Anonymous.** 2003. Industry at a glance. *World Oil* **224(5):**99.
3. **Atlas, R. M.** 1977. Stimulated petroleum biodegradation. *Crit. Rev. Microbiol.* **5:**371–386.
4. **Atlas, R. M.** 1981. Microbial degradation of petroleum hydrocarbons: an environmental perspective. *Microbiol. Rev.* **45:**180–209.
5. **Atlas, R. M.** 1984. *Petroleum Microbiology*. Macmillan, New York, N.Y.
6. **Atlas, R. M., and M. C. Atlas.** 1991. Biodegradation of oil and bioremediation of oil spills. *Curr. Opin. Biotechnol.* **2:**440–443.
7. **Atlas, R. M., and R. Bartha.** 1973. Stimulated biodegradation of oil slicks using oleophilic fertilizers. *Environ. Sci. Technol.* **7:**538–541.
8. **Atlas, R. M., and R. Bartha.** May 1976. Biodegradation of oil on water surfaces; polluting petroleum hydrocarbons. U.S. patent 3,959,127.
9. **Atlas, R. M., G. S. Sayler, R. Burlage, and A. K. Bej.** 1992. Molecular approaches for environmental monitoring of microorganisms. *BioTechniques* **12:**706–717.
10. **Bahn, P. G.** 1992. The making of a mummy. *Nature* **356:**109.
11. **Barakat, A. O., A. R. Mostafa, J. Rullkotter, and A. R. Hegazi.** 1999. Application of a multi-molecular marker approach to fingerprint petroleum pollution in the marine environment. *Mar. Pollut. Bull.* **38:**535–544.
12. **Beller, H. R., and A. M. Spormann.** 1997. Benzylsuccinate formation as a means of anaerobic toluene activation by sulfate-reducing strain PRTOL1. *Appl. Environ. Microbiol.* **63:** 3729–3731.
13. **Bellier, P., and G. Massart.** 1979. The *Amoco Cadiz* oil spill cleanup operation—an overview of the organization, control and evaluation of the cleanup techniques employed, p. 141–146. *In Proceedings of the 1979 International Oil Spill Conference.* American Petroleum Institute, Washington, D.C.
14. **Biegert, T., G. Fuchs, and F. Heider.** 1996. Evidence that anaerobic oxidation of toluene in the denitrifying bacterium *Thauera aromatica* is initiated by formation of benzylsuccinate from tol-

uene and fumarate. *Eur. J. Biochem.* **238:** 661–668.

15. **Boeda, E., J. Connan, D. Dessort, S. Muhesen, N. Mercier, H. Valladas, and N. Tisnerat.** 1996. Bitumen as a hafting material on middle Paleolithic artifacts. *Nature* **380:**336–338.
16. **Bohannon, J., X. Bosch, and J. Withgott.** 2002. Scientists brace for bad tidings after spill. *Science* **298:**1695–1696.
17. **Borden, R. C., C. A. Gomez, and M. T. Becker.** 1995. Geochemical indicators of intrinsic bioremediation. *Ground Water* **33:**180–189.
18. **Bost, F. D., R. Frontera-Suau, T. J. McDonald, K. E. Peters, and P. J. Morris.** 2001. Aerobic biodegradation of hopanes and norhopanes in Venezuelan crude oils. *Org. Geochem.* **32:**105–114.
19. **Boufadel, M. C., P. Reeser, M. T. Suidan, B. A. Wrenn, J. Cheng, X. Du, T. H. L. Huang, and A. D. Venosa.** 1999. Optimal nitrate concentration for the biodegradation of *n*-heptadecane in a variably-saturated sand column. *Environ. Technol.* **20:**191–199.
20. **Bragg, J. R., and E. H. Owens.** 1994. Clay-oil flocculation as a natural cleansing process following oil spills. Part 1. Studies of shoreline sediments and residues from past spills, p. 1–24. *In Proceedings of the Seventeenth Arctic and Marine Oilspill Program (AMOP) Technical Seminar.* Environment Canada, Ottawa, Canada.
21. **Bragg, J. R. and E. H. Owens.** 1995. Shoreline cleansing by interactions between oil and fine mineral particles, p. 219–227. *In Proceedings of the 1995 International Oil Spill Conference.* American Petroleum Institute, Washington, D.C.
22. **Bragg, J. R., R. C. Prince, E. J., Harner, and R. M. Atlas.** 1994. Effectiveness of bioremediation for the *Exxon Valdez* oil spill. *Nature* **368:**413–418.
23. **Bressler, D. C., and P. M. Fedorak.** 2000. Bacterial metabolism of fluorene, dibenzofuran, dibenzothiophene, and carbazole. *Can. J. Microbiol.* **46:**397–409.
24. **Buist, I.** 2003. Window-of-opportunity for *in situ* burning. *Spill Sci. Technol. Bull.* **8:**341–346.
25. **Bundy, J. G., A. W. J. Morriss, D. G. Durham, C. D. Campbell, and G. I. Paton.** 2001. Development of QSARs to investigate the bacterial toxicity and biotransformation potential of aromatic heterocyclic compounds. *Chemosphere* **42:**885–892.
26. **Burdon, J.** 2001. Are the traditional concepts of the structures of humic substances realistic? *Soil Sci.* **166:**752–769.
27. **Burns, K. A., S. Codi, and N. C. Duke.** 2000. Gladstone, Australia field studies: weathering and degradation of hydrocarbons in oiled mangrove and salt marsh sediments with and without the application of an experimental bioremediation protocol. *Mar. Pollut. Bull.* **41:** 392–402.
28. **Caldwell, M. E., R. M. Garrett, R. C. Prince, and J. M. Suflita.** 1998. Anaerobic biodegradation of long-chain *n*-alkanes under sulfate-reducing conditions. *Environ. Sci. Technol.* **32:**2191–2195.
29. **Canadian Coast Guard.** 1995. *Oil Spill Response Field Guide.* Canadian Coast Guard, Ottawa, Canada.
30. **Cervantes, F. J., W. Dijksma, T. Duong-Dac, A. Ivanova, G. Lettinga, and J. A. Field.** 2001. Anaerobic mineralization of toluene by enriched sediments with quinones and humus as terminal electron acceptors. *Appl. Environ. Microbiol.* **67:**4471–4478.
31. **Chakrabarty, A. M.** March 1981. Microorganisms having multiple compatible degradative energy-generating plasmids, and preparation thereof. U.S. patent 4,259,444.
32. **Chang, Y. J., J. R. Stephen, A. P. Richter, A. D. Venosa, J. Bruggemann, S. J. Macnaughton, G. A. Kowalchuk, J. R. Haines, E. Kline, and D. C. White.** 2000. Phylogenetic analysis of aerobic freshwater and marine enrichment cultures efficient in hydrocarbon degradation: effect of profiling method. *J. Microbiol. Methods* **40:**19–31.
33. **Charles Darwin Foundation for the Galapagos Islands.** 2001. *Galapagos Oil Spill: Biological Impacts of the* Jessica *Oil Spill on the Galapagos Environment.* [Online.] http://www.darwinfoundation.org/oilspill.html.
34. **Claxton, L. D., V. S. Houk, R. Williams, and F. Kremer.** 1991. Effect of bioremediation on the mutagenicity of oil spilled in Prince William Sound, Alaska. *Chemosphere* **23:**643–650.
35. **Clayton, J. R., J. R. Payne, and J. S. Farlow.** 1992. *Oil Spill Dispersants: Mechanisms of Action and Laboratory Tests.* Lewis Publishers, Boca Raton, Fla.
36. **Coates, J. D., J. Woodward, J. Allen, P. Philp, and D. R. Lovley.** 1997. Anaerobic degradation of polycyclic aromatic hydrocarbons and alkanes in petroleum-contaminated marine harbor sediments. *Appl. Environ. Microbiol.* **63:** 3589–3593.
37. **Cunningham, J. A., H. Rahme, G. D. Hopkins, C. Lebron, and M. Reinhard.** 2000. Enhanced *in situ* bioremediation of BTEX contaminated groundwater by combined injection of nitrate and sulfate. *Environ. Sci. Technol.* **35:** 1663–1670.
38. **Denissenko, M. F., A. Pao, M. Tang, and**

G. P. Pfeifer. 1996. Preferential formation of benzo[a]pyrene adducts at lung cancer mutational hotspots in P53. *Science* **274:**430–432.

39. **Dibble, J. T., and R. Bartha.** 1976. Effect of iron on the biodegradation of petroleum in seawater. *Appl. Environ. Microbiol.* **31:**544–550.
40. **Douglas, G. S., K. J. McCarthy, D. T. Dahlen, J. A. Seavey, W. G. Steinhauer, R. C. Prince, and D. L. Elmendorf.** 1992. The use of hydrocarbon analyses for environmental assessment and remediation. *J. Soil Contam.* **1:** 197–216.
41. **Downing, J. A., C. W. Osenberg, and O. Sarnelle.** 1999. Meta-analysis of marine nutrient-enrichment experiments: variation in the magnitude of nutrient limitation. *Ecology* **80:** 1157–1167.
42. **Duke, N. C., K. A. Burns, R. P. J. Swannell, O. Dalhaus, and R. J. Rupp.** 2000. Dispersant use and a bioremediation strategy as alternate means of reducing impacts of large oil spills on mangroves: the Gladstone field trials. *Mar. Pollut. Bull.* **41:**403–412.
43. **Dutta, T. K., and S. Harayama.** 2001. Biodegradation of *n*-alkylcycloalkanes and *n*-alkylbenzenes via new pathways in *Alcanivorax* sp. strain MBIC 4326. *Appl. Environ. Microbiol.* **67:** 1970–1974.
44. **Edgar, G. J., H. L. Snell, and L. W. Lougheed.** 2003. Impacts of the *Jessica* oil spill: an introduction. *Mar. Pollut. Bull.* **47:**273–275.
45. **Eriksson, M., A. Swartling, and G. Dalhammar.** 1998. Biological degradation of diesel fuel in water and soil monitored with solid-phase micro-extraction and GC-MS. *Appl. Microbiol. Biotechnol.* **50:**129–134.
46. **Etkin, D. S.** 1999. Historical overview of oil spills from all sources, p. 1097–1102. *In Proceedings of the 1999 International Oil Spill Conference*. American Petroleum Institute, Washington, D.C.
47. **Fiocco, R. J., and A. Lewis.** 1999. Oil spill dispersants. *Pure Appl. Chem.* **71:**27–42.
48. **Foght, J., K. Semple, C. Gauthier, D. W. S. Westlake, S. Blenksopp, G. Sergy, Z. Wang, and M. Fingas.** 1999. Effect of nitrogen source on biodegradation of crude oil by a defined bacterial consortium incubated under cold, marine conditions. *Environ. Technol.* **20:** 839–849.
49. **Forsberg, C.** 1995. The large-scale flux of nutrients from land to water and the eutrophication of lakes and marine waters. *Mar. Pollut. Bull.* **29:** 409–413.
50. **Frontera-Suau, R., F. D. Bost, T. J. McDonald, and P. J. Morris.** 2002. Aerobic biodegradation of hopanes and other biomarkers by crude oil-degrading enrichment cultures. *Environ. Sci. Technol.* **36:**4585–4592.
51. **Garrett, R. M., I. J. Pickering, C. E. Haith, and R. C. Prince.** 1998. Photooxidation of crude oils. *Environ. Sci. Technol.* **32:**3719–3723.
52. **GeorgeAres, A., and J. R. Clark.** 2000. Aquatic toxicity of two Corexit dispersants. *Chemosphere* **40:**897–906.
53. **Goodman, R.** 2003. Tar balls: the end state. *Spill Sci. Technol. Bull.* **8:**117–121.
54. **Grossman, M. J., R. C. Prince, R. M. Garrett, K. K. Garrett, R. E. Bare, K. R. O'Neil, M. R. Sowlay, S. M. Hinton, K. Lee, G. A. Sergy, E. H. Owens, and C. C. Guénette.** 2000. Microbial diversity in oiled and unoiled shoreline sediments in the Norwegian Arctic, p. 775–789. *In* C. R. Bell, M. Brylinski, and P. Johnson-Green (ed.), *Proceedings of the 8th International Symposium on Microbial Ecology*. Atlantic Canada Society for Microbial Ecology, Halifax, Nova Scotia, Canada.
55. **Gundlach, E.** 1997. Comparative photographs of the *Metula* spill site, 21 years later, p. 1042–1044. *In Proceedings of the 1997 International Oil Spill Conference*. American Petroleum Institute, Washington, D.C.
56. **Harayama, S., H. Kishira, Y. Kasai, and K. Shutsubo.** 2000. Petroleum biodegradation in marine environments, p. 123–137. *In* D. H. Bartlett (ed.), *Molecular Marine Microbiology*. Horizon Scientific Press, Wymondham, United Kingdom.
57. **Head, I., and R. J. P. Swannell.** 1999. Bioremediation of petroleum hydrocarbon contaminants in marine habitats. *Curr. Opin. Biotechnol.* **10:**234–239.
58. **Heider, J., A. M. Spormann, H. R. Beller, and F. Widdel.** 1999. Anaerobic bacterial metabolism of hydrocarbons. *FEMS Microbiol. Rev.* **22:**459–473.
59. **Hernaez, M. J., W. Reineke, and E. Santero.** 1999. Genetic analysis of biodegradation of tetralin by a *Sphingomonas* strain. *Appl. Environ. Microbiol.* **65:**1806–1810.
60. **Herrick, J. B., K. G. Stuart-Keil, W. C. Ghiorse, and E. L. Madsen.** 1997. Natural horizontal transfer of a naphthalene dioxygenase gene between bacteria native to a coal tar-contaminated field site. *Appl. Environ. Microbiol.* **63:** 2330–2337.
61. **Holden, C.** 1990. Gulf slick a free lunch for bacteria. *Science* **249:**120.
62. **Hozumi, T., H. Tsutsumi, and M. Kono.** 2000. Bioremediation on the shore after an oil spill from the *Nakhodka* in the Sea of Japan. I. Chemistry and characteristics of heavy oil loaded on the *Nakhodka* and biodegradation tests by a

bioremediation agent with microbiological cultures in the laboratory. *Mar. Pollut. Bull.* **40:** 308–314.

63. **Hunt, J. M.** 1996. *Petroleum Geochemistry and Geology*, 2nd ed. W. H. Freeman, New York, N. Y.

64. **Hutchins, S. R., D. E. Miller, and A. Thomas.** 1998. Combined laboratory/field study on the use of nitrate for *in situ* bioremediation of a fuel-contaminated aquifer. *Environ. Sci. Technol.* **32:**1832–1840.

65. **Ibrahim, Y. A., M. A. Abdelhameed, T. A. Al-Sahhaf, and M. A. Fahim.** 2003. Structural characterization of different asphaltenes of Kuwaiti origin. *Petrol. Sci. Technol.* **21:**825–837.

66. **International Agency for Research on Cancer.** 1983. *IARC Monograph on the Evaluation of the Carcinogenic Risk of Chemicals to Humans. Polynuclear Aromatic Hydrocarbons.* Part 1. *Chemical, Environmental and Experimental Data,* vol. 32. World Health Organization, Geneva, Switzerland.

67. **Iwabuchi, N., M. Sunairi, M. Urai, C. Itoh, H. Anzai, M. Nakajima, and S. Harayama.** 2002. Extracellular polysaccharides of *Rhodococcus rhodochrous* S-2 stimulate the degradation of aromatic components in crude oil by indigenous marine bacteria. *Appl. Environ. Microbiol.* **68:** 2337–2343.

68. **Iwamoto, T., and M. Nasu.** 2001. Current bioremediation practice and perspective. *J. Biosci. Bioeng.* **92:**1–8.

69. **Jackson, W. A., and J. H. Pardue.** 1999. Potential for enhancement of biodegradation of crude oil in Louisiana salt marshes using nutrient amendments. *Water Air Soil Pollut.* **109:**343–355.

70. **Juck, D., T. Charles, L. G. Whyte, and C. W. Greer.** 2000. Polyphasic microbial community analysis of petroleum hydrocarbon-contaminated soils from two northern Canadian communities. *FEMS Microbiol. Ecol.* **33:**241–249.

71. **Juhasz, A. L., and R. Naidu.** 2000. Bioremediation of high molecular weight polycyclic aromatic hydrocarbons: a review of the microbial degradation of benzo[a]pyrene. *Int. Biodeter. Biodegrad.* **45:**57–88.

72. **Kanaly, R. A., and S. Harayama.** 2000. Biodegradation of high-molecular-weight polycyclic aromatic hydrocarbons by bacteria. *J. Bacteriol.* **182:**2059–2067.

73. **Kasai, Y., H. Kishira, K. Syutsubo, and S. Harayama.** 2001. Molecular detection of marine bacterial populations on beaches contaminated by the *Nakhodka* tanker oil-spill accident. *Environ. Microbiol.* **13:**246–255.

74. **Kasai, Y., H. Kishira, T. Sasaki, K. Syutsubo, K. Watanabe, and S. Harayama.** 2002. Predominant growth of *Alcanivorax* strains in oil-contaminated and nutrient-supplemented sea water. *Environ. Microbiol.* **4:**141–147.

75. **Kato, T., M. Haruki, T. Imanaka, M. Morikawa, and S. Kanaya.** 2001. Isolation and characterization of long-chain-alkane degrading *Bacillus thermoleovorans* from deep subterranean petroleum reservoirs. *J. Biosci. Bioeng.* **91:**64–70.

76. **Keith, L. H., and W. A. Telliard.** 1979. Priority pollutants. I. A perspective view. *Environ. Sci. Technol.* **13:**416–423.

77. **Kvenvolden, K. A., F. D. Hostettler, P. R. Carlson, J. B. Rapp, C. N. Threlkeld, and A. Warden.** 1995. Ubiquitous tar balls with a California-source signature on the shorelines of Prince William Sound, Alaska. *Environ. Sci. Technol.* **29:**2684–2694.

78. **Latha, K., and D. Lalithakumari.** 2001. Transfer and expression of a hydrocarbon-degrading plasmid pHCL from *Pseudomonas putida* to marine bacteria. *World J. Microbiol. Biotechnol.* **17:**523–528.

79. **Leahy, J. G., and R. R. Colwell.** 1990. Microbial degradation of hydrocarbons in the environment. *Microbiol. Rev.* **54:**305–315.

80. **Lee, K.** 1999. Bioremediation of oil impacted shorelines, p. 69–85. *In* T. Murphy and M. Munawar (ed.), *Aquatic Restoration in Canada.* Ecovision World Monograph Series. Backhuys Publishers, Leiden, The Netherlands.

81. **Lee, K.** 2003. Oil-particle interactions in aquatic environments: influence on the transport, fate, effect and remediation of oil spills. *Spill Sci. Technol. Bull.* **8:**3–8.

82. **Lee, K., and S. deMora.** 1999. *In situ* bioremediation strategies for oiled shoreline environments. *Environ. Technol.* **20:**783–794.

83. **Lessard, R. R., and G. DeMarco.** 2000. The significance of oil spill dispersants. *Spill Sci. Technol. Bull.* **6:**59–68.

84. **Leveille, T. P.** 1991. The *Mega Borg* fire and oil spill: a case study, p. 273–278. *In Proceedings of the 1991 International Oil Spill Conference.* American Petroleum Institute, Washington, D.C.

85. **Levy, E. M., and K. Lee.** 1988. Potential contribution of natural hydrocarbon seepage to benthic productivity and the fisheries of Atlantic Canada. *Can. J. Fish. Aquat. Sci.* **45:**349–352.

86. **Limpert, E., W. A. Stahel, and M. Abbt.** 2001. Log-normal distributions across the sciences: keys and clues. *Bioscience* **51:**341–352.

87. **Lin, Q., I. A. Mendelssohn, C. B. Henry, P. O. Roberts, M. M. Walsh, E. B. Overton, and R. J. Portier.** 1999. Effects of bioremediation agents on oil degradation in mineral and sandy salt marsh sediments. *Environ. Technol.* **20:** 825–837.

88. **Lindstrom, J. E., R. C. Prince, J. R. Clark, M. J. Grossman, T. R. Yeager, J. F. Braddock, and E. J. Brown.** 1991. Microbial populations and hydrocarbon biodegradation potentials in fertilized shoreline sediments affected by the T/V *Exxon Valdez* oil spill. *Appl. Environ. Microbiol.* **57:**2514–2522.

89. **Lovley, D. R.** 2000. Anaerobic benzene degradation. *Biodegradation* **11:**107–116.

90. **Lovley, D. R., J. C. Woodward, and F. H. Chapelle.** 1994. Stimulated anoxic biodegradation of aromatic hydrocarbons using Fe(III) ligands. *Nature* **370:**128–131.

91. **Lunel, T., J. Rusin, C. Halliwell, and L. Davies.** 1997. The net environmental benefit of a successful dispersant operation at the *Sea Empress* incident, p. 185–194. *In Proceedings of the 1997 International Oil Spill Conference.* American Petroleum Institute, Washington, D. C.

92. **Lung, W. S., J. L. Martin, and S. C. McCutcheon.** 1993. Eutrophication analysis of embayments in Prince William Sound, Alaska. *J. Environ. Eng.* **119:**811–824.

93. **Macnaughton, S. J., J. R. Stephen, A. D. Venosa, G. A. Davis, Y. J. Chang, and D. C. White.** 1999. Microbial population changes during bioremediation of an experimental oil spill. *Appl. Environ. Microbiol.* **65:**3566–3574.

94. **Margesin, R., and F. Schinner.** 2001. Biodegradation and bioremediation of hydrocarbons in extreme environments. *Appl. Microbiol. Biotechnol.* **56:**650–663.

95. **Mathew, M., J. P. Obbard, Y. P. Ting, Y. H. Gin, and H. M. Tan.** 1999. Bioremediation of oil contaminated beach sediments using indigenous microorganisms in Singapore. *Acta Biotechnol.* **19:**225–233.

96. **McMillen, S. J., A. G. Requejo, G. N. Young, P. S. Davis, P. D. Cook, J. M. Kerr, and N. R. Gray.** 1995. Bioremediation potential of crude oil spilled on soil, p. 91–99. *In* R. E. Hinchee, C. M. Vogel, and F. J. Brockman (ed.), *Microbial Processes for Bioremediation.* Battelle Press, Columbus, Ohio.

97. **McNally, D. L., J. R. Mihelcic, and D. R. Lueking.** 1998. Biodegradation of three- and four-ring polycyclic aromatic hydrocarbons under aerobic and denitrifying conditions. *Environ. Sci. Technol.* **32:**2633–2639.

98. **Melendez-Colon, V. J., A. Luch, A. Seidel, and W. M. Baird.** 1999. Comparison of cytochrome P450- and peroxidase-dependent metabolic activation of the potent carcinogen dibenzo[a,1]pyrene in human cell lines: formation of stable DNA adducts and absence of a detectable increase in apurinic sites. *Cancer Res.* **59:** 1412–1416.

99. **Michel, J., and B. L. Benggio.** 1995. Testing and use of shoreline cleaning agents during the *Morris J. Berman* spill, p. 197–202. *In Proceedings of the 1995 International Oil Spill Conference.* American Petroleum Institute, Washington, D.C.

100. **Miget, R. J., C. H. Oppenheimer, H. I. Kator, and D. A. LaRock.** 1969. Microbial degradation of normal paraffin hydrocarbons in crude oil, p. 327–331. *In Proceedings of the Joint Conference on Prevention and Control of Oil Spills.* American Petroleum Institute, Washington, D.C.

101. **Mills, M. A., J. S. Bonner, M. A. Simon, T. J. McDonald, and R. L. Autenreith.** 1997. Bioremediation of a controlled oil release in a wetland, p. 609–616. *In Proceedings of the Twentieth Arctic and Marine Oilspill Program (AMOP) Technical Seminar.* Environment Canada, Ottawa, Canada.

102. **Mueller, D. C., J. S. Bonner, R. L. Autenrieth, K. Lee, and K. Doe.** 1999. The toxicity of oil-contaminated sediments during bioremediation of a wetland, p. 1049–1052. *In Proceedings of the 1999 International Oil Spill Conference.* American Petroleum Institute, Washington, D.C.

103. **Mullin, J. V., and M. A. Champ.** 2003. Introduction/overview to *in situ* burning of oil spills. *Spill Sci. Technol. Bull.* **8:**323–330.

104. **National Research Council.** 1985. *Oil in the Sea: Inputs, Fates and Effects.* National Academy Press, Washington, D.C.

105. **National Research Council.** 1989. *Using Oil Spill Dispersants on the Sea.* National Academy Press, Washington, D.C.

106. **National Research Council.** 2002. *Oil in the Sea.* III: *Inputs, Fates and Effects.* National Academy Press, Washington, D.C.

107. **Nauman, S. A.** 1991. Shoreline clean-up: equipment and operations, p. 431–438. *In Proceedings of the 1991 International Oil Spill Conference.* American Petroleum Institute, Washington, D.C.

108. **Nixon, S. W.** 1995. Coastal marine eutrophication: a definition, social causes, and future concerns. *Ophelia* **41:**199–219.

109. **Nodar, M. A., and M. del Rosario Martino.** 1999. *San Jorge* oil spill, an experience from Uruguay, p. 1107. *In Proceedings of the 1999 International Oil Spill Conference.* American Petroleum Institute, Washington, D.C.

110. **Olagbende, O. T., G. O. Ede, L. E. D. Inyang, E. R. Gundlach, E. S. Gilfillan, and D. S. Page.** 1999. Scientific and cleanup response to the IDOHO-QIT oil spill, Nigeria. *Environ. Technol.* **20:**1213–1222.

111. **Olivieri, R., P. Bacchin, A. Robertiello, N. Oddo, L. Degen, and A. Tonolo.** 1976. Mi-

crobial degradation of oil spills enhanced by a slow-release fertilizer. *Appl. Environ. Microbiol.* **31:**629–634.

112. **Omiecinski, C. J., R. P. Remmel, and V. P. Hosagrahara.** 1999. Concise review of the cytochrome P450s and their role in toxicology. *Toxicol. Sci.* **48:**151–156.

113. **Oudot, J.** 2000. Biodegradability of the *Erika* fuel oil. *Comptes Rendus Serie III* **323:**945–950.

114. **Owens, E. (ed.).** 1996. *Field Guide for the Protection and Cleanup of Oiled Arctic Shorelines.* Environment Canada, Prairie and Northern Region, Yellowknife, Northwest Territories, Canada.

115. **Owens, E. H., J. R. Harper, W. Robson, and P. D. Boehm.** 1987. Fate and persistence of crude oil stranded on a sheltered beach. *Arctic* **40:**109–123.

116. **Owens, E. H., G. A. Sergy, L. Gusmán, Z. Wang, and J. Baker.** 1999. Long-term salt marsh recovery and pavement persistence at *Metula* spill sites, p. 847–863. *In Proceedings of the Twenty-Second Arctic and Marine Oilspill Program (AMOP) Technical Seminar.* Environment Canada, Ottawa, Canada.

117. **Owens, E. H., A. M. Sienkiewicz, and G. A. Sergy.** 1999. Evaluation of shoreline cleaning versus natural recovery: the *Metula* spill and the KOMI operations, p. 503–509. *In Proceedings of the 1999 International Oil Spill Conference.* American Petroleum Institute, Washington, D.C.

118. **Page, C. A., J. S. Bonner, P. L. Summer, and R. L. Autenrieth.** 2000. Solubility of petroleum hydrocarbons in oil/water systems. *Mar. Chem.* **70:**79–87.

119. **Page, C. A., J. S. Bonner, P. L. Summer, T. J. McDonald, R. L. Autenrieth, and C. B. Fuller.** 2000. Behavior of a chemically-dispersed oil and a whole oil on a near-shore environment. *Water Res.* **34:**2507–2516.

120. **Peters, K., and J. M. Moldowan.** 1993. *The Biomarker Guide; Interpreting Molecular Fossils in Petroleum and Ancient Sediments.* Prentice-Hall, Englewood Cliffs, N.J.

121. **Phelps, C. D., and L. Y. Young.** 2001. Biodegradation of BTEX under anaerobic conditions: a review. *Adv. Agron.* **70:**329–357.

122. **Potter, T. L., and B. Duval.** 2001. Cerro Negro bitumen degradation by a consortium of marine benthic microorganisms. *Environ. Sci. Technol.* **35:**76–83.

123. **Prince, R. C.** 1993. Petroleum spill bioremediation in marine environments. *Crit. Rev. Microbiol.* **19:**217–242.

124. **Prince, R. C.** 1998. Crude oil biodegradation, p. 1327–1342. *In* R. A. Meyers (ed.), *The Encyclopedia of Environmental Analysis and Remediation,* vol. 2. John Wiley, New York, N.Y.

125. **Prince, R. C.** 2002. Petroleum and other hydrocarbons, biodegradation of, p. 2402–2416. *In* G. Bitton (ed.), *Encyclopedia of Environmental Microbiology.* John Wiley, New York, N.Y.

126. **Prince, R. C., and J. R. Bragg.** 1997. Shoreline bioremediation following the *Exxon Valdez* oil spill in Alaska. *Bioremed. J.* **1:**97–104.

127. **Prince, R. C., D. L. Elmendorf, J. R. Lute, C. S. Hsu, C. E. Haith, J. D. Senius, G. J. Dechert, G. S. Douglas, and E. L. Butler.** 1994. 17α(H), 21β(H)-hopane as a conserved internal marker for estimating the biodegradation of crude oil. *Environ. Sci. Technol.* **28:**142–145.

128. **Prince, R. C., R. E. Bare, G. N. George, C. E. Haith, M. J. Grossman, J. R. Lute, D. L. Elmendorf, V. Minak-Bernero, J. D. Senius, L. G. Keim, R. R. Chianelli, S. M. Hinton, and A. R. Teal.** 1993. The effect of bioremediation on the microbial populations of oiled beaches in Prince William Sound, Alaska, p. 469–475. *In Proceedings of the 1993 International Oil Spill Conference.* American Petroleum Institute, Washington, D.C.

129. **Prince, R. C., J. R. Clark, J. E. Lindstrom, E. L. Butler, E. J. Brown, G. Winter, M. J. Grossman, R. R. Parrish, R. E. Bare, J. F. Braddock, W. G. Steinhauer, G. S. Douglas, J. M. Kennedy, P. J. Barter, J. R. Bragg, E. J. Harner, and R. M. Atlas.** 1994. Bioremediation of the *Exxon Valdez* oil spill: monitoring safety and efficacy, p. 107–124. *In* R. E. Hinchee, B. C. Alleman, R. E. Hoeppel, and R. N. Miller (ed.), *Hydrocarbon Remediation.* Lewis Publishers, Boca Raton, Fla.

130. **Prince, R. C., R. E. Bare, R. M. Garrett, M. J. Grossman, C. E. Haith, L. G. Keim, K. Lee, G. J. Holtom, P. Lambert, G. A. Sergy, E. H. Owens, and C. C. Guénette.** 1999. Bioremediation of a marine oil spill in the Arctic, p. 227–232. *In* B. C. Alleman and A. Leeson (ed.), In Situ *Bioremediation of Petroleum Hydrocarbon and Other Organic Compounds.* Battelle Press, Columbus, Ohio.

131. **Prince, R. C., R. E. Bare, R. M. Garrett, M. J. Grossman, C. E. Haith, L. G. Keim, K. Lee, G. J. Holtom, P. Lambert, G. A. Sergy, E. H. Owens, and C. C. Guénette.** 2003. Bioremediation of stranded oil on an Arctic shoreline. *Spill Sci. Technol. Bull.* **8:**303–312.

132. **Proctor, L. M., E. Toy, L. Lapham, J. Cherrier, and J. P. Chanton.** 2001. Enhancement of Orimulsion biodegradation through the addition of natural marine carbon substrates. *Environ. Sci. Technol.* **35:**1420–1424.

133. **Radwan, S. S., and A. S. AlMuteirie.** 2001. Vitamin requirements of hydrocarbon-utilizing soil bacteria. *Microbiol. Res.* **155:**301–307.

134. **Raghavan, P. U. M., and M. Vivekanandan.** 1999. Bioremediation of oil-spilled sites through seeding of naturally adapted *Pseudomonas putida*. *Int. Biodeter. Biodegrad.* **44:**29–32.

135. **Reilly, T. J., F. Csulak, and P. Van Cott.** 1999. Selecting a preferred restoration alternative for the *Julie N* oil spill, p. 1089–1092. *In Proceedings of the International Oil Spill Conference.* American Petroleum Institute, Washington, D.C.

136. **Rice, J. A.** 2001. Humin. *Soil Sci.* **166:**848–857.

137. **Ripp, S., D. E. Nivens, Y. Ahn, C. Werner, J. Jarrell, J. P. Easter, C. D. Cox, R. S. Burlage, and G. S. Sayler.** 2000. Controlled field release of a bioluminescent genetically engineered microorganism for bioremediation process monitoring and control. *Environ. Sci. Technol.* **34:**846–853.

138. **Robbins, W. K., and C. S. Hsu.** 1996. Petroleum composition, p. 352–370. *In* M. Howe-Grant (ed.), *Kirk-Othmer Encyclopedia of Chemical Technology*, 4th ed., vol. 18. John Wiley & Sons, New York, N.Y.

139. **Rosenberg, E., R. Legman, A. Kushmaro, E. Adler, H. Abir, and E. Z. Ron.** 1996. Oil bioremediation using insoluble nitrogen source. *J. Biotechnol.* **51:**273–278

140. **Rosenberg, E., R. Legman, A. Kushmaro, R. Taube, E. Adler, and E. Z. Ron.** 1992. Petroleum bioremediation—a multiphase problem. *Biodegradation* **3:**337–350.

141. **Ross, S. L., C. W. Ross, F. Lepine, and R. K. Langtry.** 1980. IXTOC-I oil blowout, p. 25–38. *In Proceedings of a Symposium on Preliminary Results from the September 1979 Researcher/Pierce IXTOC-I Cruise.* National Oceanic and Atmospheric Administration, Boulder, Colo.

142. **Schneider, J., R. J. Grosser, K. Jayasimhulu, W. L. Xue, B. Kinkle, and D. Warshawsky.** 2000. Biodegradation of carbazole by *Ralstonia* sp. RJGII.123 isolated from a hydrocarbon contaminated soil. *Can. J. Microbiol.* **46:**269–277.

143. **Sergy, G., C. C. Guenette, E. Owens, R. C. Prince, and K. Lee.** 1999. Treatment of oiled sediment shorelines by sediment relocation, p. 549–554. *In Proceedings of the 1999 International Oil Spill Conference.* American Petroleum Institute, Washington, D.C.

144. **Shikada, A.** 1999. The statistical analysis of manual removal of stranded oils; lessons learned from oil spills caused by the Russian tanker *Nakhodka*, p. 1119–1121. *In Proceedings of the 1999 International Oil Spill Conference.* American Petroleum Institute, Washington, D.C.

145. **Shin, W. S., J. H. Pardue, W. A. Jackson, and S. J. Choi.** 2001. Nutrient enhanced biodegradation of crude oil in tropical salt marshes. *Water Air Soil Pollut.* **131:**135–152.

146. **Simoneit, B. R. T., and P. F. Lonsdale.** 1982. Hydrothermal petroleum in mineralized mounds at the seabed of Guaymas Basin. *Nature* **295:**198–202.

147. **SolanoSerena, F., R. Marchal, M. Ropars, J. M. Lebeault, and J. P. Vandecasteele.** 1999. Biodegradation of gasoline: kinetics, mass balance, and fate of individual hydrocarbons. *J. Appl. Microbiol.* **86:**1008–1016.

148. **Sousa, S., C. Duffy, H. Weitz, L. A. Glover, and E. Bar.** 1998. Use of a lux-modified bacterial biosensor to identify constraints to bioremediation of BTEX-contaminated sites. *Environ. Toxicol. Chem.* **17:**1039–1045.

149. **Spormann, A. M., and F. Widdel.** 2000. Metabolism of alkylbenzenes, alkanes, and other hydrocarbons in anaerobic bacteria. *Biodegradation* **11:**85–105.

150. **Stephen, J. R., Y. J. Chang, Y. D. Gan, A. Peacock, S. M. Pfiffner, M. J. Barcelona, D. C. White, and S. J. Macnaughton.** 1999. Microbial characterization of a JP-4 fuel-contaminated site using a combined lipid biomarker/polymerase chain reaction-denaturing gradient gel electrophoresis (PCR-DGGE)-based approach. *Environ. Microbiol.* **1:**231–241.

151. **Sticher, P., M. C. M. Jaspers, K. Stemmler, H. Harms, A. J. B. Zehnder, and J. R. van der Meer.** 1997. Development and characterization of a whole-cell bioluminescent sensor for bioavailable middle-chain alkanes in contaminated groundwater samples. *Appl. Environ. Microbiol.* **63:**4053–4060.

152. **Swannell, R. P. J., and F. Daniel.** 1999. Effect of dispersants on oil biodegradation under simulated marine conditions, p. 169–176. *In Proceedings of the 1999 International Oil Spill Conference.* American Petroleum Institute, Washington, D.C.

153. **Swannell, R. P. J., K. Lee, A. Basseres, and F. X. Merlin.** 1994. A direct respirometric method for *in-situ* determination of bioremediation efficiency, p. 1273–1286. *In Proceedings of the Seventeenth Arctic Marine Oilspill Program Technical Seminar.* Environment Canada, Ottawa, Canada.

154. **Swannell, R. P. J., K. Lee, and M. McDonagh.** 1996. Field evaluations of marine spill bioremediation. *Microbiol. Rev.* **60:**342–365.

155. **Swannell, R. P. J., D. Mitchell, G. Lethbridge, D. Jones, D. Heath, M. Hagley, M. Jones, S. Petch, R. Milne, R. Croxford, and K. Lee.** 1999. A field demonstration of the efficacy of bioremediation to treat oiled shorelines following the *Sea Empress* incident. *Environ. Technol.* **20:**863–873.

156. **Tay, S. T. L., F. H. Hemond, L. R. Krumholz, C. M. Cavanaugh, and M. F. Polz.**

2001. Population dynamics of two toluene degrading bacterial species in a contaminated stream. *Microb. Ecol.* **41:**124–131.

157. **Teas, H., R. R. Lessard, G. P. Canevari, C. P. Brown, and R. Glenn.** 1993. Saving oiled mangroves using a new non-dispersing shoreline cleaner, p. 761–763. *In Proceedings of the 1993 International Oil Spill Conference.* American Petroleum Institute, Washington, D.C.
158. **Thouand, G., P. Bauda, J. Oudot, G. Kirsch, C. Sutton, and J. F. Vidalie.** 1999. Laboratory evaluation of crude oil biodegradation with commercial or natural microbial inocula. *Can. J. Microbiol.* **45:**106–115.
159. **Tissot, B. P., and D. H. Welte.** 1984. *Petroleum Formation and Occurrence.* Springer-Verlag, Berlin, Germany.
160. **Tsutsumi, H., M. Kono, K. Takai, T. Manabe, M. Haraguchi, I. Yamamoto, and C. Oppenheimer.** 2000. Bioremediation on the shore after an oil spill from the *Nakhodka* in the Sea of Japan. III. Field tests of a bioremediation agent with microbiological cultures for the treatment of an oil spill. *Mar. Pollut. Bull.* **40:** 320–324.
161. **U.S. Department of Transportation.** 1992. 33 CFR Part 155, Vessel Response Plans; Proposed Rule. *Fed. Regist.* **57:**27514–27553.
162. **van Beilen, J. B., M. G. Wubbolts, and B. Witholt.** 1994. Genetics of alkane oxidation in *Pseudomonas oleovorans. Biodegradation* **5:**161–174.
163. **VanHamme, J. D., J. A. Odumeru, and O. P. Ward.** 2000. Community dynamics of a mixed-bacterial culture growing on petroleum hydrocarbons in batch culture. *Can. J. Microbiol.* **46:**441–450.
164. **Varadaraj, R., M. L. Robbins, J. Bock, S. Pace, and D. MacDonald.** 1995. Dispersion and biodegradation of oil spills on water, p. 101–106. *In Proceedings of the 1995 International Oil Spill Conference.* American Petroleum Institute, Washington, D.C.
165. **Venosa, A. D., J. R. Haines, W. Nisamaneepong, R. Govind, S. Pradhan, and B. Siddique.** 1992. Efficacy of commercial products in enhancing biodegradation in closed laboratory reactors. *J. Ind. Microbiol.* **10:**13–23.
166. **Venosa, A. D., M. T. Suidan, B. A. Wrenn, K. L. Strohmeier, J. R. Haines, B. L. Eberhart, D. King, and E. Holder.** 1996. Bioremediation of an experimental oil spill on the shoreline of Delaware Bay. *Environ. Sci. Technol.* **30:**1764–1775.
167. **Watkinson, R. J., and P. Morgan.** 1990. Physiology of aliphatic hydrocarbon-degrading microorganisms. *Biodegradation* **1:**79–92.
168. **Weise, A. M., C. Nalewajko, and K. Lee.** 1999. Oil-mineral fine interactions facilitate oil biodegradation in seawater. *Environ. Technol.* **20:** 811–824.
169. **Widdel, F., and R. Rabus.** 2001. Anaerobic biodegradation of saturated and aromatic hydrocarbons. *Curr. Opin. Biotechnol.* **12:**259–276.
170. **Wilhelms, A., S. R. Larter, I. M. Head, P. Farrimond, R. Di-Primio, and C. Zwach.** 2001. Biodegradation of oil in uplifted basins prevented by deep-burial sterilization. *Nature* **411:**1034–1037.
171. **Wrabel, M. L., and P. Peckol.** 2000. Effects of bioremediation on toxicity and chemical composition of no. 2 fuel oil: growth responses of the brown alga *Fucus vesiculosus. Mar. Pollut. Bull.* **40:**135–139.
172. **Wright, A. L., R. W. Weaver, and J. W. Webb.** 1997. Oil bioremediation in salt marsh mesocosms as influenced by N and P fertilization, flooding, and season. *Water Air Soil Pollut.* **95:** 179–191.
173. **Xing, X. H., T. Tanaka, K. Matsumoto, and H. Unno.** 2000. Characteristics of a newly created bioluminescent *Pseudomonas putida* harboring TOL plasmid for use in analysis of a bioaugmentation system. *Biotechnol. Lett.* **22:** 671–676.
174. **Yaws, C. L., H. C. Yang, J. R. Hopper, and K. C. Hansen.** 1990. 232 hydrocarbons: water solubility data. *Chem. Eng.* **97(4):**177–182.
175. **Yaws, C. L., H. C. Yang, J. R. Hopper, and K. C. Hansen.** 1990. Organic chemicals: water solubility data. *Chem. Eng.* **97(5):**87.
176. **Yaws, C. L., X. Pan, and X. Lin.** 1993. Water solubility data for 151 hydrocarbons. *Chem. Eng.* **100(2):**108–111.
177. **Zhang, X., and L. Y. Young.** 1997. Carboxylation as an initial reaction in the anaerobic metabolism of naphthalene and phenanthrenes by sulfidogenic consortia. *Appl. Environ. Microbiol.* **63:**4759–4764.
178. **Zilinskas, R. A., and P. J. Balint (ed.).** 1998. *Genetically Engineered Marine Microorganisms. Environmental and Economic Risks and Benefits.* Kluwer, Boston, Mass.

BIOREMEDIATION OF METALS AND RADIONUCLIDES

Jonathan R. Lloyd, Robert T. Anderson, and Lynne E. Macaskie

8

INTRODUCTION

Metals have played a pivotal role throughout the development of human civilizations, from as early as 15000 BCE, when gold and copper were first worked, through the development of primitive smelting techniques (4000 BCE) and the widespread application of the blast furnace that ushered in the Industrial Revolution in the 18th century. Metals are not just part of our long industrial heritage. They also are finding increasing use in areas as diverse as medicine, electronics, catalysis, and the generation of nuclear power. Given our long and intimate association with metals and our continued reliance on these important natural resources, it is not surprising that their use (and abuse) can lead to significant environmental problems that need to be addressed.

Estimates of the global market for the cleanup and prevention of metal contamination vary, but conservative calculations suggest that the current market for metal bioremediation may be about $360 billion in the United States alone in 2005. The emerging market for the cleanup of radioactive contamination in the United States may already be worth as much as $300 billion (77). Current estimated costs of decontamination and safe disposal of radioactive waste in the United Kingdom have also been estimated at $86 billion, according to a recent United Kingdom Government White Paper (9). Unfortunately, existing chemical techniques are not always economical or cost-effective for the remediation of water or land contaminated with metals and radionuclides. Current strategies for land contaminated with metals include the use of "dig and dump" approaches that only move the problem to another site and are expensive and impractical for large volumes of soil or sediment. Likewise, soil washing, which removes the smallest particles that bind most of the metals, is useful but can be prohibitively expensive for some sites. "Pump and treat" technologies rely on the removal of metals from the site in an aqueous phase which is treated ex situ (e.g., above land). These approaches can cut down on excavation costs but are still expensive, and metal removal can be inefficient. A potentially economical alternative is to develop biotechnological ap-

Jonathan R. Lloyd, School of Earth, Atmospheric and Environmental Sciences, Williamson Building, University of Manchester, Manchester M13 9PL, United Kingdom. *Robert T. Anderson,* Office of Biological and Environmental Research SC-75 / Germantown Building, U.S. Department of Energy, 1000 Independence Ave., S.W., Washington, DC 20585-1290. *Lynne E. Macaskie,* School of Biosciences, The University of Birmingham, Birmingham B15 2TT, United Kingdom.

Bioremediation: Applied Microbial Solutions for Real-World Environmental Cleanup
Edited by Ronald M. Atlas and Jim C. Philp © 2005 ASM Press, Washington, D.C.

proaches that could be used in the sediment or soil (in situ) to either extract the metals or stabilize them in forms that are immobile or nontoxic.

In addition to development of biotechnologies to treat contaminated land, there is also considerable interest in more effective techniques that can be used to treat water contaminated with metals from a range of industrial processes. Problems inherent in currently used chemical approaches include a lack of specificity associated with some ion-exchange resins and the generation of large quantities of poorly settling sludge through treatment with alkali or flocculating agents. This chapter gives an overview of metal-microbe interactions and describes how they could be harnessed to clean up metal-contaminated water, soil, and land. Although this is a comparatively young and maturing field, there are already several examples of successful uses of biotechnological approaches for metal bioremediation. Key industrial-scale (ex situ) and field-scale (in situ) processes are highlighted as special case studies.

METAL AND RADIONUCLIDE CONTAMINATION

The U.S. Environmental Protection Agency lists among the priority pollutants trace metals that are recoverable and dissolved, including antimony, chromium, mercury, silver, arsenic, copper, nickel, thallium, beryllium, lead, selenium, zinc, and cadmium. In addition to these, toxic metals and radionuclides are priority contaminants at nuclear installations worldwide. For example, the U.S. Department of Energy's 120 sites contain 1.7 trillion gallons of contaminated groundwater and 40 million cubic meters of contaminated soil and debris. More than 50% of the sites are contaminated with radionuclides, with the priority radionuclides being cesium-137, plutonium-239, strontium-90, technetium-99, and uranium-238 and U-235, in addition to toxic heavy metals including chromium, lead, and mercury (77). The U.S. Department of Energy Natural and Accelerated Bioremediation (NABIR) program is notable in its support of emerging biotechnological approaches for dealing with this legacy waste. In Europe, the European Union Water Framework Directive (2000/60/EC) is the principal driver for environmental legislation and also recognizes metals among the 32 substances on the priority list. These include lead, nickel, mercury, cadmium, and tin and their compounds. Attempts to remediate sites contaminated with radioactive waste in Europe are not so advanced.

METAL-MICROBE INTERACTIONS

Microorganisms can interact with metals and radionuclides via many mechanisms, some of which may be used as the basis of potential bioremediation strategies. The major types of interaction are summarized in Fig. 8.1. In addition to the mechanisms outlined, accumulation of metals by plants (phytoremediation) warrants attention but is not yet an established route for the bioremediation of metal contamination. The reader is referred to the work of Salt, Raskin, and colleagues (92, 98, 99), who give a detailed description of (i) phytoextraction, the use of metal-accumulating plants to remove toxic metals from soil; (ii) rhizofiltration, the use of plant roots to remove toxic metals from polluted waters; and (iii) phytostabilization, the use of plants to eliminate the bioavailability of toxic metals in soils. The role of the microbiota associated with the plant root system in metal accumulation by plant tissue remains relatively poorly studied. In this connection, the use of reed bed technologies and artificial wetlands for bioremediation (41, 42, 50) should also be mentioned, since these, too, would implicate the involvement of rhizosphere microorganisms. This is also discussed later in this chapter in a case study for the treatment of acid mine drainage waters, alongside other competing biotechnological approaches for passive in situ remediation.

Biosorption

The term biosorption is used to describe the metabolism-independent sorption of heavy metals and radionuclides to biomass. It encompasses both adsorption, the accumulation of substances at a surface or interface, and absorption, the almost uniform penetration of atoms or molecules

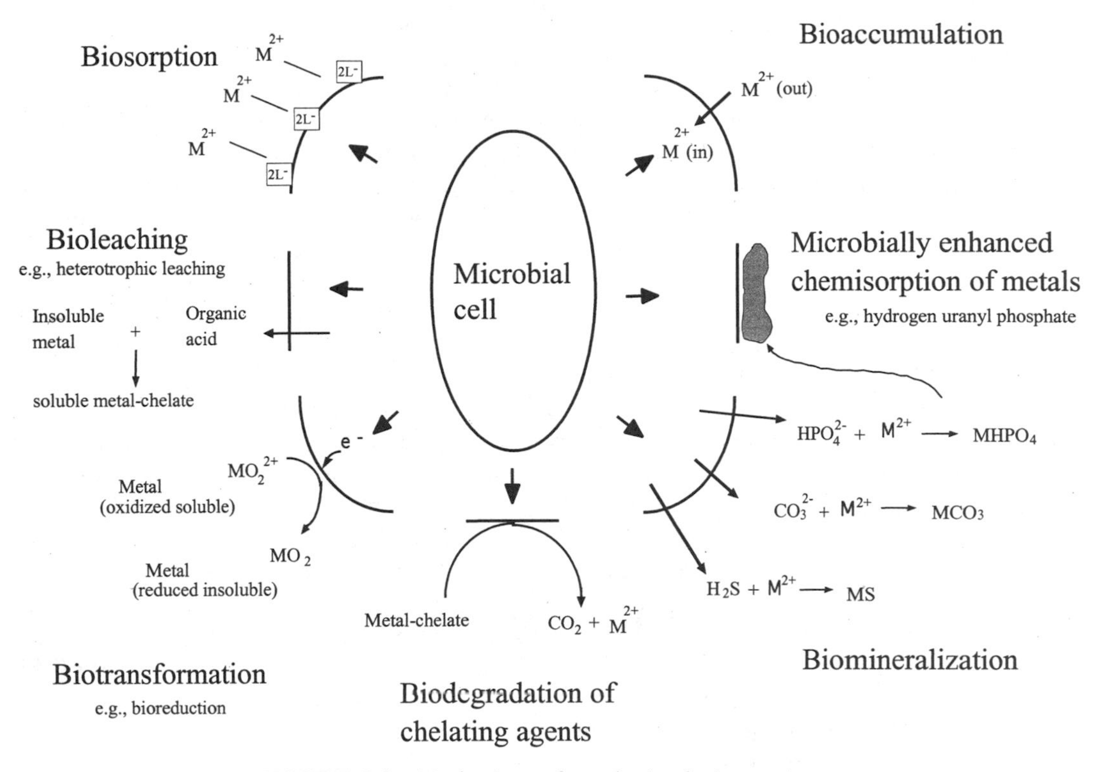

FIGURE 8.1 Mechanisms of metal-microbe interactions.

of one phase forming a solution with a second phase (37). Both living and dead biomasses are capable of biosorption, and ligands involved in metal binding include carboxyl, amine, hydroxyl, phosphate, and sulfhydryl groups. Biosorption of metals has been reviewed extensively (55, 78, 110, 118), and this chapter will note only some salient points and recent developments of interest. Volesky and Holan (118) give an excellent overview of metal biosorption and a numerical assessment of metal (uranium and thorium) biosorption along with some more recent data. Beveridge et al. (14), in addition to reviewing metal-microbe interactions, give a very useful guide to the various methods which are now available for their study and a good introduction to the problems of metal speciation, which will always influence the data and their interpretation and makes accurate comparisons between the reports of different laboratories very difficult (91). However, some generalizations are possible.

Dead biomass often sorbs more metal than its live counterpart (20, 119) and thus may be particularly suited to treatment of highly toxic waste streams. In many cases, the sorbing biomass (biosorbent) comprises the waste from another process, adding considerably to the economic attractiveness, even though the absolute metal sorption capacity may be less than that of a more attractive biomass that has to be grown for the purpose. This is not always the case; Avery and Tobin (10) noted that a laboratory strain of *Saccharomyces cerevisiae* removed less strontium from solution than a brewery strain, possibly because of properties of the extracellular polymer that are desirable for good flocculation of the yeast after fermentation. Postharvest treatments, including powdering of dry biomass (109) and the application of detergents (97), have been reported to improve biosorption, exposing additional metal-binding sites. Autoclaving after fermentation can also affect the biomass surface properties (102), as can washing (93). New approaches for improving metal uptake have included the application of electrical pulses, which increased both the initial rate of binding of uranyl ions to yeast biomass and its capacity, from 70 to 140 mg of uranium/g of biomass (22), a value comparable to that of the filamentous fungi (118).

Biosorption is generally rapid and unaffected over modest temperature ranges and in many cases can be described by isotherm models such as the Langmuir and Freundlich isotherms (118). Gadd and White (37), however, noted that more complex interactions are difficult to model because the adsorption of solutes by solids is affected by factors including diffusion, heterogeneity of the surface, and pH. An additional isotherm, the Brunauer-Emmett-Teller (BET) isotherm, which assumes multilayer binding at constant energy, has also been used to describe metal biosorption (8, 28). This model assumes that one layer need not necessarily be completely filled before another is commenced. Further insight is offered by Andres et al. (8), who summarize: "each adsorption layer of the BET model can be reduced to Langmuir behavior with homogeneous surface energy, in contrast to the adsorption energy requirements of the Freundlich isotherm." Other studies have used complex multistage kinetic approaches to model biosorption (112, 127).

Ultimately, however, the amount of residual metal remaining in solution at equilibrium is governed by the stability constant of the metal-ligand complex (69), and the only way to change the equilibrium position is to modify the binding ligand to one which has a greater binding affinity for the given metal or to transform the metal from a poorly sorbing species to one which has a higher ligand-binding affinity, e.g., by a change of metal valence. Alternatively, some new studies have applied the tools of molecular biology to enhance metal sorption, and they do warrant attention. For example, a mouse metallothionein was targeted to the outer membrane of a metal-resistant *Ralstonia eutropha* isolate (116). The engineered strain accumulated more Cd^{2+} than its wild-type counterpart and offered tobacco plants some protection from Cd^{2+} when inoculated into contaminated soil (116). A surface display technique has also been used successfully to generate ZnO-binding peptides fused to fimbriae on the surface of cells of *Escherichia coli* (51). Finally, gram-positive bacteria (staphylococci) have also been engineered to produce surface-exposed peptides able to bind Hg^{2+} and Cd^{2+} (100). Enhanced uptake of cadmium and mercury by *E. coli* expressing a metal-binding motif has also been reported (87). Future studies could usefully compare the performance of such engineered systems with that of other, more traditional biosorbents. However, it has been noted that despite the relatively long time during which biosorption has been examined, and despite considerable early interest, there has been reduced interest in commercializing this type of technology, and we are aware of no current commercial applications of this type of technology.

Metabolism-Dependent Bioaccumulation

Energy-dependent metal uptake has been demonstrated for most physiologically important metal ions, and some toxic metals and radionuclides enter the cell as chemical "surrogates" using these transport systems. Monovalent cation transport, for example, K^+ uptake, is linked to the plasma membrane-bound H^+-ATPase via the membrane potential and is therefore affected by factors that inhibit cell energy metabolism. These include the absence of substrate, anaerobiosis, incubation at low temperatures, and the presence of respiratory inhibitors such as cyanide (129). The requirement for metabolically active cells may therefore limit the practical application of this mode of metal uptake to the treatment of metals and radionuclides with low toxicity and radioactivity. For example, in the study of White and Gadd (129), increasing metal concentrations inhibited H^+ pumping, potentially de-energizing the cell membrane and reducing cation uptake. Although the presence of multiple transport mechanisms of differing affinities may cause added complication, metal influx frequently conforms to Michaelis-Menten kinetics (18).

Once in the cell, metals may be sequestered by cysteine-rich metallothioneins (44, 114) or, in the case of fungi, compartmentalized into the vacuole (37, 81). In this context, it should be emphasized that the uptake of higher-mass radionuclides, e.g., the actinides, into microbial cells has been reported sporadically and remains poorly characterized (54, 55). Metabolism-dependent bioaccumulation of metals by microorganisms has not been used commercially for bioremediation purposes.

Enzymatically Catalyzed Biotransformations

Microorganisms can catalyze the transformation of toxic metals to less soluble or more volatile forms. For example, the microbial reduction of Cr(VI) to Cr(III), Se(VI) to Se(0), V(V) to V(III), Au(III) to Au(0), Pd(II) to Pd(0), U(VI) to U(IV), and Np(V) to Np(IV) results in metal precipitation under physiological conditions and has been reviewed in detail recently (52). In many cases, the high-valence metal can be used as an electron acceptor under anoxic conditions. A good example here is the reduction of soluble U(VI) to insoluble U(IV) by Fe(III)-reducing bacteria (66), which has been harnessed recently to remediate sediments contaminated with uranium in situ (see Box 8.3). Metal detoxification is also possible through biotransformations of toxic metals. The bioreduction of Hg(II) to relatively nontoxic Hg(0) is perhaps the best-studied example and is discussed in further detail below. Finally, in addition to redox biotransformations, biomethylation may also increase the volatility of metals, with the methylation of mercury, cadmium, lead, tin, selenium, and tellurium recorded (31). These mechanisms could also offer potential use for the in situ remediation of contaminated soil (59).

In common with some examples of bioaccumulation, enzymatic biotransformations of metals can be described by Michaelis-Menten kinetics. This allows, for example, the description of a flowthrough bioreactor using an integrated form of the Michaelis-Menten equation (73, 111, 132). It is therefore possible to predict the degree of biomass loading or the operational temperature needed to maintain metal removal at a given efficiency within the constraints set by the reactor volume available, the background ionic matrix, and flow rates required (e.g., for bioprecipitation [70] of U [73] and La [111] and for bioreduction of Tc [53]).

Another advantage of using a single enzyme-mediated transformation over, for example, energy-dependent processes (e.g., bioaccumulation) is that nongrowing or even nonliving biomass with enzymatic activity may be utilized to treat radiotoxic effluents, yielding a waste material with a low organic content. The high metal and radioresistance of several enzymes, potentially useful for the bioremediation of toxic waste streams, has been confirmed (53; L. F. Strachan, M. R. Tolley, and L. E. Macaskie, paper presented at the 201st Meeting of the American Chemical Society, Atlanta, Ga., symposium on biotechnology for wastewater treatment, 1991). For many biological reductions [e.g., Cr(VI), U(VI), or Tc(VII) reduction], simple, cheaply available electron donors such as hydrogen, acetate, or formate can be supplied, negating the requirement for cofactor regeneration.

As mentioned above, microbes have evolved sophisticated approaches to deal with toxic metal (21), often involving redox transformations of the toxic metal. Perhaps the best-studied metal resistance system is encoded by genes of the *mer*, or mercury resistance, operon (45). Recent studies have confirmed the biotechnological potential of this widespread resistance determinant. Hg(II) is bound in the periplasm of gram-negative bacteria by the MerP protein, transported into the cell via the MerT transporter, and detoxified by reduction to relatively nontoxic volatile elemental mercury by an intracellular mercuric reductase (MerA). Mer proteins are expressed under the regulation of the activator protein, MerR, which binds Hg(II) and activates gene expression. Mercury-resistant bacteria and the proteins that they encode have been used recently for the bioremediation of Hg-contaminated water (Box 8.1, case study) and in the development of biosensors for bioavailable concentrations of Hg(II) (16). Such sensors may prove very useful in identifying the need for metal remediation, which will be dictated in many cases by the concentration of bioavailable toxic metals in a given soil or water matrix, and also in defining the end point for bioremediation efforts. With a focus on Hg biosensors, several different components of the *mer* system have been used in prototype sensors, including the NADPH-dependent mercuric reductase (MerA) in an enzyme-linked biosensor (30), the *mer* regulatory region in a whole-cell biosensor (38), and the MerR protein in a capacitance biosensor (17).

BOX 8.1
Case Study: Using Mercury-Resistant Bacteria To Treat Chloralkali Wastewater

Mercury-resistant bacteria have been used recently to detoxify Hg(II)-contaminated water at laboratory and pilot scale (Box Fig. 8.1). Wagner-Döbler and coworkers at the German Research Centre for Biotechnology in Braunschweig, Germany, captured reduced elemental Hg in a 20-ml immobilized cell bioreactor, inoculated with a mercury-resistant *Pseudomonas putida* strain and subsequently colonized with other mercury-resistant strains (124). A companion study demonstrated successful removal of Hg^{2+} from chloralkali electrolysis water at laboratory scale (121), prior to development of a pilot plant for Hg(II) removal using this technology (125). In the latter study, a 700-liter reactor was packed with pumice granules (particle size, 4 to 6 mm) and inoculated with seven mercury-resistant *Pseudomonas* species. Acidic wastewater from a chloralkali factory was neutralized and amended with sucrose and yeast extract prior to introduction into the bioreactor. Concentrations of up to 10 mg of Hg liter^{-1} were successfully treated with a retention efficiency of 95%, although influent spikes above this concentration had a deleterious (if reversible) effect on reactor performance. When the reactor was operated in combination with an activa-

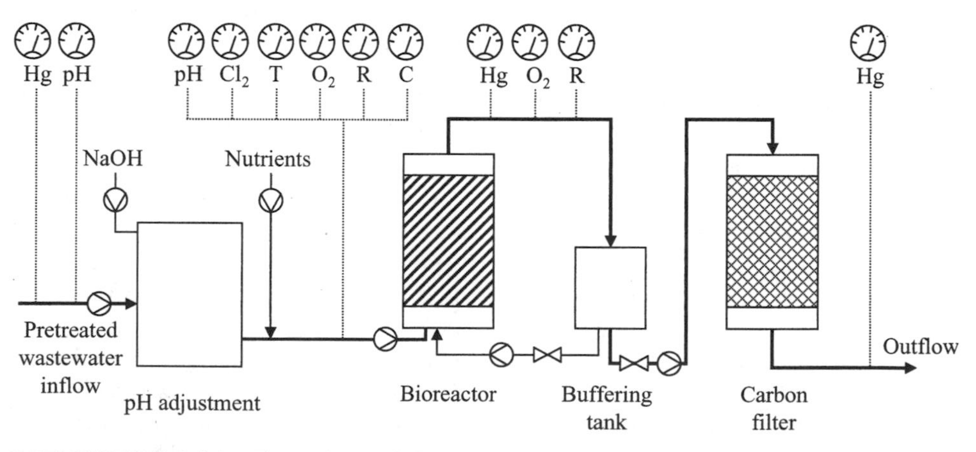

BOX FIGURE 8.1.1 Flow scheme (left) and photograph (right) of a pilot plant for removal of mercury from wastewater by mercury-resistant bacteria. The plant includes pH adjustment to pH 7, nutrient amendment, the bioreactor (volume, 1 m^3), a buffering tank, and a polishing carbon filter. Continuous automated Hg measurement is performed at the inflow, after the bioreactor, and at the outflow. pH is measured before and after adjustment to pH 7. Other parameters determined continuously are chlorine concentration (Cl_2), oxygen concentration (O_2), redox potential (R), conductivity (C), and temperature (T). Figures were kindly provided by Dr. Irene Wagner-Döbler.

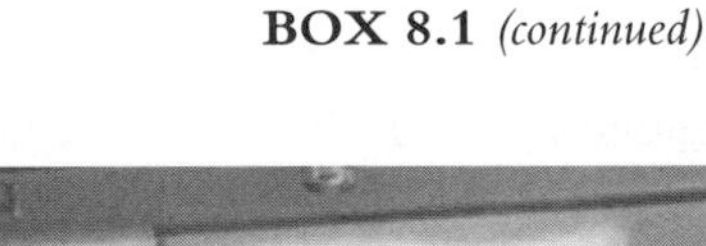

BOX 8.1 *(continued)*

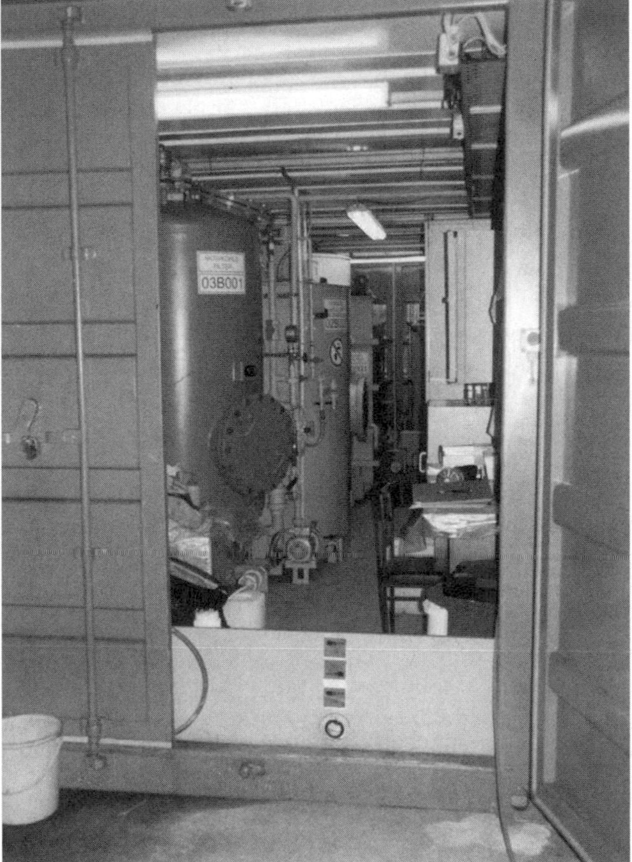

BOX FIGURE 8.1.1 *(continued)*

ted carbon filter, which also became colonized by bacteria, further removal of Hg to below 10 μg $liter^{-1}$ was reported. Very high loadings of Hg were retained in the reactor, conservatively estimated at 31.5 kg for the 700-liter vessel.

Long-term performance of the reactors has been studied, with no loss of the entrapped Hg(0) from the system over 16 months (123). Although the reactors were sensitive to mechanical and physical stresses (e.g., shear from gas bubbles and increased temperature over 41°C, respectively), the system seems robust and able to adjust to elevated Hg(II) concentrations (up to 7.6 mg $liter^{-1}$) within several days (123). With a continuous selection pressure for mercury resistance, a stable and highly active mercury-reducing microbial community is established within the bioreactors; this was confirmed by PCR-based techniques targeting the intergenic spacer region of DNA coding for 16S to 23S rRNA (16S to 23S rDNA) and a functional gene target for Hg(II) reduction, *merA* (122). The performance of the reactor system has also been studied in response to the oscillation of the mercury concentration in the bioreactor inflow (120). At low mercury concentrations, maximum Hg(II) reduction occurred near the inflow at the bottom of the bioreactor. At higher concentrations, the zone of maximum activity migrated to the upper horizons. Molecular analysis of the microbial communities showed an increasing microbial diversity along a gradient of decreasing mercury concentrations (120).

Another area where direct bioreduction of high-valence metals may prove useful in the future is in the biorecovery of precious metals from spent automotive catalysts. Platinum group metals are routinely used in automotive catalysts for atmospheric protection, but recycle technology lags behind demand. There is no available "clean technology," and leach solutions (e.g., aqua regia) to solubilize noble metals from scrap are usually highly aggressive. A microwave-assisted leaching method has been developed which gave 80% metal recovery, with the leach time reduced from 2 h to 15 min using 50% (aqueous) diluted aqua regia to give potentially a more biocompatible leachate (136). *Desulfovibrio desulfuricans* reduces soluble platinum group metals to cell-bound insoluble base metals [e.g., Pd(II) → Pd(0)] via the activity of hydrogenase (58, 134, 135), which is, surprisingly, stable at pH 2 to 3 for several hours (I. Mikheenko, unpublished data). The role of the hydrogenase is to mediate the formation of nucleation sites of Pd(0) on the cell surface. After nucleation, reduction of more Pd(II) to Pd(0) occurs autocatalytically via the deposited Pd(0) clusters (135). For use, biofilm was immobilized on a Pd–23% Ag solid alloy membrane which delivered H· to the cells with H supplied via an electrochemical chamber at the back side (136). The biomass-coated Pd-Ag alloy electrode was used in a flowthrough reactor for recovery of Pd, Pt, and Rh from aqua regia leachates (pH 2.5) of spent automotive catalysts with up to 90% efficiency at a flow residence time of 15 min. Free cells did not reduce platinum group metals from the leachates, but the electrobioreactor did so using biofilm cells preloaded with Pd(0). Reactors lacking biomass or reactors with heat-killed biofilm removed less platinum group metal, via electrochemically synthesized H· reductant alone. The use of an active biofilm layer in a flowthrough electrobioreactor provides a simple, clean, and rapid potential recycle technology. Furthermore, the biorecovered Pd(0) [Bio-Pd(0)] has high catalytic activity, since metal is deposited on the cells as supported nanoclusters. As examples, hydrogen release was promoted rapidly from hypophosphite at room temperature (135), Cr(VI) was reduced to Cr(III) under conditions where biomass alone or chemically reduced Pd(0) was ineffective (68), and Bio-Pd(0) reductively dehalogenated polychlorinated biphenyls (PCBs), which are recalcitrant to degradation and not attacked by chemically reduced Pd(0) alone (12). The latter is noteworthy since PCBs are not water soluble and the Bio-Pd(0) was able to access the suspension of nonaqueous PCB in water. The high potential of the new biomaterial in industry is also illustrated by a test hydrogenation reaction, in which the addition of hydrogen across the double bond of itaconic acid to form methyl succinic acid (an industrial test reaction) was comparable to that of a 5% Pd on carbon commercial supported palladium catalyst (N. J. Creamer, J. Wood, and L. E. Macaskie, unpublished data). The latter are excellent examples of a sustainable technology, where the materials recovered from a bioremediation process have applications in other industrial or environmental sectors, which would greatly increase the economic competitiveness of the new biotechnologies.

Biomineralization via Microbially Generated Ligands

Microorganisms are able to precipitate metals and radionuclides as carbonates and hydroxides via plasmid-borne resistance mechanisms, whereby proton influx countercurrent (antiport) to metal efflux results in localized alkalinization at the cell surface (29, 117). Alternatively, metals can precipitate with enzymatically generated ligands, e.g., sulfide (11) or phosphate (72). The concentration of residual free metal at equilibrium is governed by the solubility product of the metal complex (e.g., 10^{-20} to 10^{-30} for the sulfides and phosphates, higher for the carbonates). Most of the metal should be removed from solution if an excess of ligand is supplied. This is difficult to achieve when chemical precipitation methods and dilute solutions are used. The advantage of microbial ligand generation is that high concentrations of ligand are achieved in juxtaposition to the cell surface, which can also provide nucleation foci for the rapid onset of metal precipitation; effectively, the metals are concentrated "uphill" against a concentration gradient. This was demonstrated by using the gamma isotope ^{241}Am supplied at an input concentration of approximately 2.5 ppb; approximately 95% of the metal was removed as biomass-bound phosphate (74), with the use of gamma counting permitting detection at levels below those of most analytical methods. In many cases, the production of the ligand can also be fine-tuned by the application of Michaelis-Menten kinetics.

Sulfide precipitation, catalyzed by a mixed culture of sulfate-reducing bacteria, has been utilized first to treat water cocontaminated by sulfate and zinc (11) and also soil leachate contaminated with sulfate alongside metal and radionuclides (T. Kearney, H. Eccles, D. Graves, and A. Gonzalez, paper presented at the 18th Annual Conference of the National Low-Level Waste Management Program, Salt Lake City, Utah, 20 to 22 May, 1996). Ethanol was used as the electron donor for the reduction of sulfate to sulfide in both examples. A later study has confirmed the potential of integrating the action of sulfur-cycling bacteria (130). In the first step of a two-stage process, sulfur-oxidizing bacteria were used to leach metals from contaminated soil via the generation of sulfuric acid, and in the second step the metals were stripped from solution in an anaerobic bioreactor containing sulfate-reducing bacteria. The ubiquitous distribution of sulfate-reducing bacteria in acid, neutral, and alkali environments (89) suggests that they have the potential to treat a variety of effluents, while the ability of the organisms to metabolize a wide range of electron donors may also allow cotreatment of other organic contaminants. Recent work by Paques BV of The Netherlands has confirmed the potential of sulfate-reducing bacteria to treat metal waste in ex situ bioreactors for the treatment of a wide range of metal-contaminated waters (Box 8.2).

Bioprecipitation of metal phosphates via hydrolysis of stored polyphosphate by *Acinetobacter* spp. is dependent upon alternating aerobic (polyphosphate synthesis) and anaerobic (polyphosphate hydrolysis and phosphate release) periods (19). This group of obligately aerobic organisms is fairly restricted in the range of carbon sources utilized, but the preferred substrates (acetate and ethanol) are widely and cheaply available. The best-documented organism for metal phosphate biomineralization, a *Citrobacter* sp. (now reclassified as a *Serratia* sp. on the basis of molecular methods, the presence of the *phoN* phosphatase gene, and the production of pink pigment under some conditions [86]), grows well on cheaply available substrates, and viable cells are not required for metal uptake since it relies on hydrolytic cleavage of a supplied organic phosphate donor (72). The expense of adding organic phosphate was calculated to be the single factor which limited the economic viability of this approach (95); moreover, organophosphorus compounds are often highly toxic. A possible alternative phosphate donor, tributyl phosphate (TBP), is used widely as a cheap solvent and plasticizer, but its degradation is more difficult because this compound is a phosphate triester, with three cleavage events per mole required to liberate 1 mol of phosphate. TBP is biodegradable; this activity was unstable (108), but biogenic phosphate from TBP hydrolysis was harnessed to the removal of uranium from solution in a flowthrough system (107). This followed an earlier study which suggested that TBP could support metal removal by the metal-accumulating *Serratia* sp. (79). However, in this case the total amount of phosphate recovered was small (85), a result which could be attributed to phosphotransferase activity which recycled the liberated phosphate back onto the liberated alcohol (3 mol of alcohol coproduct per mol of phosphate [49]). The enzymatic activity responsible for TBP hydrolysis remains obscure, but a recent study of TBP hydrolysis by enzymatic degradation used cell extracts of *Serratia odorifera* (13). Jeong et al. (49) also identified a new class of TBP-hydrolyzing (photosynthetic) organisms, the potential of which for metal bioremediation has not yet been examined.

BOX 8.2
Case Study: Ex Situ Bioremediation of Metals Using Sulfate-Reducing Bacteria

The ability of sulfate-reducing bacteria to precipitate metals as insoluble metal sulfides has been used by Paques BV of The Netherlands in ex situ bioreactors for the treatment of metal-contaminated water. The patented reactor configurations, marketed under the registered trademark Thiopaq, can also be adapted to treat other waste streams containing sulfur compounds, including hydrogen sulfide.

Early development work focused on the Budel Zinc BV refinery at Budel-Dorplein in The Netherlands. Over 200,000 tons of zinc is produced annually at the refinery, which has been operated since 1973. However, zinc was refined by various companies at this site for more than 100 years, resulting in contamination of soil and groundwater with heavy metals and sulfate. In 1992, Paques designed and installed a system to treat water extracted from strategically located wells around a hydrogeological containment system installed to protect local drinking water supplies. The bioreactor system and a flow sheet of the process are shown in Box Fig. 8.2.1. In the first stage, water is passed to an anaerobic bioreactor containing sulfate-reducing bacteria that couple the oxidation of ethanol to the reduction of sulfate to sulfide. This leads to the precipitation of insoluble metal sulfides:

$$H_2S + ZnSO_4 \rightarrow ZnS\ (\text{precipitate}) + H_2SO_4$$

Excess toxic sulfide is then oxidized to elemental sulfur in an aerobic reactor, and tilted plate settlers and sand filters are used as final polishing steps to remove solids. Metal sulfides and elemental sulfur are returned to the plant for metal recovery and sulfuric acid production, respectively. Performance of this system is summarized in Box Table 8.2.1.

Since 1999, this type of technology has also been employed at Budel-Dorplein to treat process streams containing sulfate and zinc produced by the conventional roast-leach-electrowin process operated at this site. These streams were previously treated conventionally by neutralization with lime, resulting in the production of 18 tons of gypsum day^{-1}. However, recent legislation prohibited further production of solid residues from July 2000. The high-rate Thiopaq biological sulfate reduction bioreactor, supplied with hydrogen as the electron donor, was, however, able to convert zinc and sulfate into zinc sulfide (10 tons day^{-1}), which is recycled at the refinery. Paques notes that this is now the first gypsum-free zinc refinery.

Paques has also used Thiopaq to remove metals from an alkaline slag dump leachate at Kovohute Pribram lead waste recycling facility in the Czech Republic. An alkaline carbonate-buffered sodium sulfate leachate, containing lead, zinc, tin, and high concentrations of arsenic and antimony, is treated with hy-

BOX FIGURE 8.2.1 Aerial photograph (right) and flow sheet (facing page) of the Paques BV Thiopaq zinc-sulfate treatment process at the Budel Zinc BV refinery at Budel-Dorplein in The Netherlands. TPS, tilted plate settler. With permission from Paques BV.

BOX 8.3 *(continued)*

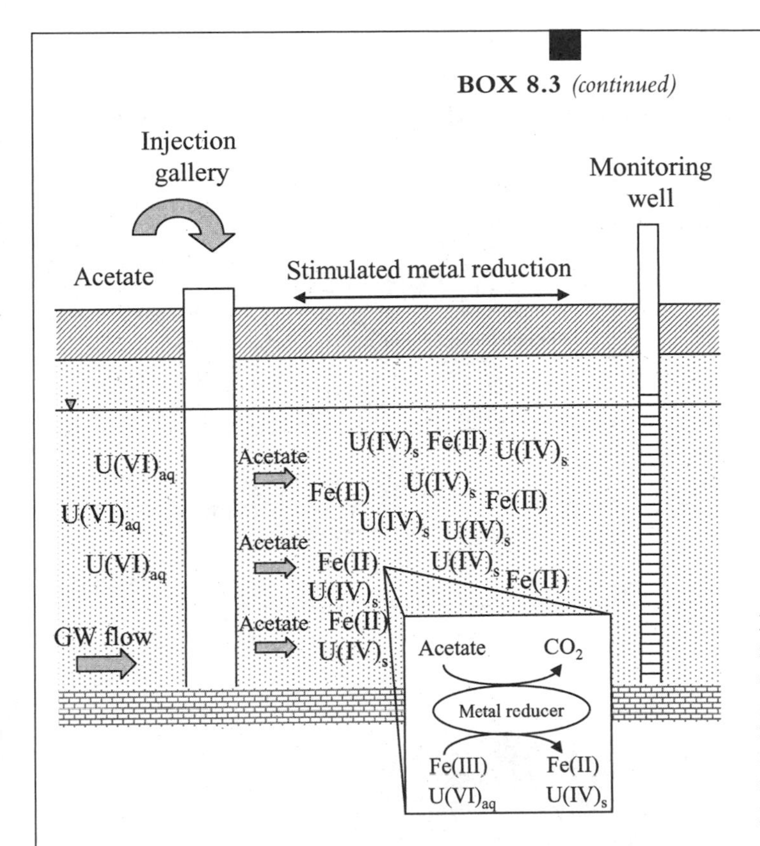

BOX FIGURE 8.3.1 Schematic showing the removal of soluble U(VI) from contaminated groundwater under Fe(III)-reducing conditions, stimulated by the addition of acetate to the subsurface. aq, aqueous phase; s, solid phase.

min^{-1}, which corresponded to a calculated volume addition to the aquifer of 1 to 3% per day (in situ acetate concentration, 1 to 3 mM). Upon the initiation of acetate injection, soluble uranium concentrations decreased rapidly within the monitoring well field, resulting in removal percentages averaging 70% of initial concentrations over a period of approximately 50 days. Loss of soluble U(VI) occurred coincident with the arrival of acetate and the production of Fe(II) and prior to any observed loss of sulfate. Furthermore, 16S rDNA-based analyses of the groundwater indicated a microbial community greatly enriched in *Geobacteraceae*, up to 89% of the detected bacterial community. Phospholipid fatty acid analyses of groundwater using lipids specific for *Geobacteraceae* also indicated an increase in *Geobacter* biomass. These results are consistent with the previous laboratory studies indicating a stimulated removal of soluble U(VI) from groundwater via the in situ stimulation of Fe(III)- and U(VI)-reducing *Geobacteraceae* (7, 33, 46).

Metal-reducing conditions were not sustained within the Old Rifle site over 50 days, and it was thought that acetate-oxidizing sulfate-reducing bacteria became dominant when Fe(III) was depleted in the vicinity of the injection gallery and the terminal electron accepting process shifted to sulfate reduction. Indeed, a complete loss of acetate (limiting under sulfate-reducing conditions in this aquifer) was accompanied by a nearly stoichiometric loss of sulfate from the groundwater. Analyses of the microbial community detected within the groundwater also indicated a shift from a community dominated by Fe(III)-reducing organisms to a community dominated by organisms known to reduce sulfate (*Desulfobacteraceae*) (7). The results stress the importance of maintaining metal reduction within the subsurface or encouraging the growth and activity of sulfate-reducing bacteria capable of U(VI) reduction; acetate-oxidizing sulfate-reducing bacteria have not been shown to reduce U(VI), although there is ample evidence that lactate-oxidizing, sulfate-reducing bacteria are able to reduce U(VI) using lactate or hydrogen as electron donors (60, 67). Thus, addition of these electron donors to the subsurface may stimulate U(VI) reduction in situ.

BOX 8.3 continues

BOX 8.3
Case Study: In Situ Uranium Bioremediation through Bioreduction *(continued)*

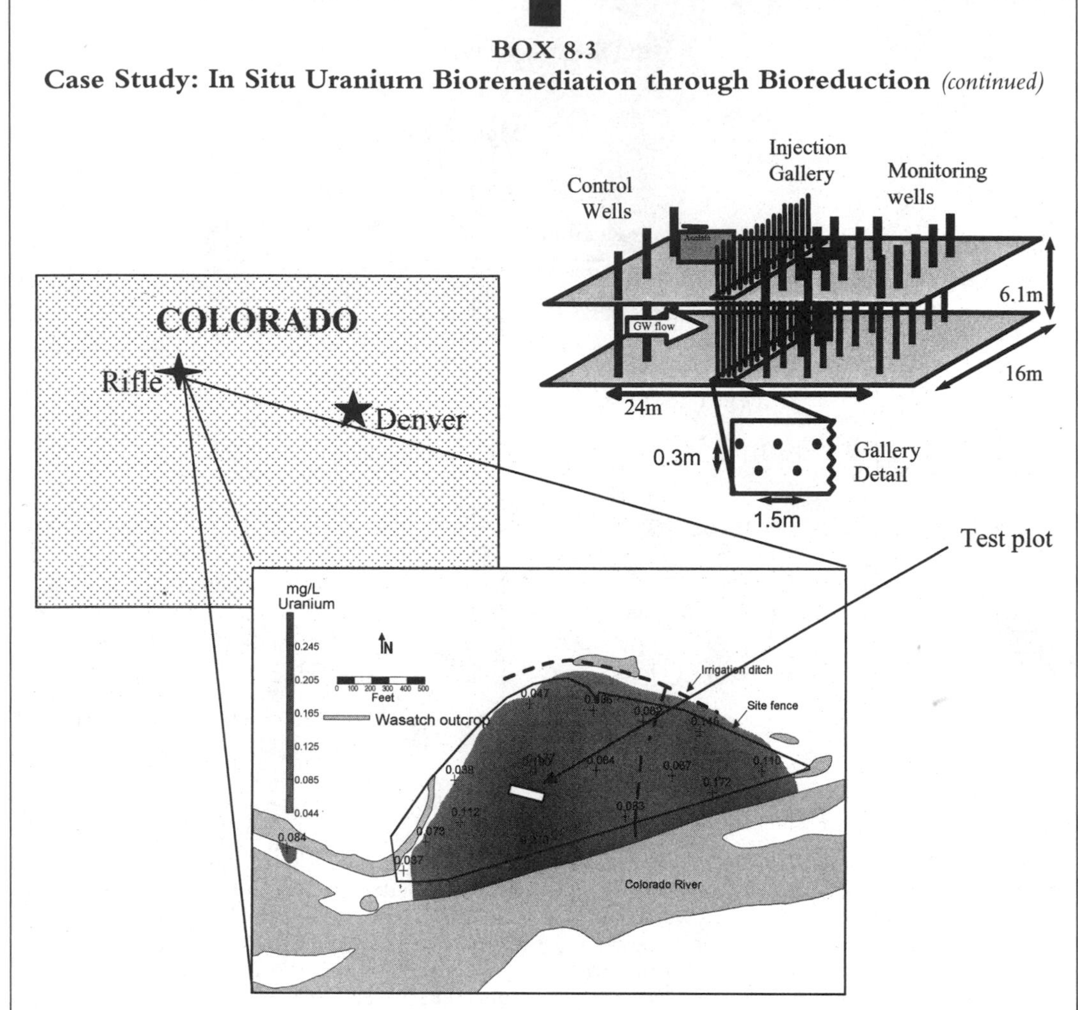

BOX FIGURE 8.3.2 Test plot for U(VI) remediation at the Old Rifle UMTRA site, consisting of an acetate injection gallery composed of 20 injection wells and 18 monitoring wells installed within a 16- by 24-m area.

nant metals such as uranium, chromium, technetium, and vanadium (5, 25, 57, 82). Additionally, the product of dissimilatory Fe(III) reduction, Fe(II), is a potential abiotic reduction for contaminant metals such as Tc(VII), V(V), and Cr(VI) (33, 57, 82). The demonstrated enrichment of *Geobacteraceae* within an aquifer during an in situ trial of stimulated uranium reduction and the potential abiotic benefits of in situ Fe(II) production suggest *Geobacteraceae* to be a target group of microorganisms within the subsurface for removing metals from groundwater. Further investigation of stimulated subsurface metal reduction will likely reveal other groups of bacteria, such as sulfate reducers, that could also play an important role in contaminant metal reduction and stabilization within subsurface environments. See Box 8.4 for another example.

FUTURE PROSPECTS

The bioremediation of metals is a rapidly maturing research topic, although at present technological advances in this area do lag behind

BOX 8.4
Case Study: Water Treatment at the Wheal Jane Mine, United Kingdom—Active (Chemical) versus Passive (Biological) Treatment

The Wheal Jane Mine is located within the Carnon River Valley in Cornwall, United Kingdom, and was operated as a tin mine from the early 18th century until it was closed and abandoned in 1991. Mine dewatering operations halted, and in January 1992 there was a buildup and sudden release of metal-contaminated mine water, which contaminated 6.5 $\times$ 10^6 m^2 of the Carnon River and Fal estuary. Emergency pump-and-treatment remediation of the site was followed in 1994 by the construction of a pilot passive treatment system and later, in 2000, by an active treatment system. The initial performance of these two contrasting systems has been reported by CL:AIRE (Contaminated Land: Applications in Real Environments) (http://www.claire.co.uk), providing evidence of the effectiveness of a sequence of discrete biotreatment approaches for the treatment of metal-contaminated mine waters versus chemical alternatives. Aerial photographs of the site and a schematic of the passive treatment processes are shown in Box Fig. 8.4.1, while the average composition of the mine water is shown in Box Table 8.4.1.

Active Chemical Treatment System

An active treatment plant was commissioned in 2000 based on a three-step high-density sludge system (mixing of water and sludge, aeration, and then clarification). After clarification, the treated water is diverted

BOX FIGURE 8.4.1 Aerial photograph (above) of the Wheal Jane Mine in Cornwall, United Kingdom, and a schematic of the passive treatment processes (next page) installed for the bioremediation of mine wastewaters (with permission from CL:AIRE).

BOX 8.4 continues

BOX 8.4
Case Study: Water Treatment at the Wheal Jane Mine, United Kingdom—Active (Chemical) versus Passive (Biological) Treatment *(continued)*

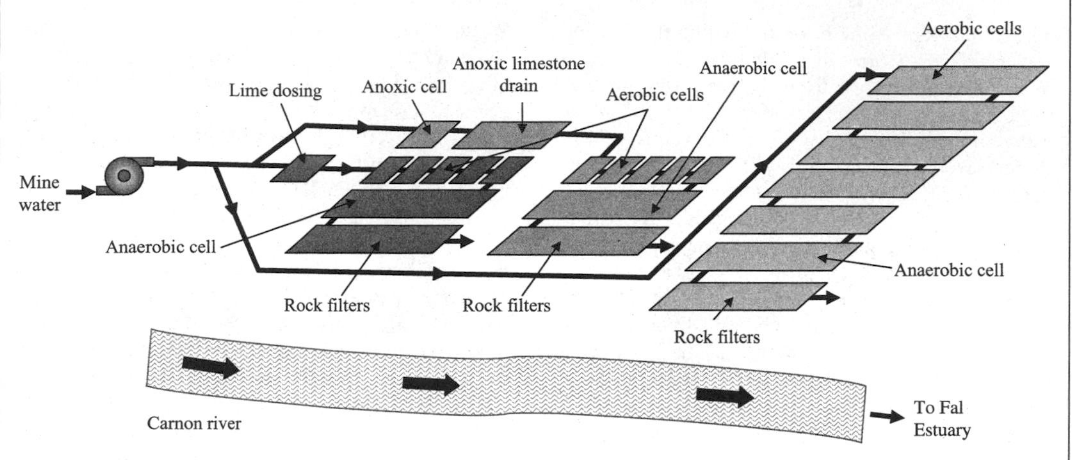

BOX FIGURE 8.4.1 *(continued)*

to the Carnon River, and the metal-bearing sludge is removed to a holding tank for disposal. With pH control (raising the pH to 8.5 in the initial step), this active treatment system was capable of treating 12×10^6 m^3 of water in the first 22 months of operation. During this time, approximately 3,200 tons of metal was removed at an efficiency of 99.2%. It is against this benchmark that bioremediation processes must compete.

Passive Biological Treatment Systems

Three pilot passive processes were installed, allowing a range of biotreatment strategies to be tested in series.

BOX TABLE 8.4.1 Average composition of Wheal Jane tin mine water (1995 to 1998)

Parameter	Value
Fe	150 mg $liter^{-1}$
Zn	2.5 mg $liter^{-1}$
As	2.5 mg $liter^{-1}$
Cd	0.12 mg $liter^{-1}$
Cu	0.5 mg $liter^{-1}$
Mn	20 mg $liter^{-1}$
Al	50 mg $liter^{-1}$
SO_4^{2-}	300 mg $liter^{-1}$
pH	3.9

The processes consisted of flowthrough aerobic artificial reed beds, anaerobic cells supporting sulfate-reducing bacteria, and finally shallow rock filters supporting algal growth. Pretreatment regimens to raise the pH of the influent water were compared, including lime dosing (to pH 4.5) and passage through an anoxic limestone drain, versus a no-pretreatment control stream.

Aerobic Cells (Artificial Reed Beds)

Five aerobic cells containing artificial reed beds were used to facilitate the initial precipitation of Fe(III) hydroxide/oxydroxide, which in turn sorbed arsenic present in the waters. The cells were planted with a 50:50 mix of *Phragmites* and *Typha*, with *Scirpus* used in some additional cells. Growth of the plants was good with the anoxic limestone drain pretreatment (influent to aerobic cells, pH 6.0), and the Fe(III) deposits were associated with the plant root systems in these cells. Here, the average Fe removal rate was 7.6 g m^{-2} day^{-1}, with 90% removal in the first cell alone. The more acid streams from the lime dosing step inhibited plant growth dramatically, but there was still iron removal in all aerobic cells. With the lime dosing pretreatment, the average rate of Fe removal was 4 g m^{-2} day^{-1}, while no pretreatment supported a rate of 5.8 g m^{-2} day^{-1}. In all cases, microbial oxidation of Fe(II) by moderately acidophilic bacteria played an important role in precipitation of the Fe(III) hydroxide-oxydroxides.

BOX 8.4 *(continued)*

Anaerobic Cells (Compost Bioreactors)
Underground chambers were filled with straw and sawdust as long-term sources of organic carbon to promote sulfate reduction, while manure was added as an inoculum for sulfate-reducing bacteria. This form of anaerobic metabolism was used to increase alkalinity and precipitate dissolved metals as insoluble sulfides. This proved successful if a conditioning phase was used to allow the reactors to stabilize, as identified during a 12-month shutdown of the compost bioreactors when the main mine water feed pipe fractured in 2001. After this shutdown, the pH of the effluent was consistently between 6 and 7, with concentrations of sulfide, Zn, and Fe below levels of detection, even without pretreatment of the mine water. Without the initial conditioning phase, the anaerobic cells did not operate effectively, even in the lime dosing or anoxic lime drain pretreatment streams.

Shallow Rock Filters (Algal Biofilms)
A series of 10 shallow pools filled with small granite pebbles was also tested as a final polishing step, to promote algal growth and precipitate metals via the generation of alkalinity (driven by consumption of CO_2 during oxygenic photosynthesis). These shallow rock pools were used downstream of the anaerobic cells. Again, the low pH of the effluent prevented algal growth, even in the streams pretreated by lime dosing or passage through the anoxic lime drain. Interestingly, following the conditioning shutdown described above, this final polishing step was noted to shift the pH of waters exiting the compost reactors from 6.5 to 7.5 and to remove Mn to below 0.5 mg $liter^{-1}$.

To conclude, this pilot study has demonstrated the potential of a passive treatment system comprised of a series of discrete bioprocesses to treat very aggressive toxic mine water with high loadings of metals. However, it is apparent from these studies that each of the three components of the operation described above has to function optimally for the successful passive treatment of acid mine drainage water, and this may require the introduction of suitable conditioning periods, especially for the anaerobic sulfate reduction steps in this process.

developments for the bioremediation of organics discussed in other chapters of this book. However, driven by the realization that large areas of land contaminated with metal and radionuclides cannot be economically remediated by conventional chemical approaches, significant resources have become available for this research area. Of particular note have been subsurface research programs (e.g., the U.S. Department of Energy NABIR program) that have supported fundamental studies on metal-microbe interactions. Supported by genomics-enabled studies ongoing in many laboratories worldwide (61), we can expect this research area to develop further in the near future, delivering more robust technologies for the bioremediation of metal-contaminated waters and land. Exciting developments in the use of microorganisms for the recycling of metal waste, with the formation of novel biominerals with unique properties, are also predicted in the near future.

REFERENCES

1. **Abdelouas, A., Y. L. Lu, W. Lutze, and H. E. Nuttall.** 1998. Reduction of U(VI) to U(IV) by indigenous bacteria in contaminated ground water. *J. Contam. Hydrol.* **35:**217–233.
2. **Abdelouas, A., W. Lutze, and E. H. Nuttall.** 1998. Chemical reactions of uranium in groundwater at a mill tailings site. *J. Contam. Hydrol.* **34:**343–361.
3. **Abdelouas, A., W. Lutze, and H. E. Nuttall.** 1999. Uranium contamination in the subsurface: characterization and remediation, p. 433–473. *In* P. C. Burns and R. Finch (ed.), *Uranium: Mineralogy, Geochemistry and the Environment*, vol. 38. Mineralogical Society of America, Washington, D.C.
4. **Ahmann, D.** 1997. Bioremediation of metal-contaminated soil. *Soc. Ind. Microbiol. News* **47:** 218–233.
5. **Anderson, R. T., and D. R. Lovley.** 2002. Microbial redox interactions with uranium: an environmental perspective, p. 205–233. *In* F. R. Livens and M. Keith-Roach (ed.), *Microbiology and Radioactivity*. Elsevier, Amsterdam, The Netherlands.
6. **Anderson, R. T., J. Rooney-Varga, C. V. Gaw, and D. R. Lovley.** 1998. Anaerobic benzene oxidation in the Fe(III)-reduction zone of petroleum-contaminated aquifers. *Environ. Sci. Technol.* **32:**1222–1229.
7. **Anderson, R. T., H. A. Vrionis, I. Ortiz-Bernad, C. T. Resch, P. E. Long, R. Day-**

vault, K. Karp, S. Marutzky, D. R. Metzler, A. Peacock, D. C. White, M. Lowe, and D. R. Lovley. 2003. Stimulating the in situ activity of *Geobacter* species to remove uranium from the groundwater of a uranium-contaminated aquifer. *Appl. Environ. Microbiol.* **69:**5884–5891.

8. **Andres, Y., H. J. MacCordick, and J.-C. Hubert.** 1993. Adsorption of several actinide (Th, U) and lanthanide (La, Eu, Yb) ions by *Mycobacterium smegmatis. Appl. Microbiol. Biotechnol.* **39:**413–417.
9. **Anonymous.** 2002. *Managing the Nuclear Legacy—A Strategy for Action.* United Kingdom Government White Paper. Presented to Parliament by the Secretary of State for Trade and Industry by Command of Her Majesty. [Online.] http://www.dti.gov.uk.
10. **Avery, S. A., and J. M. Tobin.** 1992. Mechanisms of strontium uptake by laboratory and brewing strains of *Saccharomyces cerevisiae. Appl. Environ. Microbiol.* **58:**3883–3889.
11. **Barnes, L. J., F. J. Janssen, J. Sherren, J. H. Versteegh, R. O. Koch, and P. J. H. Scheeren.** 1991. A new process for the microbial removal of sulphate and heavy metal from contaminated waters extracted by a geohydrological control system. *Chem. Eng. Res. Des.* **69A:**184–186.
12. **Baxter-Plant, V. S., I. P. Mikheenko, and L. E. Macaskie.** 2003. Sulphate-reducing bacteria, palladium and the reductive dehalogenation of chlorinated aromatic compounds. *Biodegradation* **14(2):**83–90.
13. **Berne, C.** 2004. Etude des mecanismes de degradation du phosphate de tributyle par des bacteries. Ph.D. dissertation. Universite de la Mediterranee, Aix-Marseille, France.
14. **Beveridge, T. J., M. N. Hughes, H. Lee, K. T. Leung, R. K. Poole, I. Savvaidis, S. Silver, and J. T. Trevors.** 1997. Metal-microbe interactions: contemporary approaches. *Adv. Microb. Physiol.* **38:**177–243.
15. **Bonthrone, K. M., G. Basnakova, F. Lin, and L. E. Macaskie.** 1996. Bioaccumulation of nickel by intercalation into polycrystalline hydrogen uranyl phosphate deposited via an enzymatic mechanism. *Nat. Biotechnol.* **14:**635–638.
16. **Bontidean, I., E. Csoregi, P. Corbisier, J. R. Lloyd, and N. L. Brown.** 2002. Bacterial metal-responsive elements and their use in biosensors for monitoring of heavy metals, p. 647–680. *In* B. Sankar (ed.), *The Handbook of Heavy Metals in the Environment.* Marcel Dekker, Inc., New York, N.Y.
17. **Bontidean, I., J. R. Lloyd, J. L. Hobman, J. R. Wilson, E. Csoregi, B. Mattiasson, and N. L. Brown.** 2000. Bacterial metal-resistance proteins and their use in biosensors for the detection of bioavailable heavy metals. *J. Inorg. Biochem.* **79:**225–229.
18. **Borst-Pauwels, G. W. F. H.** 1981. Ion transport in yeast. *Biochim. Biophys. Acta* **650:**88–127.
19. **Boswell, C. D., R. E. Dick, H. Eccles, and L. E. Macaskie.** 2001. Phosphate uptake and release by *Acinetobacter johnsonii* in continuous culture and coupling of phosphate release to heavy metal accumulation. *J. Ind. Microbiol. Biotechnol.* **26:**333–340.
20. **Brady, D., A. Stoll, and J. R. Buncan.** 1994. Biosorption of heavy metal cations by non-viable yeast biomass. *Environ. Technol.* **15:**429–439.
21. **Bruins, M. R., S. Kapil, and F. W. Oehme.** 2000. Microbial resistance to metals in the environment. *Ecotoxicol. Environ. Safety* **45:**198–207.
22. **Bustard, M., A. Donnellan, A. Rollan, L. McHale, and A. P. McHale.** 1996. The effect of pulsed field strength on electric field stimulated biosorption of uranium by *Kluyveromyces marxianus* IMB3. *Biotechnol. Lett.* **18:**479–482.
23. **Caccavo, F., Jr., D. J. Lonergan, D. R. Lovley, M. Davis, J. F. Stolz, and M. J. McInerney.** 1994. *Geobacter sulfurreducens* sp. nov., a hydrogen- and acetate-oxidizing dissimilatory metal-reducing microorganism. *Appl. Environ. Microbiol.* **60:**3752–3759.
24. **Carpentier, W., K. Sandra, I. De Smet, A. Brige, L. De Smet, and J. Van Beeumen.** 2003. Microbial reduction and precipitation of vanadium by *Shewanella oneidensis. Appl. Environ. Microbiol.* **69:**3636–3639.
25. **Chen, J. M., and O. J. Hao.** 1998. Microbial chromium(VI) reduction. *Crit. Rev. Environ. Sci. Technol.* **28:**219–251.
26. **Clearfield, A.** 1988. Role of ion exchange in solid-state chemistry. *Chem. Rev.* **88:**125–148.
27. **Cummings, D. E., O. L. Snoeyenbos-West, D. T. Newby, A. M. Niggemyer, D. R. Lovley, L. A. Achenbach, and R. F. Rosenzweig.** 2003. Diversity of *Geobacteraceae* species inhabiting metal-polluted freshwater lake sediments ascertained by 16S rDNA analyses. *Microb. Ecol.* **46:**257–269.
28. **deRome, L., and G. M. Gadd.** 1991. Use of pelleted and immobilized yeast and fungal biomass for heavy metal and radionuclide recovery. *J. Ind. Microbiol.* **7:**97–104.
29. **Diels, L., Q. Dong, D. van der Lelie, W. Baeyens, and M. Mergeay.** 1995. The *czc* operon of *Alcaligenes eutrophus* CH34: from resistance mechanism to the removal of heavy metals. *J. Ind. Microbiol.* **14:**142–153.
30. **Eccles, H., G. Garnham, C. Lowe, and N. Bruce.** March 1996. Biosensors for detecting metal ions capable of being reduced by reductase

enzymes. U.S. patent 5,500,351. http://patft.uspto.gov.

31. **Ehrlich, H. L.** 1996. *Geomicrobiology*, 3rd ed. Marcel Dekker, Inc., New York, N.Y.

32. **Ellwood, D. C., M. J. Hill, and J. H. P. Watson.** 1992. Pollution control using microorganisms and metal separation, p. 89–112. *In* J. C. Fry, G. M. Gadd, R. A. Herbert, C. W. Jones, and I. A. Watson-Craik (ed.), *Microbial Control of Pollution*, vol. 48. Cambridge University Press, Cambridge, United Kingdom.

33. **Fendorf, S., B. W. Wielinga, and C. M. Hansel.** 2000. Chromium transformations in natural environments: the role of biological and abiological processes in chromium(VI) reduction. *Int. Geol. Rev.* **42:**691–701.

34. **Finneran, K. T., R. T. Anderson, K. P. Nevin, and D. R. Lovley.** 2002. Potential for bioremediation of uranium-contaminated aquifers with microbial U(VI) reduction. *Soil Sed. Contam.* **11:**339–357.

35. **Francis, A. J.** 1994. Microbial transformations of radioactive wastes and environmental restoration through bioremediation. *J. Alloys Compounds* **213/214:**226–231.

36. **Gadd, G. M.** 2000. Microbial interactions with tributyltin compounds; detoxification, accumulation, and environmental fate. *Sci. Total Environ.* **258:**119–127.

37. **Gadd, G. M., and C. White.** 1989. Heavy metal and radionuclide accumulation and toxicity in fungi and yeasts, p. 19–38. *In* R. K. Poole and G. M. Gadd (ed.), *Metal-Microbe Interactions*. Society for General Microbiology, IRL Press, Oxford, England.

38. **Geiselhart, L., M. Osgood, and D. S. Holmes.** 1991. Construction and evaluation of a self-luminescent biosensor. *Ann. N.Y. Acad. Sci.* **646:**53–60.

39. **Gorby, Y. A., F. Caccavo, Jr., and H. Bolton, Jr.** 1998. Microbial reduction of cobalt(III) EDTA in the presence and absence of manganese(IV) oxide. *Environ. Sci. Technol.* **32:**244–250.

40. **Gorby, Y. A., and D. R. Lovely.** 1992. Enzymatic uranium precipitation. *Environ. Sci. Technol.* **26:**205–207.

41. **Gray, K. R., and A. J. Biddlestone.** 1995. Engineered reed-bed systems for wastewater treatment. *Trends Biotechnol.* **13:**248–252.

42. **Gray, N. F.** 1992. *Biology of Wastewater Treatment*. Oxford University Press, Oxford, United Kingdom.

43. **Greiner, R., U. Konietzny, and K. D. Jany.** 1993. Purification and characterization of two phytases from *Escherichia coli*. *Arch. Biochem. Biophys.* **303:**107–113.

44. **Higham, D. P., P. J. Sadler, and M. D. Scawen.** 1984. Cadmium-resistant *Pseudomonas putida* synthesizes novel cadmium binding proteins. *Science* **225:**1043–1046.

45. **Hobman, J. L., J. W. Wilson, and N. L. Brown.** 2000. Microbial mercury reduction, p. 177–197. *In* D. R. Lovely (ed.), *Environmental Microbe-Metal Interactions*. ASM Press, Washington, D. C.

46. **Holmes, D. E., K. T. Finneran, R. A. O'Neil, and D. R. Lovely.** 2002. Enrichment of members of the family *Geobacteraceae* associated with stimulation of dissimilatory metal reduction in uranium-contaminated aquifer sediments. *Appl. Environ. Microbiol.* **68:**2300–2306.

47. **Hunsberger, L. R., and A. B. Ellis.** 1990. Excited-state properties of lamellar solids derived from metal complexes and hydrogen uranyl phosphate. *Coord. Chem. Rev.* **97:**209–224.

48. **Inoue, H., O. Takimura, H. Fuse, K. Murakami, K. Kamimura, and Y. Yamaoka.** 2000. Degradation of triphenyltin by a fluorescent pseudomonad. *Appl. Environ. Microbiol.* **66:**3492–3498.

49. **Jeong, B. C., H.-W. Kim, S. J. Owen, R. E. Dick, and L. E. Macaskie.** 1994. Phosphoesterase activity and phosphate release from tributyl phosphate by a *Citrobacter* sp. *Appl. Biochem. Biotechnol.* **47:**21–32.

50. **Kalin, M.** 2000. Biogeochemical and geological considerations in designing wetland treatment systems in post-mining landscapes. *Waste Manage.* **21:**191–196.

51. **Kjaergaard, K., J. K. Sorenson, M. A. Schembri, and P. Klemm.** 2000. Sequestration of zinc oxide by fimbrial designer chelators. *Appl. Environ. Microbiol.* **66:**10–14.

52. **Lloyd, J. R.** 2003. Microbial reduction of metals and radionuclides. *FEMS Microbiol. Rev.* **27:**411–425.

53. **Lloyd, J. R., C. L. Harding, and L. E. Macaskie.** 1997. Tc(VII) reduction and accumulation by immobilized cells of *Escherichia coli*. *Biotechnol. Bioeng.* **55:**505–510.

54. **Lloyd, J. R., and L. E. Macaskie.** 2002. The biochemical basis of radionuclide-microbe interactions, p. 313–342. *In* F. R. Livens and M. Keith-Roach (ed.), *Microbiology and Radioactivity*. Elsevier, Amsterdam, The Netherlands.

55. **Lloyd, J. R., and L. E. Macaskie.** 2000. Bioremediation of radioactive metals, p. 277–327. *In* D. R. Lovely (ed.), *Environmental Microbe-Metal Interactions*. ASM Press, Washington, D. C.

56. **Lloyd, J. R., and L. E. Macaskie.** 1996. A novel PhosphorImager-based technique for

monitoring the microbial reduction of technetium. *Appl. Environ. Microbiol.* **62**:578–582.

57. **Lloyd, J. R., V. A. Sole, C. V. Van Praagh, and D. R. Lovely.** 2000. Direct and Fe(II)-mediated reduction of technetium by Fe(III)-reducing bacteria. *Appl. Environ. Microbiol.* **66**: 3743–3749.
58. **Lloyd, J. R., P. Yong, and L. E. Macaskie.** 1998. Enzymatic recovery of elemental palladium by using sulfate-reducing bacteria. *Appl. Environ. Microbiol.* **64**:4607–4609.
59. **Losi, M. E., and W. T. Frankenberger.** 1997. Bioremediation of selenium in soil and water. *Soil Sci.* **162**:692–702.
60. **Lovley, D., and E. J. Phillips.** 1992. Reduction of uranium by *Desulfovibrio desulfuricans*. *Appl. Environ. Microbiol.* **58**:850–856.
61. **Lovely, D. R.** 2003. Cleaning up with genomics: applying molecular biology to bioremediation. *Nat. Rev. Microbiol.* **1**:36–44.
62. **Lovley, D. R.** 1993. Dissimilatory metal reduction. *Annu. Rev. Microbiol.* **47**:263–290.
63. **Lovley, D. R.** 2001. Reduction of iron and humics in subsurface environments, p. 193–217. *In* J. K. Frederickson and M. Fletcher (ed.), *Subsurface Microbiology and Biogeochemistry*. Wiley-Liss, Inc., New York, N. Y.
64. **Lovley, D. R., and J. D. Coates.** 2000. Novel forms of anaerobic respiration of environmental relevance. *Curr. Opin. Microbiol.* **3**:252–256.
65. **Lovley, D. R., and E. J. P. Phillips.** 1992. Bioremediation of uranium contamination with enzymatic uranium reduction. *Environ. Sci. Technol.* **26**:2228–2234.
66. **Lovley, D. R., E. J. P. Phillips, Y. A. Gorby, and E. Landa.** 1991. Microbial reduction of uranium. *Nature* **350**:413–416.
67. **Lovley, D. R., E. E. Roden, E. J. P. Phillips, and J. C. Woodward.** 1993. Enzymatic iron and uranium reduction by sulfate reducing bacteria. *Mar. Geol.* **113**:41–53.
68. **Mabbett, A. N., and L. E. Macaskie.** 2002. A new bioinorganic process for the remediation of Cr(VI). *J. Chem. Technol. Biotechnol.* **77**: 1169–1175.
69. **Macaskie, L. E.** 1991. The application of biotechnology to the treatment of wastes produced from nuclear fuel cycle: biodegradation and bioaccumulation as a means of treating radionuclide-containing streams. *Crit. Rev. Biotechnol.* **11**:41–112.
70. **Macaskie, L. E.** 1990. An immobilized cell bioprocess for the removal of heavy metals from aqueous flows. *J. Chem. Technol. Biotechnol.* **49**: 357–379.
71. **Macaskie, L. E., and A. C. R. Dean.** 1990. Trimethyl lead degradation by free and immobilized cells of an *Arthrobacter* sp. and by the wood decay fungus *Phaeolus schweinitzii*. *Appl. Microbiol. Biotechnol.* **33**:81–87.
72. **Macaskie, L. E., R. M. Empson, A. K. Cheetham, C. P. Grey, and A. J. Skarnulis.** 1992. Uranium bioaccumulation by a *Citrobacter* sp. as a result of enzymically-mediated growth of polycrystalline HUO_2PO_4. *Science* **257**: 782–784.
73. **Macaskie, L. E., R. M. Empson, F. Lin, and M. R. Tolley.** 1995. Enzymatically-mediated uranium accumulation and uranium recovery using a *Citrobacter* sp. immobilised as a biofilm within a plug-flow reactor. *J. Chem. Technol. Biotechnol.* **63**:1–16.
74. **Macaskie, L. E., B. C. Jeong, and M. R. Tolley.** 1994. Enzymically-accelerated biomineralization of heavy metals: application to the removal of americium and plutonium from aqueous flows. *FEMS Microbiol. Rev.* **14**:351–368.
75. **Macaskie, L. E., J. R. Lloyd, R. A. P. Thomas, and M. R. Tolley.** 1996. The use of microoorganisms for the remediation of solutions contaminated with actinide elements, other radionuclides and organic contaminants generated by nuclear fuel cycle activities. *Nucl. Energy* **35**: 257–271.
76. **Macaskie, L. E., P. Yong, T. C. Doyle, M. G. Roig, M. Diaz, and T. Manzano.** 1997. Bioremediation of uranium-bearing wastewater: biochemical and chemical factors influencing bioprocess application. *Biotechnol. Bioeng.* **53**: 100–109.
77. **McCullough, J., T. C. Hazen, S. M. Benson, F. B. Metting, and A. C. Palmisano.** 1999. *Bioremediation of Metals and Radionuclides . . . What It Is and How It Works*. Lawrence Berkeley National Laboratory, Berkeley, Calif. http://www.lbl.gov/NABIR.
78. **McHale, A. P., and S. McHale.** 1994. Microbial biosorption of metals: potential in the treatment of metal pollution. *Biotechnol. Adv.* **12**: 647–652.
79. **Michel, L. J., L. E. Macaskie, and A. C. R. Dean.** 1986. Cadmium accumulation by immobilized cells of a *Citrobacter* sp. using various phosphate donors. *Biotechnol. Bioeng.* **28**:1358–1365.
80. **National Academies of Science.** 1999. *Groundwater and Soil Cleanup: Improving Management of Persistent Contaminants*. National Academies Press, Washington, D.C. http://www.nap.edu.
81. **Okorov, L. A., L. P. Lichko, V. M. Kodomtseva, V. P. Kholodenko, V. T. Titovsky, and I. S. Kulaev.** 1977. Energy-dependent transport of manganese into yeast cells and distri-

bution of accumulated ions. *Eur. J. Biochem.* **75:** 373–377.

82. **Ortiz-Bernad, I., R. T. Anderson, H. A. Vrionis, and D. R. Lovley.** 2004. Vanadium respiration by *Geobacter metallireducens*: a novel strategy for the in situ removal of vanadium from groundwater. *Appl. Environ. Microbiol.* **70:** 3091–3095.
83. **Ostanin, K., E. H. Harms, P. E. Stevis, R. Kuciel, M.-M. Zhou, and R. L. Van Etten.** 1992. Overexpression, site-directed mutagenesis and mechanism of *E. coli* acid phosphatase. *J. Biol. Chem.* **267:**22830–22836.
84. **Ostanin, K., and R. Van Etten.** 1993. Asp304 of *Escherichia coli* acid phosphatase is involved in leaving group protonation. *J. Biol. Chem.* **268:** 20778–20784.
85. **Owen, S., B. C. Jeong, P. S. Poole, and L. E. Macaskie.** 1992. Tributyl phosphate degradation by immobilized cells of a *Citrobacter* sp. *Appl. Biochem. Biotechnol.* **34/35:**693–707.
86. **Pattanapipitpaisal, P., A. N. Mabbett, J. A. Finlay, A. J. Beswick, M. Paterson-Beedle, A. Essa, J. Wright, M. R. Tolley, U. Badar, N. Ahmed, J. L. Hobman, N. L. Brown, and L. E. Macaskie.** 2002. Reduction of Cr(VI) and bioaccumulation of chromium by Gram positive and Gram negative microorganisms not previously exposed to Cr-stress. *Environ. Technol.* **23:** 731–745.
87. **Pazirandeh, M., B. M. Wells, and R. L. Ryan.** 1998. Development of bacterium-based heavy-metal biosorbants: enhanced uptake of cadmium and mercury by *Escherichia coli* expressing a metal-binding motif. *Appl. Environ. Microbiol.* **64:**4068–4072.
88. **Pham-Thi, M., and P. Columban.** 1985. Cationic conductivity, water species motions and phase transitions in $H_3OUO_2PO_4{\cdot}3H_2O$ (HUP) and MUP related compounds ($M^+ = Na^+, K^+, Ag^+, Li^+, NH_4^+$). *Solid State Ion* **17:**295–306.
89. **Postgate, J. R.** 1979. *The Sulphate Reducing Bacteria.* Cambridge University Press, Cambridge, United Kingdom.
90. **Pozas-Tormo, R., L. Moreno-Real, M. Martinez-Lara, and S. Bruque-Gamez.** 1987. Intercalation of lanthanides into $H_3OUO_2PO_4{\cdot}3H_2O$ and $C_4H_9NH_3UO_2PO_4{\cdot}3H_2O$. *Inorg. Chem.* **26:**1442–1445.
91. **Pumpel, T.** 1997. Metal biosorption: a structured data space? *Res. Microbiol.* **148:**514–515.
92. **Raskin, I., R. D. Smith, and D. E. Salt.** 1997. Phytoremediation of metals: using plants to remove pollutants from the environment. *Curr. Opin. Biotechnol.* **8:**221–226.
93. **Riordan, C., M. Bustard, R. Putt, and A. P. McHale.** 1997. Removal of uranium from solution using residual brewery yeast: combined biosorption and bioprecipitation. *Biotechnol. Lett.* **19:**385–387.
94. **Rodriguez, H., and R. Fraga.** 1999. Phosphate solubilizing bacteria and their role in plant growth promotion. *Biotechnol. Adv.* **17:**319–339.
95. **Roig, M. G., J. F. Kennedy, and L. E. Macaskie.** 1995. *Biological Rehabilitation of Metal Bearing Wastewaters.* Final report. EC contract EV5V-CT93-0251. Programme: Environment. The European Commission, Brussels, Belgium.
96. **Röling, W. F. M., B. M. van Breukelen, M. Braster, B. Lin, and H. W. van Verseveld.** 2001. Relationships between microbial community structure and hydrochemistry in a landfill leachate-polluted aquifer. *Appl. Environ. Microbiol.* **67:**4619–4629.
97. **Ross, I. S., and C. C. Townsley.** 1986. The uptake of heavy metals by filamentous fungi, p. 49–58. *In* H. Eccles and S. Hunt (ed.), *Immobilisation of Ions by Biosorption.* Ellis Horwood, Chichester, United Kingdom.
98. **Salt, D. E., M. Blaylock, N. P. B. A. Kumar, V. Dushenkov, B. D. Ensley, I. Chet, and I. Raskin.** 1995. Phytoremediation: a novel strategy for the removal of toxic metals from the environment using plants. *Biotechnology* **13:** 468–474.
99. **Salt, D. E., R. D. Smith, and I. Raskin.** 1998. Phytoremediation. *Annu. Rev. Plant Phys. Plant Mol. Biol.* **49:**643–668.
100. **Samuelson, P., H. Wernerus, M. Svedberg, and S. Stahl.** 2000. Staphylococcal surface display of metal-binding polyhistidyl peptides. *Appl. Environ. Microbiol.* **66:**1243–1248.
101. **Satroutdinov, A. D., E. G. Dedyukhina, T. I. Chistyakova, M. Witschel, I. G. Minkevich, V. K. Eroshin, and T. Egli.** 2000. Degradation of metal-EDTA complexes by resting cells of the bacterial strain DSM 9103. *Environ. Sci. Technol.* **34:**1715–1720.
102. **Simmons, P., J. M. Tobin, and I. Singleton.** 1995. Considerations on the use of commercially available yeast biomass for the treatment of metal-containing effluents. *J. Ind. Microbiol.* **14:** 240–246.
103. **Snoeyenbos-West, O., K. P. Nevin, R. T. Anderson, and D. R. Lovley.** 2000. Enrichment of *Geobacter* species in response to stimulation of Fe(III) reduction in sandy aquifer sediments. *Microb. Ecol.* **39:**153–167.
104. **Tebo, B. M., and A. Y. Obraztsova.** 1998. Sulfate-reducing bacterium grows with Cr(VI), U(VI), Mn(IV), and Fe(III) as electron acceptors. *FEMS Microbiol. Lett.* **162:**193–198.
105. **Thomas, R. A. P., A. J. Beswick, G. Basnakova, R. Moller, and L. E. Macaskie.** 2000.

Growth of naturally-occurring microbial isolates in metal-citrate medium and bioremediation of metal-citrate wastes. *J. Chem. Technol. Biotechnol.* **75:**187–195.

106. **Thomas, R. A. P., K. Lawlor, M. Bailey, and L. E. Macaskie.** 1998. The biodegradation of metal-EDTA complexes by an enriched microbial population. *Appl. Environ. Microbiol.* **64:** 1319–1322.
107. **Thomas, R. A. P., and L. E. Macaskie.** 1996. Biodegradation of tributyl phosphate by naturally occurring microbial isolates and coupling to the removal of uranium from aqueous solution. *Environ. Sci. Technol.* **30:**2371–2375.
108. **Thomas, R. A. P., and L. E. Macaskie.** 1998. The effect of growth conditions on the biodegradation of tributyl phosphate and potential for the remediation of acid mine drainage waters by a naturally-occurring mixed microbial culture. *Appl. Microbiol. Biotechnol.* **49:**202–209.
109. **Tobin, J. M., D. G. Cooper, and R. J. Neufeld.** 1984. Uptake of metal ions by *Rhizopus arrhizus* biomass. *Appl. Environ. Microbiol.* **47:** 821–824.
110. **Tobin, J. M., C. White, and G. M. Gadd.** 1994. Metal accumulation by fungi: applications in environmental biotechnology. *J. Ind. Microbiol.* **13:**126–130.
111. **Tolley, M. R., L. F. Strachan, and L. E. Macaskie.** 1995. Lanthanum accumulation from acidic solutions using *Citrobacter* sp. immobilized in a flow through bioreactor. *J. Ind. Microbiol.* **14:**271–280.
112. **Treen-Sears, M. E., B. Volesky, and R. J. Neufeld.** 1984. Ion exchange/complexation of the uranyl ion by *Rhizopus* biosorbent. *Biotechnol. Bioeng.* **26:**1323–1329.
113. **Tucker, M. D.** 1996. *Technical Considerations for the Implementation of Subsurface Microbial Barriers for Restoration of Ground Water at UMTRCA Sites.* SAND96–1459. Sandia National Laboratories, Albuquerque, N.M.
114. **Turner, J. S., and N. J. Robinson.** 1995. Cyanobacterial metallothioneins: biochemistry and molecular genetics. *J. Ind. Microbiol.* **14:**119–125.
115. **U.S. Department of Energy, Grand Junction Office.** 1999. Final site observational work plan for the UMTRA project Old Rifle site GJO-99–88-TAR. U. S. Department of Energy, Grand Junction, Colo.
116. **Valls, M., S. Atrian, V. de Lorenzo, and L. A. Fernandez.** 2000. Engineering a mouse metallothionein on the cell surface of *Ralstonia eutropha* CH34 for immobilization of heavy metals in soil. *Nat. Biotechnol.* **18:**661–665.
117. **Van Roy, S., K. Peys, T. Dresselaers, and L. Diels.** 1997. The use of an *Alcaligenes eutrophus* biofilm in a membrane bioreactor for heavy metal recovery. *Res. Microbiol.* **148:**526–528.
118. **Volesky, B., and Z. R. Holan.** 1995. Biosorption of heavy metals. *Biotechnol. Prog.* **11:** 235–250.
119. **Volesky, B., and H. A. May-Phillips.** 1995. Biosorption of heavy metals by *Saccharomyces cerevisiae. Appl. Microbiol. Biotechnol.* **42:**797–806.
120. **von Canstein, H., Y. Li, J. Leonhäuser, E. Haase, A. Felske, W.-D. Deckwer, and I. Wagner-Döbler.** 2002. Spatially oscillating activity and microbial succession of mercury-reducing biofilms in a technical scale bioremediation system. *Appl. Environ. Microbiol.* **68:** 1938–1946.
121. **von Canstein, H., Y. Li, K. N. Timmis, W. D. Deckwer, and I. Wagner-Döbler.** 1999. Removal of mercury from chloralkali electrolysis wastewater by a mercury-resistant *Pseudomonas putida* strain. *Appl. Environ. Microbiol.* **65:** 5279–5284.
122. **von Canstein, H. F., Y. Li, and I. Wagner-Döbler.** 2001. Long term stability of mercury reducing microbial biofilm communities analysed by 16S-23S rDNA interspacer region polymorphism. *Microb. Ecol.* **42:**624–634.
123. **von Canstein, H. F., Y. Li, and I. Wagner-Döbler.** 2001. Long-term performance of bioreactors cleaning mercury contaminated wastewater and their response to temperature and mercury stress and mechanical perturbation. *Biotechnol. Bioeng.* **74:**212–219.
124. **Wagner-Döbler, I., H. Lunsdorf, T. Lubbehusen, H. F. von Canstein, and Y. Li.** 2000. Structure and species composition of mercury-reducing biofilms. *Appl. Environ. Microbiol.* **66:** 4559–4563.
125. **Wagner-Döbler, I., H. von Canstein, Y. Li, K. N. Timmis, and W. D. Deckwer.** 2000. Removal of mercury from chemical wastewater by microoganisms in technical scale. *Environ. Sci. Technol.* **34:**4628–4634.
126. **Watson, J. H. P., and D. C. Ellwood.** 1994. Biomagnetic separation and extraction process for heavy metals from solution. *Minerals Eng.* **7:** 1017–1028.
127. **Weidemann, D. P., R. D. Tanner, G. W. Strandberg, and S. E. Shumate II.** 1981. Modelling the rate of transfer of uranyl ions onto microbial cells. *Enzyme Microb. Technol.* **3:** 33–40.
128. **Whicker, F. W., T. G. Hinton, M. M. MacDonell, J. E. Pinder III, and L. J. Habegger.** 2004. Avoiding destructive remediation at DOE sites. *Science* **303:**1615–1616.
129. **White, C., and G. M. Gadd.** 1987. Inhibition of H^+ efflux and K^+ uptake and induction of

K^+ efflux in yeast by heavy metals. *Tox. Assess.* **2**:437–447.

130. **White, C., A. K. Sharman, and G. M. Gadd.** 1998. An integrated microbial process for the bioremediation of soil contaminated with toxic metals. *Nat. Biotechnol.* **16**:572–575.

131. **Yong, P., and L. E. Macaskie.** 1997. Effect of substrate concentration and nitrate inhibition on product release and heavy metal removal by a *Citrobacter* sp. *Biotechnol. Bioeng.* **55**:821–830.

132. **Yong, P., and L. E. Macaskie.** 1995. Enhancement of uranium bioaccumulation by a *Citrobacter* sp. via enzymatically-mediated growth of polycrystalline $NH_4UO_2PO_4$. *J. Chem. Technol. Biotechnol.* **63**:101–108.

133. **Yong, P., and L. E. Macaskie.** 1999. The role of sulfate as a competitive inhibitor of enzymatically-mediated heavy metal uptake by *Citrobacter* sp.: implications in the bioremediation of acid mine drainage water using biogenic phosphate precipitant. *J. Chem. Technol. Biotechnol.* **74**: 1149–1156.

134. **Yong, P., N. A. Rowson, J. P. G. Farr, I. R. Harris, and L. E. Macaskie.** 2002. Bioaccumulation of palladium by *Desulfovibrio desulfuricans*. *J. Chem. Technol. Biotechnol.* **77**:593–601.

135. **Yong, P., N. A. Rowson, J. P. G. Farr, I. R. Harris, and L. E. Macaskie.** 2002. Bioreduction and biocrystallization of palladium by *Desulfovibrio desulfuricans* NCIMB 8307. *Biotechnol. Bioeng.* **80**:369–379.

136. **Yong, P., N. A. Rowson, J. P. G. Farr, I. R. Harris, and L. E. Macaskie.** 2003. A novel electrobiotechnology for the recovery of precious metals from spent automotive catalysts. *Environ. Technol.* **23(3)**:289–297.

PREEMPTIVE BIOREMEDIATION: APPLYING BIOTECHNOLOGY FOR CLEAN INDUSTRIAL PRODUCTS AND PROCESSES

Mike Griffiths and Ronald M. Atlas

9

INTRODUCTION

A combination of economic, environmental, and social pressures, the so-called "triple bottom line," is driving companies in a wide spectrum of industries to explore and use alternative manufacturing processes and to develop novel, environmentally friendly products. In essence, industry is turning to preemptive bioremediation as a strategy to reduce potential environmental problems and liabilities and make industrial products and processes more sociably acceptable in an age of ecoactivism. Preemptive bioremediation aims at producing materials that are less likely to persist in the environment and cause harm, processes and products that are based upon renewable resources, and processes that require less energy, i.e., using biotechnology for clean products and processes that result in the formation of fewer pollutants. There have been several reviews of the current and potential uses of biocatalysis to reduce pollution and to produce chemical products of industrial value (52, 59, 69).

Although the implementation of preemptive bioremediation strategies has been slower than expected, biotechnology is providing a rapidly expanding tool kit so that new product and process creation can be expected to accelerate. In this chapter, we describe some of these new developments, identify the barriers faced by the new technology, and indicate how they might be overcome. In addition, we will highlight the contributions to be made by the latest research and development. Later in the chapter, we will examine how companies assess their various options and make decisions concerning the introduction of novel products and processes.

The applications of so-called "white" biotechnology (25, 26) (application of biotechnology to industrial processes, as distinct from "red" [the application to medicine] and "green" [the application to agriculture]) fall conveniently into two distinct groups:

- the replacement of fossil fuel raw materials by renewable (biomass) raw materials, and
- the replacement of a conventional, nonbiological process by one based on biological systems, such as whole cells or enzymes, used as reagents or catalysts.

Renewable Materials

George Washington Carver is remembered as a pioneering developer of industrial products

Mike Griffiths, Mike Griffiths Associates, St. Non, Pleasant Valley, Stepaside, Narberth, Pembrokeshire SA67 8NY, United Kingdom. *Ronald M. Atlas*, Graduate School, University of Louisville, Louisville, KY 40292.

Bioremediation: Applied Microbial Solutions for Real-World Environmental Cleanup
Edited by Ronald M. Atlas and Jim C. Philip

from plants. Carver's work resulted in the creation of 325 products from peanuts, more than 100 products from sweet potatoes, and hundreds more from a dozen other plants native to the southern United States. These products contributed to rural economic improvement by offering alternative crops to cotton that were beneficial for the farmers and for the land (35).

In 1941, Henry Ford, who was reputed to wear clothes made from soybean-based fibers (6), demonstrated a handmade car with a body made entirely from plant-derived plastic. Unfortunately, he never managed to make it compete economically with conventional plastics. There are those who believe we are entering a "carbon-constrained world" where the production of carbon dioxide will have a significant cost, but in the short term the commercial fate of bioproducts will hinge on the price of crude oil.

Many industrial chemicals and products of chemical manufacture in common use are carbon based and derive from fossil carbon sources, primarily crude oil and natural gas, which are finite resources. When these manufactured materials reach the end of their useful life, if they cannot be recycled, they are often incinerated or left to degrade, giving rise to CO_2 and other polluting compounds.

The parallel universe of chemical and material production, based on biomass raw materials and separate entirely from that based on fossil carbon, is being explored to a much lesser extent, in part because of the complex chemistry involved but also because of the relative costs of the raw materials. Nonetheless, a wide range of nonfood products are currently derived from biomass. To give just a few examples: production of cotton for fibers exceeds that of all synthetic fibers; natural rubber competes in quantity with synthetics; starch is used as an additive in paper and textiles and as an adhesive; vegetable oils are made into soaps and detergents; and biomass-based ethanol is a significant replacement for gasoline.

Bioprocesses

Currently, bioprocesses account for commercial production of more than 15 million tons of chemical products per year, including organic and amino acids, antibiotics, industrial and food enzymes, fine chemicals, active ingredients for crop protection, pharmaceutical products, and fuel ethanol. These products account for almost $10 billion in annual sales at the bulk level but are largely unrecognized by consumers, who buy them as ingredients or components of products with other names. In addition, biotechnology is finding uses in industries as diverse as pulp and paper, textiles, and metal extraction.

Building upon an Organization for Economic Cooperation and Development (OECD) study, the Biotechnology Industry Organization (BIO) (12) has produced a report, with primary focus on the United States, entitled "New Biotech Tools for a Cleaner Environment." The BIO report is intended to inspire corporate leaders and policy makers but will be of relevance to nongovernmental organizations and the general public.

By way of illustration, Table 9.1 from BIO (12) gives just some applications of industrial biotechnology. The implementation of microbial technology in the decades 1960 to 1990 has led to the development of the current applications of industrial enzymes in detergent compositions and the manufacture of high-fructose corn syrup and the like.

Perhaps a few hundred different enzymes are now used commercially, derived primarily from microorganisms that can be cultured in the laboratory. Enzymes can be remarkably specific or very nonspecific: proteases, for example, may operate on an individual peptide bond in but one protein or, like the enzymes involved in protein degradation, can attack a wide range of substrates. The other characteristic is the high levels of rate acceleration they achieve—the protease-catalyzed reactions can be as much as 10^{16} times faster than the uncatalyzed equivalent reactions (67). Many enzymes can accept unnatural substrates, and genetic engineering can alter their stability, broaden their substrate specificity, and increase their specific activity.

The primary impact of genetic engineering

TABLE 9.1 Some industrial biotechnology applications by industrial sector[a]

Biological fuel cells	Oil and gas desulfurization	Metal ore heap leaching
Fine and bulk chemicals	Leather degreasing	Electroplating/metal cleaning
Chiral compound synthesis	Biohydrogen	Rayon and other synthetic fibers
Synthetic fibers for clothing	Biopolymers for plastic packaging	Metal refining
Pharmaceuticals	Coal bed methane water treatment	Vitamin production
Food-flavoring compounds	Chemical/biological warfare agent decontamination	Sweetener production (high-fructose syrup)
Biobased plastics	Pulp and paper bleaching	Oil well drill hole completion (nontoxic cake breakers)
Biopolymers for automobile parts	Biopulping (paper industry)	Road surface treatment for dust control
Bioethanol transportation fuel	Specialty textile treatment	Textile dewatering
Nutritional oils	Enzyme food processing aids	Vegetable oil degumming

[a] Taken from reference 12.

so far has been to make improvements in the economics of enzyme production and enhancement of enzyme functionality. Much attention in biocatalyst design is on increasing the useful lifetime of biocatalysts and having them operate under conditions not usually found in nature. The process engineer is unlikely to favor a delicate catalyst that must be replaced every few hours.

Problems slowing the development of these biotechnological processes include the expensive nature of current raw materials which are food and feed alternatives, while the enzymes and enzymatic processes require considerable research and development expenditure and long lead times. The first difficulty may be overcome by using the large amount of waste biomass left over after processing for higher-added-value products. Nevertheless, stimulated by the increasing use of biofuels, biorefineries may take their place alongside petroleum refineries. This additional raw material base will change the prospects of agriculture, especially in developed countries.

Sustainability

Biotechnology can be used to increase the environmental friendliness of industrial processes and to encourage a shift in companies' emphasis from end-of-pipe cleanup to inherently clean processes. There are many examples of companies moving back up the pipe by, for example, introducing closed-loop water supply systems; it is a smaller step than replacing chemical conversion with biocatalysis but is still a useful contribution.

One factor limiting the potential of biotechnology has been the absence of techniques for comparing the overall long-term sustainability of alternative processes. In the 1970s and earlier, sustainability was one-dimensional; it was equated with the profit necessary for the long-term survival of the company. Later, environmental concerns were added, and in the 1990s a third dimension was added: societal concerns. Hence, the triple bottom line. More and more companies are beginning to adopt the principles of sustainable development in their everyday activities and to see that doing so does not generate extra cost but, rather, can be an economic advantage (see the Cargill Dow LLC website,http://www.cargilldow.com/corporate/life-cycle). Environmental considerations are thus not being addressed in isolation but are being integrated with the economic and social aspects of the business.

Until recently, two major questions posed by industrialists have remained unanswered: could biotechnology provide a cheaper option to conventional industrial processes, and could economic improvements and environmental friendliness go hand in hand (50, 51)? A steady stream of innovation is emerging from academic biotechnology laboratories, but it will not necessarily be taken up by industry unless it is clearly demonstrable that there is a cost advantage. Cost reduction can be direct (lower material and/or energy inputs, waste treatment costs, reduced capital expenditure) or indirect (lower risk to the general public, fewer obligations in the way of eventual cleanup, reduced global pollution levels, improved downstream recycling).

Environmental assessment should be applied to all products and processes, large or small, and in companies of all sizes. All stages of the life cycle of a product or process may affect the environment, and consequently the design of industrial processes must take into consideration everything from choice and quantities of raw materials utilized to reuse of wastes. Environmentally friendly processes will consume less energy and raw material and markedly reduce or even eliminate wastes. Biotechnology is capable of providing tools that help achieve these goals and in the process ensure that industrial sustainability is in fact being achieved.

RENEWABLE RESOURCES

Plant and animal sources have provided industrial raw materials, such as timber, paper, vegetable oils, tallow, and many others, for centuries. Following the first production of crude oil, however, this abundant and cheap resource quickly became the principal raw material for a burgeoning chemical industry. Only in times of conflict, when resources take on a strategic importance, has there been any significant return to natural raw materials. Today, with increasing attention being paid to greenhouse gases (GHGs) and global warming, renewable raw materials may once again have a role to play.

Perhaps the most attractive feature of plant biomass, which is a lignocellulosic material, is that it is a product of photosynthesis and as such is a CO_2-neutral, renewable resource. As plants grow, they take in CO_2 from the atmosphere, and this carbon is largely retained (sequestered) until the plant is used. It is this principle that makes the planting of forests a "sink" for atmospheric carbon. When plant raw material is used in any industrial process, it can give off only as much CO_2 as it took in in the first place. Only the fossil fuels that may also be used in the process, in transport, for example, contribute to the buildup of greenhouse gases.

Chemicals from Plants

Extracts from plants, such as sugars, starch, and oils, are the major raw materials for both foodstuffs and chemicals. Sucrose from sugarcane and sugar beet is a major resource for ethanol production; dextrose (glucose) from cornstarch is a basis for chemicals such as lysine and lactic acid. Oils from soybean and rapeseed (canola) have a wide range of industrial uses. Cargill, a major corn wet-miller, has seen the possibilities in widening the use of its resources and has based several joint ventures, with CSM of The Netherlands to make lactic acid, with Mitsubishi to make erythritol, and with Degussa to make lysine (Box 9.1), at its major facilities in Nebraska. The company's latest development is a plant to make polylactide (PLA) in a joint venture with Dow (see below). In addition, Cargill has a joint venture with Roche to make vitamin E from a soy oil waste product.

Soy is a particularly important renewable resource. Several hundred industrial products made from soybeans were developed in the 1930s and 1940s, including adhesives, rubber substitutes, printing inks, and plastics. Amides, esters, and acetates of biohydrocarbons are currently being used as plasticizers and blocking or slip agents and mold release agents for synthetic polymers. Biohydrocarbons linked to amines, quaternary ammonium ligands, alcohols, phosphates, and sulfur ligands are used as fabric softeners, surfactants, emulsifiers, corrosion inhibitors, antistatic agents, hair conditioners, ink carriers, biodegradable solvents, cosmetic bases, and perfumes. Complexes with aluminum, magnesium, or other metal compounds have produced greases and marine lubricating oils.

Soy protein and sugars may be used in the production of polyurethane foams for packing,

BOX 9.1
Lysine Feed Additive

Midwest Lysine LLC, a joint venture between Cargill and Degussa-Hüls, has built a plant in Blair, Nebr., to produce 75,000 metric tons of the amino acid lysine per year. Based on dextrose as raw material, the lysine will be used as a feed additive to increase the nutritional value of plant proteins.

Lysine has been produced for many years by fermentation, using *Corynebacterium* or *Brevibacterium* spp. The conventional product is L-lysine-HCl, which is produced by a multistep process. When Degussa decided to become a producer, it was realized that the "conventional" process would be very expensive both because of the large amounts of waste and bacterial biomass that are made as by-products and because of the loss of product during downstream processing.

A new product, Biolys 60, was developed, and a new process which reduces the by-products and the wastes almost to zero was invented and patented by Degussa. Degussa changed raw materials and fermentation process so that the fermentation broth contains lysine and by-products in such a ratio that the product has 60% lysine when dried. Because such a fermentation broth is very difficult to dry, a special technique that results in a granulated, dust-free product had to be developed. In comparison to the conventional process, the new process is very environmentally friendly because no wastes are produced. This is an example of a low-value bulk product which would never have been economic without such savings.

The $100 million plant, which employs 70 people, began operations in June 2000.

Source: Degussa

insulation, and padding. Using these materials to replace expensive polyalcohols improves the biodegradability of these materials and also reduces the need for environmentally undesirable fluorocarbon foaming agents. Coextrusion of soy protein with polyvinyl chloride produces a silk-like fiber, with greatly increased wet/dry tensile strength, while retaining wearing comfort, as a result of the hydrophilic nature of the protein fibers. A DuPont company is using soy protein for coated paper and paperboard, adhesives, and inks (see soypolymers.dupont.com).

A number of steroid hormones are derived via bioprocesses from the natural phytosterols in soybean, conifers, and rapeseed. Schering has made androstendione and androsta-dienedione on a 200-m^3 scale using *Mycobacterium* mutants (71).

Use of a larger number of plant species is also occurring. For example, natural rubber is currently derived from only one tree genus, *Hevea*. There are, however, more than 300 other genera producing isoprenes, many yet to be exploited.

The structural components of biomass vary from source to source, but lignin, cellulose, and hemicellulose are always present. The U.S. pulping industry produces 20 million tons of lignin as a by-product of papermaking. Virtually all of this lignin—the second most abundant biological polymer on Earth—gets burned as waste. Wood scientists, who think of wood as three-dimensional biopolymer composites, want to see this resource become a supply of high-technology materials, including plastics. Lignin's complicated and only partially understood chemical structure has so far discouraged researchers from developing a routine chemical basis for its exploitation. However, the lignin that emerges from the pulping process appears to follow some structural rules. For example, lignin components of specific molecular sizes link and dissociate in a particular order, and with this knowledge it has been possible to develop methods for casting films made of the lignin biopolymer.

A number of strategic developments, especially those sponsored by the U.S. Department of Energy (DOE), are occurring in this area (62) (Box 9.2). Three- and four-year projects, the result of an April 2002 solicitation for projects for "Biomass Research and Development for the Production of Fuels, Power, Chemicals and Other Economical and Sustainable Products," are part of a 4-year, $20-million-per-year program intended as a major component of sugar platform research for the next several years. Projects were chosen to be minimum-50%-cost-shared, multidisciplinary plans leading to commercialization of biomass fuel, power, or product technologies. Six pro-

BOX 9.2
Biorefineries

On 6 October 2003, DuPont and the U.S. DOE's NREL announced a joint research agreement leading towards the development of the world's first integrated "biorefinery" that uses corn or other renewable resources, rather than traditional petrochemicals, to produce a host of valuable fuels and value-added chemicals.

The $7.7 million Cooperative Research and Development Agreement calls for DuPont and the NREL to collaboratively develop, build, and test a biorefinery pilot process that will make fuels and chemicals from the entire corn plant, including the fibrous material in the stalks, husks, and leaves and the starchy material in the kernels.

The agreement is part of the larger $38 million DuPont-led consortium known as the Integrated Corn-Based Bioproducts Refinery (ICBR) project. The ICBR project—which includes DuPont, NREL, Diversa Corp., Michigan State University, and Deere & Co.—was awarded $19 million in matching funds from the DOE in 2002 to design and demonstrate the feasibility and practicality of alternative energy and renewable resource technology.

The initiative will develop the world's first fully integrated biorefinery, which will be capable of producing a range of products from a variety of plant material feedstocks. Several biorefineries currently produce a range of products mainly from starch-rich or protein-rich biomass, while other biorefineries start with a variety of vegetable oils.

Source: DuPont

jects were selected from 23 full proposal submittals invited from 190 preproposals, of which two are as follows.

- The $26 million Cargill Dow LLC project entitled "Making Industrial Biorefining Happen." In addition to official team members Iogen, Shell Global Solutions, and CNH Global NV, Cargill Dow plans to involve agricultural grower organizations, national laboratories, universities, and environmental and social nongovernment organizations to develop and build a pilot-scale biorefinery that produces sugars and chemicals such as lactic acid and ethanol from grain.
- The $18.2 million DuPont project entitled "Integrated Corn-Based Biorefinery." With help from Diversa, the U.S. DOE's National Renewable Energy Laboratory (NREL), Michigan State University, and Deere & Co., DuPont will lead the development of a biorefinery concept that converts both starch (such as corn) and lignocellulose (such as corn stover) to fermentable sugars for production of value-added chemicals (such as 1,3-propanediol [PDO]) and fuel ethanol.

For some time, there has been increased interest and very substantial research in the production of chemicals using renewable feedstocks, particularly in the United States. In addition to the environmental attractions of using renewable resources, this has been driven by concerns about the dependence on imported oil. The United States is rich in the supply of renewable agricultural feedstocks such as corn, which can be used to produce low-cost starch raw materials, the cost of which has traditionally been low. On the negative side, concern that assumptions for industrial biotechnology may not take into account what is realistic in terms of the use of farmland and the consequent effect on the delivered price of biomass raw material has been expressed. For example, in the United States, soybean consumption outstrips domestic supply in most years. Further, farmers have been educated to return plant residues to the soil by tillage to preserve long-term soil structure and health and might therefore not be willing to see this material used as an industrial raw material.

While the conventional chemical industry is often reluctant to adopt biotechnology because of perceived risk and expense, contrary to common belief, some renewable raw material costs are now reaching a crossover with their fossil equivalents (34). The only limitation is the technology to convert these materials into useful products, and governments are increasingly willing to support research efforts to overcome these hurdles. The U.S. government recently published draft rules to encourage federal purchasing of biobased industrial products, and it subsidizes the production of bioethanol. In 2003, the U.S. government pledged $400 million in grants to support industrial biotechnology.

There has been a collaborative project between Pacific Northwest National Laboratory and the NREL to identify the value-added chemicals suitable as targets for biorefineries. From a list of more than 300 potential chemicals, a shortlist of 30 was chosen, and these were narrowed further to 12 which are three- to six-carbon molecules with multiple functional groups with high potential to be converted to new families of compounds.

Genencor and Eastman Chemical, which holds a 42.5% stake in Genencor, have developed a one-step fermentation for the ascorbic acid intermediate 2-ketogluconic acid from glucose, replacing four steps in the conventional synthesis. Replacement of one of the four enzymes with an NADP-dependent dehydrogenase recycles cofactors and results in no emission of CO_2 from the process. The firms declared their intention to commercialize the ketogluconic acid bioprocess several years ago; capital costs were estimated to be half those for the existing process, and low costs could also open up new markets, such as use of ascorbic acid as a reducing agent. It should be noted, however, that during the period of development there has been a significant reduction in the price of conventionally produced ascorbic acid.

Genencor has also been collaborating with DuPont on a bioprocess for the production of PDO directly from glucose (27). In 1998, the companies succeeded in genetically engineering a microorganism capable of synthesizing PDO from glucose in a single step. The technology involves inserting four genes taken from various species of bacteria and yeast into industrial strains of *Escherichia coli*. The proteins produced from these four genes then subvert *E. coli*'s biochemical pathways, feeding raw materials that would normally be destined for other purposes to an enzyme that synthesizes PDO.

The following year, Genencor and DuPont achieved a second milestone by developing a second-generation organism capable of much higher yield, rate, or product concentration in water. Yields from the process are such that the researchers believed they needed only to double the efficiency of the prototype that they created in order to have a commercial product. The two companies now claim to have far exceeded their targets and to be nearing performance suitable for commercialization. In late 2002, DuPont and Genencor announced that their scientists had used pathway engineering to combine DNA from several different microorganisms into one production strain, resulting in a 500-fold increase in bioprocessing productivity. The PDO monomer is used to make DuPont Sorona 3GT polymer (a polyester copolymer of PDO and terephthalic acid), the company's newest and most advanced polymer platform (14 October 2002 press release, http://www.genencor.com).

DuPont, jointly with sugar producer Tate & Lyle, has now demonstrated large-scale feasibility of DuPont's bio-PDO process (56) in a 200,000-lb year^{-1} bio-PDO pilot plant at Tate & Lyle subsidiary A. E. Staley Manufacturing's corn wet-mill in Decatur, Ill. Meanwhile, DuPont is using chemically synthesized PDO to build a market for the polyester. DuPont predicts that lowering the cost of PDO will broaden the commercial appeal of 3GT and also make PDO an attractive feedstock for polyols used in polyurethane elastomers and synthetic leathers.

ChemSystems reviewed the alternative processes for PDO in late 1998 and concluded that the biological route could compete with petrochemical routes if it was back-integrated to glucose production from corn. DuPont says further improvements have taken the process "well beyond the most optimistic case described in that study."

Genetic technology will dramatically change the materials available from biomass. Recent advances in recombinant genetic biotechnology of soybeans have led to ways of altering the lipid composition in order to increase the variety of biohydrocarbons available. Additionally, the yield, structure, and degree of saturation of the oils from soy and other vegetable sources will be modified.

Bioengineering of crop plants will improve the markets for oils and fatty acids. DuPont, Monsanto, and Dow are all marketing vegetable oils enriched in oleic acids. Crop developers hope to manufacture specialty oils for industrial applications, although limited funding for product development and higher-than-expected costs are slowing development.

DuPont is exploring application of its high-oleic-acid soybean oil, which can be chemically epoxidized to form nine-carbon diacids for plasticizers, and has cloned the genes needed to epoxidize fatty acids into the plant. It has also cloned the metabolic machinery to conjugate fatty acids for coatings or hydroxylate them for lubricants.

Monsanto engineered rapeseed oil for industrial uses, enriching the oil with lauric acid for surfactants, myristate for making soaps and detergents, and medium-chain fatty acids for lubricants. However, Monsanto has now given these applications a low priority while it concentrates on food and agricultural applications.

Crop enhancement may eventually cut the cost of making a wide range of chemical products. Several firms are seeking to make high-value proteins in crops. Prodigene, for example, has brought to market a plant-derived bovine trypsin. The company has also commercialized another maize-based protein, a bovine protease inhibitor, aprotinin, used to prevent protein degradation during cell culture. Large-scale production in corn can greatly lower the price, since adding capacity is relatively easy. Prodigene is also working with Genencor to make industrial enzymes in plants. The companies are particularly hopeful about applications in which an enzyme-enriched plant could be added directly into an industrial process, eliminating costly purification steps.

In addition to production of new chemicals from biomass, there exists the possibility of using photosynthesis directly, the aim being to create an artificial system mimicking photosynthesis to convert CO_2 to hydrocarbons and other organics. Microbes, both land and sea based, remain a huge untapped resource of metabolic diversity. In the long term, atmospheric CO_2 may be a significant raw material for these products.

While not strictly a plant-derived material, a spider silk protein is being produced from goat's milk by Nexia (41; also see http://nexiabiotech.com). This synthetic material (brand name BioSteel) has uses ranging from body armor to microsurgical sutures. Spinning of BioSteel proteins into nanometer-diameter fibers has been achieved, and Nexia is now determining the product specifications for medical and microelectronic applications.

Biodegradable Plastics

The route to sustainable development lies in optimizing the recycling of all materials. Plastics in particular are so widely used that considerable interest and effort have gone into both direct recycling and the development of varieties that can be "recycled" by biodegradation.

The structure of a molecule is key to its biodegradability. In general, polymers with mixed backbone linkages (carbon-oxygen or carbon-nitrogen) show greater susceptibility to hydrolysis than carbon-carbon backbone polymers. Polymers with aromatic components or branched regions tend to be more resistant to attack than straight-chain aliphatic components. The primary biological mechanism for degradation of high-molecular-weight polymers such as plastics is hydrolysis by extracellular enzymes produced by microorganisms. To be biodegradable, the polymer chain must be flexible and have a stereo configuration that allows it to fit into the active site of a degradative enzyme.

A general class of biodegradable plastics are the microbially produced polyesters which have ester bonds that are susceptible to enzymatic attack. These compounds include poly (betahydroxyalkanoates), or PHAs, an example of which is polyhydroxybutyrate (PHB). PHAs are synthesized by bacteria as a reserve material, and nitrogen and phosphorus limitation can be used to enhance intracellular PHA accumulation. Under the appropriate conditions, the bacterium *Alcaligenes eutrophus*, for example, can accumulate an astounding 96% of its dry weight as PHA.

In the late 1980s, the applied microbiology laboratory at Massachusetts Institute of Technology became the first research facility to bioengineer PHA biopolymers using recombinant DNA, and their technology is licensed to Metabolix (16 May 2001 press release, http://www.monsanto.co.uk). Other researchers have successfully transferred the genes for the three enzymes involved in the synthesis of PHB into *E. coli*, in which they were expressed. A completely biodegradable copolymer, poly(3-hydroxybutyrate-3 hydroxyvalerate), or PHBV, is produced by some types of bacteria. In the early 1980s, Zeneca (formerly part of ICI) developed a process for growing large batches of these bacteria and harvesting the PHBV. PHBV is renewable, biodegrades rapidly and completely (to carbon dioxide and water) in soil or in a landfill yet remains stable in storage, is biocompatible for medical devices and personal hygiene products, and can be processed into films and containers by existing technology. The material, known as Biopol, is so far one of only a very few commercially produced bioplastics. Zeneca sold its Biopol business to Monsanto, who in May 2001 sold the brand to Metabolix.

Metabolix has developed the most advanced, high-throughput microbial production systems in existence today for the production of PHA polymers. These microbial production systems can express pathways encoded by upwards of nine genes from a number of different species, stably integrated into the chromosome with specific productivities severalfold greater than those of wild-type systems. These integrated pathways enable the production of a range of copolymers varying widely in properties and serving diverse applications with costs well under $1/lb.

A novel approach to making plastics is either to have the plant produce the raw materials or, more radically, to make it grow the finished product. In 1999, a team at Monsanto used rape and cress plants to synthesize a PHA by adding bacterial genes from a bacterium, *Ralstonia eutropha*, chosen because it produces high levels of PHAs, into their experimental plants. Workers at Michigan State University have grown PHB in *Arabidopsis thaliana*, a plant of the rapeseed family. The plastic-producing genes from *Alcaligenes eutrophus* have been inserted into the plant, which then makes PHB granules throughout its leaves, stems, and roots. These granules can be recovered for use as a raw material for bioplastic production. Current research is attempting the transfer to the rape plant itself.

While bacterial PHAs are too expensive to be commercially viable, those produced in plants should be cheaper. In 2001, Metabolix received a $15 million cost-share grant from the U.S. DOE to produce PHAs directly in green tissue plants, such as switchgrass, tobacco, and alfalfa. PHAs produced directly in plants will provide cost-competitive, sustainable alternatives to the large-volume, general-purpose plastics now in prevalent use. Metabolix is developing the technology to enable this, producing PHAs in specific plant tissues, such as seeds or leaves, directly by photosynthesis using carbon dioxide and water as the raw materials. PHAs produced this way will not only be low-cost alternatives for many of the large-volume plastics such as polyethylene, polystyrene, polypropylene (PP), and polyethylene terephthalate (PET) now in use, but will also be sustainable raw materials for a number of important chemicals currently based on petrochemical sources.

BASF has also looked at a related material, polyhydroxybutanoic acid from transgenic canola (rape), and although it is competitive with polypropylene on an ecoefficiency basis, the net present value was regarded as too low and the scientific risks in development were seen as too high.

The development of PLAs is a good example of a new process based on renewable resources. PLAs are biodegradable plastics suitable for packaging applications and fibers. They are made by the polymerization of a lactide which is produced from lactic acid. For many years, lactic acid has been produced by both fermentation and chemical routes. More recently, developments in the fermentation process and particularly in downstream recovery appear to have given the bioprocess an overall economic advantage as well as the environmental benefit of being based on renewable raw materials. Cargill Dow's 140,000-tons-per-annum PLA plant began operation in late 2001. For background to Cargill's joint venture with Dow, see their website (http://www.cargill.com and http://www.cargilldow.com).

The first-generation PLA is being produced from corn (maize) in the United States, while in other parts of the world local crops such as rice or sweet potatoes will be used. However, Cargill Dow is developing conversion technologies to utilize lignocellulosic waste materials—straw and bagasse, for example.

To compete with polyester and other conventional petroleum-based polymers, Cargill Dow has located its commercial-scale plant next to a low-cost dextrose supply, Cargill's corn wet-milling complex. Cargill Dow is fermenting Cargill's dextrose to lactic acid and then chemically cracking the lactic acid into three chiral isomers of lactide. Finally, various combinations of the lactides are made to generate a range of polymers. Cargill Dow currently makes Nature Works brand packaging and Ingeo brand fibers (44).

Relying on dextrose ties bioprocesses to corn wet-mills in North America and, in Europe, to wheat processors, but the ability to use a wider range of sugars is developing rapidly. Cargill Dow is exploring novel processes that would allow the use of cheaper feedstocks than dextrose, a capability that would cut the cost of making PLA as well as novel products. Cargill Dow's next plant will not be so limited. The enzyme-converting technology and the ability to adjust fermentations to use a wider variety of sugars have advanced to the point where corn wet-mills will not be needed.

Processing technology to use sucrose from sugar cane, which costs about 5¢ lb^{-1} compared to 10 to 12¢ lb^{-1} for dextrose, is already available. Corn fiber, which corn wet-mills sell locally as animal feed for as little as 3¢ lb^{-1}, may be the next major raw material in the United States. Corn fiber consists of a range of five- and six-carbon sugars, but research and development on bioprocesses to ferment these sugars is well developed.

Farm groups in the United States believe PLAs are an important new market, given slumping commodity prices and concerns over the safety of genetically modified foods. Although Cargill Dow's process uses fermentation, it does not depend on transgenic organisms because many microorganisms already have the capacity to make lactic acid.

Toyota Motor Corporation (60) became the first automaker in the world to use bioplastics in the manufacture of auto parts, employing them in the cover for the spare tire on a model that went on sale in 2003. The bioplastic used is made from polylactic acid, which can be used in the manufacture of products after being heated and shaped. Toyota is building a plant to undertake test production of bioplastic at a factory in Japan, with production beginning in 2004. The company expects to produce 1,000 tons of bioplastic annually, which will be used not just in car parts but in many other plastic products.

Mitsubishi Plastics has already succeeded in raising the heat resistance and strength of polylactic acid by combining it with other biodegradable plastics and filler, and the result was used to make the plastic casing of a new version of Sony Corp.'s Walkman released last fall. NEC Corp., meanwhile, is turning its attention to kenaf, a type of fibrous plant native to tropical areas of Africa and Asia. A mixture of polylactic acid and kenaf fiber that is 20% fiber by weight produces a plastic that is strong enough and heat resistant enough to be used in electronic goods. At present, approximately 14 million tons of plastic is produced in Japan annually. Though bioplastics and other environment-friendly plastics account for only about 10,000 tons, the market for bioplastics is expected to grow by 400% by 2005.

Two key public concerns are fossil fuel use and GHG emissions. The importance of these mark a shift away from emphasis on the biodegradability of plastic materials. Less fossil fuel use and reduced GHG emissions are also increasingly correlated with better economic performance.

Renewable Fuels

Since the 1970s, a number of countries have been involved in the manufacture of liquid fuels based on plant raw materials. Production of bioethanol (Box 9.3) continues on a large scale in Brazil and the United States, with more recent interest in Canada, while a wider range of countries are exploring the potential of biodiesel (Box 9.4).

BOX 9.3
Bioethanol

A combination of national security and the need to meet targets agreed under the Kyoto Agreement is driving a third wave of interest in biofuels and in particular bioethanol. Low-carbon-emissions scenarios reflect the emergence of ethanol as a significant source of fuel for both the transportation and the industrial sectors. In the longer term, a zero-emission ethanol fuel could be produced from sustainable agricultural and biomass sources. Cornstarch (United States) and sugarcane (Brazil) are currently the major sources of ethanol, which is either blended with gasoline or used on its own.

Meanwhile, the traditional ethanol fuel industry continues to grow. Across the United States, 75 ethanol plants are now operating and are able to produce more than 10.6 billion liters of ethanol per year. The 13 plants now under construction will add another 500 million gallons in production capacity (65). There are those who believe this could rise as high as 75 billion liters by 2020 (5).

The European Union wants biofuel, currently 0.3% of transportation fuel, to rise to 5.75% by 2010 to meet Kyoto Agreement targets. This figure represents 9.3 million tons of ethanol, the yield of 3.7 million ha of wheat or sugar beet, but although this may seem a huge area, some 5.6 million ha are currently idle.

All major vehicle manufacturers warrant their cars for use of E-10 fuel (10% ethanol plus 90% gasoline). Many manufacturers are now producing flexible fuel vehicles with engines capable of accepting blends of up to 85% ethanol.

The use of cornstarch will always have to compete with alternative food and feed uses, and consequently most interest is now directed towards the use of cellulose from waste biomass from forest industries or grain production. In the United States, the primary potential raw material is corn stover, while in Canada wheat straw may be the major source.

Corn ethanol plants use coal or natural gas to fuel their distillation process. The CO_2 produced by this combustion has to be taken into account when emissions of GHGs in the transportation sector are estimated. The levels of CO_2 emissions fall dramatically when the waste lignin from lignocellulosic raw materials is used as fuel. In its 1997 scenarios, the U.S. DOE made estimates of CO_2 emissions from transportation fuel production (Box Table 9.3.1).

The Government of Canada's Action Plan 2000 on Climate Change expressed its intention to invest $394 million over the next 5 years. This, together with $625 million which is in the 2000 budget, represented a commitment of over $1 billion in specific actions to reduce GHG emissions by 65 megatons a year. The initiatives outlined in the Action Plan will take Canada one-third of the way to achieving the target established in the Kyoto Protocol.

A key sector targeted by Canada is transportation, which is currently the largest source (25%) of GHGs. Without further action, GHGs from this sector could be 32% above 1990 levels by 2010. Canada's current annual gasoline consumption is 25 to 30 billion liters, 5% of which is E-10. Measures in the Action Plan include increasing Canada's ethanol production from 250 million liters to 1 billion liters, allowing 25% of the total gasoline supply to contain 10% ethanol.

The province of Saskatchewan in Canada estimates that it has enough waste biomass at present, some 22

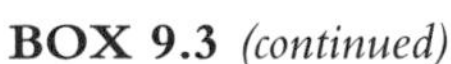

BOX 9.3 *(continued)*

BOX TABLE 9.3.1 Comparative full-cycle CO_2 emissions[a]

Fuel	CO_2 emission (kg/gal^{-1})
Gasoline	11.8
Ethanol from corn (assumes coal-fired boiler)	10.2
Ethanol from corn (assumes natural gas as fuel)	7.0
Ethanol from cellulose (assumes lignin as fuel)	0.06

[a] Data are estimates from the U.S. DOE.

million tons, to produce 8.7 billion liters of fuel ethanol. However, using hybrid poplars and other agricultural cellulose, this could rise to 50 billion liters, without any reduction of food grain production.

Neither corn-based ethanol nor ethanol from cellulose is economically competitive with gasoline. Before the introduction of organisms capable of fermenting multiple sugars, ethanol from biomass was projected to cost $1.58/gal (1980s). Today, the cost has fallen to $1.16/gal. The program forecasts a fall to $0.82 gal in this decade and, as production rises, to compete with gasoline. According to a U.S. DOE analysis (in 2000 figures), if the enzymes necessary to convert biomass to ethanol can be bought for less than 10¢ per gallon of ethanol, the cost of making ethanol could drop as low as 75¢ a gallon, a figure approaching the production cost of gasoline. Genencor believes the enzyme cost could be reduced to 5¢ per gallon of ethanol. Iogen estimates that its product could be competitive based on a raw material price of $28 per ton, a figure acceptable to Saskatchewan farmers at recent seminars.

Emissions of volatile organic compounds (VOCs) react with nitrogen oxides in sunlight to form ground-level ozone, the cause of smog. Because ethanol contains oxygen, it reduces smog and local air pollution. According to the U.S. Environmental Protection Agency (EPA), every 1% increase in oxygenate use decreases toxic emissions by 4.5%.

Chicago has some of the worst levels of air quality in the United States, and strategies for reducing smog in this region have focused largely on VOCs since 1970. The leading approach since 1990 has been the use of reformulated gasoline (RFG). By a wide margin, RFG has been the largest single source of emission reduction in the Chicago area. A number of other U.S. regions have chosen to use RFG with the consequence that, according to the U.S. EPA, one-third of all gasoline sold in the United States is RFG. RFG contains various compounds containing oxygen (known as oxygenates). In the Chicago area, over 90% of the oxygenate is supplied as ethanol. In addition to reducing emissions, RFG oxygenates displace the carcinogen benzene, which is found in conventional gasoline. Total VOC emissions in metropolitan Chicago fell from ca. 2,000 tons day^{-1} in 1970 to 801 tons day^{-1} in 1996. Between 1990 and 1996, RFG contributed 27% of this drop in emissions.

In the 1990s, the U.S. DOE National Biofuels Program (48) focused on developing new, more versatile microorganisms to extract more ethanol from biomass. The program's mission is to develop cost-effective, environmentally friendly technologies for production of alternative transportation fuel additives from plant biomass. The goal is to develop technology which can utilize nonfood sources of sugars for ethanol production. Additionally, the program has collected rigorous material and energy balance data to give increased confidence to projected performance and cost figures.

Recent research has focused on cellulase enzymes. Work is also targeted at organisms capable of converting all the sugars in biomass, especially the pentose sugars. Alternative strategies include the use of the *E. coli* workhorse by adding the capability to make ethanol to strains which can metabolize a range of sugars and by adding sugar metabolism to yeasts which produce alcohol. The program is supporting work at the Universities of Wisconsin and Toronto which is evaluating both a yeast strain and a recombinant form of the organism *Zymomonas* developed by the U.S. DOE.

BOX 9.4
Biodiesel

Biodiesel is made by transesterifying animal fat or vegetable oil with ethanol and can be directly substituted for diesel either as neat fuel (B100) or as an oxygenate additive (typically 20% [B20]) (64). B20 earns credits for alternative fuel use under the Energy Policy Act of 1992, and it is the only fuel that does not require the purchase of a new vehicle. In Europe, the largest producer and user of biodiesel, the fuel is usually made from rapeseed (canola) oil. In the United States, the second largest producer and user of biodiesel, the fuel is usually made from soybean oil or recycled restaurant grease. In 2002, 15 million gallons of biodiesel was consumed in the United States.

E-diesel is a fuel that uses additives in order to allow blending of ethanol with diesel. It includes ethanol blends of 7.7 to 15% with up to 5% special additives that prevent the ethanol and diesel from separating at very low temperature or if water contamination occurs. While the use of E-diesel reduces mileage by some 7 to 10%, there are environmental benefits such as reductions in particulates, CO and NOx.

The U.S. Clean Air Act specifically required the use of oxygenated fuels during winter months in areas exceeding standards for carbon monoxide and RFG in areas exceeding standards for ground-level ozone. In order to fulfill the oxygenate requirement, ethanol is now blended with gasoline in nearly all the carbon monoxide nonattainment areas and is added to a modest but growing portion of RFG.

Although diesel fuel regulations do not require the use of oxygenates per se, oxygen-containing renewable diesel alternatives such as biodiesel (fatty-acid methyl ester made from vegetable oil or animal fat) and E-diesel can dramatically reduce emissions from diesel engines. Biodiesel used straight or in a typical 20% blend with petroleum diesel reduces visible smoke, odor, and toxic emissions, as shown in Box Table 9.4.1.

In January 2004, Archer Daniels Midland announced a partnership with Volkswagen to develop biodiesel fuels.

BOX TABLE 9.4.1 Emissions from biodiesel fuel

Gas or parameter	% Change in emissions	
	B100	B20
Carbon monoxide	−47	−12
Hydrocarbons	−67	−20
Particulates	−48	−12
Nitrogen oxides	+10	+2
Air toxics	−60–90	−12–20
Mutagenicity	−80–90	−20

The benefits of biomass to ethanol technology (38) include increased energy security, reduction in GHG emissions, use of renewable resources, microeconomic benefits for rural economies, and development of a carbohydrate-based chemical process industry.

Unlike starch, which contains homogeneous and easily hydrolyzed polymers, lignocellulosic plant materials contain cellulose, polyphenolic lignins, hemicellulose, and other components bound together in complex ways. The main challenges are the separation (pretreatment) of this complex to make the useful components, cellulose and hemicellulose, more readily available and to reduce the major conversion process costs, including that of the enzymes.

Several physical, chemical, and biological pretreatment processes are under investigation, potentially applicable to the widest possible range of biomass, from wood waste to corn stover to domestic paper waste. The criteria for successful pretreatment are high yields of cellulose and hemicellulose, low lignin degradation, low capital and energy costs, and recoverable process chemicals. Physical treatments such as the use of high temperature can be of limited effectiveness and expensive, while biological processes such as the use of wood-rotting fungi require long treatment times. Chemical treatments, for example with dilute acid or ammonia, may have advantages. A limitation of all of these is their capital-intensive nature.

Cellulase enzymes have been, until recently, high-cost products with relatively low activities. Because of the complex nature of the substrate, many related enzymes, such as endo- and exogluconases, are required to act synergistically to achieve complete breakdown of the cellulose molecule.

The Biomass Research and Development Act allows the U.S. DOE to place the same emphasis on biomass as a source of raw sugars for chemicals as it does on lowering the cost of bioethanol fuel. The DOE expects enzyme producers to lead the cost improvements. In particular, cellulase costs must fall 10-fold, from 30 to 40¢ gal of ethanol produced^{-1} to less than 5¢ gal^{-1}, before biomass conversion becomes profitable for large-scale ethanol production. In 2000, the DOE signed 3-year contracts with Genencor and Novozymes to achieve those cost improvements.

The $17 million grant awarded to Genencor by the U.S. DOE in 2000 had the goal of achieving a 10-fold cost reduction for the enzymes, particularly cellulase, required to produce ethanol. In 3 years, the company exceeded this target and developed an enzyme system (tailored to a dilute acid pretreatment of corn stover), a production organism, and a production process for the new cellulase. New and improved enzymes were developed by protein engineering and directed evolution. In late April, Novozymes announced that it had cut the cost of the enzymes needed for producing ethanol from cellulose by a factor of 20. The gains were achieved in part by Novozymes' advances in enzyme technologies and in part by improved pretreatment processes for corn wastes that were developed by the DOE's NREL. The new pretreatment process allows the use of fewer enzymes per gallon of ethanol produced.

While glucose can easily be fermented by using existing organisms, hydrolysis of cellulose and hemicellulose produces a wide range of both C_6 and C_5 sugars. Wild organisms which convert these exist, but ethanol yields are not sufficient and many by-products which inhibit fermentation result. Research into engineering of modified microorganisms, including ones which will combine hydrolysis of the cellulose and alcohol production, and novel enzymes continues.

Iogen (http://www.iogen.ca), an enzyme company in Ottawa, Canada, is now producing fuel ethanol commercially from cellulose. Iogen makes its ethanol from wheat straw and corn stover and uses steam explosion to free the cellulose from hemicellulose and lignin digesting the cellulose with cellulase. One ton of wheat straw yields 85 gal of ethanol and 200 kg of lignin, which is burned as an energy source. The company's newly opened demonstration plant has a capacity of 260,000 gal per year, and the company plans a 42-million-gallon plant with an expected completion date of 2007. Iogen estimates that the market for cellulose alcohol will grow to $10 billion by 2012 (65). The U.S. DOE expects that ethanol could eventually supply 30% or more of the U.S. transportation market.

Like Iogen, Novozymes and Genencor make cellulase enzymes for textile and pulp processing. Both companies are trying to make the currently known cellulases more active but are also searching for novel enzymes that could assist the process. The intention is to genetically engineer all of the necessary steps into a single organism.

The Future for Renewables

In its report, BIO (12) asks a very pertinent question: what if industrial biotechnology were more widely used?

These are a few of their answers.

- Biotechnology process changes in the production and bleaching of pulp for paper **reduce the amount of chlorine chemicals necessary for bleaching by 10 to 15%.** If applied across the industry, these process changes could **reduce chlorine in water and air as well as chlorine dioxide by a combined 75 tons per year.** Biotechnology processes cut bleaching-related energy uses by 40%—a savings that has the potential to create additional pollution reductions—and **lower wastewater toxicity.**

- Biotechnology process changes in the textile finishing sector **reduce water usage by about 17 to 18%, cost associated with water usage and air emissions by 50 to 60%, and energy demand for bleaching by about 9 to 14%.**
- Biotechnology process changes in plastic production replace petrochemical feedstocks with feedstocks made from organic material such as corn or even corn stovers, thereby **reducing demand for petrochemicals by 20 to 80%.** Because these bioplastics are biodegradable, their use could also **reduce plastics in the waste stream by up to 80%**. Waste burdens are reduced partly because disposable food service items such as plates, cups, and containers can be composted along with the food waste, eliminating the need for separation. These bioplastics can be used to make products ranging from clothing to car parts, all of which can be composted instead of disposed of in landfills or incinerators.
- Biotechnology process changes allow for bioethanol production not only from corn but also from cellulosic biomass such as crop residues; **bioethanol from cellulose generates 8 to 10 times as much net energy as is required for its production. It is estimated that 1 gal of cellulosic ethanol can replace 30 gal of imported oil equivalents**. The closed-loop nature of **using cellulosic biomass to produce bioethanol can contribute substantially to the mitigation of GHG emissions** and can help provide a partial solution to global warming.
- Biotechnology process changes in the nutriceutical and pharmaceutical sector in the production of riboflavin (vitamin B_2) **reduce associated carbon dioxide emissions by 80% and water emissions by 67%**. Changes in the production of the antibiotic cephalexin **reduce carbon dioxide emissions by 50%, energy demand by 20%, and water usage by 75%**. The market share of the biotechnology method of vitamin B_2 production increased from 5% in 1990 to 75% in 2002.

According to Craig Venter, making ethanol is old-think. New-think would use fuel cells supplied with biological hydrogen. This could be produced either by organisms such as *Carboxydothermus* spp., which produce hydrogen as a waste product, or by a completely synthetic organism in which the sugar-forming pathways in photosynthesis have been turned off.

We should not always be fixated on biotechnology (63). Researchers at the University of Minnesota claimed early this year to have discovered a method of producing hydrogen from ethanol that is efficient enough to potentially serve as an economical source of hydrogen. The engineers used an automotive fuel injector to vaporize a mixture of ethanol and water and then used a catalyst to convert that vapor into a mixture of hydrogen, carbon dioxide, and other by-products. The researchers claim that a unit small enough to fit in a person's hand would be able to generate enough hydrogen to fuel a 1-kW fuel cell, capable of powering an average home.

A major advantage of the University of Minnesota invention is its use of a mixture of ethanol and water, eliminating an energy-costly step needed to separate the two for use as a combustion fuel. The efficiency of the conversion process and the fuel cell add to that benefit. "We can potentially capture 50 percent of the energy stored in sugar (in corn), whereas converting the sugar to ethanol and burning the ethanol in a car would harvest only 20 percent of the energy in sugar (65)."

BIOPROCESSING

Enzymes have been used on an industrial scale, in detergents for example, since the 1950s, but their role in biocatalysis has been fully accepted at industrial level only in the last few decades or so, with the lead coming from the pharmaceutical and fine-chemical industries. Although there is ongoing research into the catalytic properties of antibodies and also segments of DNA and RNA, none of these have found their way into industrial processes. Consequently, when we speak of bioprocesses we mean the use of whole cells, cell fragments, or purified enzymes as biocatalysts.

Since most, if not all, novel technologies go through a typical S-curve of development, it should be appreciated that industrial biotechnology is still near the foot of its growth curve. As chemical products become more diverse, the trend is from stoichiometric synthesis towards using the complexity of biological systems—moving from single-reaction biocatalysis to metabolic pathway engineering, from single synthetic steps to cascade catalysis in which a number of enzymes act in concert, without the need to add and remove protective groups. Molecules with several functional groups pose particular difficulties for traditional organic synthesis but are the natural targets for the enzymatic approach. As molecules with pharmaceutical action get more and more complicated, so their manufacture will increasingly require a biological approach.

The ability to manipulate the genetic material of microorganisms has underpinned the widening use of enzymes. Genes for individual enzymes have been transferred from one organism and expressed in another, more suitable organism; multiple copies can be made to overproduce an enzyme, and enzymes from different organisms can be assembled, in one organism, into novel synthetic pathways.

In parallel with these developments in genetic engineering have come improvements in biochemical engineering which have yielded novel commercial reactor and fermentor designs and operation, improved control techniques, and downstream separation. These have resulted in more rapid delivery of products to the marketplace. It is no longer the case that biotechnological solutions are relevant only to high-added-value products, such as pharmaceuticals. Bulk chemicals, including polymers, and heavy-duty industrial processes may have a biotechnological component.

Biocatalysts

While some 3,000 enzymes have been described, maybe only 10% of these are available on an industrial scale (39). These carry out a diverse group of chemical reactions and include enzymes that form and break carbon-carbon bonds (the ligases and lyases), oxidize and reduce organic compounds (dehydrogenases, oxidoreductases), and hydrolyze or form esters, amides, nitriles, peptides, etc., and also have more exotic activities such as the expandase enzyme, which can expand five-membered rings to six-membered rings. These enzymes may be used as free or immobilized whole cells, crude and purified enzyme preparations, bonded to membranes or in cross-linked crystals. Many are now based on recombinant organisms.

Hydrolysis reactions were among the first to be substituted by biological alternatives, but reduction reactions are also being replaced. A widely used enzyme in this respect is alcohol dehydrogenase, originally derived from yeast but with more robust, more active versions of this enzyme now available from other sources. Another functional group transformation originally catalyzed by yeast is the reduction of carbon-carbon double bonds (71).

Almost all the enzymes commercially used today come from the small fraction of organisms that can be grown under controlled conditions. Early reviews of the use of industrial enzymes are presented by Dordick and Wong and Whitesides (23, 71).

Many more organisms, sources of novel biocatalysts, grow in extreme environments, and their enzymes have potentially useful properties in that they can operate in conditions of pH and temperature well away from the norm (1, 53). Novel methods for extracting genes and expressing them in genetically modified organisms have widened the possibilities for finding new and useful activities (16). Crude enzymes normally contain cell debris, nucleic acids, inactive proteins, and pigments. Such impurities interfere with purification of the final product and make the workup difficult and expensive. Immobilized and crystalline enzymes overcome this difficulty, since they are heterogeneous catalysts and do not contaminate the product with unwanted protein. Another serious drawback of crude enzyme preparations is the presence in the mixture of several competing enzymes. These enzymes may have different stereo- and regioselectivities and thus reduce the yield and optical purity of the final product. These competing enzyme activities can be removed by crystallization processes.

Obtaining enantiomerically pure intermediates and products is of utmost importance to the pharmaceutical industry and, to an increasing extent, to the agrochemical industry as well. Biocatalytic reactions offer many advantages, particularly for the synthesis of these chemicals. Conventional processes often give rise to racemic mixtures of chiral isomers (left- and right-handed versions) of the product, only one of which is likely to be useful or chemically active. The inherent selectivity associated with many enzyme-catalyzed reactions means that they can lead to the production of a single isomer, thereby minimizing complicated separations in the downstream processing. Esterases, lipases, and proteases have been of particular importance in this area, and some of these enzymes have the added advantage of being stable and active in organic solvents.

A number of potentially useful syntheses require cofactors such as ATP or NADH. Regeneration of the cofactor from its reaction product is essential for the process to be economically feasible. Use of whole cells rather than free enzymes is one route to cofactor regeneration (54).

Restriction of enzymes to aqueous media has limited their industrial possibilities (37). Many products and their precursors are insoluble in water or react with it to give unwanted by-products. However, the belief that enzymes are denatured in organic solvents derives from experiments in aqueous-organic mixtures rather than in the pure solvent. In the absence of water, proteins are very rigid and lack the flexibility to unfold. This increased rigidity has been shown to result in improved stability, including thermal stability, in a number of enzymes.

Recent research has shown that enzymes do not need to be kept in aqueous reaction media to catalyze technologically useful reactions. Water itself is not an ideal medium for synthetic purposes, since it often participates in the side reactions and may complicate product recovery. This has led to a new form of enzyme biotechnology in which lipases, proteinases, and carbohydrases (enzymes which normally hydrolyze lipids, proteins, and carbohydrates, respectively) are controlled by water depletion and persuaded to form the chemical bonds they would normally break. For example, lipases, esterases, and proteases obviously cannot carry out their normal hydrolysis in dry solvents, but, in the presence of alternative nucleophiles such as alcohols or amines, they can catalyze transesterification and aminolysis reactions and the synthesis of esters (the reverse of hydrolysis) becomes thermodynamically favorable (72). This is the basis of emerging methods for modifying the functionality and nutritional properties of, for example, food fats and the synthesis of food emulsifiers and flavors.

Perhaps the most interesting aspect of enzyme activity in solvents other than water is that it is sometimes possible to alter significantly the enzyme selectivity merely by switching solvents. For example, the enantioselectivity of chymotrypsin acetylation can be reversed by changing from cyclohexane to acetonitrile.

Hydrolases generally display a substantial relaxation of specificity in anhydrous conditions and, as a result, accept a variety of very unusual substrates. For example, subtilisin, an enzyme whose natural function is protein hydrolysis, readily catalyzes the addition of acidic subunits to sugars in organic solvents to yield surface-active food emulsifiers.

While many enzymes do not survive in organic solvents, and a high proportion of those that do become much less active, a few enzyme-catalyzed reactions actually seem to proceed better in nonaqueous media than in water. Industrial advantages of using nonaqueous media include increased solubility of nonpolar substrates and elimination of microbial contamination. Another incentive to use enzymes that can withstand organic solvents is that these solvents make the enzymes more selective with respect to the substrates they bind. One reason for this change in selectivity is that the changed flexibility of the protein chain can affect the way the enzymes select their targets.

Possible remedies for the lower activity in solvents are being discovered. An important factor in enzyme activity in aqueous solutions is the pH, something which is meaningless in the context of organic solvents. However, it has been discovered that enzymes have a pH memory—their activity in the solvent is related to the pH of the last solution they were in. Consequently, lyophilization of an enzyme from aqueous solution at its pH optimum can increase its activity in a subsequent solvent considerably. The reduced flexibility in organic solvents is another reason for decreased activity, and the addition of solvents capable of making multiple hydrogen bonds, such as ethylene glycol, can increase the flexibility and hence the activity.

There is no natural environment that mimics an organic solvent and thus is a source for appropriate organisms. However, salt, like solvents, dehydrates enzymes, sticking to them and shielding them from water; therefore, to improve the level of activity in solvents, research groups are looking at extremozymes from halophilic (salt-tolerant) bacteria. These organisms can survive in water which has 30 times the salinity of the sea, and their enzymes actually require very salty water to function.

Current Processes

We can illustrate the rapid increase in the industrial uses of biotechnology by examples, all of which have emerged in the last few decades.

In the 1980s, Hoffman LaRoche, in Basel, Switzerland, developed a commercial process which used baker's yeast to reduce the double bond of a cyclic α,β-unsaturated ketone as the key step. This reduction gave a pure chiral ketone while making considerable cost savings over the existing chemical conversion.

One of the earliest large-scale uses of enzymes was the manufacture of high-fructose corn syrup using xylose isomerase (36).

BASF has used a lipase-based enzymatic resolution of amines since 1993. Several chiral amines are being produced in a multiproduct facility at scales up to several hundred tons.

DSM uses amidases from several bacterial species to produce a range of nonproteinogenic amino acids. The genes encoding these amidases have been isolated and cloned in *E. coli* so that overproduction occurs, resulting in much-improved biocatalysts (14).

An example of the commercial use of enzymes in organic solvents is the production of enantiopure 2-chloro- and 2-bromopropionic acids, which are intermediates in the manufacture of some herbicides. These have been obtained by an enantioselective butanolysis in anhydrous solvents. BASF has also used enzymes in solvents to manufacture enantiopure amines (31). Eli Lilly has used penicillin acylase to synthesize a pure enantiomeric antibiotic, and Bristol-Myers-Squibb has used a lipase in the stereospecific synthesis of taxol.

Whole cells are most often used for reactions in which cofactors have to be regenerated. Although this is possible in vitro, the use of live cells is usually less expensive. Kyowa Hakko Kogyo manufactures oligosaccharides using whole-cell cofactor regeneration schemes and recombinant metabolic pathways in nonviable but metabolically active microorganisms (40). Both metabolic intermediates and cofactors diffuse freely in the reaction medium. Free enzymes can have advantages over the whole-cell approach, which may be more difficult to predict and control. Degussa manufactures L-leucine in a membrane reactor using enzymatic recycling of NADH.

Although this chapter focuses primarily on the chemical industry, since this is the major user of bioprocesses, it is worth noting that biotechnology is making an impact on a wide range of other industrial sectors. The following examples, which are not an exhaustive selection, are drawn mainly from a recent OECD report (50).

BIOLOGICAL PRODUCTION OF ACRYLAMIDE

Acrylamide is produced industrially as a monomer for synthetic fibers, flocculating agents, etc. The conventional synthetic process involves hydration of the nitrile with sulfuric acid and/or the use of inorganic catalysts. The inclusion of the enzyme nitrile hydratase by Mitsubishi Rayon in an alternative route avoids several costly steps, was the first case of such a replacement, and is possibly the first use of biotechnology in the petrochemical industry.

Advantages of the bioprocess are that the production process is simpler and the reaction is carried out at ambient pressure and temperature; there is no need for a catalysis recovery and reactivation process; less energy is consumed; investment in equipment is lower; the production cost is also lower; and the high level of selectivity and low reaction temperature reduce the amount and range of by-products.

PRODUCTION OF AMMONIUM ACRYLATE

Ammonium acrylate is conventionally manufactured by the chemical conversion of acrylonitrile to acrylic acid, which is subsequently neutralized with ammonia. The first biological process, based on two sequential enzyme catalysts, nitrile hydratase and amidase, had the disadvantage that there existed the possibility of a buildup of the toxic intermediate acrylamide. A breakthrough came with the discovery of a species of *Rhodococcus* possessing a nitrilase enzyme which was capable of converting acrylonitrile directly to acrylic acid without the buildup of any intermediate acrylamide. A benefit of this catalyst was that it was an excellent scavenger of acrylonitrile.

A crucial economic point is that acrylic acid made by the conventional route has to be diluted and neutralized with ammonia before it is used. The biological route effectively adds water to acrylonitrile to make aqueous ammonium acrylate directly. This represents a substantial saving on raw materials.

The major driver for the project was to minimize the capital cost of the plant to ensure a rapid payback. A fed-batch process utilizing free cell biocatalyst was found to give the lowest capex.

BIOBLEACHING OF PULP

Xylanases are key enzymes in the biobleaching process for paper pulp. Originally, they were very expensive and very fragile, needing carefully controlled conditions of pH and temperature. Additionally, they contained cellulases, which are not desirable because they break up the cellulose polymer chains, reducing yield and pulp strength. Originally, the driver for the use of xylanase was the need to reduce chlorine at mills where there was insufficient chlorine dioxide (ClO_2) to achieve elemental chlorine-free status. In the early days of xylanase use, the mills absorbed the high cost of the enzyme to avoid having to build larger ClO_2 generators. Suppliers, such as Iogen, now produce a high-purity product which is more robust and have now reduced the cost of xylanase to the point where they are promoting it on its ability to reduce operating costs alone.

The advantage of xylanase is that it does not need any investment because it is added directly ahead of the bleaching line. Xylanase is not a delignifier; there is speculation that by breaking bonds it somehow "activates" the lignin, and consequently treated pulp requires some 10 to 15% less bleaching chemicals.

One Canadian paper mill achieved environmental compliance with a 50% margin of safety and a bleaching cost that was significantly lower in 17 months of intensive trials, having further revised brightness targets in most stages.

GYPSUM REDUCTION IN ZINC REFINING

Budel Zink BV has operated a zinc refinery at Budel-Dorplein in The Netherlands since

1973. Over 200,000 tons of zinc is produced annually. The conventional roast-leach-electrowin process produces various wastewater streams containing sulfate and zinc. Until mid-2000, these streams were treated conventionally by neutralization with milk of lime, resulting in the production of gypsum which was stored on-site. The Dutch government indicated that they would prohibit further storage of residues on-site as of 1 July 2000. For this reason, alternative wastewater treatment processes were studied in order to arrive at a process in which storage of gypsum is avoided and in which an effluent which complies with Dutch legislation can be produced.

The difficulties in producing a consistently high quality of clean gypsum for sale led to a decision not to pursue this route, and an alternative in which no gypsum is produced was chosen involving use of a high-rate biological sulfate reduction technology in which zinc and sulfate are converted into a zinc sulfide product which can be recycled to the refinery. With the successful implementation of this technology, no gypsum is produced and an improvement of the water quality has been achieved.

BIOLEACHING OF COPPER

Copper smelters are inherently heavily polluting, with 50% of global smelter capacities operating with a sulfur dioxide recovery of less than 85%. Hydrometallurgical processing, which involves transferring the metal constituents of an ore into solution, is generally considered to be more environmentally acceptable, especially for ore containing problem elements such as arsenic. Other factors driving development of hydrometallurgical processes include the high transport costs for shipment of concentrates to a remote smelter; the proven success of hydrometallurgy in treatment of gold, zinc, and nickel-cobalt sulfide concentrates; and the success of heap leach operations for copper recovery.

Most of the commercial mineral-processing bioreactors are operated with mesophilic bacteria at about 40°C. Activity can be improved by using moderate thermophiles such as *Thiobacillus caldus* and *Sulfobacillus* species, which grow optimally at about 45°C, while strains of *Sulfolobus*-like and *Metallosphaera*-like *Archaea* can be active to at least 85°C.

Based on comparative costs to achieve equally low environmental emissions, bioleaching is competitive compared to smelting for treatment of sulfide concentrates. Use of themophiles represents a major breakthrough in the bioleaching of base metal sulfide concentrates, and benefits include a significant improvement in economic performance compared to that of mesophiles.

ENZYMATIC DEGUMMING OF VEGETABLE OIL

A prerequisite for physical refining of crude vegetable oils is a low phosphatide content in the oil entering the final deacidification-deodorization stage. The content of phosphatides is reduced in a degumming step, and one way of doing this is enzymatically in a process based on the hydrolysis of the phosphatide molecule. The enzyme phospholipase A_2 catalyzes the splitting of the fatty acid ester, and the resulting lysolecithin molecule is water soluble and can be separated from the oil by centrifugation. This reaction takes place in an intensive dispersion of the oil with the liquid enzyme solution at a mild temperature of 60°C and pH 5.

A commercial process developed by Lurgi uses considerably less caustic soda, phosphoric and sulfuric acid as well as water (washing and dilution water), and steam than the conventional chemical degumming method. Although a detailed comparison of the cost of the two methods is difficult because the plants are engineered differently, the enzymatic process is cheaper in terms of operational cost. The price of the enzyme solution has a major impact, and it will be reduced further by the application of biotechnological methods. The enzymatic process requires lower capital investment, and integration of the new process into existing plants is simple.

BIODESULFURIZATION OF GASOLINE

Conventional hydrodesulfurization of gasoline is an expensive and energy-intensive process requiring high temperature and pressure (see reference 61). Expensive processes are required to generate hydrogen and to convert the main by-product, hydrogen sulfide, into elemental sulfur. The conventional process saturates olefins in the gasoline, which results in a lowered octane rating. Biodesulfurization (BDS) offers potential cost savings because the process operates at ambient temperature and pressure and produces only nontoxic by-products. Biocatalysis minimizes side reactions such as saturation of olefins. Consequently, BDS retains the quality and value of the gasoline produced.

To achieve the commercial goals for BDS, bacterial strains that produce enzymes that will utilize the sulfur in thiophenes and benzothiophenes have been isolated and characterized. The genes responsible for the sulfur biotransformation have been isolated, cloned, and overexpressed; the biochemical activity of the desulfurization enzymes must yet be optimized and, if necessary, the genes engineered into a gasoline-tolerant host organism. In addition, a suitable bioreactor system that permits contact between the bacterial cells and the gasoline must be developed and a suitable product recovery system must be designed.

It is anticipated that capital costs will be 50% lower and that operating costs will be 15 to 25% lower than for hydrodesulfurization. The application of gasoline BDS in U.S. refineries is expected to yield domestic savings of $900 million in capital equipment and $450 million in annual operating expenditures.

Canada's Prime West Energy Inc. expects to begin using sulfur-oxidizing microorganisms to process sour gas and cut sulfur emissions in 2004 at its facility at Valhalla in Alberta (7). The company plans to use the Shell-Paques process to purify its natural gas. This process was developed by Shell Global Solutions International and Paques BV. The Shell-Paques process is designed to achieve at least 99.8% sulfur recovery. This gas treatment technology uses sulfur-oxidizing microorganisms to convert hydrogen sulfide to solid, fertilizer-grade sulfur. Besides reducing emission, this process reduces costs. Thus, the Shell-Paques process satisfies environmental and economic goals.

Process Development. The decision to design and implement one manufacturing process rather than an alternative is always a complex one involving many parameters and is almost always made on the basis of a less than ideal data set. Environmental benefits alone are not sufficient incentives for the adoption of biotechnology by companies. Decisions are much more influenced by economic considerations, company strategy, and product quality. In its approach to such a decision, a company needs to decide which parameters to take into consideration: economic (cost of production, investments, etc.), occupational health, regulatory aspects (product approval), environmental, customer perception, company profile and values, and many others. It must then gather the facts together, making sure it has access to comparable data for the alternative processes.

As the applications widen, so the international marketplace for bioproducts and processes is increasing. Naturally, the lead is coming from the pharmaceutical sector, in which, in 1998, total biopharmaceutical sales reached $13 billion, an increase of 17% over the previous year. Outside the pharmaceutical sector, the industrial enzyme market is estimated to double in size from 1997 ($400 million) to 2004. Nevertheless, according to a survey conducted in 2000 by the environmental and engineering consultancy Entec (19 January 2001 press release, http://www.entecuk.com), industry still lacks a clear understanding of the meaning of sustainable development. From 104 companies surveyed in seven industrial sectors in the United Kingdom, including pharmaceuticals and oil and gas, 45% of directors and chief executives had not heard of sustainable development; 78% of respondents thought that pressure for sustainability was coming from regulators, demonstrating that any moves towards sustainable development are likely to be compliance driven; and 41% felt that the result of sustainable development would be more costs and additional work.

The implementation of sustainable biotechnology solutions has been slower than it might have been, partly because real-life experience of its application is only slowly acquired by and disseminated between companies. An underlying cause is that the change to a biotechnology solution seems to have large economic implications and the associated risks seem large to the industrial manager.

The problem of management education is still one to be faced today. As one company's president has put it, "Sustainability may well be understood at the top levels in big companies—the problem is application—and middle management has other objectives. The average manager in a pulp and paper mill, for example, joined at 18 to 23 with, perhaps, a bachelor's degree, worked in the plant for the whole of his life and is now 53 and only uses his own practical experience gained over the last 30 years. A worry he has is continuity of production process—he doesn't want to report to the board that there have been production problems because of the introduction of new technology."

Stimulus for development may come from the marketplace with increasing competition or from legislative pressure. Both of these factors have driven the move towards pure enantiomeric pharmaceutical products; environmental legislation has been a major driver in sectors such as pulp and paper and metal extraction. Where clean water is an important raw material, as in food processing, then limited availability may be sufficient stimulus for process development.

A major difficulty for specialist biotechnology companies is how to persuade chemical engineers of the advantages of the new approach. Biocatalysis is still significantly underutilized in the pharmaceutical industry, for example; the success rate is much higher than perceived. In practice, this may mean demonstrating a process at large fermentor (small pilot) scale. One such company with long-term links to major intermediate producers claims that if they know a company's ideal process parameters, they can provide an enzyme to meet these needs. The idea of adjusting a process around an enzyme tends to put off chemical companies, and therefore the enzyme should be optimized to the process. What properties—stability, specificity, activity in solvent, temperature, etc.—are important? It is now possible to search for multiple properties simultaneously.

Many of the drawbacks perceived by process engineers, i.e., low yields and throughput, high dilutions, limited enzyme availability, and low enzyme stability, have largely disappeared. It is now accepted that water may be a suitable medium for industrial processes, while at the same time enzymes are being modified in such a way that they can be used in the organic media with which chemists are more familiar. However, biological systems are usually still too sensitive to permit the use of intensification technologies such as increased turbulence or high temperatures and pressures. Such limitations may be removed by using extremophiles, especially thermophiles (17). On the other hand, bioprocesses do have perceived advantages in that they operate at lower temperature and pressure, while chemical processes require harsher conditions, and in that enzyme catalysts are biodegradable after use but inorganic catalysts are more difficult to dispose of. Integrating biochemistry with traditional chemistry is often the best solution.

Some problems can be solved by choosing a suitable process engineering technology, such as immobilization or cross-linking the catalyst or engineering the enzyme itself.

Collaboration. New participants in the bioprocessing field include such established firms as Celanese and Chevron, who are beginning work on their own bioprocesses by making agreements with small specialist companies which have developed tools for metabolic pathway engineering. Celanese, for example, has established a research and royalties agreement with Diversa (55) because the latter has the ability to "genetically engineer the metabolic processes of an entire cell to perform the desired reaction." Chevron Research and

Technology has entered a 3-year agreement with Maxygen to develop bioprocesses to replace chemical processes, including the conversion of methane to methanol, and Hercules signed on with Maxygen to gain access to the latter's gene-shuffling catalyst optimization technology after launching an extensive in-house research and development program pursuing bioprocesses for specialty chemicals. Maxygen also has commercial links with Novozymes, DSM, Pfizer, and Rio Tinto, while Diversa has had similar arrangements with Dow, Aventis, Glaxo, and Syngenta.

Diversa agreed to work with Novartis to commercialize enzymes for use as animal feed additives and to develop genes that enhance crop plants. It also optimized a heat-tolerant enzyme, discovered in a microorganism colonizing a deep-sea hydrothermal vent, for use by Halliburton subsidiary Halliburton Energy Services, to enhance oil field recovery. Diversa is producing the enzyme for incorporation in Halliburton Energy Services' fracturing fluids. Maxygen is using its gene-shuffling technology, which rapidly generates variants of gene sequences, to help Novozymes optimize industrial enzymes for detergents, food processing, and other applications and to improve antibiotic production for DSM.

Major companies such as BASF and Dow turn to specialist biochemical companies for their specific expertise. Thus, both of these companies collaborate with Integrated Genomics Inc. in Chicago, Ill., who specialize in producing complete genomes of selected microorganisms and analyzing biochemical pathways, blending genome sequencing information with biochemistry, genetics, and physiology into a model of an organism's metabolism. They have developed a database of over 300 genomes and some 3,500 pathways.

Dow AgroSciences' insecticidal product, spinosad, is made by fermentation using a naturally occurring organism. Under an agreement with Dow AgroSciences, Integrated Genomics would determine the complete DNA sequence of the bacterium that produces spinosad and predict its capabilities. This information would supplement Dow AgroSciences' ongoing efforts to improve the spinosad production process.

Statoil, the last of the commercial methane-fermenting companies, has joined forces with DuPont. Norferm, a Statoil subsidiary, produces protein for the aquaculture market from its methane-fermenting facility in Tjeldbergodden, Norway, while DuPont has developed biobased technologies that use methane as a raw material for high-value products. The target markets for the joint venture are animal feeds and the fermentation industry.

Possibilities for the Future. The next generation of bioprocesses aims to make existing big-volume chemicals and polymers and will compete head-to-head with petroleum-based production. Bioprocess plants will then need to be considerably less costly to run than conventional chemical units (28). Bioprocesses are becoming competitive with conventional chemical routes, but industry experts believe that further improvements in enzymatic catalysis and fermentation engineering may be required before other companies are prepared to announce world-scale bioprocessing plants. Competitive economics may ultimately come from the development of bioprocesses that use cheap biomass feedstocks such as agricultural wastes, rather than the dextrose that is the currently preferred renewable raw material.

More-versatile enzymes will provide the basis for improved biocatalysis and increased yield, rate, and selectivity of bioprocesses while retaining the advantages of sustainable chemistry. Improved catalysts will better target desired chemical reactions and will allow better use of lower-cost feedstocks from biomass. Achieving these advances will involve overcoming challenges in the following areas.

- New enzymes must be isolated from the currently unexplored realms of microbes being discovered through studies of biodiversity.

- Substrate specificity (a biocatalyst's preference for a specific molecular structure, from among many similar structures, as its raw material) and activity of known enzymes must be enhanced.
- Sequential enzymatic pathways (metabolic pathways) that perform multiple synthetic steps in industrial microorganisms must be engineered to make new products through lower-cost and more efficient processes.

It must not be forgotten that biotechnology does not claim inevitable superiority in each and every case. Biotechnology may well be best used as one of a series of tools and when integrated into other processes. The comparative analysis which is recommended here may well reveal that nonbiological approaches have strong support. BASF, for example, chose to make indigo via chemical synthesis rather than a bioprocess on the basis of a detailed ecoefficiency analysis. Also, ongoing research into inorganic catalysis provides strong competition. Enzymes have stimulated chemists to design organic and inorganic mimics which operate in harsher environments and are more robust. It is also the case that choice of a renewable feedstock does not in itself guarantee sustainability. This is particularly true if fossil fuels are used during the manufacturing process. Also, oil, rather than biomass, may be a more economic source of complex monomers.

For a new molecule entering the development phase, the need to produce kilo quantities quickly may outweigh economic considerations. It follows that the manufacturing process used at this stage may be modified in the course of optimizing against other parameters, such as economic cost. In a recent development of a chiral intermediate for a U.S.-based pharmaceutical company, three alternative approaches were used: biocatalysis, asymmetric hydrogenation, and crystallization, all of which gave product of acceptable quality. The enzyme process was used to make tens of kilograms for early supply, but one of the other processes is likely to be chosen as the final manufacturing process.

Many major companies are hoping that bioprocesses will enable them to produce compounds that are currently beyond the reach of industrial chemistry. DuPont, for example, has a wide range of industrial biotechnology research and development projects under way. The company, in addition to internal activities, has a $35-million, 5-year alliance with the Massachusetts Institute of Technology. DuPont says it is in the process of selecting a follow-up project for large-scale development now that bio-PDO is well on the way to commercialization. For example, it has engineered another biocatalyst for a different polymer intermediate, dodecandioic acid, which is produced directly from dodecane.

Altus Biologics Inc. has developed a proprietary platform technology (Crystalomics) that transforms crude enzyme mixtures into pure, stable, heterogeneous catalysts with high activity by crystallizing the target enzyme and then chemically cross-linking the molecules to form a CLEC (cross-linked enzyme crystal). The resulting material is insoluble yet porous, so the substrate can access all the active enzyme sites. With this technology, the material retains the catalytic activity of the enzyme while gaining the structural integrity to withstand extremes of temperature and pH, proteases, and an extensive range of solvents. These cross-linked protein crystals can be thought of as a novel type of organic zeolite. This technology is applicable to virtually any enzyme. To date, more than 15 enzymes have been produced in CLEC form. These include alcohol dehydrogenases, asparaginase, elastase, esterases, lipases, luciferase, lysozyme, penicillin acylase, subtilisin, thermolysin, and urease. CLEC catalysts are heterogeneous and can be readily isolated, recycled, and reused many times. Normally, 1 to 2 orders of magnitude in stabilization against heat and more than 3 orders of magnitude in stabilization against water-miscible organic solvents over soluble enzymes can be achieved. The synthetic crystalline enzyme is so stable that it can be recycled thousands of times, using

1/20 the amount of enzyme used in traditional methods.

In addition to carrying out reactions in organic solvents, enzymes have been shown to function in supercritical fluids and even in the gas phase. In the latter case, the enzyme is immobilized and the substrate is adsorbed onto the solid surface. This technology has been proposed for the analysis of volatile pollutants in process off-gases. There may eventually be procedures based on nonaqueous whole-cell catalysis for conducting multistep processes. A number of solvent-tolerant bacterial strains have been identified (21).

Further development activities include such interesting possibilities as enzyme binding to glass (in which the enzyme is contained in small pores in the glass) and "dry state" enzymes (the enzyme is immobilized on, e.g., glass wool and operates on molecules in the vapor phase).

In addition to being adsorbed onto chromatographic column materials, solubilized enzymes have been incorporated into growing polymers to yield immobilized preparations (68). This technique can yield polymer beads, for example, for use in packed-bed reactors.

Research and Development. In the documents emerging from the Vision 2020 process, the chemical industry has given an important lead to biotechnology research and development, believing that biobased industries utilizing biocatalysts will be imperative for carbon sequestration and energy conservation to address greenhouse concerns and environmental pollution issues. "New Biocatalysts: Essential Tools for a Sustainable 21st Century Chemical Industry" is available on the Council for Chemical Research Vision 2020 website (18). This site also allows access to the chemical industry's document "Technology Vision 2020: The U.S. Chemical Industry."

In particular, this document identifies key barriers to be overcome, including:

- a limited knowledge of enzyme-biocatalyst mechanisms;
- a poor understanding of metabolic pathways for secondary metabolites, including pathway interactions;
- a limited number of methods to engineer whole organisms, i.e., metabolic engineering; and
- the high cost of producing many enzymes and cofactors for biocatalyst application.

The proposed goals for the next two decades should be:

- to increase the temperature stability of enzymes up to 120 to 130°C;
- to increase activity by 100- to 10,000-fold over current levels (in water or organic solvents);
- to achieve enzymatic turnover rates comparable to those of current chemical catalysts;
- to achieve an enzyme lifetime durability of months to even years;
- to increase the types of enzyme in use (expand the toolbox) to include isomerases, transferases, oxidoreductases, lyases, and ligases;
- to improve robustness of enzymes and microbes under immobilized conditions; and
- to improve molecular modeling to permit de novo design of enzyme function.

The potential for discovering new biocatalysts is largely untapped, since 99% of the microbial world has been neither studied nor harnessed to date (4, 22). Recognized today through their genetic code (DNA sequence), members of the domains *Archaea* and *Eubacteria* (10, 70) are expected to provide biocatalysts of much broader utility as this microbial diversity is further understood. In the next few decades, the DNA of many industrially important microorganisms and plants will be sequenced, metabolic pathways will be more fully understood, and quantitative models will be available. Very low-cost raw materials for bioprocesses will be derived from agricultural and forestry wastes and, to an increasing extent, cultivated feedstock crops. Known biocatalysts will be improved through the application of molecular biology.

The conventional process to discover enzymes relies upon the isolation and growth of cultures of microorganisms. The obvious limitation here is the need to grow the organism, and thus only those organisms which can be grown under laboratory conditions can be used. However, estimates of the percentage of microorganisms which have been cultured vary from as little as 0.001% in seawater to 1 to 1.5% from activated sludge (3).

Computing is inevitably playing an increasing part in developing new processes. A French company, Metabolic Explorer (http://www.metabolic-explorer.com), working with a database of metabolic pathways and a computer model of *E. coli* and given a starting material and an end product, will assemble the best set of pathways, work out how to fit them into *E. coli*, and say which of the existing *E. coli* pathways should be deleted. The company has done contract work for several other firms and on its own behalf is developing a route to methionine (an animal feed supplement worth $1.4 billion a year). Among other materials, the company has developed a route from glucose to acrylate. Another company, Bio-Technical Resources, has used a similar technique to make glucosamine.

Two quite different approaches to novel enzymes exist, each with its own band of supporters. One is the rational design approach, in which knowledge of existing protein structures is used to predict and design modified enzymes. The second is directed evolution, in which many mutations and recombinations are made and screened for selected properties. The combination of these techniques, together with detailed sequencing of the genomes of a range of organisms, is giving rise to tailored microbes capable of producing many new and existing products for which only chemical routes have previously been available.

Rational Design. Protein engineering permits the alteration of the amino acid sequences of enzyme proteins in such a way that their folded three-dimensional structure acquires new properties, such as improved stability under industrial conditions, and/or confers an altered substrate preference closer to commercial needs. For example, the enzyme xylose isomerase is inactivated by a chemical reaction between glucose (its substrate) and the secondary amino groups of the amino acid lysine which are critical to its three-dimensional structure. Researchers at Gist Brocades found a way of modifying the gene for the enzyme so that the lysine units are replaced by less reactive groups (arginine units), which will still hold the enzyme structure together. The consequence is that the glucose no longer reduces the catalytic activity of the enzyme.

Measurement methods and data in the protein spectroelectrochemistry area that will lead to improved understanding of intra- and interprotein electron transfer processes have been developed. This understanding helps industrial biocatalyst development by allowing for more efficient utilization of carbon sources (e.g., renewable resources) and nutrients and in developing new ways to drive organic syntheses (47).

The metabolic pathway by which microorganisms and plants convert glucose to aromatic amino acids is a current focus of the biothermodynamic measurements. This metabolic pathway is under investigation by several large chemical companies as an environmentally friendly source of aromatic hydrocarbons.

Researchers at the California Institute of Technology in Pasadena have devised a computer technique that can investigate the possible sequences of, say, 100 amino acids and score them according to a set of rules, the sequence with the highest score being the winner (4). The first rule is that the sequence must lower the energy barrier to the formation of intermediate states, i.e., behave as a catalyst. Another rule ensures that the protein can fold up in a biologically plausible way. This would normally be an immensely complicated calculation but, using the so-called dead-end elimination theorem, a systematic reduction in the complexity of the calculation is achieved by throwing away many of the wrong answers right at the beginning.

Using thioredoxin, a protein with no catalytic effect, the researchers modified the amino acid sequence until a new protein with enzymatic activity emerged. Synthetic DNA derived from the new enzyme was inserted into a bacterium and the enzyme was expressed, allowing its activity to be measured. The next step will be to refine the computational method so as to improve the enzyme's activity and to examine more interesting reactions. These researchers have now founded a company, Xencor, to exploit this work.

The number of possible amino acid sequences is far in excess of the capability of current search techniques. The role of rational design may therefore be to narrow down the areas for subsequent combinatorial search. The ultimate step will be to specify a biocatalyst from scratch, and already the first halting steps are being taken in this direction (11).

Directed Evolution. While protein engineering based on site-specific mutagenesis has considerably increased the knowledge of enzyme catalysis, it has had little success as a basis for designing enzyme variants, largely because of lack of understanding of the exact protein structure required for a particular reaction (8).

There exist protein superfamilies, which are groups of proteins with distinct chemical functions and recognizably similar amino acid sequences and three-dimensional structures. Genomic and proteomic searches can identify these superfamilies through their amino acid homologies (33). This sometimes allows the activity of an unknown protein to be established. There may be sufficient conservation of sequence in these families of enzymes that the activity of a novel biocatalyst may be revealed by its amino acid sequence.

However, the goal of enzyme engineering is often to develop properties not found in the natural sequences, and although specialist companies have developed technologies to substitute amino acids one at a time and screen all possible variants, these properties are often not related to any small subset of amino acid residues; rather, residues widely separated in the protein molecule may be making an important contribution.

Such problems may be overcome by a combination of random mutagenesis, gene recombination, and high-throughput screening, so-called directed evolution (2, 49). Quite a number of specialist companies now have proprietary technology in this area. By specifying the screening conditions, new enzymes can be selected with multiple traits at the same time. Degussa, for example, is currently evaluating a whole-cell catalyst containing such a modified enzyme for the commercial production of L-methionine.

If it were possible to capture all the DNA in a given sample, all the enzymes would; in theory, be available. Diversa's technology allows large fragments in a sample to be cloned and amplified. DNA libraries are then constructed and used as source material for enzyme discovery (Box 9.5). Diversa claims that its library collections contain genetic material from over 2 million microorganisms, compared with the ca. 10,000 species described in the literature. Their ultrahigh-throughput screening technologies are capable of screening 1 billion genes per day.

The biggest challenge for the industrial biotechnology sector is to prove that their methods represent a commercially viable alternative to traditional chemical manufacturing (46). Accor-

BOX 9.5
A Screening Example

A DNA library based on a soil sample taken from an alkaline desert was screened for esterase-lipase activity. Some 120 unique enzymes belonging to 21 protein families, which represent only 16% of the theoretical amount of DNA in the sample, were discovered. The implied microbial diversity was confirmed by 16S RNA analysis which revealed the existence of DNA from at least 5,000 different genomes (45).

Source: Diversa

ding to Diversa, success in chemical markets means "having a robust, high-activity enzyme that is not just effective, but cost-effective . . . Green is nice, but it's just a feature."

Using computational design and directed evolution, researchers at Duke University (24) have transformed ribose-binding protein, a protein with no catalytic abilities, into a highly active enzyme. In this research, ribose-binding protein was transformed into an enzyme with triose phosphate isomerase activity. This demonstrates the power of computational biology combined with directed evolution. The latter can identify beneficial mutations far from the active site which can be difficult to find by computation but which are important for optimizing the catalytic process. In principle, the design tools are general and may be used to design many different enzymes at will.

DNA shuffling, in which gene libraries are created by recombination of parent genes, can combine genes from multiple parents and even different species to generate not only improved enzymes, but also ones with properties not present in the parents and not known to exist in nature.

Genes encoding enzymes of a complete metabolic pathway can be transferred to more suitable hosts. These pathways can then be optimized or new pathways can be created in order to synthesize novel products. This technique underpins the new process developed by DuPont and Genencor to make PDO.

Directed evolution of enzymes using random genetic mutation and recombination followed by screening for a desired trait has been a more successful approach to the modification of enzyme properties. In this technique, no a priori knowledge of protein structure-function relationships is required. For example, an aldolase from *E. coli* which synthesizes only D-sugars and requires a phosphorylated substrate was evolved into a new form which does not require a phosphate and can synthesize both D- and L-sugars. It is worth noting that none of the six mutations that arose in order to reach the new form is in the enzyme's active site (29).

Gene shuffling, in which DNA is denatured and then annealed in novel recombinations, can give unexpected results. For example, starting with 26 sources of a protease enzyme, shuffling has given rise to a library of 654 variants, 5% of which are better than the best parent. In one case, DNA shuffling has given increases in enzyme activity of nearly 200-fold, and in another example a 30-fold increase in yield of an antibiotic was achieved in just 9 months. Activity is one target, but solubility, production conditions, substrate cost, and substrate specificity are also important. The principle is to develop a library of enzymes—Maxygen, for example, claims to have some 400 variants of nitrilases. In another case, shuffling produced a progeny enzyme with properties possessed by none of the parents, in this case a heat-stable lipase.

The most exciting example took the genes for just two enzymes differing by only nine amino acids. In the recombinant library produced from these two were enzymes with activities increased by 2 orders of magnitude and some entirely novel catalytic activity.

For a further discussion on directed evolution, see the work of Arnold and Georgiou (9).

Extremozymes

Organisms with alien biochemistries have been collected from the very harshest environments on earth—in the depths of oil wells, in arctic ice, in desiccating salt marshes, and from thermal vents on the deep ocean floor. The enzymes, known as extremozymes, from these organisms are of interest because while we have yet to devise a conventional catalyst as selective as an enzyme, the operational stability of enzymes under practical conditions still requires enhancement. Process engineering is looking to these extremozymes to combine exquisite precision with the robustness needed to survive in industrial processes.

As Diversa has shown, extremophile organisms are very useful sources for the raw material of genetic manipulation. Reduced cooling costs, reduced medium viscosity, increased solubility of reactants, reduced contamination by mesophiles, greater resistance to solvents and detergents, increased protein stability, and possible extra resistance to mechanical stress are just some of the improved properties which are emerging from studies of just one class of extremophiles, the thermo- and hyperthermophiles (17, 19, 20). (Thermophiles grow optimally between 65 and 85°C, while hyperthermophiles grow at temperatures in excess of 85°C.) Very few hydrothermal vent sites have been sampled for microorganisms, and most of them are chemically unique, suggesting the existence of equally unique organisms with widely different metabolic capabilities (58).

The unusual properties of extremozymes may lead to entirely novel chemical processes. For example, while attempting to express the gene for an enzyme from a heat vent organism in *E. coli*, it was found that the gene coded for a much larger protein which was inactive at low temperatures but which split apart, spontaneously producing new enzymes and three other fragments that spliced themselves when heated to form a further enzyme.

Where might biotechnology go next? The possibilities opened up by the new techniques of genetic direction and manipulation are endless. However, there is also the possibility of entirely new chemistry. For example, there is increasing interest in the biochemistry of silicon (13) and the emergence of a silicon biotechnology, one of the targets of which is the development of new materials and synthetic methods. These new materials might be polymers with surface-active properties, nanostructured composites such as a chitosan-silica hybrid with uses in enzyme immobilization, porous ceramics, and electrochemical sensors, and small molecules which may display unusual inhibitory activity towards enzymes significant in disease mechanisms. Even further down the line is the combination of biotechnology with nanotechnology, an area in which Genencor is already collaborating with Dow Corning.

In 1986, it was discovered that DNA coded for a 21st amino acid (selenocysteine), and in 2002 a group from Ohio State University (32) discovered a 22nd genetically encoded amino acid. These discoveries suggest that the genetic code may be more versatile than once thought and that it may be further manipulated to make new industrial enzymes.

PROCESS ASSESSMENT AND SUSTAINABLE DEVELOPMENT

Deciding whether or not to adopt a new industrial process, be it based on biotechnology or conventional physics and chemistry, requires a number of important decisions. However, the point at which these decisions are made is only the crossroads at which many different pieces of information converge and from which the process will travel down one of a number of alternative routes.

An essential rationale for the use of biotechnology is that it is thought to bring increased sustainability and lower environmental impacts to industrial processes. However, this raises the joint problems of how to demonstrate that these changes actually occur and to compare and contrast alternative processes while they are still on the drawing board. Ultimately, what is required is a framework or methodology, preferably internationally accepted, to evaluate biotechnology and bioprocess technologies with respect to economic and environmental costs and benefits (i.e., their contribution to industrial sustainability).

While environmental considerations are an important subset of the parameters to be considered in any process analysis, they are just that—a subset. Other areas such as operating costs or process control are in principle equally important, although, starting in the early 1990s, the weighting of risks began increasing.

Techniques adopted for comparing alternative products and processes need to address economic considerations such as capital expenditure and operating costs, supplies of raw materials (availability and security of supply), processing considerations such as the ease with which a new process element can be integrated into an existing operation or onto an existing site, the nature of the marketplace, and the activities of the competition. Is there, for example, a need for world-scale plants, or could the market be better served by smaller modules more conveniently located?

Process profile analysis, as used by DSM in The Netherlands, is one example of a technique which may be used in an early stage of process development, permitting the winnowing of maybe 10 ideas to give two or three for further development. This technique allows analysis of development possibilities for an existing process or rating of one's own process against that of a competitor. It can equally be used for several alternatives on one site or for the same process at different locations.

Process profile analysis requires that a set of agreed parameters, which are given different weightings for each market sector, be chosen. These might be, for example, operating costs, capex, process control, and internal and external risks. Each of these can be further subdivided; internal risks, for example, might be waste streams and health risks, while external risks might be availability of key materials, new laws and regulations, and patentability of ideas.

BASF has developed a similar process, called ecoefficiency analysis (http://www.corporate.basf.com), which is used to compare related processes and products. So far, the company has conducted over 100 analyses, including 50 in conjunction with their customers. This technique takes into account the views of the end user, life cycle assessment (LCA) (see below), total costs and environmental burden and affects strategy, marketing, research and development, and even politics.

One example compared the manufacture of polyhydroxybutanoic acid from transgenic canola with bacterial fermentation and also with chemically synthesized PP. The evaluation factors and their respective weightings were as follows: energy consumption, 20%; materials, 20%; land, 10%; emissions, 20% (air, 50% [GHG, 50%; ozone depletion, 20%; photochemical ozone, 20%; acidification, 10%]; water, 15%; waste, 15%). On the emissions side, the fermentation had the worst score; comparing overall environmental burden, the PP and canola-sourced material had high scores and fermentation had the lowest. On a cost basis, the PP was just better than canola. BASF did not proceed with the canola route because the net present value was too low, the scientific risk was high, and the time to market was too long.

In any analysis, all alternatives should have a level playing field, so that like is compared with like. Thus, alternatives should be imagined as being on the same site manufacturing the same quantities of product. The principles are simple and are relevant to all activities and sizes of company. It is the level of detail that should always be adjusted to ensure fitness for purpose.

Economic Comparisons

Biotechnology products must succeed by competing economically. Being environmentally preferable is not enough. PLA, for example, is being brought to market strictly on price and performance since Cargill Dow cannot rely on the public to pay a premium for the greener product.

Genencor learned the limits of green marketing while researching bioroutes to indigo, which is conventionally produced via a harsh chemical process that generates carcinogens and toxic wastes. Genencor successfully modified the metabolic pathways in *E. coli* by adding a gene from another bacterium to make the enzyme naphthalene dioxygenase. However, by the time bio-indigo was ready for market, competition from China had eroded the price of indigo by more than 50%. Fabric mills were not willing to pay the premium price Genencor needed to justify investment in a commercial-scale operation. Genencor has included this type of price deflation into its future projects, including the bioprocess for ascorbic acid.

While it is clear that all the major companies recognize that products must succeed by competing economically, advances in genomics and genetic engineering, coupled with increasing environmental pressures, mean that the competitive position of bioprocessing will continue to improve. Perhaps even bio-indigo will return to the marketplace.

Case histories of the development of a number of biotechnological processes are contained in a recent OECD report (50). The key messages of the case studies in this report are that the use of bioprocesses is invariably more environmentally friendly than the conventional processes they replace and that they are also more economic. However, the economic aspects were always given priority by their developers.

A simple first estimate of process feasibility might be based on overall production levels and the volumetric activity of the enzyme. Thus, a rule of thumb is that for products valued at less than \$20 kg^{-1}, production should exceed 1,000 tons per year. Improving biocatalyst stability and activity reduces medium costs and the size of reactor vessels. Even products valued at \$2 to 5 kg^{-1} are worth considering if the target production is more than 50,000 tons per year. A detailed analysis of production costs for the bio-oxidation of *n*-alkanes has been published, which showed that product costs, whether manufactured in batch or continuous process, would be on the order of \$8 kg^{-1} based on a production scale of 10,000 tons per year (42).

There are processes in which the enzyme is cheap enough or the final product is of sufficiently high added value that a single use without recovery is economically acceptable, but most biocatalysts are used in an immobilized form so that they can be recovered and reused.

Other Analysis Techniques

Several tools exist for evaluating the influence of technical products and processes on the environment. If both the damage to the local environment and the total pollution load on the global environment are to be assessed, appropriate tools have to be applied.

Environmental management systems focus on auditing management systems and to some extent on the environmental performance of organizations and companies.

Risk assessment calculates the likelihood that environmental safety limits may be exceeded or that adverse effects may occur. It usually combines the degree of sensitivity of the environment with the intensity of the disturbance to which this environment is exposed. It is often used to assess the risk of adverse effects on human health and the environment which are associated with specific hazardous activities or substances, e.g., use of toxic chemicals. Risk assessment does not assess global environmental problems, such as the greenhouse effect, and it does not cover the entire life cycle of a product or service.

Technology assessment reveals the likelihood of social, economic, and environmental effects associated with technologies, e.g., nuclear power generation, and is an attempt to establish an early warning system to detect, control, and direct technological changes and developments in order to maximize the public good while minimizing the public risks. Technology assessment is a tool to support the decision-making process at a political level.

Environmental impact assessment identifies and evaluates local impacts on the physical and social environment of a specific project or activity, e.g., the building of a new production plant. As a project-oriented tool, the environmental impact assessment does not include all the steps of the life cycle of a product or service.

Life Cycle Analysis (LCA)

Of all of the methodological approaches to assessment, one of the most promising has been identified as that of LCA, also called life cycle analysis. LCA is a way of evaluating the environmental impact of alternative products and processes in terms of their energy and materials, taking into account the entire life cycle of a product or process "from the cradle to the grave." A good example of such an assessment has been published for the Cargill Dow PLA product, Nature Works (66).

The LCA process consists of four steps: definition of aims and scope, inventory analysis (collection of the relevant inputs and outputs), impact assessment (including weighting, which is the most disputed aspect of LCA because it requires judging of the relative importance of different factors, e.g., CO_2 emissions versus mercury pollution of soil), and interpretation.

There is a high degree of consensus on the methodological framework, which has been published by the International Standards Institute (see below).

This type of analysis offers a way to:

- decide whether a product, process, or service is in fact reducing the environmental load or merely transferring it upstream to resource suppliers or downstream to treatment or disposal stages;
- determine where in a process the most severe environmental impact is felt; and
- make quantitative comparisons of alternative process options and competing technologies.

The analysis is often used to reveal the most hazardous steps in the life cycle of a product and to decide which of several product alternatives is associated with the least potential impact on the environment. Potential effects are recorded, as it is not possible to obtain the data required to predict actual effects on the environment. If industrial products and processes are to be subjected to a critical appraisal from the viewpoint of sustainability, the attractiveness of LCA lies in its:

- use of the life cycle concept for products and systems;
- description of impacts on the ecological system;
- opportunity for ecological optimization, including feedback between parts of the life cycle chain;
- possibility of objective or fair comparisons of ecological systems; and
- easier objective communication of ecological problems.

LCA studies also have limitations, which should not be disregarded. They cannot:

- determine the overall impact of a product or a group of products on the environment;
- evaluate parameters, such as availability and renewability of raw materials;
- compare products produced for different purposes and/or under different conditions;
- provide generalizations on methods of disposal; or
- determine decisions (although they can support decision makers).

Although LCA was developed a quarter of a century ago, it has so far had little impact in the bioprocess and bioproduct sectors. Only recently have concerted efforts been made to harmonize the methods involved to give greater credibility to and confidence in LCA. The use of LCA, particularly by small companies, may be hindered by cost and time constraints and by the amount of resources required to analyze complex products. Inventory calculations are made difficult because the energy and raw material inputs and waste outputs are usually not controlled by a single company, and it may be impossible to obtain the relevant data.

The environmental impact of a product or process is of primary concern. To assess this, it is necessary to identify what information is required, how it can be collected, and how reliable it is. Goals that need to be borne in mind are the preservation or efficient use of natural resources and the reduction or optimization of pollution by emissions and waste. These goals concern the environment as a source and a sink, and they comprise both the minimization of pollution and its optimization in cases where the minimization of one parameter is coupled with the increase of another. It is precisely with regard to such considerations that the importance of LCA studies emerges.

By its very nature, the use of biotechnology, and especially of renewable raw materials, gives rise to a number of specific problems. Factors such as the use of a dedicated crop for manufacture rather than food use and the effect of widespread monoculture on biodiversity need to be considered. Any detailed analysis may need to include production inputs to agriculture, such as seeds, fertilizers, pesticides, cultivation, crop storage, and farm waste management. Possible damage to soil by removal of corn stover is not something that would occur, for example, to the average chemical process engineer.

Although LCA offers great promise, the

unique environmental and social issues peculiar to biotechnology require special consideration. While ethical issues, risk assessment, and the economic aspects of decisions are not strictly part of an LCA, any analytical tool, if it is to be useful, must include these considerations.

While the ideal way of comparing the environmental impact of different processes is by elaborating detailed assessments for each alternative, at the early stages of process development the level of accuracy and detail required by a traditional LCA can be too high and therefore too costly to generate, and much of the information required may not be available. Instead, a qualitative approach using a relatively short list of parameters can provide valuable feedback.

The Society of Environmental Toxicology and Chemistry (57) has an advisory group and many publications on LCA.

LCA TECHNIQUES

An LCA can be conducted at different levels, depending on the purpose and application. At its simplest, an assessment of environmental aspects can be based on limited and usually qualitative inventory. A so-called screening LCA or streamlined LCA covers the whole life cycle, with the aim of providing essentially the same results as a detailed LCA, but with a significant reduction in expense and time. In order to ensure that the overall result gives a true picture of the impact, a quality check of the data is necessary.

Roche has carried out a simplified LCA on their product riboflavin (Box 9.6).

Nowadays, computer software packages or programs, such as SimaPro 4.0 (a software tool developed by PRé Consultants BV, Amersfoort, The Netherlands, to simplify the work of LCA [http://www.pre.nl/]), are available to ease the task of performing a full LCA. The International Standards Organization (ISO) has published a number of documents relating to detailed LCA methodology (ISO 14040 to 14043 cover principles and framework, goal and scope definition and life cycle inventory analysis, life cycle impact assessment, and life cycle interpretation) (http://www.iso.org).

BOX 9.6
LCA of Riboflavin Manufacture

Roche has carried out an in-house LCA of the chemical and biological processes for the manufacture of vitamin B_2 (riboflavin).

Raw Materials
The biological process requires 1.5 times as much raw material overall but only a quarter of the nonrenewable raw materials.

Water Consumption
The amount of process and waste treatment water in the chemical process is seven times as large. The biological process requires about double the water, but the greater part of this is for cooling and is not purified.

Energy
The two processes use about the same quantity of energy. The proportion of high-value electricity for stirring, cooling, and evaporation is about double in the biological process, but the steam and natural gas consumption is reduced and hence the CO_2 emissions from fossil fuel combustion are also reduced.

Emissions
Particulates from product formulation are comparable. Solvents are emitted at each stage of the chemical process, and in total these are double the ethanol emitted by the biological process. The latter also gives off odors which are reduced to the necessary level by scrubbing and adsorption. The chemical process gives rise to three times the emissions in water of the biological process. The wastewater from the biological process contains inorganic salts and residues from easily biodegradable biomass, while the waste from the chemical process additionally contains organic chemicals.

Solid Waste
The solid waste from the biological process is exclusively biomass, which is returned to the soil as a nutrient after composting. The compost contains the predominant part of the nitrogen and phosphorus nutrients used in fermentation. The chemical process produces, in addition to a reduced amount of biomass, solid chemical wastes (distillation and filter residues) which have to be incinerated.

Source: Roche

There are a number of other software programs for LCA available, each with its own strengths and weaknesses. They are useful tools for mapping out the overall environmental impacts of production from cradle to grave and can help to reveal the crucial steps in the production process for environmental improvement. The databases, however, may include data from only one region and may reflect the practices of that region.

It must be kept in mind that the calculations for a complete life cycle are done on the basis of a range of assumptions and different data sources. In particular, comparisons assume that alternative manufacturing routes are investigated and compared at the same level of detail.

Sometimes, no data on the environmental loading of the process are available. They are in practice assumed not to exist. This can be important for the results and conclusions of a comparative LCA, if the gaps of one alternative are more important than those of the other. The important gaps then may be filled either by additional data gathering or by obtaining the mass and energy balance through process simulation.

LCA BOUNDARIES

The beginning and end points in the LCA process are called the LCA boundaries. Their position is a matter of choice when an analysis is made and may be determined by any number of external factors. This is particularly relevant for biological processes, such as forestry, which may be considered either as part of the environment or as part of the economic system. When considered as part of the economic system, sunlight, CO_2, H_2O, etc., are environmental inputs, but biomass is the input when biological processes are considered as part of the environment. The main question is which activities and processes are to be considered part of the production system, and which are part of the environment?

Questions arise as to the extent to which the assessment should look upstream of the process; assessment of natural feedstocks or their production may reveal unsuspected problems at the cradle stage, as in the case of the production of biodiesel from oilseed rape (15). Apart from product cleanliness in use and the utilization of by-products (oil cake and glycerol), very large-scale cultivation raises two potential upstream problems: widespread pollen allergenicity and, because brassicaceous plants rarely form mycorrhizal associations, possible soil impoverishment and loss of mycorrhizal inocula from the soil following prolonged cropping.

LCA is confined to the material and energy balances of an activity; sociopolitical and economic criteria, which are also important for decision makers, do not fall within its scope. However, because of its quantitative nature, LCA may enable a trade-off between environmental impact and other elements, including cost components (30). Clean technology is often equated with reduced risk as well as resource efficiency and waste minimization. To what extent, then, should risk assessment be incorporated into LCA? LCA does not cover the intrinsic risks of processes. Risk assessment, in contrast, deals with low-frequency, high-impact events.

Products and processes can be hazardous; that is, they may be injurious to human health and/or the environment. All process steps, including extraction of materials, processing of materials, manufacturing of a product, use of a product, and waste management following the use of a product, must be taken into account in assessing the impact on the environment.

An exhaustive analysis of a product's life cycle has to take into consideration not only the mass and energy flows linked to raw material acquisition and production processes, but also those related to transportation, use, and disposal.

ASSESSMENT IN PRACTICE

Bleach Cleanup of Textiles. Novozymes has published data on their process for removing residual hydrogen peroxide after bleaching prior to dyeing (bleach cleanup). This step uses the enzyme catalase, which has the following advantages: it has no adverse effects on dyestuffs, there is no need to heat or rinse prior to dyeing, there is no risk of harmful overdosing, and there is no formation of byproducts in wastewater. Catalase is biodegradable and is not toxic for aquatic organisms.

A comparison of the amount of water used in the bleach cleanup process shows that by using the enzyme, the dye house saves 6,300 to 19,000 liters of water per ton of textiles. By substituting the enzyme for a reducing agent in a hot rinse, additional energy savings of 1.6 to 1.8 GJ ton of textiles^{-1} can be made, and, owing to the reduced energy consumption, release of CO_2 is lowered by 100 to 120 kg ton of textiles produced^{-1}.

Stonewashing of Jeans. The "stonewashed look" is one of the many available appearances of jeans. This is achieved by locally removing the indigo color by a process in which pumice stone is added to the washing drum to abrade the garment. Enzymes can be used to facilitate the abrasion of the indigo dye from the yarn surface. In practice, three washing methods are used: stonewashing with pumice only, stone-free washing with enzymes only ("biostoning"), and washing with a combination of pumice and enzymes.

LCA was performed on these methods. Of the various process steps used in a jeans finishing center, three specifically relate to the production of stonewashed jeans, namely; stonewashing, wash off (removal of the stones), and wash (cleanup of the garment). Only these three steps were covered by the LCA concerned. The boundaries for the LCA included the mining of the pumice, the production of the cellulases, transport, and finally the jeans finishing process. The main inputs and outputs for the LCA were defined as the energy, raw materials, aiding compounds, and releases that contribute to the product's environmental impact.

The LCA results show that the biostoning method scores best in almost all respects. On the basis of the LCA, it seems justified to consider the biostoning process more environmentally friendly than the pumice stoning process. However, this was not the driving force behind the industrial success of this process; instead, the traditional process was replaced for reasons of economics and quality.

Bioethanol Production. The production of bioethanol has been compared with that of synthetically produced ethanol. Synthetic ethanol is produced from ethylene by catalytic hydration with sulfuric acid, and the resulting ester is hydrolyzed. The data for synthetic ethanol production have been obtained from industry. Bioethanol is made from sugar, which may be derived from crops such as sugarcane, sugar beet, or maize. When made from sugarcane, the cane is crushed and the sugar is extracted with water. The residue, known as bagasse, may be used to generate steam for the subsequent processes. Yeast is used to ferment the sugar, and the ethanol is removed by distillation. Burning the bagasse for fuel means that the part of the ethanol production process from the cane-crushing stage to distillation is self-sufficient in energy.

Ethanol production from renewable raw materials requires very large amounts of energy which are, however, predominantly renewable. In the case of sugarcane, the energy supply was assumed to be self-sufficient and to require only small quantities of fossil energy. The demand for fertilizer, transportation, and machinery is approximately 6 MJ kg of ethanol^{-1}. Synthetic ethanol production uses crude oil and natural gas as the carbon source. The process steps—refinery, steam cracker for ethylene production, and actual synthesis—consume 62 MJ of fossil energy per kg of ethanol. In terms of CO_2 emissions, biotechnological production has major advantages: bioethanol acts as a CO_2 sink, with sugarcane even more than for grain.

In a study on the sustainability of bioethanol, the U.S. NREL concluded that the process needs nonfood biomass and new technology to hydrolyze cellulase and hemicellulose and ferment unusual sugars. An LCA made by NREL was used to compare bioethanol based on corn stover in Iowa and petroleum, the result of which was an 86% drop in fossil energy input and a 69% drop in petroleum use when bioethanol was used. This fossil energy reduction translates directly into reduction of GHGs. However, bioethanol is not cost-effective: it needs the new enzymes from Novozyme or Genencor; fluctuations in the price of oil are important, and collecting a too-high proportion of the stover will damage soil.

L-Carnitine by Chemical and Biochemical Routes. A comparison of L-carnitine by chemical and biochemical routes was made by Lonza (43), whose need for a new process arose from intellectual property and quality issues and cost aspects. The chemical route, which is based on epichlorhydrin with a final-stage optical resolution, is inefficient because D-carnitine cannot be racemized. The biosynthetic route to the optically pure material starts with 4-butyrobetaine, which in turn is made from γ-butyrolactone, which is cheap and available, using a four-step (but one-pot) reaction.

The first-generation process was continuous, one stage only, and had high productivity but only 95% conversion. The second generation was also continuous, had two stages, and had high productivity and high conversion (99.5%) but high process complexity. The industrial solution was a simple fed-batch process, with high conversion and lower capex. This version was robust but had lower productivity.

In the economic comparison, the bio route is better by more than 20% and the purity is far higher. On a sustainability basis, wastewater, total organic carbon, and waste for incineration are all significantly reduced.

Scouring of Cotton Fiber. The cotton fiber is a single cell. In the scouring process, the outer wax and cuticular layers are removed, leaving the pure cellulose. Novozymes developed a new product known as BioPreparation containing the enzyme pectic lyase. In practical tests of the enzyme route at two mills making yarn or knit cloth, water use is down 35 to 50%, energy is down by 20 to 45%, effluent cost is reduced, product yield and plant throughput are up, and there are improvements in product quality.

CONCLUDING REMARKS

The drive for cleaner production and the parallel rise of biotechnology as a production industry are still in relative infancy. However, there is a major economic driver in that production methods based upon biotechnology can reduce energy costs. Also, biodegradable products are attractive to an ecologically conscious public. Thus, one expects reasonable growth of environmental biotechnology-based industries that can reduce the need for environmental cleanup by producing cleaner products and by using cleaner production processes.

An interesting comparison can be drawn between development of the electronics and biotechnology industries. Before the mass sales of the personal computer in the 1970s, it was predicted that biotechnology and the electronics industries would enjoy relatively similar development to market. However, the biotechnology industry has lagged far behind. The issue centers on the use of living material. Defining the limits of performance of a silicon chip is relatively straightforward compared to the same analysis for a biocatalyst. Sensitivity to the external environment, short lifetime and shelf life, greater variability in tolerances, greater variability in responses, and need for cofactor regeneration all conspire to make biotechnology products more difficult to construct and deploy than electronic products. The manufacturing process for electronic goods is also simpler: writing source code for a computer program is a different proposition from purifying the genetic code of a gene for use in a bioproduct. On top of this are the ethical and safety issues surrounding the release of biological materials to the environment.

Nevertheless, the examples given in this chapter show that large corporations have seen the benefits of a lean, green biotechnology industry to the extent that they have been willing to put their money behind promising biotechnologies. That this will continue in the future, in spite of technical, regulatory, and stock market variables, is a given. The lure of, for example, replacing complex and difficult organic synthetic chemistry with a highly specific biotechnology solution with a small number of steps, all at low temperature and pressure, is too tantalizing for big industry to ignore.

REFERENCES

1. **Adams, M. W., and R. M. Kelly.** 1998. Finding and using hyperthermophilic enzymes. *Trends Biotechnol.* **16:**329–332.
2. **Affholter, J., and F. Arnold.** 1999. Engineering a revolution: chemists are applying methods of "unnatural selection' to build better catalysts. *Chem. Britain* **35:**48.
3. **Amann, R. I., W. Ludwig, and K. H. Schleifer.** 1995. Phylogenetic identification and in situ detection of individual microbial cells without cultivation. *Microbiol. Rev.* **59:**143–169.
4. **Anonymous.** 22 March 2001. Survey: designer enzymes. *Economist* **358:**10.
5. **Anonymous.** 27 March 2003. Survey: reinventing yesterday. *Economist* **366(8317):**S18–S22. http://www.economist.com.
6. **Anonymous.** 17 April 2004. Field of dreams. *Economist* **371(8370):**53.
7. **Anonymous.** 2004. Removing hydrogen sulphide the natural way. *Impact* **1:**14. http://www.shellglobalsolutions.com.
8. **Arnold, F. H.** 2001. Combinatorial and computational challenges for biocatalyst design. *Nature* **409:**253.
9. **Arnold, F. H., and G. Georgiou (ed.).** 2003. *Directed Enzyme Evolution: Screening and Selection Methods.* Humana Press, Fredericksburg, Pa.
10. **Barns, S. M., R. E. Fundyga, M. W. Jeffries, and N. R. Pace.** 1994. Remarkable archaeal diversity detected in a Yellowstone National Park hot spring environment. *Proc. Natl. Acad. Sci. USA* **91:**1609–1613.
11. **Benson, D. E., M. S. Wisz, and H. W. Hellinga.** 2000. Rational design of nascent metalloenzymes. *Proc. Natl. Acad. Sci. USA* **97:**6292.
12. **Biotechnology Industry Organization.** 2004. New biotech tools for a cleaner environment. [Online.] http://www.bio.org.
13. **Bond, R., and J. C. McAuliffe.** 2003. Silicon biotechnology: new opportunities for carbohydrate science. *Aust. J. Chem.* **56(1):**7–11.
14. **Broxterman, R., T. Sonke, H. Wories, and W. van den Tweel.** 2000. Biocatalytic production of unnatural amino acids. *Pharm. Manuf. Int.* vol. 61.
15. **Bull, A. T.** 1996. Biotechnology for environmental quality: closing the circles page. *Biodivers. Conserv.* **5:**1–27.
16. **Bull, A. T., A. C. Ward, and M. Goodfellow.** 2000. Search and discovery strategies for biotechnology: the paradigm shift. *Microbiol. Mol. Biol. Rev.* **64:**573–606.
17. **Bustard, M. T., J. G. Burgess, V. Meeyoo, and P. C. Wright.** 2000. Novel opportunities for marine hyperthermophiles in emerging biotechnology and engineering industries. *J. Chem. Technol. Biotechnol.* **75:**1095–1109.
18. **Council for Chemical Research.** Technology vision 2020: the US chemical industry. [Online.] http://www.ccrhq.org.
19. **Cowan, D. A.** 1995. Hyperthermophilic enzymes: biochemistry and biotechnology, p. 351–364. *In Hydrothermal Vents and Processes.* Geological Society publication no. 87. Geological Society Publishing House, London, U.K.
20. **Cowan, D. A., and J. A. Littlechild.** 1996. High temperature enzymes; sources of information for engineering protein stability, p. 197–237. *In* L. M. Savage (ed.), *Enzyme Technology for Industrial Applications.* IBC Biomedical Library Series, Southbridge, Mass.
21. **deBont, J. A. M.** 1998. Solvent-tolerant bacteria in biocatalysis. *Trends Biotechnol.* **16:**493–499.
22. **DeLong, E. F., K. Y. Wu, B. B. Prezelin, and R. V. Jovine.** 1994. High abundance of Archaea in Antarctic marine picoplankton. *Nature* **371:**695–697.
23. **Dordick, J. S.** 1991. *Biocatalysts for Industry.* Plenum Press, New York, N.Y.
24. **Dwyer, M. A., L. L. Looger, and H. W. Hellinga.** 2004. Computational design of a biologically active enzyme. *Science* **304:**1967–1971.
25. **European Association for Bioindustries.** 2003. White biotechnology position paper: clean, sustainable, and white. [Online.] http://www.europabio.org.
26. **European Association for Bioindustries.** 2004. EuropaBio: white biotech. [Online.] http://www.europabio.org/white_biotech.htm.
27. **Fairley, P.** 2000. Directed evolution. *Chem. Week* **162:**29–33.
28. **Fairley, P.** 2001. Bio-processing comes alive: no longer a field of dreams. *Chem. Week* **163:**23–26.
29. **Fong, S., T. D. Machajewski, C. C. Mak, and C.-H. Wong.** 2000. Directed evolution of D-2-keto-3-deoxy-6-phosphogluconate aldolase to new variants for the efficient synthesis of D- and L-sugars. *Chem. Biol.* **7:**873–883.
30. **Guinee, S. B., H. Heijungs, U. de Haes, and G. Huppes.** 1993. Quantitative life cycle assessment of products. 1. Goal definition and inventory. *J. Cleaner Prod.* **1:**3–13.

31. **Gutman, A. L., E. Meyer, E. Kalerin, F., Polyak, and J. Sterling.** 1992. Enzymatic resolution of racemic amines in a continuous reactor using inorganic solvents. *Biotechnol. Bioeng.* **40:** 760–767.
32. **Hao, B., W. Gong, T. K. Ferguson, C. M. James, J. A. Krzycki, and M. K. Chan.** 2002. A new UAG-encoded residue in the structure of a methanogen methyltransferase. *Science* **296:** 1462–1466.
33. **Henrissat, B., and A. Bairoch.** 1993. New families in the classification of glycosyl hydrolases based on amino acid sequence similarities. *Biochem. J.* **293:**781–788.
34. **Herrera, S.** 2004. Industrial biotechnology—a chance at redemption. *Nat. Biotechnol.* **22:** 671–675.
35. **Iowa State University Special Collections Department.** 1998. The legacy of George Washington Carver. [Online.] http://www.lib.iastate.edu.
36. **Jensen, V. J., and S. Rugh.** 1987. Industrial-scale production and application of immobilized glucose isomerase. *Methods Enzymol.* **136:** 356–370.
37. **Klibanov, A. M.** 2001. Improving enzymes by using them in organic solvents. *Nature* **409:** 241–246.
38. **Knauf, M., and M. Moniruzzaman.** 2004. Lignocellulose biomass processing: a perspective. *Int. Sugar J.* **106**(1263)**:**147–150.
39. **Koeller, K. M., and C. H. Wong.** 2001. Enzymes for chemical synthesis. *Nature* **409:** 232–240.
40. **Koizumi, S., T. Endo, K. Tabata, and A. Ozaki.** 1998. Large-scale production of UDP-galactose and globotriose by coupling metabolically engineered bacteria. *Nat. Biotechnol.* **16:** 847–850.
41. **Lazaris, A., S. Arcidiacono, Y. Huang, J. F. Zhou, F. Duguay, N. Chretien, E. A. Welsh, J. W. Soares, and C. N. Karatzas.** 2002. Spider silk fibers spun from soluble recombinant silk produced in mammalian cells. *Science* **295:**472–476.
42. **Mathys, R. G., A. Schmid, and B. Witholt.** 1999. Integrated two-liquid phase bioconversion and product-recovery processes for the oxidation of alkanes: process design and economic evaluation. *Biotechnol. Bioeng.* **64:**459–478.
43. **McCoy, M.** 2001. Making drugs with little bugs. *Chem. Eng. News* **79(21):**37–43. [Online.] http://echo.louisville.edu.
44. **McCoy, M.** 2003. Starting a revolution. *Chem. Eng. News* **81(50):**17–18.
45. **Miller, C. A.** 2000. Advances in enzyme discovery technology: capturing diversity. *INFORM* **11:** 489–496.
46. **Mullin, R.** 2004. Biotech uptick move into March. *Chem. Eng. News* **82(10):**24. [Online.] http://pubs.acs.org.
47. **National Institute of Standards and Technology.** 2003. Bioprocess engineering measurements. [Online.] http://www.nist.gov.
48. **National Renewable Energy Laboratory.** 2004. Introduction to biofuels. [Online.] http://www.nrel.gov.
49. **Ness, J. E., S. B. del Cardayre, J. Minshull, and W. P. C. Stemmer.** 2000. Molecular breeding: the natural approach to protein design. *Adv. Protein Chem.* **55:**261–292.
50. **Organization for Economic Cooperation and Development.** 2001. *The Application of Biotechnology to Industrial Sustainability.* Organization for Economic Cooperation and Development, Paris, France. http://www1.oecd.org.
51. **Organization for Economic Cooperation and Development.** 1998. Biotechnology for clean and industrial products and processes: towards industrial sustainability. [Online.] http://www.oecd.org/pdf/.
52. **Roberts, S. M., N. J. Turner, A. J. Willetts, and M. K. Turner.** 1995. *Introduction to Biocatalysis Using Enzymes and Microorganisms.* Cambridge University Press, Cambridge, United Kingdom.
53. **Rondon, M. R., R. M. Goodman, and J. Handelsman.** 1999. The Earth's bounty: assessing and accessing soil microbial diversity. *Trends Biotechnol.* **17:**403–409.
54. **Schmid, A., J. S. Dordick, B. Hauer, A. Kiener, M. Wubbolts, and B. Witholt.** 2001. Industrial biocatalysis today and tomorrow. *Nature* **409:**258–268.
55. **Scott, A.** 2000. Celanese collaborates with Diversa. *Chem. Week* **162:**32.
56. **Sissell, K.** 2000. DuPont, Tate & Lyle link to develop renewable polymers. *Chem. Week* **162:** 14.
57. **Society of Environmental Toxicology and Chemistry.** 2004. SETAC life-cycle assessment. [Online.] http://www.setac.org/lca.html.
58. **Takami, H., A. Inoue, F. Fuji, and K. Horikoshi.** 1997. Microbial flora in the deepest sea mud of the Marianas Trench. *FEMS Microbiol. Lett.* **152:**279–285.
59. **Thayer, A. M.** 2001. Biocatalysis. *Chem. Eng. News* **79:**27.
60. **Trends in Japan.** 2003. Bioplastic: eco-friendly material has a bright future. [Online.] http://web-japan.org/trends/science/sci031212.html.
61. **U.S. Department of Energy: Energy Efficiency and Renewable Energy.** 2004. Industrial Technologies Program. [Online.] http://www.eere.energy.gov/industry/.
62. **U.S. Department of Energy: Energy Effi-**

ciency and Renewable Energy. 2004. Sugar platform biorefineries. [Online.] https://www.eere.energy.gov/biomass/.

63. Reference deleted.
64. **U.S. Department of Energy: Energy Efficiency and Renewable Energy.** 2004. Renewable diesel fuel. http://www.eere.energy.gov/biomass/.
65. **U.S. Department of Energy: Energy Efficiency and Renewable Energy.** 2004. Biomass program—commercial status. http://www.eere.-energy.gov/biomass/.
66. **Vink, E. T. H., K. R. Rabago, D. A. Glassner, and P. R. Gruber.** 2003. Applications of life cycle assessment to Nature Works™ polylactide (PLA) production. *Poly. Degrad. Stabil.* **80(3):**403–419.
67. **Walsh, C.** 2001. Enabling the chemistry of life. *Nature* **409:**226–231.
68. **Wang, P., M. V. Sergeeva, L. Lim, and J. S. Dordick.** 1997. Biocatalytic plastics as active and stable materials for biotransformations. *Nat. Biotechnol.* **15:**789–793.
69. **Webster, L. C., P. T. Anastas, and T. C. Williamson.** 1996. Environmentally benign production of commodity chemicals through biotechnology, p. 198–211. *In* P. T. Anastas and T. C. Williamson (ed.), *Green Chemistry: Designing Chemistry for the Environment.* American Chemical Society, Washington, D.C.
70. **Woese, C. R., O. Kandler, and M. L. Wheelis.** 1990. Towards a natural system of organisms: proposal for the domains Archaea, Bacteria, and Eucarya. *Proc. Natl. Acad. Sci. USA* **87:**4576–4579.
71. **Wong, C.-H., and G. M. Whitesides.** 1994. *Enzymes in Synthetic Organic Chemistry.* Pergamon Press, Oxford, United Kingdom.
72. **Zaks, A., and A. M. Klibanov.** 1985. Enzyme-catalyzed processes in organic solvents. *Proc. Natl. Acad. Sci. USA* **82:**3192–3196.

SUBJECT INDEX